W9-BFP-889

The Biology of Neuropeptide Y
and Related Peptides

Contemporary Neuroscience

The Biology of Neuropeptide Y and Related Peptides, edited by **William F. Colmers** and **Claes Wahlestedt,** 1993

Psychoactive Drugs: Tolerance and Sensitization, edited by **A. J. Goudie** and **M. W. Emmett-Oglesby,** 1989

Experimental Psychopharmacology, edited by **Andrew J. Greenshaw** and **Colin T. Dourish,** 1987

Developmental Neurobiology of the Autonomic Nervous System, edited by **Phyllis M. Gootman,** 1986

The Auditory Midbrain, edited by **Lindsay Aitkin,** 1985

Neurobiology of the Trace Elements, edited by **Ivor E. Dreosti** and **Richard M. Smith**

Volume 1: *Trace Element Neurobiology and Deficiencies,* 1983

Volume 2: *Neurotoxicology and Neuropharmacology,* 1983

The Biology of Neuropeptide Y and Related Peptides

Edited by

William F. Colmers
University of Alberta, Edmonton, Alberta, Canada

and

Claes Wahlestedt
Cornell University Medical College, New York, NY

Humana Press • Totowa, New Jersey

999 Riverview Drive, Suite 208
Totowa, New Jersey 07512

Printed in the United States of America. 10 9 8 7 6 5 4 3 2 1

Library of Congress Cataloging-in-Publication Data

Main entry under title:

The biology of neuropeptide Y and related peptides / edited by William F. Colmers and Claes Wahlestedt.
p. cm. — (Contemporary Neuroscience)
Includes bibliographical references and index.
ISBN 0-89603-241-8
1. Neuropeptide Y. 2. Neuropeptides. I. Colmers, William F. II. Wahlestedt, Claes. III. Series.
QP552.N38B56 1993
599'.0188—dc20 92-41416
CIP

Preface

The subject of peptides, their receptors, and actions is not novel, but has received intense scrutiny from investigators representing many diverse fields of biomedical sciences. With the growing realization that peptides pervade the physiology of every organ system came attempts to explain the purpose of these apparently abundant signaling systems. With time, it became clear that at least some peptide systems *coordinated* responses of the organism for single purposes. For instance, the vasopressin system in mammals appears to be intimately involved in volume homeostasis, and therefore coordinates changes in peripheral blood flow and blood pressure, renal outflow, body temperature, and other measures to conserve volume in response to fluid loss. Whenever a new peptide appears on the scene, eventually an attempt is made to ascribe a global "purpose" to its actions.

Neuropeptide tyrosine (neuropeptide Y, NPY) and the structurally related peptides, i.e., peptide YY (PYY) and pancreatic polypeptide (PP), of the so-called PP-fold family, or, perhaps better, NPY family, are abundant and ubiquitous. Indeed, many investigators have suggested that NPY is the most abundant mammalian neuropeptide identified to date. The coexistence of NPY with so-called classical transmitters has attracted much interest and many attempts have been made to assess the role of NPY in cotransmission. The structure and parts of its precursors are extraordinarily conserved throughout evolution, matching any other neurohormonal peptide, indicating strong selection pressure for its conservation, and perhaps implying an important role(s) of the peptide. Much work has been done in the ten years since NPY and PYY were first isolated from porcine brain and gut, respectively, and these peptides have been found to be active in virtually every physiological system studied. The true physiological (and pathophysiological) importance of NPY and its chemical relatives is, however, far from well understood, in large part because receptor antagonists have not been available. Moreover, possible links between NPY and human disease have been proposed based on alterations in peptide levels, as has been the case also with other peptides and other putative signaling molecules; the uniqueness of NPY in pathophysiological contexts thus remains to be established.

This book is an attempt to present an integrated examination of the research to date on NPY and its congeners. The authors have striven to examine the impact that the studies of these peptides have had on their respective fields, to provide an overview of the present state of knowledge, and to speculate on the larger questions of their mechanisms of action, what their "purpose(s)" in a given system may be, and to assess the possibility of causal links to human disease. In this, we have attempted to make this book very different from the two preceding works on NPY, both of which are symposium proceedings, with chapters of necessarily restricted scope. Hence, this volume is not meant to to celebrate NPY and its congeners, but rather to critically evaluate their various proposed roles in the perspective of other signaling molecules.

Finally, we also wish to remark on the very high level of cooperation and collegiality that has marked the development of the NPY/PYY/PP field. It is our belief that it is exactly this commitment to the advancement of knowledge rather than to individual benefits that has enabled the field to progress so rapidly. It has been the great privilege of both editors to make friends with many researchers in the field. It is our fervent hope that the level of collegiality that has marked the early and rapid development of the field will continue now that it is approaching its maturity.

W. F. Colmers
C. Wahlestedt

Contents

Structure and Expression of the Neuropeptide Y Gene
Janet M. Allen and Domingo Balbi

Organization of Neuropeptide Y Neurons in the Mammalian Central Nervous System
Stewart H. C. Hendry

PP, PYY, and NPY: *Occurrence and Distribution in the Periphery*
F. Sundler, G. Böttcher, E. Ekblad, and R. Håkanson

Characterization of Receptor Types for Neuropeptide Y and Related Peptides
Lars Grundemar, Sören P. Sheikh, and Claes Wahlestedt

Origin and Actions of Neuropeptide Y in the Cardiovascular System

Zofia Zukowska-Grojec and Claes Wahlestedt

Central Cadiovascular Actions of Neuropeptide Y

Moira A. McAuley, Xiaoli Chen, and Thomas C. Westfall

Neuropeptide Y in Relation to Behavior and Psychiatric Disorders: *Some Animal and Clinical Observations*

Markus Heilig

Contributors

JANET M. ALLEN • *Physiological Laboratory, University of Cambridge, Cambridge, UK*
DOMINGO BALBI • *Physiological Laboratory, University of Cambridge, Cambridge, UK*
DAVID BLEAKMAN • *Department of Pharmacological and Physiological Sciences, University of Chicago, Chicago, IL*
ANDERS G. BLOMQVIST • *Department of Medical Genetics, Uppsala University, Uppsala, Sweden*
G. BÖTTCHER • *Departments of Medical Cell Research and Pharmacology, University of Lund, Lund, Sweden*
XIAOLI CHEN • *Department of Pharmacological and Physiological Science, Saint Louis University School of Medicine, St. Louis, MO*
WILLIAM F. COLMERS • *Department of Pharmacology, University of Alberta, Edmonton, Alberta, Canada*
HELEN M. COX • *Department of Pharmacology, Royal College of Surgeons of England, London, UK*
E. EKBLAD • *Departments of Medical Cell Research and Pharmacology, University of Lund, Lund, Sweden*
LARS GRUNDEMAR • *Department of Clinical Pharmacology, Lund University Hospital, Lund, Sweden*
R. HÅKANSON • *Departments of Medical Cell Research and Pharmacology, University of Lund, Lund, Sweden*
MARKUS HEILIG • *Departments of Psychiatry and Neurochemistry, University of Göteborg, Mölndal, Sweden*
STEWART H. C. HENDRY • *Krieger Mind-Brain Institute, Johns Hopkins University, Baltimore, MD*
JAMES I. KOENIG • *Department of Physiology and Biophysics, Georgetown University Medical Center, Washington, DC*
DAN LARHAMMAR • *Department of Medical Genetics, Uppsala University, Uppsala, Sweden*
MOIRA A. MCAULEY • *Department of Pharmacological and Physiological Science, Saint Louis University School of Medicine, St. Louis, MO*

JOHN K. MCDONALD • *Department of Anatomy and Cell Biology, Emory University School of Medicine, Atlanta, GA*

RICHARD J. MILLER • *Department of Pharmacological and Physiological Sciences, University of Chicago, Chicago, IL*

SÖREN SHEIKH • *Laboratory for Molecular Endocrinology, Biotechnology Centre for Neuropeptide Research, Copenhagen, Denmark*

CHARLOTTE SÖDERBERG • *Department of Medical Genetics, Uppsala University, Uppsala, Sweden*

B. GLENN STANLEY • *Department of Psychology and Department of Neuroscience, University of California, Riverside, CA*

FRANK SUNDLER • *Departments of Medical Cell Research and Pharmacology, University of Lund, Lund, Sweden*

CLAES WAHLESTEDT • *Department of Neurology, Cornell University Medical College, New York, NY*

THOMAS C. WESTFALL • *Department of Pharmacological and Physiological Science, Saint Louis University School of Medicine, St. Louis, MO*

ZOFIA ZUKOWSKA-GROJEC • *Department of Physiology and Biophysics, Georgetown University Medical Center, Washington, DC*

Evolution of the Neuropeptide Y Family of Peptides

Dan Larhammar, Charlotte Söderberg, and Anders G. Blomqvist

1. Introduction

Neuropeptide Y (NPY) is a member of a peptide family that also includes peptide YY (PYY), pancreatic polypeptide (PP), and fish pancreatic peptide Y (PY). This family of peptides is sometimes referred to as the pancreatic polypeptide family, since PP was the first of these to be discovered *(1)*. However, we show here that NPY has remained much more conserved during evolution than PP. This is why this peptide family should be more appropriately called the NPY family.

The members of the NPY family display extensive sequence identity as shown in Fig. 1. All members of the family consist of 36 amino acids and have a carboxyterminal amide. The amide group was in fact the feature that led to the discovery of PYY *(2)* and NPY *(3)*.

The three peptides found in mammals seem to occur in anatomically distinct compartments, since NPY is found in the nervous system, PYY in the intestine, and PP in the pancreas. However, NPY is present also in adrenal medulla (*see* 4 and references therein) and in rat megakaryocytes *(5)*, and can be released from platelets *(6)*. Recently, the presence of NPY mRNA in the pancreas of dexamethasone-treated rats was reported *(6b)*. PYY has been detected in some parts of the CNS *(7–10)* as well as in the pancreas *(8,11–16)*.

From: *The Biology of Neuropeptide Y and Related Peptides;*
W. F. Colmers and C. Wahlestedt, Eds. © 1993 Humana Press Inc., Totowa, NJ

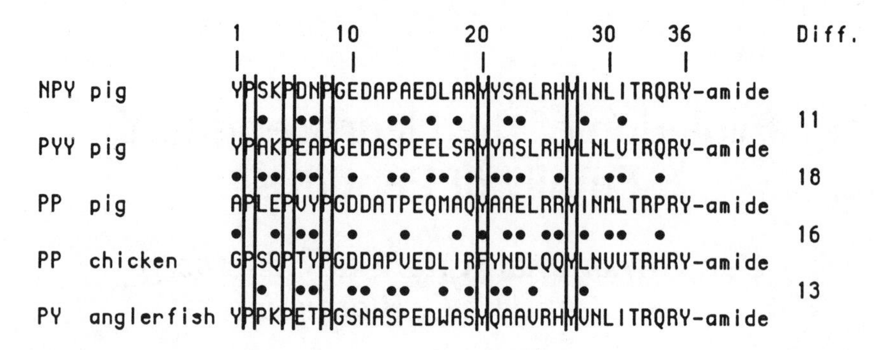

Fig. 1. Sequence alignment of porcine NPY, PYY, and PP, chicken PP, and anglerfish PY. Dots mark differences to the top sequence. Boxes enclose positions assumed to be important for the tertiary structure. For references, *see* Table 2.

Table 1
Sequence Identities Between the Members of the NPY Family[a]

	NPY pig	PYY pig	PP pig	PP chick	PY anglerfish
NPY pig	–	69	50	56	64
PYY pig	*25*	–	50	53	64
PP pig	*18*	*18*	–	44	47
PP chick	*20*	*19*	*16*	–	36
PY anglerfish	*23*	*23*	*17*	*13*	–

[a]Sequence identities between the NPY family members aligned in Fig. 1. The table shows both the number of identical positions (italics) and the percent identity.

Porcine NPY and PYY show 69% sequence identity (*see* Table 1), and numerous studies have shown that NPY and PYY can act on the same receptors with similar potencies (*17; see also* Wahlestedt et al., this vol.). In contrast, the pancreatic peptides differ considerably, not only to NPY and PYY, but also between species; PP of pig and chicken and PY of anglerfish display only 36–47% identity to each other, and are actually less conserved among themselves than to NPY and PYY (Table 1).

To try to resolve the evolutionary relationships between the various members of the NPY family of peptides, we decided to isolate DNA clones from several vertebrate species representing all major branches of the vertebrate tree. DNA clones isolated by crosshybridization would allow prediction of the peptide sequences. Of particular interest to us was how NPY has changed during vertebrate evolution and whether fish have a true NPY peptide in addition to their PY. The aim of this chapter is to review the structural information available in the literature on the various members of the NPY family of peptides, as well as our own recent molecular genetic studies of NPY and related peptides in several species of vertebrates.

2. History of Discovery

2.1. Pancreatic Polypeptide (PP)

Originally isolated as a byproduct of insulin purification from chicken pancreas *(1)*, pancreatic polypeptide (PP) was found to be localized to specific cells in the pancreas *(18)* and to consist of 36 amino acids with a carboxyterminal amide group *(19)*. Pancreatic peptides with sequence similarity to chicken PP have been described in three additional species of birds and nine mammalian species as well as American alligator and bullfrog (*see* Fig. 8 and Table 2). DNA clones encoding prepro-PP have been reported for human, rat, mouse, and guinea pig (Table 2). Physiological studies have shown that PP inhibits pancreatic secretion *(57–59)*.

2.2. Peptide YY (PYY)

PYY was discovered in porcine intestine *(2)* using an identification procedure designed to detect peptides with carboxyterminal amides *(60)*. It was named peptide YY, since it begins and ends with a tyrosine (single-letter code Y). PYY has been sequenced in its entirety from pig, rat, and human (Fig. 7 and Table 2). The sequence of rat prepro-PYY has been deduced from a cDNA clone (Table 2).

PYY has vasocontrictor action *(15)*, and inhibits pancreatic secretion and gut motility *(32,61)*. PYY has been found to act on some of the same types of receptors as NPY (Wahlestedt et al., this vol.).

2.3. Neuropeptide Y (NPY)

NPY was first purified from porcine brain *(3)* using the same approach as described above for PYY. It was named neuropeptide Y, since it begins and ends with a tyrosine. The NPY sequence has been reported for pig *(20)* and six other mammals (Fig. 6 and Table 2). DNA clones encoding prepro-NPY have been reported for

Table 2
NPY-Family Sequence References

Sequence	Species	Seq. obtained from			References	
		Peptide	cDNA	Gene	Author(s)	Year
NPY	Pig	X			20	1982
	Cow	X			21	1989
	Sheep	X			22	1988
	Human	X			23	1988
			X		24	1984
				X	25	1986
	Rat	X			23	1988
		X			26	1988
			X		27	1987
			X		28	1988
				X	29	1987
				X	30	1989
	Rabbit	X			23	1989
	Guinea-pig	X			23	1988
	Chicken		X	X	61a	1992
	Frog	X			31	1991
	Goldfish		X		61a	1992
	Torpedo marmorata				61a	1992
PYY	Pig	X			32	1982
	Human	X			33	1988
	Rat	X			26	1988
	Rat		X		14	1987
	Rat			X	13	1991
PP	Pig	X			34	1979
	Cow	X			35	1979
	Sheep	X			35	1979
	Dog	X			35	1979
	Cat	X			36	1986
	Human		X		37	1984
			X		38	1984
			X		39	1985
				X	40	1985
	Rat	X			41	1984
	Rat		X		42	1986
	Rat			X	43	1988
	Mouse		X		43	1988
	Guinea-pig	X			44	1987
			X		45	1988

	Chicken	X	19	1975
			46	1984
	Turkey	X	47	1981
	Goose	X	48	1984
	Ostrich	X	49	1987
	Alligator	X	46	1984
	Bullfrog	X	50	1988
PY	Anglerfish	X	51	1985
			52	1989
	Daddy sculpin	X	53	1986
		X	54	1987
	Coho salmon	X	55	1986
	Alligator gar	X	56	1987

human and rat (Table 2). Using NPY probes, we have isolated crosshybridizing DNA clones from several nonmammalian vertebrates; chicken, goldfish, the ray *Torpedo marmorata,* horned shark, *Xenopus laevis,* zebrafish, and river lamprey. The first three of these have been completely sequenced *(61a)* and are included in this chapter.

NPY occurs abundantly throughout the mammalian central nervous system as well as in the peripheral nervous system (*see* elsewhere in this vol.). NPY influences many physiological parameters, notably blood pressure, food intake, circadian rhythms and sexual behavior (reviewed in *62; see also* chapters by Stanley; McAuley et al.; Zukowska-Grojec; McDonald and Koenig; and Sundler in this vol.).

2.4. Fish Pancreatic Polypeptide (PY)

A nonamidated 37 amino acid peptide isolated from anglerfish pancreas by Andrews et al. *(51)* was found to be equally identical to NPY and PYY (64%), whereas it was less similar to porcine PP (47% identity). This peptide was named APY or aPY for anglerfish peptide tyrosine(Y), sometimes also referred to as peptide YG for tyrosine(Y)-glycine(G) relating to its first and last amino acids. Similar but amidated peptides were subsequently isolated from the pancreas of daddy sculpin, coho salmon, and alligator gar (Table 2). Although these were named peptide YY or pancreatic polypeptide, we prefer to use the designation PY, since these fish peptides, like anglerfish PY, were all extracted from the pancreas, and their relationship to the mammalian peptides NPY, PYY, and PP is still unclear (*see* Sections 5.5. and 5.6.). Anglerfish PY has recently been reported to occur also without Gly-37, but instead with a carboxyterminal amide *(29)*. The function of fish PY is unknown.

3. Phylogenetic Studies with Antisera

3.1. Vertebrates

Antisera raised against chicken and bovine PP were found to identify immunoreactivity in the brains of various mammalian species (reviewed in *8*), the bullfrog *Rana catesbeiana (63)*, and the African lungfish, *Protopterus annectens (64)*. This immunoreactivity was initially assumed to correspond to PP. However, histochemical studies and radioimmunoassays with an antiserum to NPY, in combination with high-pressure liquid chromatography, showed that the PP immunoreactivity in the rat CNS actually corresponds to NPY *(8)*, as it presumably does also in the other mammals. NPY-like immunoreactivity has been reported in the brains of several mammals (for references, *see 65*) and the CNS of pigeon *(65)*, lizard *(66)*, turtle *(67)*, frog *(68)*, newt *(69)*, the bony fishes anglerfish, goldfish, sea bass, salmon, and trout *(70–74)*, a cartilaginous fish *(75)*, and lampreys *(76,77)*. NPY-like immunoreactivity has been reported to occur also in sympathetic ganglia of bullfrog *(78)*. PYY-like immunoreactivity has been identified in some neuronal populations in rat brain *(7–9)*. These latter immunoreactivities have been proposed to constitute oxidized NPY *(79)*. However, *in situ* hybridizations with oligonucleotide probes specific for NPY and PYY show that mRNA for both peptides is present in rat brainstem neurons *(10)*.

Nerve fibers immunoreactive with an antiserum to PYY were found in the spinal cord of the lamprey Ichthyomyzon unicuspis *(82a)*. Immunocytochemical studies of brainstem and spinal neurons in the river lamprey, *Lampetra fluviatilis*, have revealed two distinct cell populations that differ in their pattern of recognition by antisera *(83)*. One cell type was labeled by antisera against porcine NPY, chicken PP, and bovine PP, whereas a different cell type was labeled by antisera against porcine PYY, rat PP, and bovine PP. (Thus, the antiserum against bovine PP recognized both cell types.) We have recently isolated two distinct NPY-like clones from a lamprey brain cDNA library (Söderberg et al., submitted), which may correspond to the two peptides detected in these two cell populations.

Antisera against PP, NPY, and PYY have been used in phylogenetic studies of the vertebrate pancreas *(80)*. With the antisera against bovine and avian PP, immunoreactivity was detected in the pancreas of mammals, birds, reptiles, amphibians, a bony fish (daddy sculpin, *Cottus scorpius)*, and a cartilaginous fish (spiny dogfish, *Squalus acanthias)*. However, no immunoreactivity was detected in islets of a lower vertebrate, the Atlantic hagfish, *Myxine glutinosa*. The antisera against porcine NPY and PYY detected immunoreactivity in the pancreas of tetrapods only.

NPY-like immunoreactivity in gut nerves has been investigated in an extensive comparative study that included several species of teleost and elasmobranch fishes and a hagfish (*see 81*, and references therein). NPY-like material was found in nerves of the stomach in all teleosts possessing a stomach except the eel. NPY-like immunoreactivity was found also in nerves of the intestine or rectum of some teleosts. All four species of ray had NPY-like material throughout the gut. The hagfish, *Myxine glutinosa*, showed no immunoreactivity.

In the teleost fish *Poecilia reticulata*, NPY-like immunoreactivity has been found both in gut nerves and endocrine cells *(82)*. It is possible that the material seen in the endocrine cells could correspond to a PYY-like or PY-like peptide.

3.2. Invertebrates

The initial studies with antibodies to avian or bovine PP were extended to include also several invertebrates. Immunoreactivity was detected in the nervous system of the earthworm *Lumbricus terrestris (84)* and the pond snail *Lymnaea stagnalis (85)*. In insects, PP-like immunoreactivity was found in the brain and suboesophageal ganglion of the blowfly *Calliphora erythrocephala (85a)*. This immunoreactivity was purified (but not sequenced) from the brain of the closely related species *Calliphora vomitoria (86)*. In *Calliphora vomitoria*, immunoreacitivity was observed in brain, ganglia, nerve fibers, and gut endocrine cells *(87)*. The cockroach *Periplaneta americana* displayed a similar distribution of PP-like immunoreactivity *(63,87a,88–90)*. In ascidians (tunicates), immunoreactivity was observed in the cerebral ganglion and the alimentary tract epithelium of *Ciona intestinalis (91)*.

Subsequently, antisera more specific to PYY and NPY were employed in studies of various species of invertebrates. Immunoreactivity to a PYY antiserum was observed in the brain of the tobacco hornworm moth *Manduca sexta (92)*. NPY immunoreactivity was found in the eyestalks of the lobster *Homerus gammarus (92a)*. In insects, NPY immunoreactivity was found in the locust *Locusta migratoria* in the cephalic and thoracic nervous systems *(93)* as well as in the nervous system and endocrine midgut cells *(94)* and also in the brain of the gray fleshfly *Sarcophaga bullata (94)*. Immunoreactivity was detected in neurons of the nematode *Ascaris lumbricoides* using an antiserum to porcine PP with documented crossreactivity to NPY *(95)*. In the latter study, an antiserum raised to NPY was negative. A study of several invertebrates *(96)* reported the presence of PP-like immunoreactivity, but absence of NPY-like immunoreactivity in the nervous system of the Platyhelminth *Planaria gonocephala*, the mollusc *Aplysia kurodai*, and the insects silkworm, *Bombyx mori*, and cricket, *Gryllus bimaculatus*. A different mollusc, *Fusitriton oregonensis*, contained PP-reactive neurons as well as some NPY-reactive neurons. Both the silkworm and the cricket had gut endocrine cells with PP-like immunoreactivity.

The immunoreactivities detected in the nervous system of these invertebrates presumably corresponds to NPY, since NPY has remained much more conserved during evolution than PP (*see* Section 5.5.). It is possible that the immunoreactivity detected in endocrine gut cells described in several of the studies cited above may correspond to mammalian PYY.

4. Sequence and Structure Analyses

4.1. Peptide Sequences

NPY, PYY, PP, and PY have been isolated and sequenced from several vertebrate species. All of the 33 known sequences, including those derived from DNA sequences (*see* Section 4.2.), are shown in Figs. 6–9. The sequence references have been compiled in Table 2.

4.2. DNA Sequences and Gene Structures

The human NPY precursor was initially cloned as DNA complementary to mRNA *(24)* by screening with synthetic oligonucleotides deduced from the porcine peptide sequence. Prepro-NPY was found to consist of a signal peptide of 28 amino

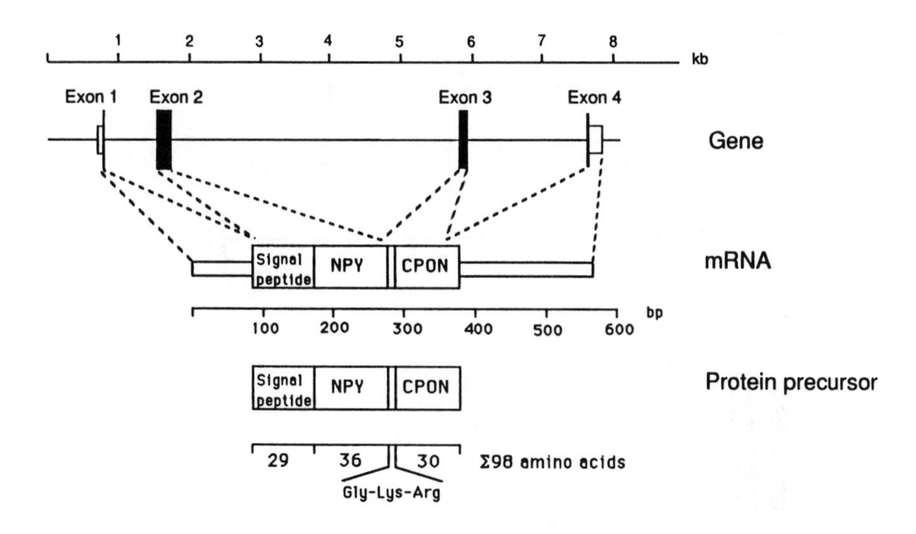

Fig. 2. Schematic outline of the rat NPY gene, mRNA, and protein precursor (taken from *29*). Note that each structure has a separate scale. In the gene, filled boxes denote coding parts and open boxes show noncoding parts of the mature mRNA. Gly-Lys-Arg constitutes the proteolytic processing site in the precursor.

acids, mature NPY, a Gly-Lys-Arg proteolytic processing site, and a carboxyterminal extension called CPON (for carboxyterminal peptide of NPY). The human cDNA clone was subsequently used to isolate genomic clones that allowed determination of the exon–intron organization of the human gene *(25)* and the rat gene *(29)*. Rat cDNA clones were isolated using synthetic oligonucleotides predicted from the human cDNA sequence *(27)* and using the human cDNA as probe *(28)*.

The relationship between the mature rat NPY peptide and its precursor mRNA, and gene is shown in Fig. 2. The exon–intron organization has been found to be identical for the human *(25)*, rat *(29)*, and chicken genes *(61a)* as shown in Fig. 3. The only major difference is that the rat gene has its initiation codon for protein translation at the end of exon 1, whereas the human and chicken genes have their initiation codons at the beginning of exon 2.

The exon–intron organization is known also for the genes of rat PYY *(13)*, human PP *(40)*, and rat PP *(43)* (*see* Fig. 3). Both the rat PYY gene and the human PP gene display identical exon–intron organization to the NPY gene, which provides further evidence

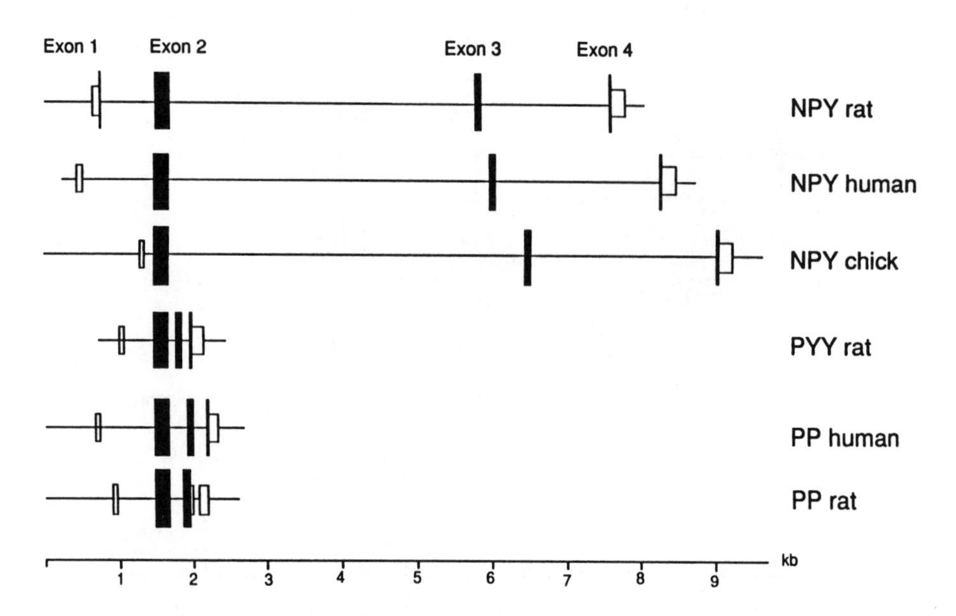

Fig. 3. Exon–intron organization of all known genes belonging to the NPY family. Genes are aligned at exon 2. The coding parts of the exons are shown as filled boxes, whereas the noncoding parts are shown as open boxes. For references, *see* Table 2.

for a shared evolutionary ancestry of these genes. However, both PYY and PP have considerably smaller introns as shown in Fig. 3. Rat PP, as well as mouse PP, has a slightly different exon organization because of a shifted splice donor site at the end of exon 3 *(97)*. This causes the last few amino acids of the rat PP precursor, those encoded by exon 4, to be translated in a different reading frame. Also the guinea pig PP precursor deviates from the majority of PP precursors in having an extended carboxyterminus *(45)*. This may also be because of altered processing of the primary gene transcript.

4.3. Zoo Blots

Our cloning of the rat NPY gene *(29)* revealed perfect identity between the rat and human mature NPY peptides. This finding prompted us to extend our structural analyses based on DNA sequences to more distantly related species. In order to obtain an indication of the degree of conservation of NPY during vertebrate evolution, we used a fragment of the rat NPY gene in hybridiza-

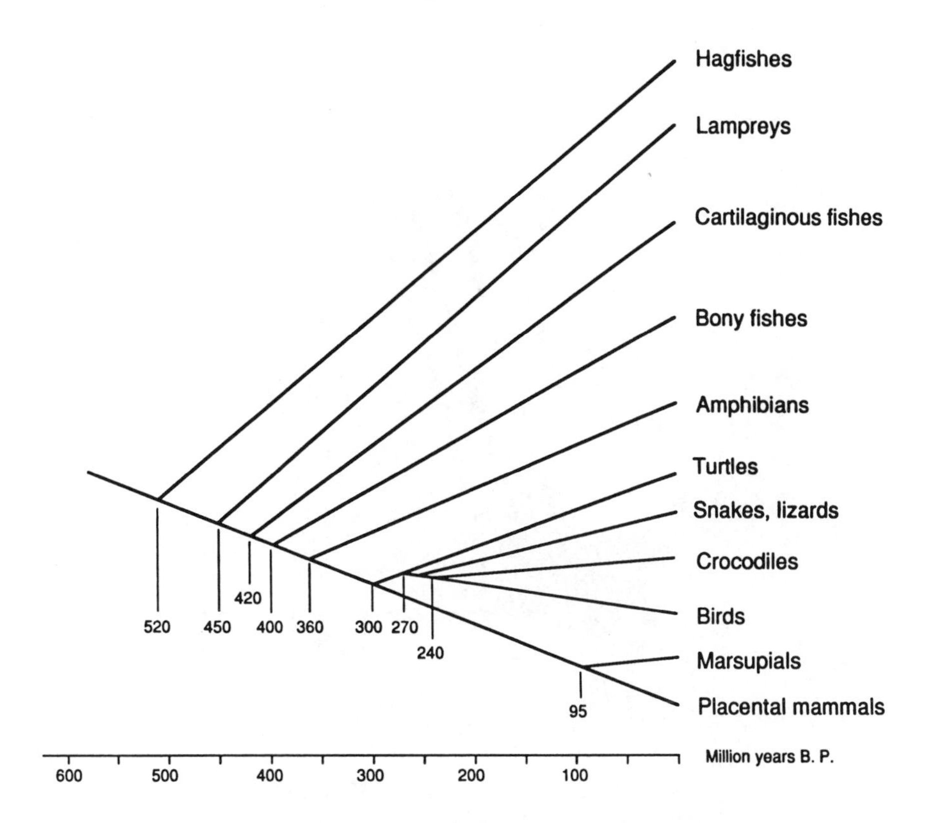

Fig. 4. Vertebrate evolutionary tree. Adapted from ref. *98*.

tions to genomic DNA isolated from a variety of species representing all major branches of the vertebrate evolutionary tree as well as urochordates (tunicates) (Fig. 4).

The DNA fragment used as probe contained only exon 2 of the rat NPY gene. It was essential to use a single-exon probe in order to minimize the complexity of bands in the zoo blot. Exon 2 encodes the signal peptide and most of mature NPY. Because the signal peptides are likely to have diverged considerably during evolution, the segment expected to crosshybridize corresponds only to NPY amino acids 1–34, the first two nucleotides of codon 35, plus the splice donor site GT, in total $34 \times 3 + 2 + 2 = 106$ nucleotides. Therefore, in order to allow crosshybridization, the hybridizations were carried out using low stringency conditions (*see* legend to Fig. 5).

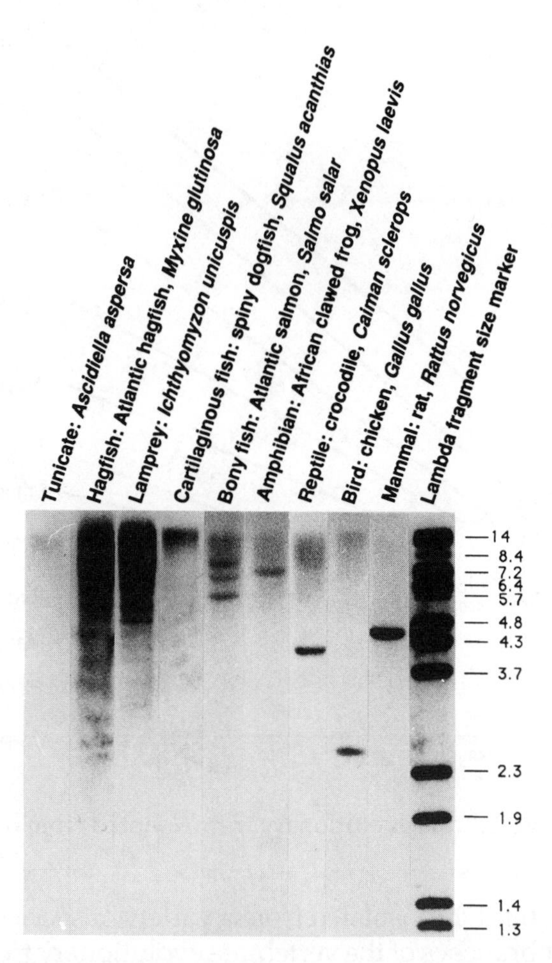

Fig. 5. Zoo blot with a rat NPY probe on genomic DNA from a panel of vertebrate species. Methods: Genomic DNA was extracted from liver (muscle from tunicate, spleen from shark), digested with the restriction enzyme *Bgl II*, electrophoresed on 0.9% agarose gel, and transferred to a nitrocellulose membrane. Hybridization was carried out for 16 h at 42°C in 25% (vol/vol) formamide, 1*M* NaCl, 10% (wt/vol) dextran sulfate, 1% $NaDodSO_4$, and salmon sperm DNA at 0.1 mg/mL. Filters were washed twice for 5 min at room temperature in 2x SSC/0.2% $NaDodSO_4$ (1x SSC = 0.15*M* NaCl/15 m*M* sodium citrate, pH 7.0) and twice for 30 min at 42°C in 2x SSC/0.5% $NaDodSO_4$. The probe was a 287-bp *Xba I-Ava I* fragment containing exon 2 of the rat NPY gene *(29)* labeled with the radionuclide ^{32}P with nick translation. The fragment size marker is bacteriophage λ DNA digested with *Bst EII*. Lanes 1–4 are from a 5-d exposure, whereas the remaining lanes are from a 3-d exposure.

```
                1       10        20        30    36            Diff.
                |        |         |         |     |
NPY man         YPSKPDNPGEDAPAEDMARYYSALRHYINLITRQRY-amide

NPY rat         YPSKPDNPGEDAPAEDMARYYSALRHYINLITRQRY-amide

NPY guinea-pig  YPSKPDNPGEDAPAEDMARYYSALRHYINLITRQRY-amide

NPY rabbit      YPSKPDNPGEDAPAEDMARYYSALRHYINLITRQRY-amide
                                •                               1
NPY pig         YPSKPDNPGEDAPAEDLARYYSALRHYINLITRQRY-amide
                                •                               1
NPY cow         YPSKPDNPGEDAPAEDLARYYSALRHYINLITRQRY-amide
                         •      •                               2
NPY sheep       YPSKPDNPGDDAPAEDLARYYSALRHYINLITRQRY-amide
                      •                                         1
NPY chicken     YPSKPDSPGEDAPAEDMARYYSALRHYINLITRQRY-amide
                                  •                             1
NPY frog        YPSKPDNPGEDAPAEDMAKYYSALRHYINLITRQRY-amide
                  •       •    •• •                             5
NPY goldfish    YPTKPDNPGEGAPAEELAKYYSALRHYINLITRQRY-amide
                          •     • •                             3
NPY Torpedo     YPSKPDNPGEGAPAEDLAKYYSALRHYINLITRQRY-amide
```

Fig. 6. Alignment of all known NPY sequences. Dots mark differences to the top sequence. Rana is the frog *Rana ridibunda.* The amide group has not been formally shown to occur in chicken, goldfish, and *Torpedo,* but is merely inferred from the presence of a Gly residue in the precursor and the presence of an amide group in all other peptides in the NPY family. For references, *see* Table 2.

A representative result is shown in Fig. 5. Both rat and chicken genomic DNAs reveal a single band that agrees in size with the structures of the isolated genes *(29; 61a).* The reptilian and amphibian samples in the blot both display one strong and one weak band (the weak reptile band is 3.7 kb and the weak amphibian band is 1.7 kb). One strong and one faint band were observed also when the genomic DNA was digested with other restriction enzymes (not shown). The two bands may correspond to the NPY gene and a related gene, possibly PYY. In the case of the amphibian DNA, it is also possible that the weak band corresponds to an additional NPY gene or pseudogene, since the species used, *Xenopus laevis*, presumably has doubled its genome *(99).* The bony fish representative, the Atlantic salmon *(Salmo salar)*, displays no less

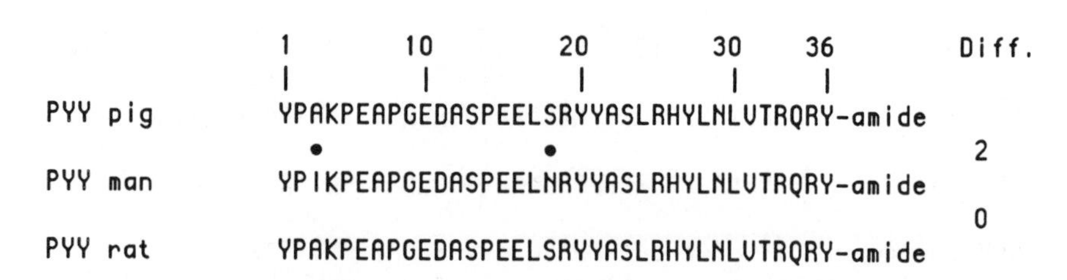

Fig. 7. Alignment of all known PYY sequences. Dots mark differences to the top sequence. For references, *see* Table 2.

than four bands. The fact that salmons are tetraploid *(100)* may account for two of the bands. The two additional bands may indicate polymorphism at both of the NPY loci or may represent a related locus, such as PY or possibly PYY. The cartilaginous fish shows no crosshybridizing bands. This may be because of suboptimal DNA quality, since we have recently been able to isolate NPY clones both from ray *(Torpedo marmorata; 61a)* and horned shark *(Heterodontus francisci;* D. Larhammar et al., unpublished). The lamprey *(Ichthyomyzon unicuspis)* DNA shows two very weak bands that may correspond to the two NPY-related peptides recently cloned as cDNA in the related species *Lampetra fluviatilis* (C. Söderberg et al., submitted). Neither the hagfish *(Myxine glutinosa)* nor the tunicate *(Ascidiella aspersa)* revealed any crosshybridizing bands.

Although a zoo blot is a rather crude way to determine whether a DNA sequence has remained conserved during evolution, our results indicated that NPY, in spite of the shortness of the probe used, had indeed been sufficiently preserved to allow crosshybridization between rat and lamprey, i.e., during almost the entire evolutionary history of vertebrates. We have subsequently been able to isolate and sequence DNA clones encoding NPY or NPY-like peptides from several vertebrate species, namely chicken, goldfish, *Torpedo marmorata (61a)*, horned shark, and river lamprey. Clones hybridizing to chicken NPY have also been isolated from *Xenopus laevis* and zebrafish, but have not yet been confirmed by nucleotide sequence determination. We have also used a rat PYY probe to isolate clones from a zebrafish genomic library. These clones do not hybridize to the goldfish NPY probe.

```
                1        10        20        30    36         Diff.
                |        |         |         |     |
PP  pig         APLEPVYPGDDATPEQMAQYAAELRRYINMLTRPRY-amide
                     •    •                                   2
PP  cow         APLEPEYPGDNATPEQMAQYAAELRRYINMLTRPRY-amide
                 •   •    •                                   3
PP  sheep       ASLEPEYPGDNATPEQMAQYAAELRRYINMLTRPRY-amide
                          •           •                       2
PP  man         APLEPVYPGDNATPEQMAQYAADLRRYINMLTRPRY-amide
                                                              0
PP  dog         APLEPVYPGDDATPEQMAQYAAELRRYINMLTRPRY-amide
                          •                                   1
PP  cat         APLEPVYPGDNATPEQMAQYAAELRRYINMLTRPRY-amide
                     •    •  •  •   •••      •                8
PP  rat         APLEPMYPGDYATHEQRAQYETQLRRYINTLTRPRY-amide
                     •    •         •••      •                6
PP  mouse       APLEPMYPGDYATPEQMAQYETQLRRYINTLTRPRY-amide
                              •        •                      2
PP  guinea-pig  APLEPVYPGDDATPQQMAQYAAEMRRYINMLTRPRY-amide
                • •• •      •• •••••••• •• • ••  •            20
PP  chicken     GPSQPTYPGDDAPVEDLIRFYNDLQQYLNVVTRHRY-amide
                • •• •      •• •••••••• •• • ••  •            20
PP  turkey      GPSQPTYPGDDAPVEDLIRFYDNLQQYLNVVTRHRY-amide
                • •• •   •  •• •••••••• •• ••••• •            23
PP  goose       GPSQPTYPGNDAPVEDLXRFYDNLQQYRLNVFRHRY-amide
                • •• •      •• •••••••• •• • ••  •            20
PP  ostrich     GPAQPTYPGDDAPVEDLVRFYDNLQQYLNVVTRHRY-amide
                •  • •    • •• ••• •••• •• • ••    •          19
PP  alligator   TPLQPKYPGDGAPVEDLIQFYNDLQQYLNVVTRPRF-amide
                  •  ••   •   • •   ••• ••  •••    •          15
PP  bullfrog    APSEPHHPGDQATPDQLAQYYSDLYQYITFITRPRF-amide
```

Fig. 8. Alignment of all known PP sequences. Dots mark differences to the top sequence. Ostrich is *Struthio camelus*, alligator is the American alligator *Alligator mississippiensis*, and bullfrog is *Rana catesbeiana*. For references, *see* Table 2.

4.4. Peptide Sequence Alignments

Figure 6 shows all NPY sequences reported in the literature as well as those recently deduced from DNA sequences determined in our laboratory, i.e., chicken, goldfish, and the ray *Torpedo marmorata (61a)*. NPY shows perfect identity among the mammals, except the three artiodactyls, i.e., cow, sheep, and pig. Chicken has only one difference from the human sequence. The largest

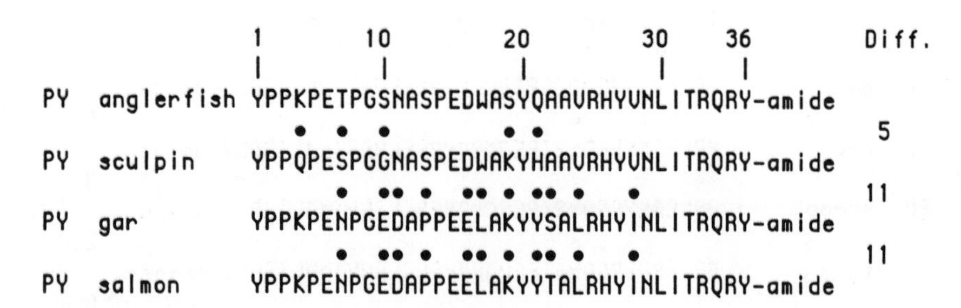

Fig. 9. Alignment of all known PY sequences. Dots mark differences to the top sequence. Anglerfish is *Lophius americanus*, sculpin is the daddy sculpin *Cottus scorpius*, gar is the alligator gar *Lepisosteus spatula*, and salmon is the Pacific coho salmon *Oncorhynchus kisutch*. For references, *see* Table 2.

number of differences between any two NPY sequences is scored for the goldfish–mammal comparison (five replacements). The *Torpedo marmorata* sequence differs from the human sequence at three positions and from the goldfish sequence at two. Since the *Torpedo* sequence has no unique positions, it is likely to be identical to the NPY of the common ancestor of cartilaginous fishes, bony fishes, and mammals *(61a)*. Analogously, the frog NPY sequence has probably remained constant since amphibians diverged from the tetrapod ancestor. After the goldfish had diverged, it probably acquired two amino acid replacements (positions 3 and 16) and the chicken–mammal lineage acquired three (positions 11, 17, and 19).

Figure 7 shows the three PYY sequences reported in the literature, all representing mammalian species, namely pig, rat, and human. The sequence information is still too limited to allow any conclusions to be drawn regarding the degree of conservation of PYY during vertebrate evolution. The sequence divergence of the human sequence (two replacements as compared to the rat and porcine sequences) might indicate that PYY has undergone more sequence divergence than NPY. We have recently isolated PYY candidate clones from a zebrafish genomic library (Söderberg and Larhammar, unpublished) using the rat PYY cDNA clone as probe.

Figure 8 shows the known PP sequences. Mature PP shows zero to nine differences in comparisons between mammals. The most divergent sequences are those of rat and mouse. The other mammals differ from each other at only zero to five positions. Among the four bird sequences, the closely related chicken and turkey differ at two positions, whereas the largest number of differences, six, is scored for the goose–ostrich comparison. The divergence between mammals and birds (and frog) is dramatic and much more extensive than for NPY. In fact, mammalian and chicken PPs are no more similar to each other than either is to NPY or PYY (Figure 12 and Table 4). Chicken and alligator differ from each other at six positions, whereas these differ from mammals at 20 and 19 positions, respectively. This comparison supports a closer relationship of birds to reptiles than to mammals (*see* Fig. 4; *101*), in agreement with the majority of anatomical characters (*see 102* for review), but in contrast to sequence comparisons of other proteins, such as myoglobin and β-hemoglobin (reviewed by *101*) as well as 18S ribosomal RNA *(103)*.

Figure 9 shows an alignment of the PY sequences of four species of fish, namely anglerfish, daddy sculpin, salmon, and gar. These species are not closely related inasmuch as the first three represent different superorders, and gar even belongs to a different division *(104)*. Surprisingly, all four of the fish PY sequences show higher identity scores to mammalian NPY and PYY (61–86%) than to mammalian PP (47–53%) (*see* Table 3). Furthermore, there are no amino acid positions that are shared by PY and PP, but different in NPY and PYY. Thus, there is no structural evidence that the pancreatic peptides PY and PP are more closely related to each other than to NPY or PYY.

4.5. Peptide Three-Dimensional Structure

Pancreatic polypeptide of turkey has been used to determine the tertiary structure by X-ray crystallography *(47,126)*. A schematic drawing is shown in Fig. 11. The peptide was found to be U-shaped with one stem of the U having an extended polyproline helix (positions 1–8) and the other stem consisting of an α helix (positions 14–31). The proline helix has Pro residues at positions 2, 5, and 8. These interdigitate with two aromatic side chains extending from the α helix at positions 20 (Phe) and 27 (Tyr), thus holding together the two stems of the U. The carboxyterminus is oriented away from the α-helix axis because of a turn at residues 33 and 34. It is unusual that small peptides form stable three-dimensional structures. Indeed, the PP molecule is considered the smallest that can

Table 3
Sequence Identities of Various Neuroendocrine Peptides Between Humans and Fish

Peptide[a]	Human sequence compared to: Taxon	Species	Identities	%Identity	References: Human sequence Ref. no.	Year	Fish sequence Ref. no.	Year
Somatostatin	Cartilaginous fish	*Torpedo marmorata*	14/14	100	105	1982	115	1985
CRF	Bony fish	*Catostomus commersoni*	39/41	95	106	1983	116	1988
	Mammal	Sheep	34/41	83	106	1983	117	1981
NPY	Cartilaginous fish	*Torpedo marmorata*	33/36	92	24	1984	61a	1985
α-MSH	Cartilaginous fish	Spiny dogfish	12/13	92	107	1973	118	1974
Vasopressin	Bony fish	White sucker (vasotocin)	8/9	89	108	1958	119	1989
Glucagon	Cartilaginous fish	Lesser spotted dogfish	25/29	86	109	1983	120	1987
VIP	Cartilaginous fish	Lesser spotted dogfish	23/28	82	110	1983	121	1987
Oxytocin	Bony fish	White sucker (isotocin)	7/9	78	109	1958	119	1989
ACTH	Cartilaginous fish	Spiny dogfish	28/39	72	111	1972	122	1974
					107	1973		
Insulin	Cartilaginous fish	Spiny dogfish	36/51	71	112	1979	123	1983
Calcitonin	Bony fish	Salmon (calcitonin 2)	18/32	56	113	1968	125	1971
Relaxin	Cartilaginous fish	Spiny dogfish	21/51	41	114	1984	125	1986

[a]MSH stands for melanocyte-stimulating hormone, ACTH stands for adrenocorticotropic hormone, and CRF stands for corticotropin (ACTH)-releasing factor.

Table 4
Sequence Identities Between the Various Pancreatic Peptides, NPY, and PYY[a]

	PY gar	PY salmon	PY anglerfish	PY sculpin	PP pig	PP chicken	PP bullfrog	NPY Torpedo	PYY rat
PY gar	–	97	69	69	53	47	50	86	75
PY salmon	*35*	–	69	69	53	47	47	83	75
PY anglerfish	*25*	*25*	–	86	47	36	36	64	64
PY sculpin	*25*	*25*	*31*	–	47	39	36	64	61
PP pig	*19*	*19*	*17*	*17*	–	44	58	47	50
PP chicken	*17*	*17*	*13*	*14*	*16*	–	44	50	53
PP bullfrog	*18*	*17*	*13*	*13*	*21*	*16*	–	50	39
NPY Torpedo	*31*	*30*	*23*	*23*	*17*	*18*	*18*	–	64
PYY rat	*27*	*27*	*23*	*22*	*18*	*19*	*14*	*23*	–

[a]Sequence identities between the peptide sequences aligned in Fig. 12. The table shows both the number of identical positions (italics) and the percent identity.

```
                          1        10        20        30     36        Total
                          |        |         |         |      |
NPY Torpedo               YPSKPDNPGEGAPAEDLAKYYSALRHYINLITRQRY-amide
NPY family invariant       P  P  PG  A             Y    TR R -amide       9
NPY family highly cons.    P  P  PG  A  E         L  Y N  TR RY-amide     13
NPY-PYY-PY invariant      YP  P  PG  A  E    Y     RHY NL TRQRY-amide     18
NPY-PYY-PY highly cons.   YP KP  PGE A  E    YY   LRHY NL TRQRY-amide     24
```

Fig. 10. Alignment showing positions that are invariant or highly conserved among the peptides belonging to the NPY family or to the NPY-PYY-PY subset. Pro-2 is scored as invariant, although it differs in sheep PP. Highly conserved means that fewer than three sequences differ out of a total of 33 in the NPY family and a total of 17 in the NPY-PYY-PY subset. Also, an amino acid designated as highly conserved must be represented in all subsets.

form a regular tertiary structure *(128)*. PPs of bird, cow, and dog have been found to occur in solution as homodimers *(129)*, a structure expected to occur also in vivo in secretory granules.

Subsequent studies have shown that the PPs of other species may adopt the same tertiary structure *(130)*. Computer modeling has suggested also that NPY may fit in the same tertiary structure *(27,131)*. In fact, prolines 2, 5, and 8, as well as Gly at position 9, have remained conserved in all peptides belonging to the NPY family (except that Pro-2 has been replaced by Ser in sheep PP). Also, all peptides have Tyr or Phe at position 20 and Tyr at positon 27 (*see* Figs. 6–10). Surprisingly, recent NMR (nuclear magnetic resonance) analyses of porcine NPY in water solution *(132)* have shown that there is no contact between these two stems of the molecule. The proline-rich amino terminus was found to adopt many different conformations owing to *cis/trans*-isomerizations of the prolines. Furthermore, the NMR studies showed that the α helix encompasses residues 11–36. Nevertheless, the virtually perfect conservation of prolines 2, 5, and 8 indicates that these residues do play an important role. The NMR studies seem to confirm a dimeric structure.

Analyses of peptide fragments corresponding to portions of NPY have shown that the α-helix fragment 13–36 is sufficient for biological activity on the presynaptic type of NPY receptors called Y2 (*see* Grundemar et al. and Bleakman et al. this vol.). In contrast, the postsynaptic Y1 receptors require both the proline-rich segment and the α helix. The segment 7–17 of NPY has been found to be

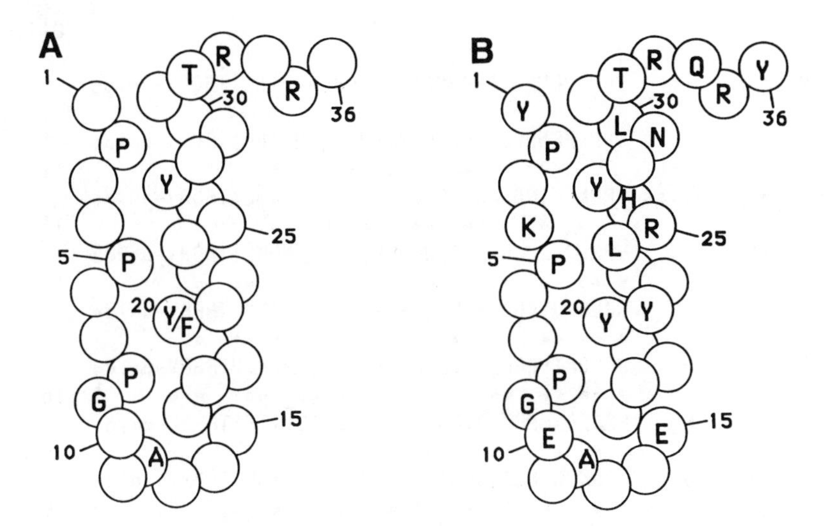

Fig. 11. Schematic structure of the NPY family of peptides adapted from Fuhlendorff et al. *(127)*. (A) shows invariant positions in the NPY family (*see* Fig. 10) and (B) shows amino acids that are highly conserved in the NPY-PYY-PY subset (*see* Fig. 10). Pro-2 is scored as invariant, although it differs in sheep PP. Highly conserved means that fewer than three sequences differ out of a total of 17 and that the amino acid is represented in all three subsets.

dispensable and could in fact be omitted in synthetic NPY analogs, which instead had the two parts (segments 1–6 and 18–36) linked together by a disulfide bridge *(133)*. Synthetic peptide constructs with more extensive central deletions showed reduced affinity to receptors *(133,134)*. The conservation between NPY and PP of the main structural features is further emphasized by receptor studies using a synthetic hybrid molecule *(127)* composed of porcine PP(1–30) and porcine NPY(31–36). This hybrid construct was almost as potent as intact NPY on Y2 receptors, although intact PP is inactive on this receptor *(135)*.

4.6. Implications of Sequence Analyses for Three-Dimensional Structure

As shown in Figs. 10 and 11A, nine positions are identical in all of the 33 sequences (disregarding position 2 in sheep PP), and another nine positions are identical in the 17 NPY, PYY, and PY sequences. Thus, 50% of the positions have remained perfectly con-

```
               1        10        20        30    36          Diff.
               |        |         |         |     |
PY  gar        YPPKPENPGEDAPPEELAKYYSALRHYINLITRQRY-amide
                                    ●                            1
PY  salmon     YPPKPENPGEDAPPEELAKYYTALRHYINLITRQRY-amide
                     ●  ●● ●  ●● ● ●● ●   ●                     11
PY  anglerfish YPPKPETPGSNASPEDWASYQAAVRHYVNLITRQRY-amide
                  ●  ●  ●● ●  ●●   ●● ●   ●                     11
PY  sculpin    YPPQPESPGGNASPEDWAKYHAAVRHYVNLITRQRY-amide
               ● ●● ●●  ●  ●  ●● ● ●●●  ●   ●●  ●               17
PP  pig        APLEPVYPGDDATPEQMAQYAAELRRYINMLTRPRY-amide
               ● ●● ●●  ●   ● ● ●●● ●● ●● ● ●●  ●               19
PP  chicken    GPSQPTYPGDDAPVEDLIRFYNDLQQYLNVVTRHRY-amide
               ● ●● ●●  ●● ● ●●  ●   ● ●●  ●●   ● ●             18
PP  bullfrog   APSEPHHPGDQATPDQLAQYYSDLYQYITFITRPRF-amide
                 ●  ●    ●  ● ●                                  5
NPY Torpedo    YPSKPDNPGEGAPAEDLAKYYSALRHYINLITRQRY-amide
                 ●   ●     ●    ●●  ●●    ●  ●                   9
PYY rat        YPAKPEAPGEDASPEELSRYYASLRHYLNLVTRQRY-amide
```

Fig. 12. Alignment of the various pancreatic peptides from fish and tetrapods, as well as *Torpedo* NPY and rat PYY. Dots mark differences to the top sequence. For references, *see* Table 2.

served in the NPY-PYY-PY peptides. Among these positions are the three prolines of the Pro helix, which are assumed to interdigitate with the two residues with aromatic side chains at positions 20 and 27. Figure 11B shows the 22 positions in the NPY-PYY-PY subset, which have remained highly conserved (as defined in the figure legend). There is not a single member of this peptide family that has undergone any insertions or deletions. Furthermore, all peptides possess a carboxyterminal amide in agreement with numerous studies showing that the amide group is essential for binding to receptors and for physiological activity. It has been shown that this amide is important for formation of the α helix *(128)*.

Positions 17, 21, 24, 28, and 31 face the same side of the α helix (see Fig. 11) and interact with the same region of a second monomer to form a dimer *(130)*. These positions always accommodate hydrophobic residues (except Gln-21 in anglerfish PY, His-21 in sculpin PY, and Arg-28 in goose PP). Most peptides have a leucine at position 17. A few PP sequences and half of NPY sequences have a methionine at this position, which may cause the dimers to become less stable *(see 130)*. Thus, the propensity of NPY to form dimers may differ slightly among species *(see also 61a)*. Two positions on the hydrophilic face of the amphiphilic α helix, i.e., Arg-19 and His-26 (*see* Fig. 11) have been proposed to be important for activity on the Y2 receptor studied on rat vas deferens *(91)*. Since all three of the synthetic NPY variants that were studied had replacements at both of these positions, it was not possible to assess whether position 19 or 26 or both were of importance in this context. The fact that His-26 is identical in all NPY-PYY-PY sequences, whereas position 19 varies even among the NPY sequences (Lys in goldfish and *Torpedo)* indicates that His-26 may be a crucial amino acid for Y2 activity. In agreement with this, the short synthetic fragment NPY(25-36) binds to receptor *(137)*, and both NPY(25-36) and PYY(25-36) do show activity on rat vas deferens *(138)*, although at high concentrations.

The positions 32–36 are not involved in either tertiary or quaternary interactions according to Glover et al. *(130)*, who found that the α helix ends at position 31 in turkey PP. In contrast, the NMR studies of porcine NPY *(132)* indicated that the α helix extends all the way to the carboxyterminus. Whatever the secondary structure, these positions are presumably important for receptor interactions, since they harbor three perfectly conserved positions, 32, 33, and 35, and two highly conserved positions, 34 and 36 (Figs. 10 and 11). Indeed, it has been found that replacement of Gln-34 in NPY with Pro (as in the mammalian PP sequences) abolishes binding to Y2 receptors with little effect on Y1-receptor binding *(139,140)*.

Positions 15–17 at the beginning of the α helix seem to tolerate only fairly conservative replacements (positions 15 and 16: Asp, Glu, or Gln; and position 17: Leu, Met, or Trp). Nevertheless, PYY and NPY differ from each other considerably in this region with five replacements out of six in positions 13–18. The fact that NPY and PYY are able to compete virtually equipotently at the same receptor sites suggests that this region is not involved in direct interactions with receptors in agreement with binding studies using peptide analogs lacking this region *(133)*. On the other hand, this segment has remained conserved among the known NPY sequences. This may suggest also that this part of the peptide is involved in interactions with other molecules (nonreceptors or receptor subtypes) and therefore needs to be preserved.

5. Evolution of the NPY Family

5.1. NPY and PYY Are Distinct Subsets

Sequence comparison of the various members of the NPY peptide family clearly identifies the NPY sequences and the PYY sequences as two distinct, but closely related subsets (approx 70% identity). In contrast, the pancreatic peptides of both fish (PY) and tetrapods (PP) display extensive sequence diversity, which makes it difficult to assess their relationship to each other as well as to NPY and PYY. Gar and salmon PY are most similar to *Torpedo* NPY (86 and 83% identity, respectively). Anglerfish and sculpin PY are both almost equally identical to NPY (64%) and to gar PY (69%) (Fig. 12 and Table 4). In fact, it has been questioned whether bony fish have a true NPY peptide in addition to their pancreatic PY peptide.

5.2. The Goldfish Has a True NPY Peptide

Our cloning work in goldfish provides persuasive evidence that bony fish do have a true NPY peptide in addition to their pancreatic peptide PY. The peptide predicted from our goldfish clone is likely to correspond to NPY because:

1. It shows higher sequence identity to mammalian NPY than do the fish PY peptides;
2. It shows higher identity to mammalian NPY than to fish PY;
3. The clone was derived from a neuronal source, i.e., retina; and
4. It has a carboxyterminal extension that is considerably more similar to mammalian CPON than to the anglerfish extension, and also more similar to CPON than is the anglerfish extension *(see 61a)*.

Because of the strong similarity among the mature NPY, PYY, and PY peptides, we anticipate that comparisons of prepro-peptide sequences may turn out to be informative in tracing the evolutionary interrelationships of these peptides. Also the carboxyterminal extension of the peptide predicted from our *Torpedo* cDNA clone is more similar to the extension of mammalian NPY than to those of PYY and PY *(61a)*. Our structural evidence for a goldfish NPY peptide that is distinct from PY is in agreement with reports describing the occurrence of both NPY-like and PY-like peptides in anglerfish pancreas *(141)* and brain *(70,142)* based on immunoreactivity and HPLC. NPY-like immunoreactivity has also been detected in the brain of goldfish *(71)* and Atlantic salmon, *Salmo salar (73)*. Goldfish retina has been reported to contain two NPY-like peptides *(143)*. Also the brain of a cartilaginous fish, the lesser spotted dogfish, *Scyliorhinus canicula,* contains NPY-like immunoreactivity *(75)*.

5.3. NPY Is One of the Most Conserved Neuroendocrine Peptides

We show here that NPY has remained extremely well conserved during the major part of vertebrate evolution (Fig. 6). NPY can be identified even in cartilaginous fish, as demonstrated by our cDNA clone from the ray *Torpedo marmorata (61a)* as well as partial sequence information from the horned shark *Heterodontus francisci* (Larhammar et al., unpublished). NPY shows 92% sequence identity between *Torpedo* and mammals, an evolutionary distance of more than 400 million years. The significance of this finding is illustrated in Table 3, which summarizes several sequence comparisons of neuroendocrine peptides between humans and fish. Only somatostatin has a higher degree of conservation than NPY. However, short neuroendocrine peptides like somatostatin would be expected to evolve more slowly, as each amino acid replacement would have a relatively greater impact on the overall structure of such a short peptide. In the light of this consideration, the strong conservation of NPY appears even more conspicuous, and there is no neuroendocrine peptide longer than NPY that can compete with its degree of conservation. The strong sequence conservation of NPY suggests that it subserves very important functions and/or has multiple points of interaction with other molecules.

5.4. PYY May Be Conserved, Too

Because of the limited PYY sequence data available, it remains unclear how long PYY and NPY have existed as separate entities. All mammals that have been investigated seem to have neuroendocrine cells in the gut with PYY-like immunoreactivity, i.e., humans, monkey, dog, rat, and mouse *(15,144–146)*. In fact, it is still unknown whether PYY exists in nonmammalian vertebrates. It is an interesting possibility that the PP-like immunoreactivity that has been detected in the alimentary tract epithelium of the tunicate *Ciona intestinalis (91)*, as well as the NPY-like immunoreactivity in locust endocrine midgut cells *(94)*, actually corresponds to mammalian PYY. From a zebrafish genomic library, we have isolated distinct clones with a goldfish NPY probe and a rat PYY probe, indicating that a single fish species indeed has separate homologs of NPY and PYY (Söderberg et al., submitted). However, the identities of these zebrafish clones still remain to be confirmed by nucleotide sequencing.

5.5. PP and PY Are Divergent Peptides

Whereas NPY and PYY may act on the same types of receptors, PP seems to act on a receptor that is distinct from both the Y1 and Y2 NPY/PYY receptor subtypes *(135,147,148)*. The differences in receptor actions between NPY/PYY on the one hand and PP on the other, along with the higher degree of sequence identity between NPY and PYY (70%) than between either of these with PP (50%), have been taken as evidence for a closer evolutionary relationship between NPY and PYY. It has been suggested that PP forms one evolutionary lineage and NPY/PYY a different lineage, and that the latter two peptides diverged from each other more recently *(149)*. Also, immunological phylogenetic studies of the vertebrate pancreas have been interpreted to show that PP-like immunoreactive cells appeared earlier in evolution than PYY-immunoreactive cells and NPY-immunoreactive neurons *(80)*. However, these results may include crossreactivities owing to broader spectra of specificities of the PP antisera used than of the NPY and PYY antisera.

The sequence alignments show that, in contrast to NPY, the pancreatic polypeptides of mammals, birds, a reptile, an amphibian, and teleost fish vary greatly (*see* Fig. 12 and Table 4). PY of gar and salmon are no more similar to PY of anglerfish and sculpin than mammalian NPY is to mammalian PYY. Furthermore, there are no apparent structural features that define the fish PY sequences and the mammalian, avian-crocodile, and bullfrog PP sequences as one distinct group. Thus, it cannot be formally ruled out that the various types of pancreatic peptides branched off from the NPY/PYY lineage on different occasions, possibly as many as five:

1. PY from gar and salmon;
2. PY from anglerfish and sculpin (although these peptides are slightly more identical to gar PY than to NPY);
3. PP from bullfrog;
4. PP from birds and alligator; and
5. PP from mammals.

However, since all five of these subgroups are expressed in the pancreas, and the latter three exclusively in the pancreas, it is unlikely that they all diverged separately after duplication of an NPY-like gene. Assuming that all of the pancreatic peptides share a common pancreatic ancestor, their rate of sequence divergence must have been dramatically faster than that of NPY. The tetrapod PPs have diverged particularly rapidly, because more than half of the positions have been replaced since birds and mammals diverged. We show here that NPY, in contrast, has remained virtually constant during the same period of time, since chicken differs from most mammals at only a single position. Thus, the sequence diversity among pancreatic peptides cannot be used to argue that this lineage is older than the NPY/PYY lineage, or that it branched off before NPY and PYY diverged from each other. In conclusion, the PP and PY sequences are clearly separate from the NPY and PYY subsets by structural criteria, but this cannot be taken to reflect older evolutionary age.

5.6. PP May Have Arisen Through Duplication of the PYY Gene

The pancreas has not existed as a distinct organ throughout vertebrate evolution (for reviews, *see 150,151*). In tunicates, which have a notochord only in their larval stage, there are no discrete Langerhans islet clusters. Insulin, somatostatin, and PP immunoreactivities have been localized in typical gut endocrine cells. The two cyclostome groups, hagfishes and lampreys, have a separate islet organ that consists of insulin cells and somatostatin cells. Somatostatin cells occur both in the islet organ and in the gut, whereas PP cells and glucagon cells are present in the gut mucosa only. Among the cartilaginous fishes, holocephalians, such as the ratfish, have Langerhans islets consisting of three cell types producing insulin, somatostatin, and glucagon, respectively. PP cells are found in the mucosa of the pancreatic duct. In sharks and rays, the pancreas constitutes a four-hormone islet organ as in all higher vertebrates. Thus, it seems like the PP-producing islet cells were recruited from gut endocrine cells.

The mammalian gut endocrine cells contain PYY. Since the pancreatic PP-producing cells seem to be derived from the gut endocrine cells, we propose that the PP gene arose through duplication of the PYY gene (Fig. 13), which was already committed to expres-

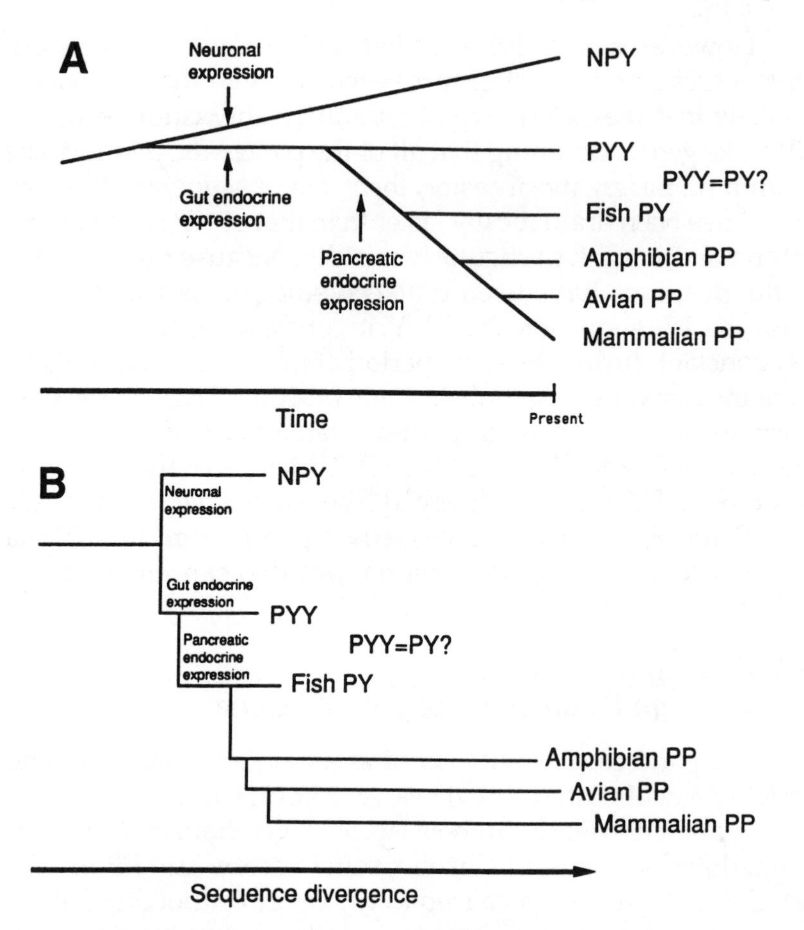

Fig. 13. Cartoons showing possible evolutionary relationships between the members of the NPY family. Note that the PYY gene is expressed also in the pancreas. Thus, fish PY may be the homolog of mammalian PYY, and the gene duplication leading to PP may have occurred after the tetrapods diverged from the bony fish. (A) Branch points schematically related to time. The cartoon only shows the hypothetical relative order of the branching events, not their distances in time. (B) Branch lengths schematically (not quantitatively) related to extent of divergence.

sion in endocrine cells. Indeed, PYY peptide and mRNA can be found also in the pancreas *(8,11,13–16)* and PYY immunoreactivity has been reported in a subset of PP-producing pancreatic cells in the mouse *(12)*. PYY mRNA has been detected in purified pancreatic islets *(13)*. These findings indicate that some gene-regulatory features may still be shared between the PYY and PP genes. After the gene duplication, the PP lineage has undergone considerable sequence divergence, probably because of weaker constraints after the duplication event. The notion that PYY and PP share a more recent common ancestor than either does with NPY lends further support from the similar exon–intron organization of the PYY and PP genes: Both the rat PYY gene and the human and rat PP genes have considerably smaller introns than the known NPY genes (Fig. 3). It will be interesting to see whether the fish genes display this striking difference in intron sizes between NPY on the one hand and PY/PYY on the other (if a separate PYY gene is present in fish).

If the postulated duplication of the PYY gene occurred before the tetrapods branched off from the bony fish, fish should also have one PYY and one PP homolog. So far, only PY has been described in fish in addition to NPY. PY has diverged less among fish than PP has among tetrapods, and furthermore PY is more similar to PYY (and NPY) than to PP. Thus, it is possible that fish PY is the homolog of mammalian PYY. If so, PY should be expected to be expressed in both gut and pancreas, like PYY in mammals. Whereas anglerfish and sculpin PY are equally similar to rat PYY and *Torpedo* NPY (Fig. 12 and Table 4), the gar and salmon PY peptides are more similar to NPY than to PYY. Gar PY actually shares the last 20 positions (17–36) with *Torpedo* NPY. This may suggest that the gar and salmon PY genes still retain many features of the ancestral duplicated NPY, whereas PY of anglerfish and sculplin as well as mammalian PYY have diverged more. The possibility that PYY is more recent than PP is unlikely, since PYY is expressed in the cells that seem to have given rise to the PP-expressing cells, yet PYY is more similar to NPY and should have arisen through duplication of the NPY gene.

Since PP has been shown to act on a receptor that is distinct from the NPY/PYY receptors *(135,147,148)*, the gene duplication of the NPY or PYY gene that gave rise to the PP gene was presumably paralleled or followed by a duplication of a gene encoding an NPY/PYY receptor. The new gene encoding the ligand PP has probably evolved in parallel with the new gene encoding the PP receptor. Future studies will show whether the structures of the NPY/PYY receptors *vis-à-vis* the PP receptor will reflect the sequence divergence of the different ligands. It is still possible that the NPY family of peptides includes additional, yet undiscovered members, although our DNA crosshybridizations have so far failed to detect any such candidates even under conditions of low stringency.

6. Summary

The neuropeptide Y family of peptides consists of neuropeptide Y (NPY), the gut endocrine peptide YY (PYY), and the pancreatic endocrine peptides called pancreatic polypeptide (PP) in tetrapods and peptide tyrosine (PY) in fish. We analyze here the interrelationships of the 33 currently available sequences belonging to this peptide family. NPY sequences are known for several mammals, chicken, goldfish, and the ray *Torpedo marmorata*. NPY of mammals is identical to NPY of *Torpedo* at 33 positions out of 36 (92%). Thus, NPY is one of the most highly conserved neuroendocrine peptides known. The sequence analyses strongly indicate that bony fish have a true NPY peptide in addition to their pancreatic peptide PY. The evolutionary relationships of the pancreatic peptides to NPY and PYY cannot be easily resolved by the available sequence information. We propose that the pancreatic peptides arose through duplication of the PYY gene. This hypothesis is based on the relationship between gut and pancreatic endocrine cells, as well as the observation that the PYY and PP genes have a much more compact exon–intron organization than the known NPY genes. After this gene duplication, the PP lineage has undergone extensive sequence divergence.

Acknowledgments

We wish to express our gratitude to Ingrid Lundell for skilful experimental assistance and to Andrew B. Leiter for unpublished data. Kent Dimberg, Göran Gezelius, Stefan Nilsson, Lars Pilström, Peter Wallén, and Jan Zabielski are gratefully acknowledged for providing animal tissues for extraction of DNA for the zoo blots. We thank Ingrid Lundell, Göran Andersson, Lennart Brodin, Lars Grundemar, and Claes Wahlestedt for comments on the manuscript. This work was supported by a grant from the Swedish Natural Science Research Council (No. B-BU 8524-308). During part of this work, D. L. was the recipient of an EMBO long-term fellowship.

References

1. Kimmel, J. R., Pollock, H. G., and Hazelwood, R. L. (1968) Isolation and characterization of chicken insulin. *Endocrinology* **83,** 1323–1330.
2. Tatemoto, K. and Mutt, V. (1980) Isolation of two novel candidate hormones using a chemical method for finding naturally occurring polypeptides. *Nature* **285,** 417–418.

3. Tatemoto, K., Carlquist, M., and Mutt, V. (1982) Neuropeptide Y—a novel brain peptide with structural similarities to peptide YY and pancreatic polypeptide. *Nature* **296,** 659–660.
4. de Quidt, M. E. and Emson, P. C. (1986) Neuropeptide Y in the adrenal gland: Characterization, distribution and drug effects. *Neuroscience* **19,** 1011–1022.
5. Ericsson, A., Schalling, M., McIntyre, K. R., Lundberg, J. M., Larhammar, D., Seroogy, K., Hökfelt, T., and Persson, H. (1987) Detection of neuropeptide Y and its mRNA in megakaryocytes: Enhanced levels in certain autoimmune mice. *Proc. Natl. Acad. Sci. USA* **84,** 5585–5589.
6a. Myers, A. K., Farhat, M. Y., Vaz, C. A., Keiser, H. R., and Zukowska-Grojec, Z. (1988) Release of immunoreactive neuropeptide Y by rat platelets. *Biochem. Biophys. Res. Comm.* **155,** 118–122.
6b. Jamal, H., Jones, P. M., Byrne, J., Suda, K., Ghatei, M. A., Kanse, S. M., and Bloom, S. R. (1991) Peptide contents of neuropeptide Y vasoactive intestinal polypeptide, and β-calcitonin gene-related peptide and their messenger ribonucleic acids after dexamethasone treatment in the isolated rat islets of Langerhans. *Endocrinology* **129,** 3372–3380.
7. Broomé, M., Hökfelt, T., and Terenius, L. (1985) Peptide YY (PYY)-immunoreactive neurons in the lower brain stem and spinal cord of rat. *Acta Physiol. Scand.* **125,** 349–352.
8. DiMaggio, D. A., Chronwall, B. M., Buchanan, K., and O'Donohue, T. L. (1985) Pancreatic polypeptide immunoreactivity in rat brain is actually neuropeptide Y. *Neuroscience* **15,** 1149–1157.
9. Ekman, R., Wahlestedt, C., Böttcher, G., Sundler, F., Håkanson, R., and Panula, P. (1986) Peptide YY-like immunoreactivity in the central nervous system of the rat. *Regul. Peptides* **16,** 157–168.
10. Pieribone, V., Brodin, L., Friberg, K., Dahlstrand, J., Söderberg, C., Larhammar, D., and Hökfelt, T. (1992) Differential expression of mRNAs for neuropeptide Y-related peptides in rat nervous tissues: Possible evolutionary conservation. *J. Neuroscience,* in press.
11. Ali-Rachedi, A., Varndell, I. M., Adrian, T. E., Gapp, D. A., Van Noorden, S., Bloom, S. R., Polak, J. M. (1984) Peptide YY (PYY) immunoreactivity is co-stored with glucagon-related immunoreactants in endocrine cells of the gut and pancreas. *Histochemistry* **80,** 487–491.
12. Böttcher G., Ahrén, B., Lundquist, I., and Sundler, F. (1989) Peptide YY: Intrapancreatic localization and effects on insulin and glucagon secretion in the mouse. *Pancreas* **4,** 282–288.
13. Krasinski, S. D., Wheeler, M. B., and Leiter, A. B. (1991) Isolation, characterization, and developmental expression of the rat peptide-YY gene. *Mol. Endocrinol.* **5,** 433–440.
14. Leiter, A. B., Toder, A., Wolfe, H. J., Taylor, I. L., Cooperman, S., Mandel, G., and Goodman, R. H. (1987) Peptide YY: Structure of the precursor and expression in exocrine pancreas. *J. Biol. Chem.* **262,** 12984–12988.
15. Lundberg, J. M., Tatemoto, K., Terenius, L., Hellström, P. M., Mutt, V., Hökfelt, T., and Hamberger, B. (1982) Localization of peptide YY (PYY) in gastrointestinal endocrine cells and effects on intestinal blood flow and motility. *Proc. Natl. Acad. Sci. USA* **79,** 4471–4475.

16. Falkmer, S., Dafgård, E., El-Salhy, M., Engström, W., Grimelius, L., and Zetterberg, A. (1985) Phylogenetical aspects on islet hormone families: a minireview with particular reference to insulin as a growth factor and to the phylogeny of PYY and NPY immunoreactive cells and nerves in the endocrine and exocrine pancreas. *Peptides 6,* **Suppl. 3,** 315–320.
17. Wahlestedt, C., Yanaihara, N., and Håkanson, R. (1985) Evidence for different pre- and post-junctional receptors for neuropeptide Y and related peptides. *Regul. Peptides* **13,** 317–328.
18. Larsson, L-I., Sundler, F., Håkanson, R., Pollock, H. G., and Kimmel, J. R. (1974) Localization of APP, a postulated new hormone, to a pancreatic endocrine cell type. *Histochemistry* **42,** 377–382.
19. Kimmel, J. R., Hayden, L. J., and Pollock, H. G. (1975) Isolation and characterization of a new pancreatic polypeptide hormone. *J. Biol. Chem.* **250,** 9369–9374.
20. Tatemoto, K. (1982) Neuropeptide Y: Complete amino acid sequence of the brain peptide. *Proc. Natl. Acad. Sci. USA* **79,** 5485–5489.
21. Tatemoto, K. (1989) Neuropeptide Y: Isolation, structure and function, in *Neuropeptide Y* (V., Mutt, K., Fuxe, T., Hökfelt, J. M., Lundberg, eds.), Raven, New York, pp. 13–21.
23. O'Hare, M. M. T., Tenmoku, S., Aakerlund, L., Hilsted, L., Johnsen, A., and Schwartz, T. W. (1988) Neuropeptide Y in guinea pig, rabbit, rat and man. Identical amino acid sequence and oxidation of methionine-17. *Regul. Peptides* **20,** 293–304.
24. Minth, C. D., Bloom, S. R., Polak, J. M., and Dixon, J. E. (1984) Cloning, characterization, and DNA sequence of a human cDNA encoding neuropeptide tyrosine. *Proc. Natl. Acad. Sci. USA* **81,** 4577–4581.
25. Minth, C. D., Andrews, P. C., and Dixon, J. E. (1986) Characterization, sequence, and expression of the cloned human neuropeptide Y gene. *J. Biol. Chem.* **261,** 11974–11979.
26. Corder, R., Gaillard, R. C., and Böhlen, P. (1988) Isolation and sequence of rat peptide YY and neuropeptide Y. *Regul. Peptides* **21,** 253–261.
27. Allen, J., Novotny, J., Martin, J., and Heinrich, G. (1987) Molecular structure of mammalian neuropeptide Y: Analysis by molecular cloning and computer-aided comparison with crystal structure of avian homologue. *Proc. Natl. Acad. Sci. USA* **84,** 2532–2536.
28. Higuchi, H., Yang, H-Y. T., and Sabol, S. L. (1988) Rat neuropeptide Y precursor gene expression. *J. Biol. Chem.* **263,** 6288–6295.
29. Larhammar, D., Ericsson, A., and Persson, H. (1987) Structure and expression of the rat neuropeptide Y gene. *Proc. Natl. Acad. Sci. USA* **84,** 2068–2072.
30. Allen, J. M. (1989) Molecular structure of neuropeptide Y and regulation of expression of its gene, in Neuropeptide Y (Mutt, V., Fuxe, K., Hökfelt, T., and Lundberg, J. M., eds.), New York, Raven, pp. 33–41.
31. Chartrel, N., Conlon, J. M., Danger, J-M., Fournier, A., Tonon, M-C., and Vaudry, H. (1991) Characterization of melanotropin-release-inhibiting factor (melanostatin) from frog brain: Homology with human neuropeptide Y. *Proc. Natl. Acad. Sci. USA* **88,** 3862–3866.
32. Tatemoto, K. (1982) Isolation and characterization of peptide YY (PYY), a candidate gut hormone that inhibits pancreatic exocrine secretion. *Proc. Natl. Acad. Sci. USA* **79,** 2514–2518.

33. Tatemoto, K., Nakano, I., Makk, G., Angwin, P., Mann, M., Schilling, J., and Go, V. L. W. (1988) Isolation and primary structure of human peptide YY. *Biochem. Biophys. Res. Comm.* **157,** 713–717.
34. Chance, R. E., Johnson, M. G., Hoffman, J. A., and Lin, T-M. (1979) Pancreatic polypeptide: a newly recognized hormone, in *Proinsulin, Insulin, C-peptide* (Baba, S., Kaneko, T. , and Yanaihara, N., eds.), Excerpta Medica, Amsterdam, pp. 419–425.
35. Chance, R. E., Moon, N. E., and Johnson, M. G. (1979b) Human pancreatic polypeptide (HPP) and bovine pancreatic polypeptide (BPP), in *Methods of Hormone Radioimmunoassay* 2nd ed. (Jaffe, B. M. and Behrman, H. R., eds.), Academic, New York, pp. 657–672.
36. Nielsen, H. V., Gether, U., and Schwartz, T. W. (1986) Cat pancreatic eicosapeptide and its biosynthetic intermediate. *Biochem. J.* **240,** 69–74.
37. Leiter, A. B., Keutmann, H. T., and Goodman, R. H. (1984) Structure of a precursor to human pancreatic polypeptide. *J. Biol. Chem.* **259,** 14702–14705.
38. Boel, E., Schwartz, T. W., Norris, K. E., and Fiil, N. P. (1984) A cDNA encoding a small common precursor for human pancreatic polypeptide and pancreatic icosapeptide. *EMBO J.* **3,** 909–912.
39. Takeuchi, T. and Yamada, T. (1985) Isolation of a cDNA clone encoding pancreatic polypeptide. *Proc. Natl. Acad. Sci. USA* **82,** 1536–1539.
40. Leiter, A. B., Montminy, M. R., Jamieson, E., and Goodman, R. H. (1985) Exons of the human pancreatic polypeptide gene define functional domains of the precursor. *J. Biol. Chem.* **260,** 13013–13017.
41. Kimmel, J. R., Pollock, H. G., Chance, R. E., Johnson, M. G., Reeve, J. R., Taylor, I. L., Miller, C., and Shively, J. E. (1984) Pancreatic polypeptide from rat pancreas. *Endocrinology* **114,** 1725–1731.
42. Yamamoto, H., Nata, K., and Okamoto, H. (1986) Mosaic evolution of prepropancreatic polypeptide. *J. Biol. Chem.* **261,** 6156–6159.
43. Yonekura, H., Nata, K., Watanabe, T., Kurashina, Y., Yamamoto, H., and Okamoto, H. (1988) Mosaic evolution of prepropancreatic polypeptide. II. Structural conservation and divergence in pancreatic polypeptide gene. *J. Biol. Chem.* **263,** 2990–2997.
44. Eng, J., Huang, C-G., Pan, Y-C. E., Hulmes, J. D., and Yalow, R. S. (1987) Guinea pig pancreatic polypeptide: Structure and pancreatic content. *Peptides* **8,** 165–168.
45. Blackstone, C. D., Seino, S., Takeuchi, T., Yamada, T., and Steiner, D. F. (1988) Novel organization and processing of the guinea pig pancreatic polypeptide precursor. *J. Biol. Chem.* **263,** 2911–2916.
46. Lance, V., Hamilton, J. W., Rouse, J. B., Kimmel, J. R., Pollock, H. G. (1984) Isolation and characterization of reptilian insulin, glucagon, and pancreatic polypeptide: Complete amino acid sequence of alligator *(Alligator mississippiensis)* insulin and pancreatic polypeptide. *Gen. Comp. Endocrinol.* **55,** 112–124.
47. Blundell, T. L., Pitts, J. E., Tickle, I. J., Wood, S. P., and Wu, C-W. (1981) X-ray analysis (1.4-Å resolution) of avian pancreatic polypeptide: Small globular protein hormone. *Proc. Natl. Acad. Sci. USA* **78,** 4175–4179.
48. Xu, Y., Lin, N., and Zhan, Y. (1984) Isolation and sequence determination of goose pancreatic polypeptide. *Scientia Sinica* (Series B) **27,** 590–592.

49. Litthauer, D. and Oelofsen, W. (1987) Purification and primary structure of ostrich pancreatic polypeptide. *Int. J. Peptide Protein Res.* **29,** 739–745.
50. Pollock, H. G., Hamilton, J. W., Rouse, J. B., Ebner, K. E., Rawitch, A. B. (1988) Isolation of peptide hormones form the pancreas of the bullfrog *(Rana catesbeiana). J. Biol. Chem.* **263,** 9746–9751.
51. Andrews, P. C., Hawke, D., Shively, J. E., and Dixon, J. E. (1985) A nonamidated peptide homologous to porcine peptide YY and neuropeptide YY. *Endocrinol.* **116,** 2677–2681.
52. Balasubramaniam, A., Andrews, P. C., Renugopalakrishnan, V., and Rigel, D. F. (1989) Glycine-extended anglerfish peptide YG (aPY) a neuropeptide Y (NPY) homologue may be a precursor of a biologically active peptide. *Peptides* **10,** 581–585.
53. Conlon, J. M., Schmidt, W. E., Gallwitz, B., Falkmer, S., and Thim, L. (1986) Characterization of an amidated form of pancreatic polypeptide from the daddy sculpin *(Cottus scorpius). Regul. Peptides* **16,** 261–268.
54. Cutfield, S. M., Carne, A., and Cutfield, J. F. (1987) The amino-acid sequences of sculpin islet somatostatin-28 and peptide YY. *FEBS Lett.* **214,** 57–61.
55. Kimmel, J. R., Plisetskaya, E. M., Pollock, H. G., Hamilton, J. W., Rouse, J. B., Ebner, K. E., and Rawitch, A. B. (1986) Structure of a peptide from coho salmon endocrine pancreas with homology to neuropeptide Y. *Biochem. Biophys. Res. Comm.* **141,** 1084–1091.
56. Pollock, H. G., Kimmel, J. R., Hamilton, J. W., Rouse, J. B., Ebner, K. E., Lance, V., and Rawitch, A. B. (1987) Isolation and structures of alligator gar *(Lepisosteus spatula)* insulin and pancreatic polypeptide. *Gen. Comp. Endocrinol.* **67,** 375–382.
57. Greenberg, G. R., McCloy, R. F., Adrian, J. E., Chadwick, V. S., Baron, J. H., and Bloom, S. R. (1978) Inhibition of pancreas and gallbladder by pancreatic polypeptide. Lancet II: 1280–1282.
58. Lin, T-M., Evans, K. C., Chance, R. E., and Spray, G. F. (1977) Bovine pancreatic polypeptide: Actions on gastric and pancreatic secretion in dogs. *Am. J. Physiol.* **232,** E311–315.
59. Taylor, I. L., Solomon, T. E., Walsh, J., and Grossman, M. I. (1979) Pancreatic polypeptide: Metabolism and effect on pancreatic secretion in dogs. *Gastroenterology* **76,** 524–528.
60. Tatemoto, K. and Mutt, V. (1978) Chemical determination of polypeptide hormones. *Proc. Natl. Acad. Sci. USA* **75,** 4115–4119.
61a. Pappas, T. N., Debas, H. T., and Taylor, I. L. (1985) Peptide YY: Metabolism and effect on pancreatic secretion in dogs. *Gastroenterology* **89,** 1387–1392.
61b. Blomqvist, A. G., Söderberg, C., Lundell, I., Milner, R. J., and Larhammar, D. (1992) Strong evolutionary conservation of neuropeptide Y: Sequences of chicken, goldfish, and *Torpedo marmorata* DNA clones. *Proc. Natl. Acad. Sci. USA* **89,** 2350–2354.
62. Gray, T. S. and Morley, J. E. (1986) Neuropeptide Y: Anatomical distribution and possible function in mammalian nervous system. *Life Sci.* **38,** 389–401.
63. Fujita, T., Yui, R., Iwanaga, T., Nishiitsutsuji-Uwo, J., Endo, Y., and Yanaihara, N. (1981) Evolutionary aspects of "brain-gut peptides": An immunohistochemical study. *Peptides* 2, **Suppl. 2,** 123–131.

64. Reiner, A. and Northcutt, R. G. (1987) An immunohistochemical study of the telencephalon of the African lungfish *Protopterus annectens.*
65. Anderson, K. D. and Reiner, A. (1990) Distribution and relative abundance of neurons in the pigeon forebrain containing somatostatin, neuropeptide Y, or both. *J. Comp. Neurol.* **299,** 261–282.
66. Marti, E., Bello, A. R., Lancha, A., and Batista, M. A. P. (1990) Neuropeptide tyrosine (NPY) and its C-terminal flanking peptide (C-PON) in the developing and adult spinal cord of a reptile. *J. Anat.* **172,** 149–156.
67. Reiner, A. and Oliver, J. R. (1987) Somatostatin and neuropeptide Y are almost exclusively found in the same neurons in the telencephalon of turtles. *Brain Res.* **426,** 149–156.
68. Danger, J-M., Guy, J., Benyamina, M., Jégou, S., Leboulenger, F., Coté, J., Toton, M. C., Pelletier, G., and Vaudry, H. (1985) Localization and identification of neuropeptide Y-like immunoreactivity in the frog brain. *Peptides* **6,** 1225–1236.
69. Perroteau, I., Danger, J-M., Biffo, S., Pelletier, G., Vaudry, H., and Fasolo, A. (1988) Distribution and characterization of neuropeptide Y-like immunoreactivity in the brain of the crested newt. *J. Comp. Neurol.* **275,** 309–325.
70. Noe, B. D., Milgram, S. L., Balasubramaniam, A., Andrews, P. C., Calka, J., and McDonald, J. K. (1989) Localization and characterization of neuropeptide Y-like peptides in the brain and islet organ of the anglerfish *(Lophius americanus). Cell Tissue Res.* **257,** 303–311.
71. Pontet, A., Danger, J-M., Dubourg, P., Pelletier, G., Vaudry, H., Calas, A., and Kah, O. (1989) Distribution and characterization of neuropeptide Y-like immunoreactivity in the brain and pituitary of the goldfish. *Cell Tissue Res.* **255,** 529–538.
72. Moons, L., Cambré, M., Ollevier, F., and Vandesande, F. (1989) Immunocytochemical demonstration of close relationships between neuropeptidergic nerve fibers and hormone-producing cell types in the adenohypophysis of the sea bass *(Dicentrarchus labrax). Gen. Comp. Endocrinol.* **73,** 270–283.
73. Vecino, E. and Ekström, P. (1990) Distribution of Met-enkephalin, Leu-enkephalin, substance P, neuropeptide Y, FMRFamide, and serotonin immunoreactivitites in the optic tectum of the Atlantic salmon (*Salmo salar* L.). *J. Comp. Neurol.* **299,** 229–241.
74. Danger, J-M., Breton, B., Vallarino, M., Fournier, A., Pelletier, G., and Vaudry, H. (1991) Neuropeptide-Y in the trout brain and pituitary: Localization, characterization, and action on gonadotropin release. *Endocrinology* **128,** 2360–2368.
75. Vallarino, M., Danger, J-M., Fasolo, A., Pelletier, G., Saint-Pierre, S., and Vaudry, H. (1988) Distribution and characterization of neuropeptide Y in the brain of an elasmobranch fish. *Brain Res.* **448,** 67–76.
76. Van Dongen, P. A. M., Hökfelt, T., Grillner, S., Verhofstad, A. A. J., Steinbusch, H. W. M., Cuello, A. C., and Terenius, L. (1985) Immunohistochemical demonstration of some putative neurotransmitters in the lamprey spinal cord and spinal ganglia: 5-hydroxytryptamine, tachykinin-, and neuropeptide-Y-immunoreactive neurons and fibers. *J. Comp. Neurol.* **234,** 501–522.
77. Negishi, K., Kiyama, H., Kato, S., Teranishi, T., Hatakenaka, S., Katayama, Y., Miki, N., and Tohyama, M. (1986) An immunohistochemical study on the river lamprey retina. *Brain Res.* **362,** 389–393.

78. Horn, J. P., Stofer, W. D., and Fatherazi, S. (1987) Neuropeptide Y-like immunoreactivity in bullfrog sympathetic ganglia is restricted to C cells. *J. Neurosci.* **7**, 1717–1727.
79. de Quidt, M., Kiyama, H., Emson, P. C. (1990) Pancreatic polypeptide, neuropeptide Y and peptide YY in central neurons, in *Handbook of Chemical Neuroanatomy*, vol. 9, Neuropeptides in the CNS, Part II (A., Björklund, T., Hökfelt, and M. J. Kuhar, eds.), Amsterdam, Elsevier, pp. 287–357.
80. El-Salhy, M., Grimelius, L., Emson, P. C., and Falkmer, S. (1987) Polypeptide YY- and neuropeptide Y-immunoreactive cells and nerves in the endocrine and exocrine pancreas of some vertebrates: an onto- and phylogenetic study. *Histochem. J.* **19**, 111–117.
81. Bjenning, C. and Holmgren, S. (1988) Neuropeptides in the fish gut: An immunohistochemical study of evolutionary patterns. *Histochemistry* **88**, 155–163.
82a. Burkhardt-Holm, P. and Holmgren, S. (1989) A comparative study of neuropeptides in the intestine of two stomachless teleosts *(Poecilia reticulata, Leuciscus idus melanotus)* under conditions of feeding and starvation. *Cell Tissue Res.* **255**, 245–254.
82b. Buchanan, J. T., Brodin, L., Hökfelt, T., Van Dongen, P. A. M., Grillner, S. (1987) Survey of neuropeptide-like immunoreactivity in the lamprey spinal cord. *Brain Res.* **408**, 299–302.
83. Brodin, L., Rawitch, A., Taylor, T., Ohta, Y., Ring, H., Hökfelt, T., Grillner, S., and Terenius, L. (1989) Multiple forms of pancreatic polypeptide-related compounds in the lamprey CNS: Partial characterization and immunohistochemical localization in the brain stem and spinal cord. *J. Neurosci.* **9**, 3428–3442.
84. Sundler, F., Håkanson, R., Alumets, J., and Walles, B. (1977) Neuronal localization of pancreatic polypeptide (PP) and vasoactive intestinal peptide (VIP) immunoreactivity in the earthworm (Lumbricus terrestris). *Brain Res. Bull.* **2**, 61–65.
85a. Schot, L. P. C., Boer, H. H., Swaab, D. F., Van Noorden, S. (1981) Immunocytochemical demonstration of peptidergic neurons in the central nervous system of the pond snail *Lymnaea stagnalis* with antisera raised to biologically active peptides of vertebrates. *Cell Tissue Res.* **216**, 273–291.
85b. Duve, H. and Thorpe, A. (1980) Localisation of pancreatic polypeptide (PP)-like immunoreactive material in neurones of the brain of the blowfly, *Calliphora erythrocephala* (Diptera). *Cell Tissue Res.* **210**, 101–109.
86. Duve, H., Thorpe, A., Neville, R., and Lazarus, N. R. (1981) Isolation and partial characterization of pancreatic polypeptide-like material in the brain of the blowfly *Calliphora vomitoria*. *Biochem. J.* **197**, 767–770.
87. Duve, H. and Thorpe, A. (1982) The distribution of pancreatic polypeptide in the nervous system and gut of the blowfly, *Calliphora vomitoria* (Diptera). *Cell Tissue Res.* **227**, 67–77.
87a. Endo, Y. and Nishiitsutsuji-Uwo, J. (1981) Gut endocrine cells in insects: ultrastructure of the gut endocrine cells of the lepidopterous species. *Biomed. Res.* **2**, 270–280.
88. Iwanaga, T., Fujita, T., Nishiitsutsuji-Uwo, J., and Endo Y. (1981) Immunohistochemical demonstration of PP-, somatostatin-, enteroglucagon- and VIP-like immunoreactivities in the cockroach midgut. *Biomedical Res.* **2**, 202–207.

89. Endo, Y., Iwanaga, T., Fujita, T., and Nishiitsutsuji-Uwo, J. (1982) Localization of pancreatic polypeptide (PP)-like immunoreactivity in the central and visceral nervous systems of the cockroach *Periplaneta. Cell Tissue Res.* **227,** 1–9.
90. Endo, Y., Nishiitsutsuji-Uwo, J., Iwanaga, T., and Fujita, T. (1982) Ultrastructural and immunohistochemical identification of pancreatic polypeptide immunoreactive endocrine cells in the cockroach midgut. *Biomed. Res.* **3,** 454–456.
91. Fritsch, H. A. R., Van Noorden, S., and Pearse, A. G. E. (1982) Gastro-intestinal and neurohormonal peptides in the alimentary tract and cerebral complex of *Ciona intestinalis* (Ascidiaceae). *Cell Tissue Res.* **223,** 369–402.
92. El-Salhy, M., Falkmer, S., Kramer, K. J., and Speirs, R. D. (1983) Immunohistochemical investigations of neuropeptides in the brain, corpora cardiaca, and corpora allata of an adult lepidopteran insect, *Manduca sexta* (L). *Cell Tissue Res.* **232,** 295–317.
92a. Charmantier-Daures, M., Danger, J-M., Netchitailo, P., Pelletier, G., and Vaudry, H. (1987) Immmunocytochemical evidence for atrial natriuretic factor- and neuropeptide Y-like peptides in the eyestalks of *Homarus gammarus* (Crustacea, Decapoda). *C. R. Acad. Sci. Paris,* **t. 305, Série III,** 479–483.
93. Rémy, C., Guy, J., Pelletier, G., and Boer, H. H. (1988) Immunohistological demonstration of a substance related to neuropeptide Y and FMRFamide in the cephalic and thoracic nervous system of the locust *Locusta migratoria. Cell Tissue Res.* **254,** 189–195.
94. Schoofs, L., Danger, J-M., Jegou, S., Pelletier, G., Huybrechts, R., Vaudry, H., and De Loof, A. (1988) NPY-like peptides occur in the nervous system and midgut of the migratory locust, *Locusta migratoria* and in the brain of the grey fleshfly, *Sarcophaga bullata. Peptides* **9,** 1027–1036.
95. Sithigorngul, P., Stretton, A. O. W., and Cowden, C. (1990) Neuropeptide diversity in *Ascaris:* An immunocytochemical study. *J. Comp. Neurol.* **294,** 362–376.
96. Yui, R., Iwanaga, T., Kuramoto, H., and Fujita, T. (1985) Neuropeptide immunocytochemistry in protostomian invertebrates, with special reference to insects and molluscs. *Peptides* 6, **Suppl. 3,** 411–415.
97. Kopin, A. S., Toder, A. E., and Leiter, A. B. (1988) Different splice site utilization generates diversity between the rat and human pancreatic polypeptide precursors. *Arch. Biochem. Biophys.* **267,** 742–748.
98. Benton, M. J. (1990) *Vertebrate palaentology.* Unwin Hyman Ltd., London.
99. Kobel, H. R. and Du Pasquier, L. (1986) Genetics of polyploid *Xenopus. Trends Genet.* **2,** 310–315.
100. Hartley, S. E. (1987) The chromosomes of Salmonid fishes. *Biol. Rev.* **62,** 197–214.
101. Larhammar, D. and Milner, R. J. (1989) Phylogenetic relationship of birds with crocodiles and mammals as deduced from protein sequences. *Mol. Biol. Evol.* **6,** 693–696.
102. Kemp, T. S. (1988) Haemothermia or Archosauria? The interrelationships of mammals, birds and crocodiles. *Zool. J. Linnean Soc.* **92,** 67–104.
103. Hedges, S. B., Moberg, K. D., and Maxson, L. R. (1990) Tetrapod phylogeny inferred from18S and 28S ribosomal RNA sequences and a review of the evidence for amniote relationships. *Mol. Biol. Evol.* **7,** 607–633.

104. Nelson, J. S. (1984) *Fishes of the World,* 2nd ed. Wiley, New York.
105. Shen, L-P., Pictet, R. L., and Rutter, W. J. (1982) Human somatostatin I: Sequence of the cDNA. *Proc. Natl. Acad. Sci. USA* **79,** 4575–4579.
106. Shibahara, S., Morimoto, Y., Furutani, Y., Notake, M., Takahashi, H., Shimizu, S., Horikawa, S., and Numa, S. (1983) Isolation and sequence analysis of the human corticotropin-releasing factor precursor gene. *EMBO J.* **2,** 775–779.
107. Bennett, H. P. J., Lowry, P. J., and McMartin, C. (1973) Confirmation of the 1-20 amino acid sequence of human adrenocorticotrophin. *Biochem. J.* **133,** 11–13.
108. Light, A. and du Vigneaud, V. (1958) On the nature of oxytocin and vasopressin from human pituitary. *Proc. Soc. Exp. Biol. Med.* **98,** 692–696.
109. Bell, G. I., Sanchez-Pescador, R., Laybourn, P. J., and Najarian, R. C. (1983) Exon duplication and divergence in the human preproglucagon gene. *Nature* **304,** 368–371.
110. Itoh, N., Obata, K-i., Yanaihara, N., and Okamoto, H. (1983) Human preprovasoactive intestinal polypeptide contains a novel PHI-27-like peptide, PHM-27. *Nature* **304,** 547–549.
111. Riniker, B., Sieber, P., Rittel, W., and Zuber, H. (1972) Revised amino-acid sequences for porcine and human adrenocorticotrophic hormone. *Nature New Biol.* **235,** 114–115.
112. Bell, G. I., Swain, W. F., Pictet, R., Cordell, B., Goodman, H. M., and Rutter, W. J. (1979) Nucleotide sequence of a cDNA clone encoding human preproinsulin. *Nature* **282,** 525–527.
113. Neher, R., Riniker, B., Rittel, W., and Zuber, H. (1968) *Helv. Chim. Acta* **51,** 1900–1905.
114. Hudson, P., John, M., Crawford, R., Haralambidis, J., Scanlon, D., Gorman, J., Tregear, G., Shine, J., and Niall, H. (1984) Relaxin gene expression in human ovaries and the predicted structure of a human preprorelaxin by analysis of cDNA clones. *EMBO J.* **3,** 2333–2339.
115. Conlon, J. M., Agoston, D. V., and Thim, L. (1985) An elasmobranchian somatostatin: Primary structure and tissue distribution in *Torpedo marmorata. Gen. Comp. Endocrinol.* **60,** 406–413.
116. Okawara, Y., Morley, S. D., Burzio, L. O., Zwiers, H., Lederis, K., and Richter, D. (1988) Cloning and sequence analysis of cDNA for corticotropin-releasing factor precursor from the teleost fish *Catostomus commersoni. Proc. Natl. Acad. Sci. USA* **85,** 8439–8443.
117. Spiess, J., Rivier, J., Rivier, C., and Vale, W. (1981) Primary structure of corticotropin-releasing factor from ovine hypothalamus. *Proc. Natl. Acad. Sci. USA* **78,** 6517–6521.
118. Bennett, H. P. J., Lowry, P. J., McMartin, C., Scott, A. P. (1974) Structural studies of α-melanocyte-stimulating hormone and a novel β-melanocyte-stimulating hormone from the neurointermediate lobe of the pituitary of the dogfish *Squalus acanthias. Biochem. J.* **141,** 439–444.
119. Heierhorst, J., Morley, S. D., Figueroa, J., Krentler, C., Lederis, K., and Richter, D. (1989) Vasotocin and isotocin precursors from the white sucker, *Catostomus commersoni:* Cloning and sequence analysis of the cDNAs. *Proc. Natl. Acad. Sci. USA* **86,** 5242–5246.

120. Conlon, J. M., O'Toole, L., and Thim, L. (1987) Primary structure of glucagon from the gut of the common dogfish *(Scyliorhinus canicula). FEBS Lett.* **214,** 50–56.
121. Dimaline, R., Young, J., Thwaites, D. T., Lee, C. M., Shuttleworth, T. J., and Thorndyke, M. C. (1987) A novel vasoactive intestinal peptide (VIP) from elasmobranch intestine has full affinity for mammalian pancreatic VIP receptors. *Biochim. Biophys. Acta* **930,** 97–100.
122. Lowry, P. J., Bennett, H. P. J., McMartin, C., and Scott, A. P. (1974) The isolation and amino acid sequence of an adrenocorticotrophin from the pars distalis and a corticotrophin-like intermediate-lobe peptide from the neurointermediate lobe of the pituitary of the dogfish *Squalus acanthias. Biochem. J.* **141,** 427–437.
123. Bajaj, M., Blundell, T. L., Pitts, J. E., Wood, S. P., Tatnell, M. A., Falkmer, S., Emdin, S. O., Gowan, L. K., Crow, H., Schwabe, C., Wollmer, A., and Strassburger, W. (1983) Dogfish insulin. Primary structure, conformation and biological properties of an elasmobranchial insulin. *Eur. J. Biochem.* **135,** 535–542.
124. Keutmann, H. T., Lequin, R. M., Habener, J. F., Singer, F. R., Niall, H. D., and Potts, J. T., Jr. (1971) in *Endocrinology 1971: Proceedings of the Third International Symposium* (S., Taylor, ed.), Heinemann Medical Books, London, pp. 316–323.
125. Büllesbach, E. E., Gowan, L. K., Schwabe, C., Steinetz, B. G., O'Byrne, E., and Callard, I. P. (1986) Isolation, purification, and the sequence of relaxin from spiny dogfish *(Squalus acanthias). Eur. J. Biochem.* **161,** 335–341.
126. Glover, I., Haneef, I., Pitts, J., Wood, S., Moss, D., Tickle, I., and Blundell, T. (1983) Conformational flexibility in a small globular hormone: X-ray analysis of avian pancreatic polypeptide at 0.98-Å resolution. *Biopolymers* **22,** 293–304.
127. Fuhlendorff, J., Langeland Johansen, N., Melberg, S. G., Thøgersen, H., and Schwartz, T. W. (1990b) The antiparallel pancreatic polypeptide fold in the binding of neuropeptide Y to Y_1 and Y_2 receptors. *J. Biol. Chem.* **265,** 11,706–11,712.
127. Sillard, R., Agerberth, B., Mutt, V., and Jörnvall, H. (1989) Sheep neuropeptide Y. A third structural type of a highly conserved peptide. *FEBS Lett.* **258,** 263–265.
128. Tonan, K., Kawata, Y., and Hamaguchi, K. (1990) Conformations of isolated fragments of pancreatic polypeptide. *Biochem.* **29,** 4424–4429.
129. Noelken, M. E., Chang, P. J., and Kimmel, J. R. (1980) Conformation and association of pancreatic polypeptide from three species. *Biochem.* **19,** 1838–1843.
130. Glover, I. D., Barlow, D. J., Pitts, J. E., Wood, S. P., Tickle, I. J., Blundell, T. L., Tatemoto, K., Kimmel, J. R., Wollmer, A., Strassburger, W., and Zhang, Y-S. (1985) Conformational studies on the pancreatic polypeptide hormone family. *Eur. J. Biochem.* **142,** 379–385.
131. MacKerell, A. D., Jr. (1988) Molecular modeling and dynamics of neuropeptide Y. *J. Computer-Aided Mol. Design* **2,** 55–63.
132. Saudek, V. and Pelton, J. T. (1990) Sequence-specific ^{1}H NMR assignment and secondary structure of neuropeptide Y in aqueous solution. *Biochem.* **29,** 4509–4515.

133. Krstenansky, J. L., Owen, T. J., Buck, S. H., Hagaman, K. A., and McLean, L. R. (1989) Centrally truncated and stabilized porcine neuropeptide Y analogs: Design, synthesis, and mouse brain receptor binding. *Proc. Natl. Acad. Sci. USA* **86,** 4377–4381.
134. Beck, A., Jung, G., Gaida, W., Köppen, H., Lang, R., and Schnorrenberg, G. (1989) Highly potent and small neuropeptide Y agonist obtained by linking NPY 1-4 via spacer to α-helical NPY 25-36. *FEBS Lett.* **244,** 119–122.
135. Schwartz, T. W., Sheikh, S. P., and O'Hare, M. M. T. (1987) Receptors on phaeochromocytoma cells for two members of the PP-fold family—NPY and PP. *FEBS Lett.* **225,** 209–214.
136. Minikata, H., Taylor, J. W., Walker, M. W., Miller, R. J., Kaiser, E. T. (1989) Characterization of amphiphilic secondary structures in neuropeptide Y through the design, synthesis, and study of model peptides. *J. Biol. Chem.* **264,** 7907–7913.
137. MacKerell, A. D., Jr., Hemsén, A., Lacroix, J. S., and Lundberg, J. M. (1989) Analysis of structure-function relationships of neuropeptide Y using molecular dynamics simulations and pharmacological activity and binding measurements. *Regul. Peptides* **25,** 295–313.
138. Grundemar, L. and Håkanson, R. (1990) Effects of various neuropeptide Y/peptide YY fragments on electrically-evoked contractions of the rat vas deferens. *Br. J. Pharmacol.* **100,** 190–192.
139. Fuhlendorff, J., Gether, U., Aakerlund, L., Johansen, N. L., Thøgersen, H., Melberg, S. G., Olsen, U. B., Thastrup, O., and Schwartz, T. W. (1990a) [Leu^{31},Pro^{34}]Neuropeptide Y: A specific Y_1 receptor agonist. *Proc. Natl. Acad. Sci. USA* **87,** 182–186.
140. Krstenansky, J. L., Owen, T. J., Payne, M. H., Shatzer, S. A., and Buck, S. H. (1990) C-terminal modifications of neuropeptide Y and its analogs leading to selectivity for the mouse brain receptor over the porcine spleen receptor. *Neuropeptides* **17,** 117–120.
141. Noe, B. D., McDonald, J. K., Greiner, F., and Wood, J. G. (1986) Anglerfish islets contain NPY immunoreactive nerves and produce the NPY analog aPY. *Peptides* **7,** 147–175.
142. Milgram, S. L., Balasumbramaniam, A., Andrews, P. C., McDonald, J. K., and Noe, B. D. (1989) Characterization of aPY-like peptides in anglerfish brain using a novel radioimmunoassay for aPY-Gly. *Peptides* **10,** 1013–1017.
143. Osborne, N. N., Patel, S., Terenghi, G., Allen, J. M., Polak, J. M., and Bloom, S. R. (1985) Neuropeptide Y (NPY)-like immunoreactive amacrine cells in retina of frog and goldfish. *Cell Tissue Res.* **241,** 651–656.
144. El-Salhy, M., Grimelius, L., Wilander, E., Ryberg, B., Terenius, L., Lundberg, J. M., and Tatemoto, K. (1983) Immunocytochemical identification of polypeptide YY (PYY) cells in the human gastrointestinal tract. *Histochemistry* **77,** 15–23.
145. Greeley, G. H., Jr., Hill, F. L. C., Spannagel, A., and Thompson, J. C. (1987) Distribution of peptide YY in the gastrointestinal tract of the rat, dog, and monkey. *Regul. Peptides* **19,** 365–372.
146. Roddy, D. R., Koch, T. R., Reilly, W. M., Carney, J. A., Go, V. L. W. (1987) Identification and distribution of immunoreactive peptide YY in the human, canine, and murine gastrointestinal tracts: species related antibody recognition differences. *Regul. Peptides* **18,** 201–212.

147. Gilbert, W. R., Frank, B. H., Gavin, J. R., III, and Gingerich, R. L. (1988) Characterization of specific pancreatic polypeptide receptors on basolateral membranes of the canine small intestine. *Proc. Natl. Acad. Sci. USA* **85,** 4745–4749.
148. Jørgensen, J. C., Fuhlendorff, J., and Schwartz, T. W. (1990) Structure-function studies on neuropeptide Y and pancreatic polypeptide—evidence for two PP-fold receptors in vas deferens. *Eur. J. Pharmacol.* **186,** 105–114.
149. Schwartz, T. W., Fuhlendorff, J., Langeland, N., Thøgersen, H., Jørgensen, J. C., and Sheikh, S. P. (1989) Y_1 and Y_2 receptors for NPY—the evolution of PP-fold peptides and their receptors, in *Neuropeptide Y* (V., Mutt, K., Fuxe, T., Hökfelt, J. M., Lundberg, eds.), Raven, New York, pp. 143–151.
151. Gapp, D. A. (1987) V. Endocrine and related factors in the control of metabolism in nonmammalian vertebrates. 17. The Endocrine Pancreas. 3. Comparative aspects of islet morphology, and 4. Evolutionary considerations of morphology, in *Fundamentals of Comparative Vertebrate Endocrinology* (I., Chester-Jones, P. M., Ingleton, and J. G., Phillips, eds.), Plenum, New York, pp. 584–604.
160. Miyachi, Y., Jitsuishi, W., Miyoshi, A., Fujita, S., Mizuchi, A., and Tatemoto, K. (1986) The distribution of polypeptide YY-like immunoreactivity in rat tissues. *Endocrinology* **118,** 2163–2167.

Structure and Expression of the Neuropeptide Y Gene

Janet M. Allen and Domingo Balbi

1. Introduction

The identity of neuropeptide Y (NPY) was first reported in 1982 *(1)*. This peptide was discovered at the time of a watershed in technology. It is now more common for complete peptide sequences to be deduced from the genetic code as a result of the expansion and ease of molecular biology techniques. However, NPY was discovered and characterized by chemical purification. Two years later, the structure of the mRNA that encodes the peptide was reported *(2)*, and after an additional three years, the genomic organization was published *(3)*.

1.1. Discovery

NPY was identified using a novel chemical method *(4,5)* that detected the presence of C-terminal amide structures, a feature of biologically active peptides. This assay had been used to isolate a related peptide, peptide tyrosine tyrosine or PYY, from porcine gut *(5,6)*. This peptide, PYY, was named using a recently adopted policy of naming newly discovered peptides by their N- and C-terminal amino acids. Because many of the gut peptides are found in the brain, Tatemoto and colleagues initially planned to follow the same extraction procedure to demonstrate this novel peptide, PYY, in the brain. However, the product of their purification was a peptide that was similar to, but distinct from PYY *(1,7)*. Their starting material for this purification was 400 kg of porcine brain

From: *The Biology of Neuropeptide Y and Related Peptides;*
W. F. Colmers and C. Wahlestedt, Eds.

("without cerebellum and pituitary"), and it yielded 56 mg of peptide. This peptide was also found to possess tyrosine residues at its N and C termini and, thus, it posed a dilemma to their policy. It too should have been called peptide tyrosine tyrosine or PYY. In order to overcome the confusion, this distinct, but related peptide was called neuropeptide Y or NPY, the N being used to indicate its neural origin.

To a molecular biologist, the concept of starting with such large amounts of material to yield such seemingly small amounts of purified peptide may be difficult to grasp. However, Tatemoto has commented many times that they were "surprised" at the "high yield" of the peptide at the initial purification (personal communication) and that, from an early stage, it was considered to be an abundant neuropeptide within the brain. Indeed, in the first paper describing NPY *(1)*, the authors comment that "NPY seems to be present in large quantities in the brain at concentrations even higher than those of VIP" and that "clearly, it will be of interest to determine whether NPY is a neurotransmitter or neurohormone and to discover its physiological role(s) in the central nervous system."

Since there was a four- to five-year gap between the identification of the peptide and the subsequent determination of the genomic organization, much was determined concerning the expression of this peptide using standard radioimmunoassay and histochemical techniques. Thus:

1. The peptide appeared to be confined in its expression to cells derived from the neural crest;
2. It is abundantly present in neurons throughout the brain (except cerebellum) and spinal cord;
3. High concentrations of NPY were found in postganglionic sympathetic nerves and the adrenal medulla; and
4. NPY peptide concentrations extracted from cell lines derived from the neural crest could be altered by trophic factors, such as nerve growth factor and dexamethasone.

Many of these features of NPY will be the subject of subsequent chapters appearing in this book. In this chapter, we will confine the discussion to the structure of cDNA that encodes NPY, its genomic organization, and factors that regulate gene expression.

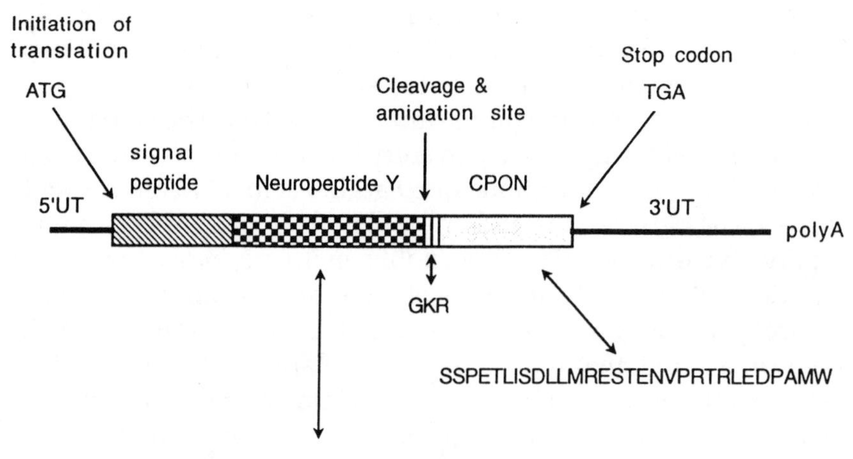

Fig. 1. Schematic representation of the mRNA and open reading frame of human NPY. The 5' and 3' untranslated regions are shown as a line (5'UT and 3'UT). The region of the open reading frame is shown as shaded boxes. The derived amino acid sequences are shown below the relevant boxes in single letter code. (CPON = C flanking Peptide Of NPY)

2. cDNA Structure

2.1. Determination of the Structure of the cDNA-Encoding NPY

The cDNA that encodes NPY was reported in 1984 *(2)*. RNA was isolated from a human phaeochromocytoma, a tumor of the adrenal medulla, known to contain exceptionally high concentrations of NPY. RNA extracted from the tumor was shown, by in vitro translation, to contain an mRNA that encoded a protein of 10.5 kDa, which was immunoprecipitated by antisera directed to NPY. The mRNA was converted into cDNA and probed by a degenerate oligonucleotide probe designed to five amino acids of the porcine sequence. The isolated cDNA consisted of 591 nucleotides, comprising 86 bases in the 5' untranslated region, 291 bases in the open reading frame, 174 bases in the 3' untranslated region, and a stretch of 40 adenosyl residues. The mRNA contained a single open reading frame that predicted a fairly simple precursor for NPY (Fig. 1), comprising 97 amino acids with a predicted molecular size of

10,839 Da. The predicted precursor comprised a hydrophobic signal peptide of 28 amino acids, necessary to cause the nascent peptide chain to enter the lumen of the endoplasmic reticulum and, thus, enter the secretory compartment of the cell. The signal peptide should be rapidly cleaved on entry into the endoplasmic reticulum yielding the prohormone of 69 amino acids. No amino acids sequence lie between the end of the signal peptide and the amino terminal tyrosine of NPY. Thus, unlike many peptides, there is no peptide flanking the N terminus of NPY, and the amino region of the prohormone comprises the 36 amino acids of human NPY. In the prohormone, the mature peptide is flanked at its C-terminal end by 33 amino acids. The amino acids immediately C terminal to the tyrosine 36 of NPY are the glycine-lysine-arginine motif necessary for posttranslational processing of the prohormone. The lysine-arginine is the proteolytic site, and the glycine residue is required to donate its amide to the tyrosine at the C terminus of NPY. The 30 amino acids that comprise the remainder of the prohormone were designated as CPON, an acronym for C flanking Peptide Of NPY.

The most striking feature from this paper *(2)* was the similarity between the predicted amino acid sequence of human NPY to that of the original porcine isolate reported by Tatemoto *(1,6)*. The two peptides differ in only one amino acid; at position 17, a leucine residue found in the porcine isolate was predicted to be replaced by a methionine residue in the human sequence (Fig. 2). This difference thus confirmed studies using reverse-phase high-pressure liquid chromatography that had suggested that porcine NPY and NPY immunoreactivity extracted from human brain were similar in nature, but not identical *(8)*.

Studies have been performed to investigate the processing of the prohormone to the mature version of NPY. In these, NPY peptide is expressed from its cDNA in AtT 20 cells, a clonal cell line derived from mouse pituitary cells. These cells, although not normally containing NPY, were capable of expressing the cDNA, and they processed the prohormone correctly to yield NPY and its flanking peptide, CPON *(9)*. Mutant cDNAs were constructed to alter the paired basic amino acids at the cleavage site. This study demonstrated a hierarchy in preferred amino acids at the cleavage site. It seems that the arginine residue present in the second position of the dibasic pair determines the efficiency of cleavage, since both the wild-type (lysine-arginine) and the mutant, arginine-arginine,

```
COMPARISON BETWEEN PORCINE AND HUMAN NPY

            1        10        20        30
Porcine     YPSKPDNPGEDAPAEDLARYYSALRHYINLITRNRY*

Human       ----------------M-------------------


COMPARISON BETWEEN HUMAN AND RAT CPON

            1        10        20        30
Human       SSPETLISDLLMRESTENVPRTRLEDPAMW

Rat         -------------------A--------S--
```

Fig. 2. Comparison of the amino acid sequences of porcine NPY with the predicted sequence of human NPY derived from the cDNA sequence (human and rat NPY are predicted to be identical), and comparison of the predicted amino acid sequences of the C-flanking peptides for NPY (CPON) derived from the human and rat cDNAs.

were cleaved at equal rates. The mutant encoding an arginine-lysine was cleaved less rapidly and efficiently, whereas the mutant encoding lysine-lysine was cleaved rapidly, but least efficiently *(10)*.

The nucleotide sequence that encodes rat NPY was described in 1987 *(11)* after the cDNA was isolated from a library constructed from mRNA extracted from rat hypothalamus. A striking level of homology was observed between the sequences reported for rat compared to those for human. Thus, the overall homology between the two cDNAs is on the order of 60%, rising to 85% when comparing nucleotide sequences in the open reading frame. Indeed, the primary structure of NPY was predicted to be identical for rat NPY compared to human NPY. Rat NPY shares with human NPY the methionine residue at position 17, this being the only difference in sequence to the original porcine isolate. NPY is, thus, highly conserved across species. This level of sequence conservation across species is remarkable and has been suggested to indicate conservation of structure for important functional considerations.

The cDNA that encodes rat NPY was found to consist of 559 bases. It terminated at the 3' end with 20 adenosine residues that were preceded by a typical polyadenylation motif (AATAAA). The

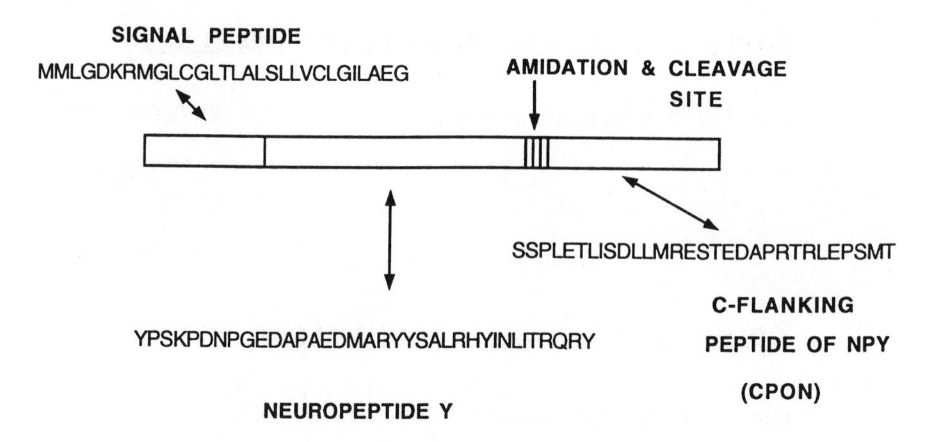

Fig. 3. Schematic representation of the predicted precursor for rat NPY. The amino acids are shown in single letter code.

cDNA predicted an open reading frame that encodes a 98 amino acid precursor of NPY (Fig. 3) similar in overall structure to the 97 amino acid human precursor. The additional amino acid predicted for the rat precursor resulted from the presence of two contiguous initiation codons (ATG) encoding methionine residues. The signal peptide was, thus, predicted to comprise 29 amino acids rather than the 28 for the human precursor. Again, the signal peptide is likely to be cleaved on entry of the nascent peptide into the lumen of the endoplasmic reticulum. The amino terminal of proNPY, thus, comprises NPY itself. Then, there occurs the motif (glycine-lysine-arginine) required to generate mature amidated NPY from the prohormone. As in the human sequence, NPY is flanked on its C-terminal side by a peptide that comprises 30 amino acids and is conserved in primary sequence with human CPON. There are only two amino acid substitutions between the two species, and these changes are very conservative in nature. At position 19 of the flanking peptide, an alanine residue in the rat sequence is substituted by a valine in the human sequence, and at position 28, the rat sequence predicts a serine residue, whereas the human sequence predicts an alanine. This degree of conservation is not observed between species for another member of this family of peptides, namely pancreatic polypeptide (PP), where very diverse sequences have been reported in the flanking peptides. Again, the conservation of structure tends to point to a functional role for the flanking peptide, CPON, but

```
PEPTIDE : Pancreatic polypeptide (PP) - Pancreatic Islets
          Peptide YY (PYY) -Endocrine cells of lower gut
          Neuropeptide Y (NPY) - Neural crest derived tissues

          All 36 amino acids with C terminal tyrosine amide
          50% homology between PP and PYY/NPY
          75% homology between NPY and PYY

PRECURSOR : Identical precursor form - Signal peptide
                                       Peptide
                                       Post-translational processing site
                                       Flanking peptide

GENE : Identical sites of intron/exon boundaries
```

Fig. 4. Homology within the pancreatic polypeptide family of peptides at the peptide, cDNA, and genomic levels of organization.

although the peptide has been shown to be costored and coreleased with NPY *(12)*, no biological activity has been shown to reside in this peptide as yet.

2.2. NPY As a Member of the PP Family

At the time of the isolation of NPY, it was suggested that this peptide represented the third member of a family of peptides. Indeed, the discovery of NPY presented an explanation for observations made by workers that some antisera raised to the avian form of PP (APP) recognized an immunoreactive material in neural tissue. It has become clear that this APP-like immunoreactivity was in fact NPY *(8)*, resulting from the immunostaining of neurons by antisera raised to APP crossreacting with endogenous NPY. There are now three members of this family of peptides (Fig. 4) (*see also* Larhammar et al., this vol.). The first discovered member of this family, pancreatic polypeptide (PP), is found in pancreatic islets of Langerhans, where it is found in endocrine cells distinct to those secreting the other three principal hormones found in pancreatic islets, namely insulin, glucagon, and somatostatin. PP acts as a hormone being released into plasma, and its main biological effect appears to be the inhibition of pancreatic exocrine secretion.

The second member of the family, PYY, was identified in 1980, following its purification from porcine gut *(5)*. This peptide is found in endocrine cells of the gastrointestinal tract *(6)*, particularly the ileum and colon, where it is in part costored with glucagon-related

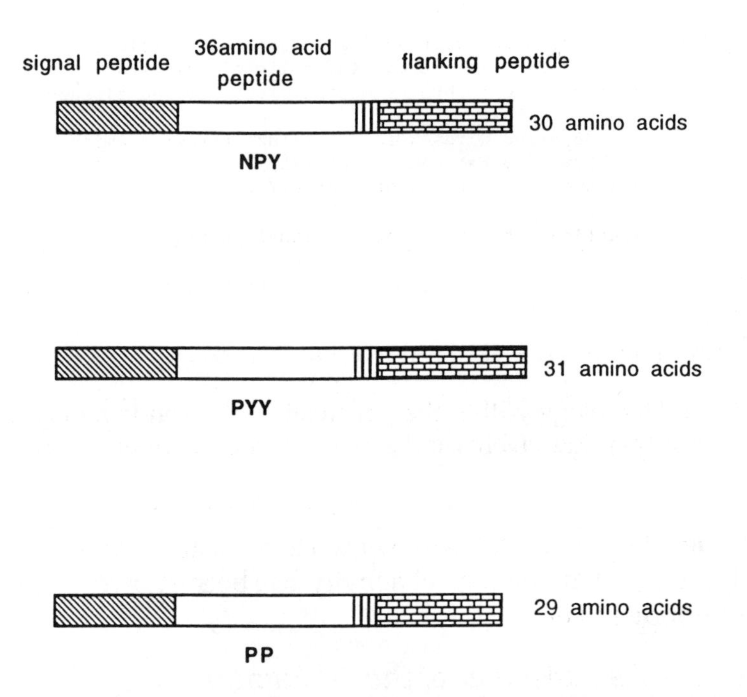

Fig. 5. Schematic representation of the prepropeptide precursors predicted from cDNAs for NPY, PYY, and PP. Each precursor contains sequences that comprise a signal peptide, the 36 amino acid mature peptide, the three amino acids required for appropriate posttranslational processing (glycine-lysine, arginine), and a peptide at their C-flanking region of 30, 31, or 29 amino acids.

peptides *(13)*. The peptide is released into plasma in response to the presence of fat in the intestine *(14)*, and its actions include inhibition of gastrointestinal motility *(15)*, pancreatic exocrine secretion, and acid secretion from the stomach *(16)*.

The cDNAs that encode the precursors for human PP *(17)* and rat PYY *(18)* have been isolated, and these reveal a conserved structure that is shared with NPY (Fig. 5). Each precursor consists of a signal peptide of 28 or 29 amino acids, the 36 amino acid mature member of the family (NPY, PYY, or PP) followed by the amino acids required for posttranslational processing (glycine-lysine-arginine) and a C-terminal peptide consisting of approx 30 residues. The marked similarities in precursor structures for NPY, PYY, and PP are strongly suggestive that each of these peptides arose by

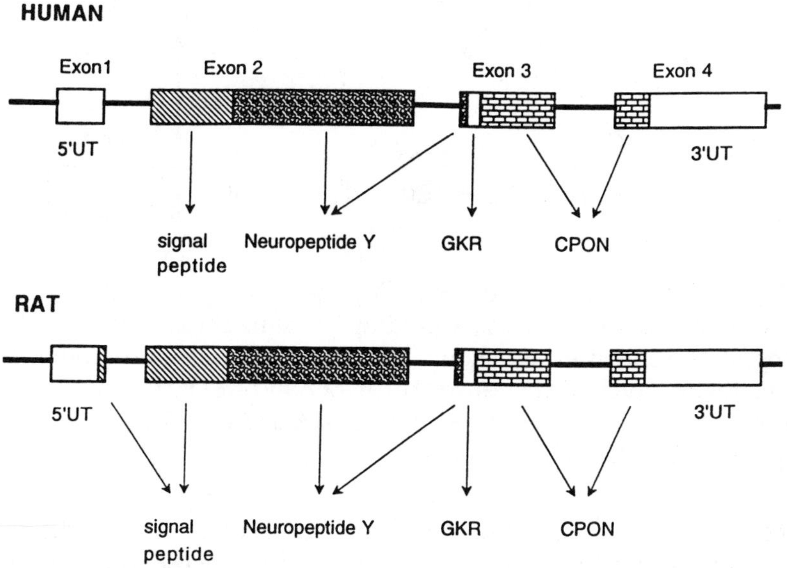

HUMAN

Exon1 Exon 2 Exon 3 Exon 4

5'UT 3'UT

signal peptide Neuropeptide Y GKR CPON

RAT

5'UT 3'UT

signal peptide Neuropeptide Y GKR CPON

Fig. 6. Schematic representation of the genomic organization of the human and rat neuropeptide Y gene designed to show the structures encoded by the four exons. The exons are shown as boxes linked by lines representing introns. The relationship of the length of exons to introns is not drawn to scale. 5'UT = 5' untranslated region of mRNA. Signal = signal peptide. GKR = glycine-lysine-arginine. CPON = C-flanking peptide of Neuropeptide Y. 3' UT = 3' untranslated region of mRNA.

duplication of a common ancestral gene *(19)*. This suggestion is supported by a finding that the genes for all three members of this family of peptides have similar overall architecture.

3. NPY Gene

3.1. Organization of the NPY Gene

The human NPY gene was isolated in 1986 *(20)*. The transcription unit spans approx 8 kb, and analysis of the gene sequence revealed that the transcribed sequences of NPY are divided into four exons separated by three introns of approx 965, 4300, and 2300 bp (Fig. 6). The exons, in general, define functional domains of NPY. Thus, the first exon encodes the 5' untranslated sequences of the mRNA. The second exon starts with the initiation codon, ATG, and

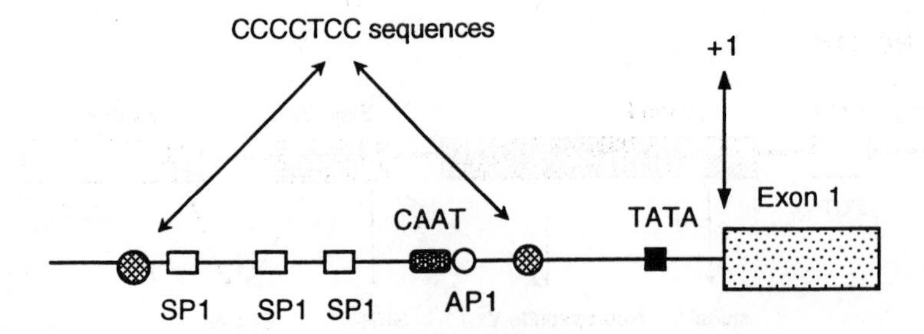

Fig. 7. 5' Flanking region of the NPY gene showing the regions known to be consensus sites for DNA binding protein. Three of the five SP-1 recognition sequences are shown at -111, -99, and -88. (The other two SP-1 sites are much further upstream at -607 and -181). The one AP-1 site is shown together with the sites of the TATA and CAAT boxes. The two sequences, CCCCTCC, are shown.

encodes the entire signal sequence and majority of NPY itself, since it ends with the first two bases of the codon for the arginine residue found as the 35th amino acid of NPY. The third exon encodes the C terminal tyrosine of NPY, the three amino acids necessary for posttranslational processing, and 23 amino acids of CPON. The fourth exon encodes the C-terminal seven amino acids of CPON and the 3' untranslated region of the mRNA, including the polyadenylation signal.

The region of DNA upstream from the transcription initiation site should be expected to be rich in *cis*-acting elements that regulate transcription directed by RNA polymase II. The *cis*-acting sequences are recognized by sequence specific DNA binding proteins that either repress or activate transcription. Analysis of the 5' flanking sequences has identified consensus sequences for several DNA binding proteins (Fig. 7). The "TATA" box is 25 bp upstream of the transcription start site, and there is a proposed "CAAT" box further upstream. The 600 bp of DNA at the 5' flanking region of the NPY gene contain five predicted SP-1 GC boxes that are located at -607, -181, -111, -99, and -88. In addition, there are two sites, CCCCTCC present at -118 and -51. There is also one AP-1 binding site. The significance of these consensus sites in controlling gene expression will be discussed later in this chapter.

The overall architecture of the rat NPY gene *(3)* is very similar to that described for the human gene. The rat gene spans 7.2 kb, and again, the transcription unit is divided into four exons (Fig. 6)

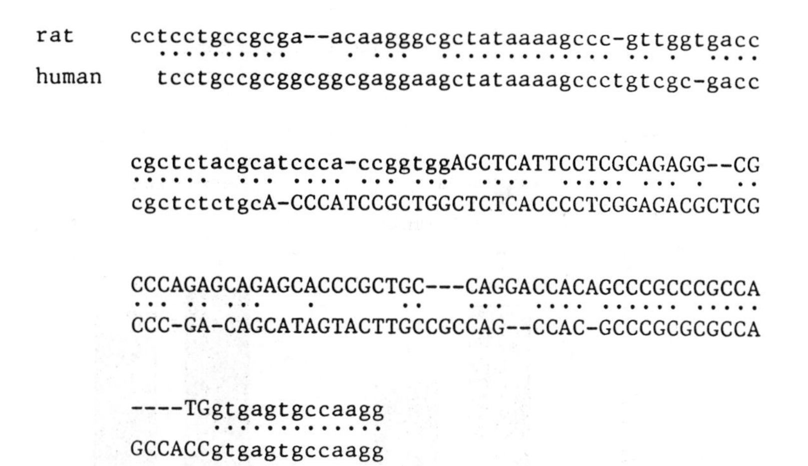

Fig. 8. Sequence of the human and rat genes in the region immediately upstream of the site of initiation of transcription. Nucleotides of the shared with the cDNA sequences are shown in capital letters. Nucleotides common to the rat and human genes are marked by a dot.

by three introns of 790, 4100, and 1850 bp, respectively. Remarkable homology was observed between the nucleotides in the 5' flanking region of the genes for human and rat NPY (Fig. 8). The exon organization is identical to that observed for the human gene. The two initiation codons (ATG) found in the rat cDNA sequence are split by the first intron in the gene. Thus, the first exon terminates in the first ATG, and the second exon starts with the second ATG. The transcription start site is preceded 30 nucleotides upstream by an AT-rich "TATA"-like element, ATAAAA.

Two rat genomic libraries were screened, and although no differences were found between the two genes within the exons, significant differences were observed in the sequences obtained from the two libraries in a region of the gene that spans 2.5 kb downstream from -670. This region comprises 670 nucleotides of 5' flanking region, exon I, intron I, exon 2, and 780 nucleotides of intron 2. Six single-base substitutions were observed together with three insertions or deletions. The changes were very suggestive of polymorphic allelic changes. However, one of the substitutions was located 80 nucleotides 5' to the "TATA"-like promoter, and it was thought that such a substitution may influence the level of expression of the gene. However, no further evidence for allelic alterations in NPY gene expression has been defined.

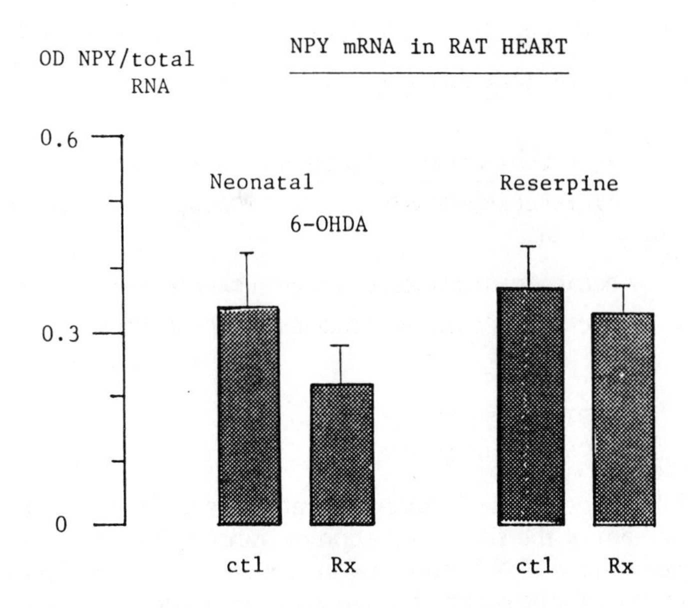

Fig. 9. Measurement of NPY specific mRNA by Northern analysis in rats treated with 6-hydroxydopamine as neonates or reserpine as adults. The animals treated with 6-hydroxydopamine had evidence of a total chemical sympathectomy and no NPY peptide could be detected within extracts of heart tissue.

This study revealed for the first time that NPY mRNA is expressed in some unusual sites, namely spleen and heart *(3)*. Both these tissues receive a dense sympathetic innervation and are known to be richly supplied by NPY-immunoreactive nerves. However the NPY immunoreactivity is almost completely localized to postganglionic sympathetic nerve fibers. Stored peptide concentrations in these tissues can be depleted by agents that damage sympathetic nerves, such as 6-hydroxydopamine *(21)* and reserpine. However, in view of the generally accepted view that mRNAs are not present in nerve terminals, surprisingly high levels of NPY mRNA were found in the heart and spleen *(3)*. Subsequent studies have shown that the NPY mRNA levels are unaltered by treatment with agents that effectively abolish detectable levels of NPY peptide in these tissues (Fig. 9) *(22)*. In the mouse spleen and rat bone marrow, NPY mRNA was demonstrated in nonneuronal megakaryocytes *(23)*. The finding of high levels of NPY mRNA in rat and mouse megakaryocytes appears to be unique to these rodents. It represents the major exception to the rule that NPY gene expression is confined to cells derived from the neural crest.

3.2. Genes for the Related Peptides, PP and PYY

The genomic organization of the gene that encodes NPY bears remarkable resemblance to the genes described for PP *(24)* and PYY *(25)*, further reinforcing the concept that the three members of this family of peptides have arisen as a result of gene duplication (Fig. 4). The gene for human PP spans 2.8 kb of DNA and contains four exons separated by three introns. These four exons split the transcription unit into functional domains in a similar way to that observed for NPY. Thus, exon 1 encodes the 5' untranslated region, exon 2 encodes the signal peptide and 35 amino acids of PP, exon 3 encodes the C-terminal tyrosine together with the posttranslational motif and 20 amino acids of the flanking peptide, whereas exon 4 contains the sequences that encode the final seven amino acids of the flanking peptide along with the 3' untranslated region of the cDNA containing the polyadenylation site.

In contrast to all the others, the structural organization of the rat PP gene differs somewhat from the genes that encode the human form of PP and other members of this gene family *(26)*. The third exon of the rat PP gene encodes the entire C-terminal flanking peptide and a portion of the 3' untranslated region of the mRNA. The sequences that encode the flanking peptide in all others of this family are found on both the third and the fourth exon in the gene structure. An intron is positioned in the triplet codon for the arginine residue found in the flanking peptide of all members of this family, except the rat version, where a leucine residue is present. It has been proposed that this variation observed in the organization of the gene for rat PP results from the recruitment of an alternative splice-donor site at the 3' end of the third exon of the rat gene *(25,27)*. This third exon is, thus, 42 bases longer, and these additional bases encode 11 amino acids followed by a translational stop codon. These nucleotides are similar to those found at the 5' end of the third intron in the human gene from PP. The flanking peptides for human and rat PP are very different. Differences at the C-terminal region are explicable by this alteration in exon–intron structure. Additionally, there appears to be a translational frameshift at the 5' end of exon 3 for rat PP. Thus, although nucleotides may compare favorably between the two genes, the resulting peptides differ markedly. The differences in the C-terminal peptides of the rat and human precursors reflect less strict evolutionary constraints than those for the mature peptide, PP itself. The difference further highlights the remarkable conservation observed between the flanking peptides (CPON) of rat and human NPY, again suggesting that these peptides are under tight functional evolutionary control, although no function has been demonstrated as yet.

The transcription unit for the rat PYY gene is short, spanning approx 1.2 kb. The overall architecture is identical to that for NPY and PP (except rat PP) with the transcription unit split into four exons by three introns. The exons appear to encode functional domains as for NPY. Both the third and fourth domains encode the flanking peptide for PYY, and typically, the intron is placed within the triplet codon that encodes an arginine residue of the flanking peptide.

Expression of the genes that encode members of this family of peptides is tightly regulated to certain tissues. The PP gene is expressed within endocrine pancreatic islets. The PYY gene is expressed in intestinal endocrine cells of the ileum and colon *(18)*. In addition, surprisingly high levels of PYY gene expression have been reported in the pancreas *(18)*. NPY gene expression is confined to cells derived from the neural crest (with the exception of rodent megakaryocytes.)

4. Chromosomal Localization of the Genes

The human chromosomal localizations of PP and NPY were determined using human–mouse somatic cell hybrid lines. The PP gene segregates with human chromosome 17 being assigned to the p11.1-qter region. The NPY gene segregates with human chromosome 7 to the pter-q22 region of the chromosome *(28)*.

A number of gene families are localized on chromosomes 7 and 17. This finding suggested that these families arose as a result of a tandem duplication of an ancient linkage group followed by chromosomal translocation. Genes that segregate to the two chromosomes in a similar fashion to that for NPY and PP include the erb proto-oncogenes, type I collagen, actin, band 3 membrane protein, and protein kinases. Thus, chromosome 17, in addition to encoding PP, also encodes α 1 chain type I collagen, erb A, erb B-β2, Hox 1, protein kinase C and erythroid band 3. Chromosome 7 encodes NPY, erb B, Hox 2, α 2 chain type I collagen, β actin, nonerythroid band 2, and protein kinase A.

It is generally considered from evolutionary data that there was a single ancestral gene that encoded an ancient form of the PP family of peptides. It is thought by many that this gene diverged to produce two genes, one encoding a primitive form of PP and one encoding a primitive PYY/NPY hybrid *(25,29)*. This initial divergence presumably occurred at the time of the chromosomal translocation to account for their different chromosomal localizations. PYY and NPY probably arose as a result of tandem duplication of the ancestral PYY/NPY hybrid gene. It is thus predicted that PYY lies close to NPY on human chromosome 7, although this has not yet been determined.

5. Regulation of Gene Expression

The amount of free peptide stored within synaptic terminals is very tightly regulated. This regulation can occur at all levels in the synthetic pathway, including factors affecting gene transcription, and regulation of a specific mRNA pool by factors affecting stability and regulation of the rate of posttranslational processing (Fig. 10). When considering the complexity of the mammalian nervous system along with the broad distribution of NPY and its diverse functions, it becomes clear that there is a need for a tight regulation of the transcription of the gene that encodes this peptide. Complex and interconnected regulatory mechanisms appear necessary for transcriptional control of NPY expression.

Information about factors that influence transcription of the NPY gene in the nervous system can be obtained from two different types of studies. Earlier studies have investigated the factors that regulate specific mRNA levels in vivo and in tissue culture, together with direct studies of transcription rates. Later studies have involved the structural analysis of the flanking regions of the gene to identify *cis*-acting regions of DNA.

Analysis of the 5' flanking region of the NPY gene reveals the presence of consensus sequences for a number of DNA-binding proteins that could act as regulatory factors (Fig. 7). Consensus sequences present in the NPY gene include five potential GC-rich

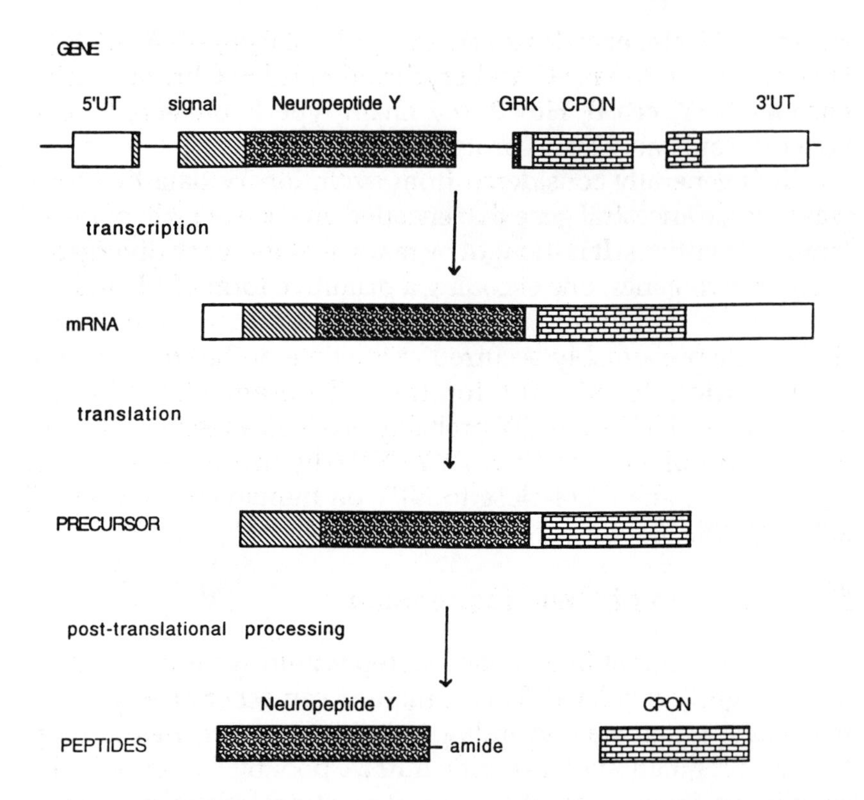

Fig. 10. Schematic representation of the NPY gene, mRNA, prohormone precursor, and mature peptide. Regulation of the amount of free peptide can occur at all stages.

SP-1 binding sites, two CCCCTC sites, a partial CAAT box, and one AP-1 binding site. Deletion analysis has revealed that the sequences necessary for expression of the NPY gene are contained between -246 and -51 *(30)*. The two regions of DNA that contain the CCCCTC consensus sequence appear to be crucial to the expression of the gene in human Lan-5 cells, a cell line established from a neuroblastoma. Deletion of the region -63 to -51, which contains one of these motifs, abolishes transcription of the gene, and deletion of the region -143 to -118, which contains the other CCCCTC sequence, results in a twofold reduction in transcription. In contrast, deletion of the CAAT box, the AP-1 binding sites, and three GC-rich SP-1 binding sites in the region -118 to -83 does not alter gene expression in this cell line *(30)*.

Gel retardation assays suggested that specific DNA/protein interactions occurred in the region of -63 to -51 of the NPY gene (containing the sequence CCCCTCC). Competition studies suggested that this region was recognized by an SP-1-like molecule, although this CT-rich site is not a typical SP-1 binding site. Although evidence that an SP-1-like protein is binding this CT-rich regions is strong, it is nevertheless not conclusive. Furthermore, although effort has been made to unravel the mechanisms of action of SP-1, little is known about the factors that control the activity of SP-1 itself and the different tissue specificity of SP-1, therefore limiting possibilities of linking the SP-1 activity with events at the cell surface. However, the absolute dependence on this region of DNA for transcription in this cell line is remarkable.

Attempts have been made to examine the presence of brain-specific DNA protein complexes related to the NPY gene, using gel retardation assays in the presence of nuclear extracts from brain *(31)*. Provisional data have demonstrated two specific DNA–protein complexes involving sequences at -50 and an 18-mer sequence that includes the TATA-like element.

AP-1 sites are present and active in genes that are activated by phorbol esters and NGF *(32)*. Thus, AP-1 sites are specifically recognized by a group of protein complexes that include among others the products of the proto-oncogenes, c-*fos* and c-*jun (33)*. The AP-1 site of the NPY gene shows no activity *(30)*. This result is surprising, since NPY expression appears to be tightly regulated by both phorbol esters *(34)* and NGF *(35)*. Obviously, this experimental approach covers only limited physiological conditions and ignores any dynamic aspects of gene regulation. It is clear that during different states of development, different factors may be switched on and off, thus actively modulating the transcription rate of many genes. There is also some evidence that specific transcription factors can display different affinities toward different consensus sequences in the presence of accessory proteins. As an example, the proto-oncogene, *jun*, in presence of *fos* shows high affinity for AP-1 sites. However, in the presence of the cyclic-AMP Response Element Binding Protein 1 (CRE-BP1), it displays greater specificity for the consensus sequence that forms part of the cyclic-AMP response element *(36)*. Therefore, it is very likely that under different conditions, sequences apart from those identified in this study *(30)* may take the control of NPY gene transcription.

Studies involved measuring levels of NPY specific mRNA have demonstrated that the gene is tightly regulated by a number of factors. Thus, Nerve Growth Factor (NGF) increases the amount of specific NPY mRNA in PC12 cells *(35)*. PC12 cells are derived from a rat phaeochromocytoma and are widely used as a model of differentiation. These cells undergo profound morphological changes following exposure to NGF, becoming differentiated and phenotypically similar to sympathetic neurons. Phorbol esters have been shown to increase NPY mRNA in PC12 cells *(34)*. Furthermore, a synergistic effect between NGF and phorbol ester on NPY gene transcription has been shown. Since both NGF and phorbol esters are known to induce increases in levels of c-*fos* and c-*jun*, it is possible that these agents affect NPY gene expression through these proto-oncogenes. It will be of interest to investigate whether the AP-1 site present in the 5' region of the NPY gene mediates the increase in NPY expression observed after NGF or phorbol ester treatment. Studies have also attempted to elucidate the intracellular route used by growth factors to stimulate transcriptional activity of the NPY gene. NGF activates several protein kinase systems *(37,38)*. Experiments using activators of cyclic-AMP and calcium or phospholipid dependent protein kinases have shown that NPY mRNA levels can be increased in PC12 cells after treatment with stimulators of both protein kinase systems *(34,39)*. Although activation of both protein kinase systems stimulated an increase in NPY mRNA levels, only phorbol esters produced a synergistic effect on NPY mRNA levels when combined with NGF. Furthermore, the NGF effect on NPY mRNA levels could be blocked by treatment of cells with inhibitors of protein kinase C, whereas inhibitors of protein kinase A had no effect on the NGF-mediated rise in NPY mRNA levels.

There are a number of studies that link electrical activity with transcription activity in the nucleus. In the rat adrenal medulla *(40,41)*, there is a marked increase in NPY mRNA levels following stimulation of splanchnic nerve activity by systemic treatment with insulin. The elucidation of the intracellular mechanisms involved in coupling electric activity with differential control of transcription in the nucleus provides an exciting new challenge.

6. Conclusions

NPY has been known for the last 10 years. It was originally discovered by chemical isolation and purification, and its amino acid sequence was reported in 1982. Since its discovery, the cDNA and gene structure have been determined, and NPY has been shown to be the third member of a family of peptides. This family demonstrates distinct tissue localization, such that PP is found in islets of Langerhans, PYY in endocrine cells of the lower gut, and NPY in cells derived from the neural crest. In addition to this tissue specific regulation, expression of the NPY gene is tightly regulated by a number of factors.

Acknowledgments

This work was supported by a grant from the Wellcome Trust. D. B. is the recipient of a scholarship from the National Academy of Sciences of Venezuela. We thank Aileen Briggs for her expert assistance in the preparation of this manuscript.

References

1. Tatemoto, K., Carlquist, M., and Mutt, V. (1982) Neuropeptide Y—a novel brain peptide with structural similarities to peptide YY and pancreatic polypeptide. *Nature* **296,** 659–660.
2. Minth, C. D., Bloom, S. R., Polak, J. M., and Dixon, J. E. (1984) Cloning, characterization and DNA sequence of a human cDNA encoding neuropeptide tyrosine. *Proc. Natl. Acad. Sci. USA* **81,** 4577–4581.
3. Larhammer, D., Ericsson, A., and Persson, H. (1987) Structure and expression of the rat neuropeptide Y gene. *Proc. Natl. Acad. Sci. USA* **84,** 2068–2072.
4. Tatemoto, K. and Mutt, V. (1978) Chemical determination of polypeptide hormones. *Proc. Natl. Acad. Sci. USA* **75,** 4115–4119.
5. Tatemoto, K. and Mutt, V. (1980) Isolation of two novel candidate hormones using a chemical method for finding naturally occurring polypeptides. *Nature* **285,** 417–418.
6. Tatemoto, K. (1982) Isolation and characterization of peptide YY (PYY), a candidate gut hormone that inhibits pancreatic exocrine secretion. *Proc. Natl. Acad. Sci. USA* **79,** 2514–2518.
7. Tatemoto, K. 1982b. Neuropeptide Y: complete amino acid sequence of the brain peptide. *Proc. Natl. Acad. Sci. USA* **79,** 5485–5489.
8. Adrian, T. E., Allen, J. M., Bloom, S. R., Ghatei, M. A., Rossor, M. N., Roberts, G. W., Crow, T. J., Tatemoto, K., and Polak, J. M. (1983) Neuropeptide Y in human brain. *Nature* **306,** 584–586.

9. Dickerson, I. M., Dixon, J. E., and Mains, R. E. (1987) Transfected human neuropeptide Y cDNA expression in mouse pituitary cells. *J. Biol. Chem.* **262,** 13,646–13,653.
10. Dickerson, I. M., Dixon, J. E., and Mains, R. E. (1990) Biosynthesis and processing of site-directed endoproteolytic mutants of proNPY in mouse pituitary cells. *J. Biol. Chem.* **265,** 2462–2469.
11. Allen, J. M., Novotny, J., Martin, J. B., and Heinrich, G. (1987) Molecular structure of mammalian neuropeptide Y: analysis by molecular cloning and computer-aided comparison with crystal structure of avian homologue. *Proc. Natl. Acad. Sci. USA* **84,** 2532–2536.
12. Allen, J. M., Yeats, J. C., Causon, R., Brown, M. J., and Bloom, S. R. (1987) Neuropeptide Y and its flanking peptide in human endocrine tumors and plasma. *J. Clin. Endocrinol. Metab.* **64,** 1199–1204.
13. Bottcher, G., Skolund, K., Eklad, E., Hakanson, R., Schwartz, T. W., and Sundler, F. (1984) Co-existence of peptide YY and glicentin immunoreactivity in endocrine cells of the gut. *Regul. Dept.* **8,** 261–266.
14. Aponte, G. W., Fink, A. S., Meyer, J. H., Tatemoto, K., and Taylor, I. L. (1985) Regional distribution and release of peptide YY with fatty acids of different chain length. *Am. J. Physiol.* **249,** G745–G750.
15. Allen, J. M., Fitzpatrick, M. L., Yeats, J. C., Darcy, K., Adrian, T. E., and Bloom, S. R. (1984) Effects of peptide YY and neuropeptide Y on gastric emptying in man. *Digestion* **30,** 255–262.
16. Pappas, T. N., Debas, H. T., Goto, Y., and Taylor, I. L. (1985) Peptide YY inhibits meal-stimulated pancreatic and gastric secretion. *Am. J. Physiol.* **248,** G118–G123.
17. Leiter, A. B., Keutmann, H. T., and Goodman, R. H. (1984) Structure of a precursor to human pancreatic polypeptide. *J. Biol. Chem.* **259,** 14,702–14,705.
18. Leiter, A. B., Toder, A., Wolfe, H. J., Taylor, I. L., Cooperman, S., Mandel, G., and Goodman, R. H. (1987) Peptide YY. Structure of the precursor and expression in exocrine pancreas. *J. Biol. Chem.* **262,** 12,984–12,988.
19. Yamamoto, H., Nata, K., and Okamoto, H. (1986) Mosaic evolution of prepropancreatic polypeptide. *J. Biol. Chem.* **261,** 6156–6159.
20. Minth, C. D., Andrews, P. C., and Dixon, J. E. (1986) Characterization Sequence and Expression of the Cloned Human Neuropeptide Y Gene. *J. Biol. Chem.* **261,** 11,974–11,979.
21. Allen, J. M., Polak, J. M., Rodrigo, J., Darcy, K., and Bloom, S. R. (1985) Localization of neuropeptide Y in nerves of the rat cardiovascular system and effect of 6-hydroxydopamine. *Cardiovascular Research* **19,** 570–577.
22. Allen, J. M., Martin J. B., and Heinrich, G. (1988) Neuropeptide Y is intrinsic to the heart, in *Advances in Atrial Peptide Research,* vol. II (Brenner, B. M. and Laragh J. H.), Raven, New York, pp. 155–160.
23. Ericsson, A., Schalling, M., McIntyre, K. R., Lundberg, J. M., Larhammar, D., Seroogy, K., Hokfelt, T., and Persson, H. (1987) Detection of neuropeptide Y and its mRNA in megakaryocytes: enhanced levels in certain autoimmune mice. *Proc. Natl. Acad. Sci. USA* **84,** 5585–5589.
24. Leiter, A. B., Montminy, M. R., Jamieson, E., and Goodman, R. H. (1985) Exons of the human pancreatic polypeptide gene define functional domains of the precursor. *J. Biol. Chem.* **260,** 13,013–13,017.

25. Krasinski, S. D., Wheeler, M. B., Kopin, A. S., and Leiter, A. B., (1990) Pancreatic polypeptide and Peptide YY Gene expression, in *Central and Peripheral Significance of Neuropeptide Y and its Related Peptides.* (Allen, J. M. and Koenig, J. I., eds.), Annals of NY Acad. Sci., vol. 611, New York, pp 73–85.
26. Yonekura, H., Nata, K., Watanabe, T., Kurashina, Y., Yamamoto, H., and Okamoto, H. (1988) Mosaic evolution of prepropancreatic polypeptide II. Structural conservation and divergence in pancreatic polypeptide gene. *J. Biol. Chem.* **263,** 2990–2997.
27. Kopin, A. S., Toder, A. E., and Leiter, A. B. (1988) Different splice site utilization generates diversity between the rat & human pancreatic polypeptide precursors. *Arch. Biochem. Biophys.* **267,** 742–748.
28. Takenchi, T., Gumucio, D .L., Yamada, T., Meister, M. H., Minth, C. D., Dixon, J. E., Eddy, R.E., and Shows, T. B. (1986) Genes encoding pancreatic polypeptide and neuropeptide Y are on chromosome 17 and 7. *J. Clin. Invest.* **77,** 1038–1041.
29. Schwartz, T. W., Fuhlendorff, J., Langeland, N., Thogersen, H., Jorgensen, J. C., and Sheikh, S. P. (1989) Y_1 and Y_2 receptors for NPY—the evolution of the PP-fold peptides and their receptors, in *Neuropeptide Y* (Mutt, V., Fuxe, K., Hokfelt, T., and Lundberg, J.M., eds.), Raven, New York, pp. 143–151.
30. Minth, C. D. and Dixon, J. E. (1990) Expression of the human neuropeptide Y gene. *J. Biol. Chem.* **265,** 12,933–12,939.
31. OoI, Y. M. and Persson H. (1990) Novel brain specific transcription factor form DNA—protein complexes with SPI and TATA regions of the rat neuropeptide Y gene promoter, in *Central and Peripheral Significance of Neuropeptide Y and Its Related Peptides* (Allen, J. M. and Koenig, J. I., eds.), Annals of the NY Acad. Sci., vol 611, New York, pp 382–387.
32. Angel, P., Imagawa, M., Chiu, R., Stein, B., Imbra, R. J., Rahmsdorf, H. J., Jonat, C., Herrlich, P., and Karim, M. (1987) Phorbol ester-inducible genes contain a common cis element recognized by a TPA-modulated *Trans*-acting factor. *Cell* **49,** 729–739.
33. Curran, T. and Franza, B. R. (1988) *Fos* and *Jun*: the AP-1 Connection. *Cell* **55,** 395–397.
34. Sabol, S. L. and Higuchi, H. (1990) Transcriptional regulation of the neuropeptide Y gene by nerve growth factor: Antagonism by glucocorticoids and potentiation by adenosine 3'-5' monophosphate and phorbol ester. *Mol. Endocrinol.* **4,** 384–392.
35. Allen, J. M., Martin, J. B., and Heinrich, G. (1987) Neuropeptide Y gene expression in PC12 cells and its regulation by nerve growth factor: a model for developmental regulation. *Mol. Brain Res.* **3,** 39–43.
36. Morgan, J. I. and T. Curran. (1991) Stimulus-transcription coupling in the nervous system: Involvement of the inducible proto-onogenes *fos* and *jun. Ann. Rev. Neurosci.* **14,** 421–451.
37. Cremis, J., Wagner, J. A., and Halegoua, S. (1986) Nerve Growth Factor Action is mediated by cyclic AMP and Ca^{++}/phospholipid-dependent protein kinases. *J. Cell. Biol.* **103,** 887–893.
38. Gomez, N. and P. Cohen. (1991) Dissection of the protein kinase cascade by which nerve growth factor activates MAP kinase. *Nature* **353,** 170–173.

39. Balbi, D. and Allen, J. M. (1992) Nerve Growth Factor may mediate the increase in neuropeptide Y mRNA levels in PC12 cells through activation of protein kinase C. *J. Physiol.* **452,** 2548.
40. Fischer-Colbrie, R., Iacangelo, A., and Eiden, L. E. (1988) Neural and humoral factors separately regulate neuropeptide Y, enkephalin, and chromogranin A and B mRNA levels in rat adrenal medulla. *Proc. Natl. Acad. Sci. USA* **85,** 3240–3244.
41. Schalling, M., Franco-Cereceda, A., Hemsén, A., Dagerlind, A., Seroogy, K., Persson, H., Hökfelt, T., and Lundberg, J. M. (1991) Neuropeptide Y and catecholamine synthesizing enzymes and their mRNAs in root sympathetic neurons and adrenal glands: studies on expression, synthesis and axonal transport after pharmacological and experimental manipulations using hybridization techniques and radioimmunoassay. *Neuroscience* **61,** 753–766.

Organization of Neuropeptide Y Neurons in the Mammalian Central Nervous System

Stewart H. C. Hendry

1. Introduction

In the discovery of most neuropeptides, function has preceded antigen. Each peptide has been identified as the mediator of at least one function, and only later has knowledge of chemical structure led to antibody production, which, in turn, has led to the discovery of peptide distribution *(1)*. Such is not the case for neuropeptide Y (NPY). Originally discovered because of its chemical structure *(2)*, NPY was quickly identified as a member of the pancreatic polypeptide family *(2,3)*, and after specific antiNPY antisera were made, it was localized to neurons throughout the central nervous system *(4–6)*. Yet only very recently have studies of function, from single-cell physiology to animal behavior, begun to determine the role of NPY as a neuroactive substance in the CNS. This process of going from localization to physiology, from antigen to function, is one that is being pursued for many CNS molecules, including proteins that are recognized with select monoclonal antibodies *(7–11)* or that are phosphorylated by particular kinases (e.g., *12,13*). It is a process that has been most successful for NPY.

There remains the fact, however, that with a few notable exceptions, particularly addressing the organization and function of the hypothalamus, studies of NPY localization have proceeded with little knowledge of what the peptide does. What has been the point? Why have literally hundreds of studies examined the distri-

From: *The Biology of Neuropeptide Y and Related Peptides;*
W. F. Colmers and C. Wahlestedt, Eds. © 1993 Humana Press Inc., Totowa, NJ

bution of NPY at a time when the function of that peptide has been a mystery? Even a very quick survey of the literature provides a very clear and compelling answer: NPY has been used as an always-convenient and often-powerful marker for identifying and studying distinct groups of neurons in the CNS. In a real sense, then, the study of NPY localization over the past decade has been the study not of one peptide, but of the brain and spinal cord, themselves.

One general statement can be made for NPY in the CNS, particularly where it has been most intensively studied: The localization of NPY allows an otherwise bewildering array of neurons to be understood as a series of chemically, morphologically, and connectionally distinct subpopulations. Within that context, two basic types of neurons have been discovered to be NPY immunoreactive: (1) short-axon cells or interneurons—these cells give rise to local connections and are, thus, elements of the intrinsic circuitry of an area, NPY interneurons are predominant in the forebrain, including the cerebral cortex, and large subcortical regions, such as the striatum and amygdala, and (2) long-projection neurons—these cells have axons that leave the region of their parent somata and extend for considerable distances to innervate specific targets. NPY projection neurons are found principally in the brainstem.

What is true for both interneurons and projection neurons is that the presence of NPY permits the examination of select neuronal populations in relative isolation. The result has been an increased understanding of the organization, development, and plasticity of regions in the forebrain and brainstem, and an understanding of the processes underlying certain neuropathologies. Perhaps most rewarding, the examination of NPY localization has come full circle in recent years, since the discovery of where NPY is and how its distribution may change during development, in adulthood, or with pathologies has led to a greater understanding of what NPY, itself, may be doing.

Most studies reviewed here employed immunological methods to localize NPY. That is, the studies are based principally on immunocytochemical methods and, to a lesser extent, on radioimmunoassay. Both are so widely used and well understood that only a simple reminder is necessary: The immunologically based methods cannot characterize a substance with absolute certainty. Even when studies have used chromatographic methods in parallel with immunological ones (e.g., *14–16*), the distinct possibility remains that antisera may be recognizing an uncharacterized, crossreacting substance. For those familiar with the history of pancreatic polypeptide localization, this reminder is hardly necessary, since it is now widely accepted that many studies that originally described avian pancreatic polypeptide (APP; *17–20*) or bovine pancreatic polypeptide (BPP; *21,22*) in the CNS were actually localizing NPY (*3,8,15,23;* cf *24*). Those studies of APP and BPP localization are included here, usually without reference to the original naming of the immunoreactive material.

Two more recently developed methods of localization are used for comparative purposes in this chapter. One is the autoradiographic localization of ligand-binding sites. A few studies have examined the binding of radioactively labeled NPY and peptide YY to localize putative NPY receptor sites. These are included at relevant points so that putative peptide and receptor distributions might be compared. It is recognized that multiple NPY receptor subtypes may exist in the CNS, but since the localization studies to date have not dealt with that subject to any great degree, the issue of where the different subtypes may exist is not discussed extensively in this chapter (*see* Grundemar et al., this vol. for a discussion). The second recently developed method is *in situ* hybridization histochemistry for the localization of mRNAs encoding for preproNPY. This very powerful method, which takes advantage of the sequencing of the mammalian preproNPY gene *(25–27)*, and the data produced from its use are included as complementary evidence for the presence of NPY somata in several CNS regions. In one region, the human cerebral cortex, NPY immunoreactivity and NPY message have been localized to the same neurons *(28)*, leaving little doubt about the true chemical nature of the cells. This is a standard for peptide and enzyme localization and characterization that will be necessary to fulfill for neurons throughout the CNS.

What follows is not intended to be an encyclopedic list of regions in the CNS where NPY has been localized by one method or another. Studies that have examined NPY-like immunoreactivity throughout the CNS of one or more species should be consulted for certain details and for detailed maps of immunoreactivity (e.g., *29–31*). (*See* chapter by Sundler et al., this vol., for the distribution of NPY in the peripheral nervous system). Instead, this chapter is intended to explore the question, "What does NPY localization tell us about the organization, development, and plasticity of specific regions in the CNS?" It is this question that has dominated research on NPY localization, and it is hoped that the quality of the original research is reflected in the present attempt to answer it.

2. Telencephalon

2.1. Cerebral Cortex

2.1.1. Cortical Organization and NPY

Neurons of the cerebral cortex form easily recognized groups possessing similar functional and connectional features. Not only is the entire cortical mantle divided into several dozen areas, but each area is also subdivided into horizontal layers and vertical columns. In addition, functionally analogous areas may differ in structure and organization among mammalian species. To varying degrees, NPY localization reveals each of these cortical characteristics: NPY neurons of the cerebral cortex are organized in ways that recognize areal, laminar, and columnar borders, and that differ among mammalian species.

Although a fundamental plan of development and organization has been attributed to most cortical areas *(32)*, studies of cortical neurotransmitters and receptors have consistently shown area-specific patterns of organization (e.g., *33–37*). Areal differences in NPY are apparent from RIA studies of peptide concentration. These studies show that in the rat *(38)*, monkey *(39)*, and human cerebral cortex *(14,40–42)*, the concentration of NPY-like immunoreactivity varies across areas by a factor of three, with the highest concentrations localized in cingulate gyrus and in association areas of temporal and the lowest concentrations found in areas of the frontal lobe *(40)* and the occipital lobe, particularly the primary visual area *(39)*. Other primary sensory areas, such as the first somatic sensory area, and the primary motor area of the human cortex have relatively high levels of NPY immunoreactivity *(40)*. Similar results have been reported for levels of NPY messenger RNA, with areas 4 and 6 of the human motor cortex containing as much as five times the amount of NPY mRNA as areas of prefrontal cortex *(43)*. The gradient in immunoreactive NPY concentration is apparent across species, although in single species, especially in rats and humans, wide variations in peptide concentration are found when data from different labs are compared (human cortex: *40,41*; rat: *16,38*). These differences may be attributed to differences in methodology or reagents, or in the case of the human material, to problems of obtaining suitable samples *(44,45)*, even though the most critical problem of variable postmortem delay does not appear to affect the measurable levels of NPY immunoreactivity *(14,46)*.

Immunocytochemical and *in situ* hybridization studies have localized NPY somata to the cerebral cortex of rodents *(47–50)*, cats *(51–54)*, pigs *(19)*, and several species of primates, including New World monkeys *(55)*, Old World monkeys *(56–62)*, and humans *(63–69)*. For all species and all areas examined, the cell bodies are present in each layer of the cortex, but are concentrated in layers II, III, V, and VI (Figs. 1A and 2A). The degree to which somata are localized to layer IV would seem to vary from a major concentration *(67–69)* to none *(51,58–60)*. However, such differences among areas and species are probably more apparent than real (i.e., they can be attributed to differences in the assignment of laminar boundaries).

The distribution of immunoreactive fiber plexuses varies across layers and, unlike the distribution of somata, also differs across areas (Fig 1B,C). The interareal differences are apparent in the relative density of the plexuses and, in some areas, by the presence of additional plexuses. These areal differences are most apparent in the monkey cerebral cortex *(56,58,59)* where primary sensory areas generally have a lower density of fibers than areas of association and limbic cortex *(58,59)*. The one exception to this pattern of low fiber density in primary sensory areas is the primary visual area (area 17), where not only does a high density of NPY-immunoreactive processes exist in the usual superficial and deep plexuses, but also two additional plexuses are present in the middle layers (layers IVB and VA). The greater density of fibers and the additional plexuses clearly mark area 17 and serve to distinguish it from the adjacent second visual area (area 18; *56,58,59,62*).

NPY-immunoreactive somata and processes are not restricted to the cortex itself, but are present in the subcortical white matter of adults in all species (Fig. 3). The cells are most numerous in primates *(56,58,59,62)*, particularly humans, where they outnumber the NPY-positive somata in the cortex, proper, by at least 2:1 and as much as 4:1 *(64)*. The white matter cells are far less numerous in the cortex of rodents (e.g., *47,56*). These neurons are usually present within 0.5–1.0 mm of the border between layer VI and the white

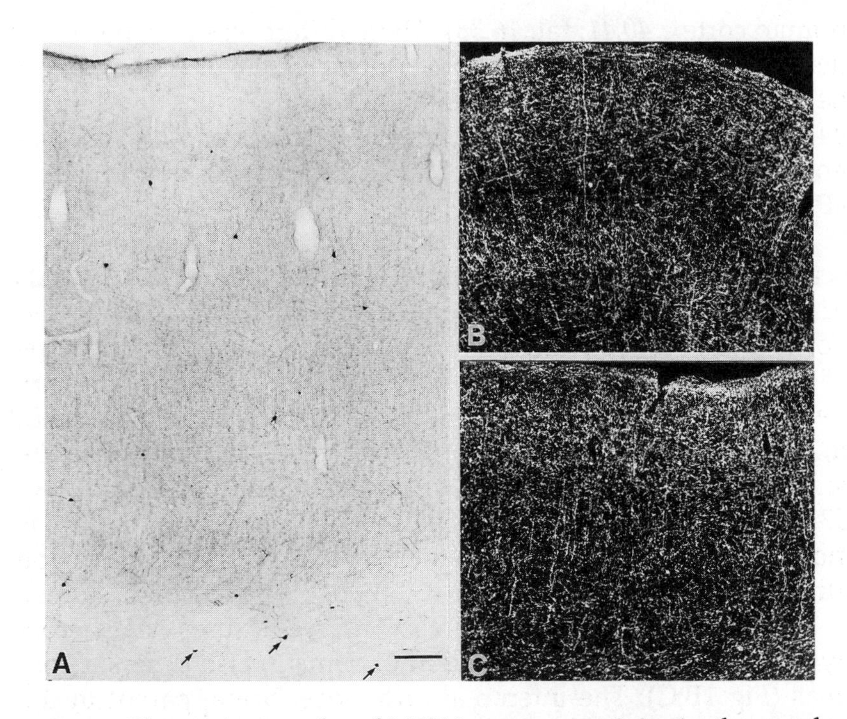

Fig. 1. Photomicrographs of NPY immunoreactivity in the monkey cerebral cortex. A. Bright field photomicrograph of immunoreactive neurons in parietal cortex (area 5), demonstrating the distribution of somata in a homotypical, six-layered area of the cortex. Somata are scattered through the depth of the cortex, but tend to be more concentrated in layers II and III and V and VI. In addition, numerous somata are present in the white matter, approx 1 mm subjacent to layer VI (arrows). B and C. Dark-field photomicrographs of NPY-immunoreactive fibers in areas of monkey temporal association cortex (B) and prefrontal cortex (C). The distribution of fibers is mainly vertical in both layers and is more rigidly layered in the temporal lobe, where a middle band, representing layer IV, contains very few fibers except those ascending or descending to neighboring layers. Bar equals 1.5 mm in A, 3.2 mm in B and C.

matter, and their processes most often take the shortest route to enter the cortex and contribute to the plexus of fibers in layers V and VI. However, in some species, immunostained processes have been found to extend over considerable distances in the white matter and to enter the corpus callosum *(17,48)*. There are some indications that within a single species of primates, and, in fact within a single individual, the numerical density of these white

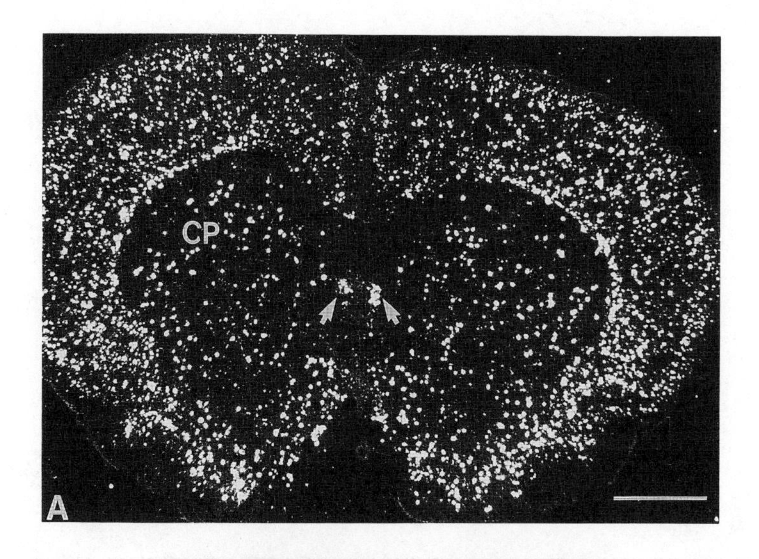

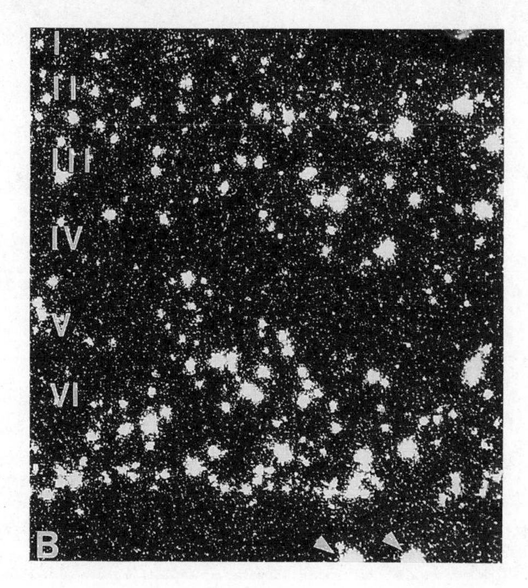

Fig. 2. Dark-field autoradiograms of rat brain processed by *in situ* hybidization histochemistry for localization of preproNPY mRNA. The distribution of labeled somata throughout the rat cortex is apparent. The clustering of NPY somata in the more superficial layers (II and III) and in deeper layers (V and VI) is also seen with this method. Fewer somata are present in the white beneath the rat cortex than are present in the monkey cortex (arrowheads in B). NPY neurons are also apparent in the caudate/putamen (CP) and in the septal nuclei (arrow). The autoradiograms are from previously unpublished experiments by C. M. Gall. Bar equals 1 mm in A and 250 μm in B.

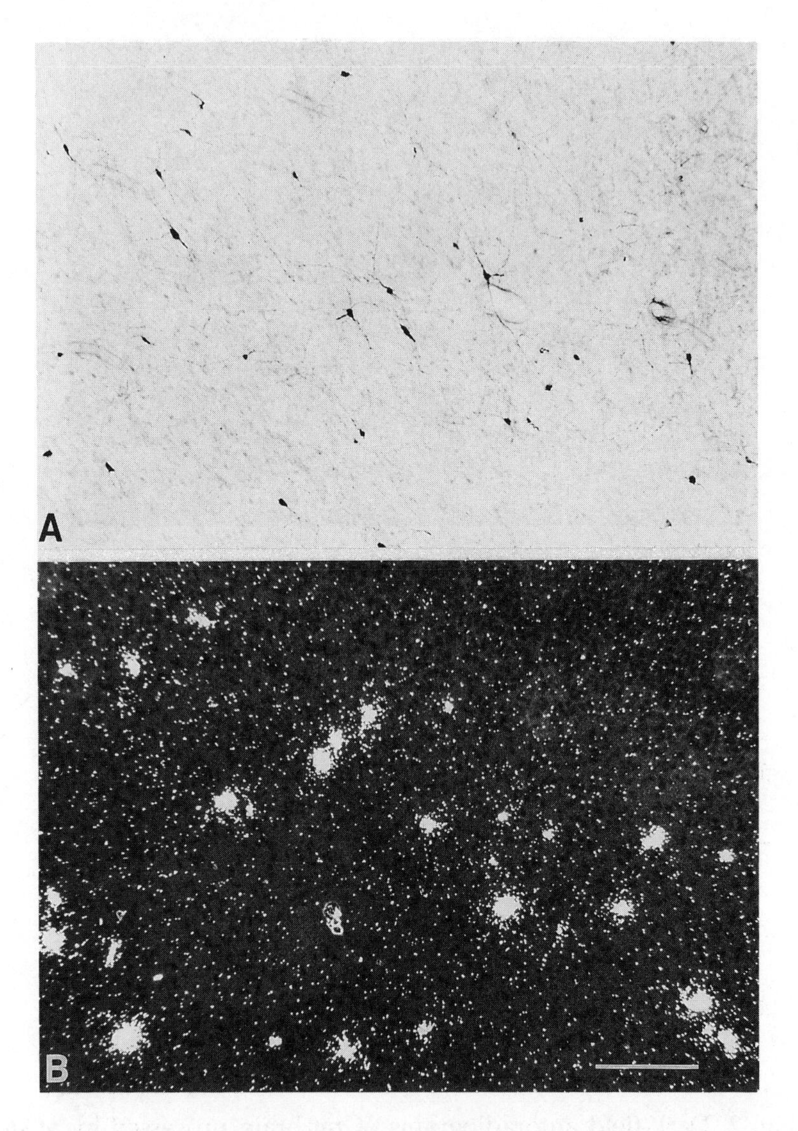

Fig. 3. Photomicrographs of NPY neurons in the white matter of monkey cortex. A. Bright-field micrograph of NPY-immunoreactive neurons in the white matter beneath the superior parietal lobule of an adult monkey. Numerous somata, giving rise to long-beaded processes are evident. B. Dark-field autoradiogram of somata labeled by *in situ* hybridization histochemistry (using an ^{35}S-labeled complementary RNA probe) in the white matter beneath area 17 of an adult monkey. Neurons are densely clustered within a few hundred microns of the border with layer VI. Bar equals 200 μm in A and 150 μm in B.

matter cells varies among areas, so that they are relatively enriched in areas of the parietal lobe and less numerous in areas of the occipital lobe *(56)*.

These data indicate that the division of the cortical mantle into functional areas and the organization of those areas into layers are reflected in the distribution of NPY somata and processes, and that some species-specific patterns are also evident. Furthermore, the distribution of NPY-immunoreactive somata follows the columnar pattern of organization in the primary visual area (area 17) of monkeys. In that area, zones of intense metabolic activity can be identified in layers II and III by high levels of histochemically localized cytochrome oxidase (CO; *70–73*). These zones form periodic patches, also referred to as "puffs" or "blobs" *(74)*, that line up in rows. The rows of CO puffs occupy the centers of eye-dominance columns and, thus, are reliable indicators of the regular, columnar organization of primate area 17 *(72,73,75)*. Kuljis and Rakic *(60)* have shown that the NPY-immunoreactive neurons in this area avoid the CO puffs: whereas these regions make up 14% of the total volume of layers II and III, fewer than 3% of the NPY-immunoreactive somata are present within them. Preliminary evidence suggests that the superficial plexus of immunoreactive fibers in monkey area is similarly organized so that the regions of the CO puffs are poorly innervated by NPY processes, whereas the regions around the puffs are richly innervated (S. Hendry, unpublished observations). Thus, both the cell bodies and processes of NPY neurons are distributed within area 17, so as to occupy select parts of the repeating, modular organization of this area. Similar findings that chemically identified neuronal elements are present selectively either in or around the CO puffs of monkey area 17 have been reported for other neurotransmitters, neuropeptides, related enzymes, and receptors *(35,76–79)*.

It should be noted, however, that in other areas of the cerebral cortex, conclusively determined to be organized into functionally and connectionally specific columns *(80–82)*, no evidence for a column-specific distribution of NPY immunoreactivity has been found. This is true even for area 17 of the cat, in which eye-dominance columns similar to those of the monkey also exist *(83,84)*. Obviously, then, NPY is not a general marker for cortical columns. Instead, the uneven distribution of NPY-immunoreactive neurons in the cortex is one indication that physiologically distinct cells, grouped into columnar units, layers, or areas, are also chemically distinct *(35–37,56,85)*. These differences in neurochemical features may contribute to the physiological differences among columns, layers, or areas, in a way that neurons innervated by NPY-containing axons would differ, for example, in receptive field size or in their response to specific stimulus features.

2.1.2. Neuronal Classes Containing NPY

Neurons of the cerebral cortex can be divided into two general populations: (1) *Pyramidal cells* have triangular somata, and prominent ascending (apical) dendrites that branch without tapering, often reach layer 1, complex systems of basal dendrites and axons that arise from somata or basal dendrites, and descend, most often into the white matter *(86)*; (2) *nonpyramidal cells* lack at least one of the pyramidal cell features. Their somata are usually round or oblong, dendrites arise from all parts of the somata, and axons may adopt one of several possible configurations *(87)*. For the most part, pyramidal cells are the efferent, or output neurons of the cerebral cortex and nonpyramidal cells are cortical interneurons (for review, *see 88*), although some examples of output cells that are nonpyramidal (e.g., *89*) and interneurons that are pyramidal have been described (e.g., *90*). Pyramidal cells can be distinguished from one another by two related characteristics: in which layer there are somata present and to which target their axons project *(91,92)*. Nonpyramidal cells, on the other hand, are reliably classified only on the basis of axonal morphology *(87)*.

All studies of NPY immunostaining agree that the great majority of NPY-positive cortical neurons are nonpyramidal cells and are, thus, interneurons. Two questions remain open: (1) Is there a population of NPY-immunoreactive pyramidal cells, and (2) can the NPY nonpyramidal cells be subdivided into distinct classes?

That NPY immunoreactive cortical neurons are exclusively nonpyramidal is commonly reported, particularly in primates (*56–60,62*; Fig. 4). However, some investigators interpret their findings to suggest that a small subpopulation of immunoreactive pyramidal neurons is present in the cortex *(14,29,51,62,67)*. The significance of such a subpopulation would be great since the presence of such cells would imply that a group of cortical projection neurons uses NPY as a signaling agent. Yet analysis of the published reports of NPY-positive pyramidal cells strongly suggests that the data on

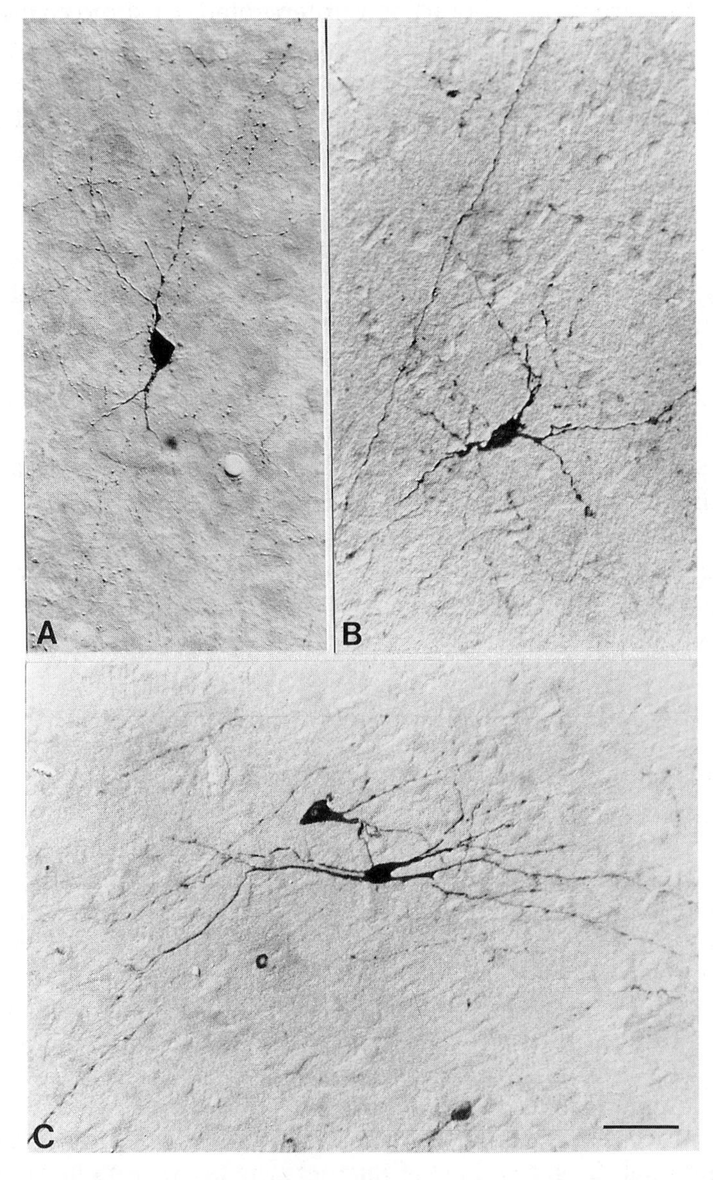

Fig. 4. Differential interference contrast photomicrographs of NPY-immunoreactive neurons in the monkey primary visual cortex (area 17). The somata give rise to many beaded dendrites that branch repeatedly. In addition, finer processes, which are presumed to be axons, are present around the somata. A. From layer III of an adult monkey. B. From layer VI of an adult monkey. C. From the white matter of a fetal (E150) monkey, 300 μm from the border with layer VI. Bar equals 20 μm.

which they are based should be reinterpreted. In some cases, the reports of pyramidal cells have focused only on the shape of the immunoreactive somata as a means of classification, but the descriptions *(67)* of the cells' processes (e.g., ascending axons) and of their laminar distribution (e.g., present in layer I) are distinctive and characteristic of nonpyramidal cells. Evidence that NPY axon terminals form symmetric synapses *(56,93)*, whereas pyramidal cell axons form exclusively asymmetric contacts *(86)* and the universally accepted finding that NPY dendrites are aspiny, whereas pyramidal cell dendrites are very spiny *(86)* also support an exclusively nonpyramidal localization of NPY. Perhaps most significantly, none of the photomicrographs or camera lucida drawings of the putative NPY-positive pyramidal neurons provides compelling documentation for this classification: The morphology of processes interpreted as apical or basal dendrites is unlike those commonly accepted from studies of Golgi-impregnated or peroxidase-injected pyramidal neurons (*see* Fig. 5) or that commonly produced with immunostaining for other antigens *(94,95)*. It is reasonable to conclude, then, from the presently available data that NPY is found only in nonpyramidal cells.

Fig. 5. *(opposite page)* Fluoresence photomicrographs of NPY immunoreactive neurons (A and C) double-labeled for GAD (B) or SRIF (D). A and B. Pair of photomicrographs from the same section through monkey visual cortex (area 17) labeled with a fluorescein label for NPY and a rhodamine label for GAD. One of the two GAD-positive neurons is also NPY-immunoreactive, indicating that NPY is present within a small subpopulation of cortical GABA neurons. The triangular shape of some NPY somata, such as the one in A, is frequently misinterpreted as evidence for NPY-immunoreactive pyramidal neurons. However, in all other morphological features, including the lack of a true apical dendrite or of a basal dendritic system, the NPY neurons are obviously nonpyramidal and, thus, most likely cortical interneurons. The presence of GAD and GABA immunoreactivity within the cells (A and B) and the symmetric synaptic contacts they form (*see* Fig. 6) confirm that the NPY neurons are nonpyramidal. C and D. Pair of micrographs taken from layer VI of monkey area 17. Three neurons (arrows) are immunoreactive for both NPY (C) and SRIF (D). That pattern of coexistence is seen for most neurons immunoreactive for the tetradecapeptide of SRIF, but neurons immunoreactive for other forms of SRIF greatly outnumber the NPY cells. Most of the NPY/SRIF cells are also GABA-positive, whereas a small proportion apparently are not GABAergic, but contain tachykinin-like immunoreactivity. Bar equals 12 μm in A and 45 μm in B.

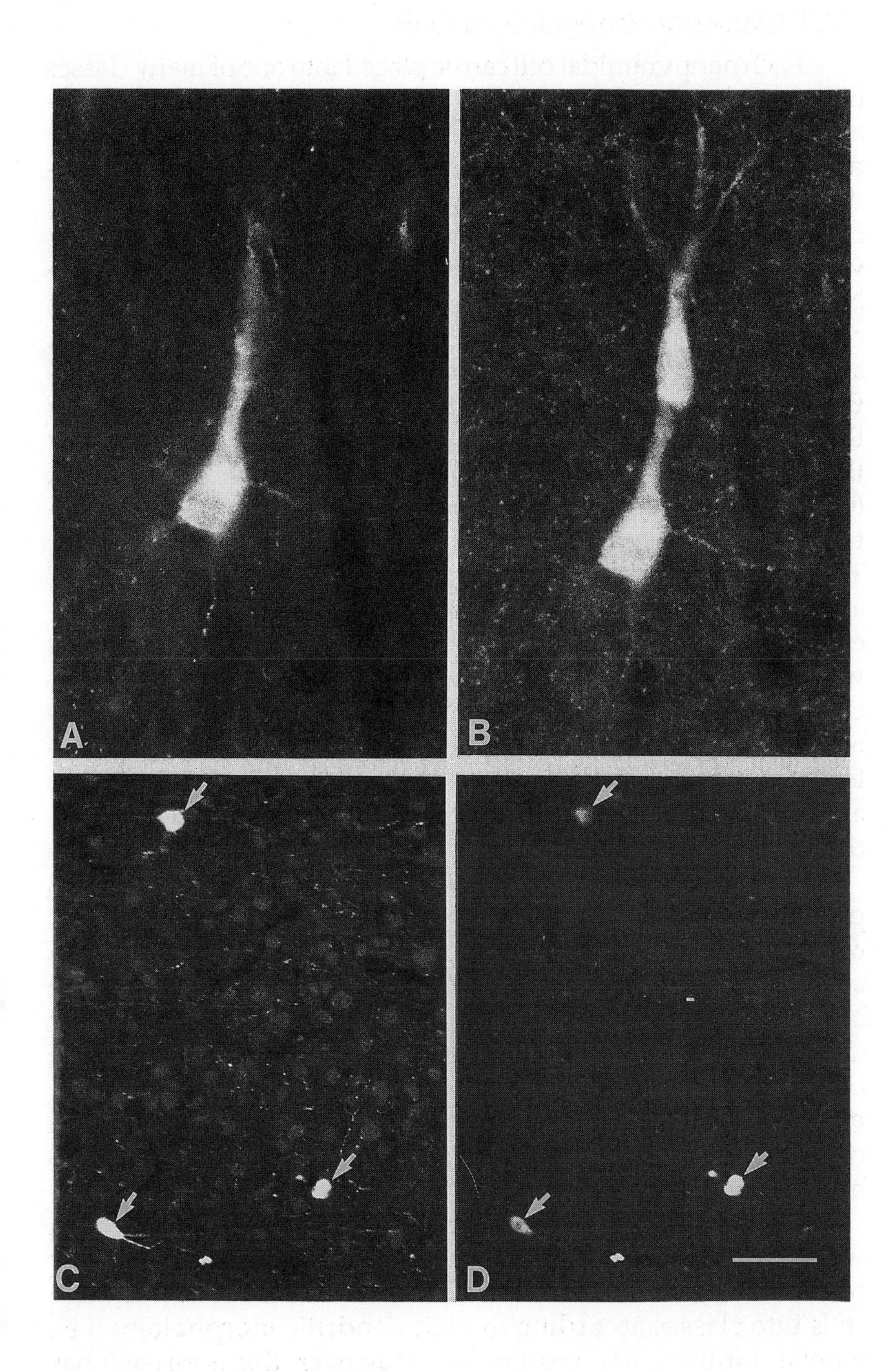
A
B
C
D

2.1.3. Classes of Nonpyramidal Cells

Each nonpyramidal cell can be placed into one of many classes based on the morphology of its axon *(87)*. Even when consideration is given only to the classes of nonpyramidal cells with aspiny dendrites, which is the case for cortical NPY cells, several candidates remain. In one study of monkey and rat cortex, it was the conclusion of the authors that the immunostaining of processes was essentially the same across all cells: No distinction was made between what might be dendrites and axons, and every cell was considered to possess long, beaded processes, which, at some point, ascended radially or obliquely through the cortical layers *(56)*. By contrast, virtually all other studies have identified processes that have the features of axons (e.g., *47,51,52,59,62*). Several of these, particularly in the cat *(52)*, monkey *(59)*, and human cortex *(66)*, have made strong cases that certain morphologically characterized classes of nonpyramidal cells are NPY-immunoreactive. These may include neurons in the deep cortical layers with long, ascending axons (Martinotti cells) and neurons with radially oriented bundles of axons and dendrites (double-bouquet cells), as well as other classes with densely intertwined axons *(66)*. Two classes of aspiny cells that are demonstrably *not* NPY-immunoreactive in normal animals are those whose axons terminate in clusters around pyramidal cell somata (basket cells; *96*) and those whose axons terminate on the initial segments of pyramidal cells (chandelier cells; *97*).

There are indications that NPY-like immunoreactivity is present in some members of a morphologically characterized class of cortical cells, but not in other members of the same class. In addition, it appears that certain morphological classes may be NPY-immunoreactive only in one or two of the many cortical areas *(59)*. These findings, that single classes of cortical neurons may be chemically heterogeneous both within and across cortical areas, may appear radical, but they have been confirmed in recent studies of the peptide, corticotropin-releasing factor *(98)*, and the calcium-binding protein, calbindin *(99)*.

Even when the full complement of cortical cells with well-characterized axons has been examined, many NPY cells remain unclassified. Some authors have attempted to divide these cells into classes according to their dendritic morphology (i.e., bipolar, bitufted, and multipolar). However, this approach has severe limitations for two reasons: (1) Every immunocytochemical preparation shows only a fraction of a cell's dendritic field; and (2) dendritic morphology is a very poor indicator of the nonpyramidal cell types in the cerebral cortex *(100)*. The danger of using dendritic morphology as a tool for interpreting nonpyramidal cell types is shown most clearly with a type designated "bipolar." Several studies have described such a type of NPY-immunoreactive cell in the cerebral cortex *(29,47,48,67)*. One interpretation has been that these cells are excitatory neurons, since morphologically characterized bipolar cells form asymmetric synapses in the cerebral cortex *(101)*. On this basis, excitatory actions for certain cortical cells containing NPY have been suggested *(67)*. However, as shown repeatedly for other peptide-immunoreactive cells that display apparently bipolar dendritic morphologies (e.g., *102,103*), and specifically for NPY-immunoreactive cells (*see* Section 2.1.6.), the axonal terminations of these neurons are of the symmetric type. These data strongly suggest that the peptides are present in neurons that are inhibitory *(57,104,105; see* Section 2.1.6.), not excitatory. In other words, whatever NPY may be contributing to the physiological characteristics of cortical neurons, the peptide is present in cells that would otherwise be classified as inhibitory interneurons. A similar temptation to misinterpret the functions of other NPY cells exists when the cells are grouped into classes according to the shape of their somata or their dendritic fields. What emerges, then, from the studies of NPY-immunoreactive neurons are two lessons, a general one that the classification of nonpyramidal cortical neurons by their dendritic morphology is always of dubious value and is often very misleading, and a specific one that the classification of NPY cortical neurons is not yet complete.

2.1.4. NPY Cells in the Subcortical White Matter

As mentioned above, many NPY-immunoreactive neurons are present in the white matter subjacent to all cortical areas *(29,47,56)*. They are particularly numerous in the primate cerebral cortex *(56,58,59,62,64)*. The morphological features of these cells does not serve to set them apart from the NPY neurons within the cortex, itself (Fig. 4). The somata of the white matter cells do tend to be more elongated, and processes that extend for several hundred microns among the subcortical axons have been described. However, most of these neurons send processes into the cortex, where they make up part of the dense fiber plexuses in layers V and VI. Thus, it is not clear whether these cells should be included in the same morphological classes as the cortical neurons or split off to form an additional class.

2.1.5. Chemical Characterization of NPY Cortical Cells

Coexistence of neuropeptides with classical neurotransmitters and with other neuropeptides is an established feature of CNS neurons *(106)*. A series of studies has indicated that NPY-immunoreactive cortical cells contain one other and, in some cases, three other peptides, a classical neurotransmitter, and a rare housekeeping enzyme.

Since the initial discovery that NPY immunoreactivity (localized with an antiserum to APP) and somatostatin (SRIF)-like immunoreactivity coexisted in human cortical neurons *(63)*, numerous studies have confirmed this observation in several species and have determined that as many as 90% of the cells displaying immunoreactivity for one of these two peptides display immunoreactivity for both (*29,105;* Fig. 5C,D). Accordingly, the concentration of the two peptides is highly correlated among many areas of the monkey cortex *(39)*. However, some studies report less frequent coexistence in the cerebral cortex *(36,110)*, and it should be noted that the studies of SRIF immunoreactivity in primate cortex that have used the best available antisera to that peptide *(109)* report the existence of a very large population of cortical SRIF-positive cells (e.g., *36,110*), certainly a larger population than the one recognized with NPY antisera *(56,58–60,62)*. Thus, although most NPY cells may also display SRIF-like immunoreactivity, a substantial SRIF population may contain no detectable NPY.

Little attempt has been made to identify subcellular compartments that may contain one or both neuropeptides, so it is not known whether NPY and SRIF are packaged and released together. However, the use of cysteamine as an agent to deplete SRIF *(111)* has shown that in cortical synaptosomal preparations, the two peptides probably exist in different compartments, since SRIF is depleted in a dose-dependent fashion, but NPY is left unaffected *(112)*. The distinct possibility exists, then, that under physiological conditions, NPY and SRIF are released separately from cortical terminals.

The subpopulation of cortical cells that is both NPY- and SRIF-immunoreactive also displays histochemical activity for the enzyme NADPH-diaphorase *(113,114; see 115)*. These are the solitary active cells originally described by Pearse *(116)* as being resistant to manipulations of brain metabolism that destroy most neurons *(117)*. The significance of this cell population for cortical function and the implications of peptide and enzyme coexistence within them are still mysteries.

Given the coexistence of NPY- and SRIF-immunoreactivity in cortical cells, the report of SRIF immunoreactivity in cortical neurons that are also immunoreactive for the GABA-synthesizing enzyme glutamate decarboxylase (GAD; *118*) indicated a coexistence of both peptides in GABA neurons. Simultaneous reports of SRIF immunoreactivity in GABA-positive neurons of cat visual cortex *(104)* and both SRIF and NPY immunoreactivity in GAD-positive neurons of monkey cortex *(157)* were able to document the wide range of mammalian species in which peptides are present in GABA neurons and to identify definitively NPY as one of those peptides (Fig. 5A,B). These and subsequent studies *(54,107)* have focused on the proportion of cells in the various species that display neuropeptide and GABA coexistence. The consensus of the several studies is the following:

1. A very small proportion of the total GABA population is immunoreactive for NPY and SRIF;
2. GABA/NPY coexistence is found in neurons throughout the cortex and in the subcortical white matter;
3. Other GABA subpopulations display immunoreactivity for other peptides (e.g., cholecystokinin octapeptide) but most GABA cells are immunoreactive for none of the currently recognized neuropeptides; and
4. Of the total NPY population, a small number consistently displays no GABA- or GAD-like immunoreactivity.

These are found in rat *(93)* and monkey cortex *(56)*, where they are most numerous in layer VI and the underlying white matter. The NPY-immunoreactive neurons that display no GABA immunoreactivity are, instead, immunoreactive for two members of the tachykinin neuropeptide family, substance P and substance K (neurokinin A), as well as SRIF *(119)*. Thus, NPY neurons can be split into two groups: one that is GABA immunoreactive and a second that is immunoreactive for four peptides. Whether these chemically distinct groups of NPY cells are also morphologically distinct is unknown.

2.1.6. Cortical Circuits and NPY

With the current state of knowledge concerning NPY neurons in the cerebral cortex, several conclusions can be drawn about their role in cortical circuitry. The coexistence of GABA and the lack of spines on their dendrites strongly indicate that NPY-positive cells are inhibitory interneurons. As expected for a population of inhibitory cells, electron microscopic studies have found that where the axon terminals of NPY cells form synapses, they are of the symmetric type and are principally on the dendrites and spines of cells (*56,58,93;* Fig. 6A,B) that are not, themselves, GABA-immunoreactive (*93*). These data indicate that the GABA/NPY cells terminate mainly on the processes of pyramidal cells and, by doing so, influence the activity of the cortical output neurons (*93*).

Many NPY-immunoreactive axons, serially reconstructed through their full extents, form no conventional synapses in either the monkey (*56*) or rat (*120,121*) cerebral cortex. Instead, the vesicle-filled expansions form parallel-membrane associations with many neuronal elements, including the terminals of unlabeled axons (Fig. 6C). In the rat cortex, perhaps a majority of the NPY terminals form membrane appositions with other terminals (*121*). Such a close and consistent association has been interpreted (*121*) as the physical substrate for the NPY inhibition of excitatory neurotransmitter release (e.g., *122,123; see also* Bleakman et al., this vol.).

The distribution of immunoreactive fibers suggests that the synapses formed by NPY cells occur with greater frequency on cells and processes in layers II, III, V, and VI. This distribution is significant because it demonstrates a preference for the fibers to occupy the layers most densely packed with pyramidal cells and generally to avoid the layer (layer IV) of densest terminations from the thalamus. Only in area 17 of primates is there a significant NPY innervation of a thalamocortical-recipient layer and then, only in the very thin layer IVA (*58*). The major thalamocortical termination zone (layer IVC) in area 17 is very poorly innervated by NPY fibers (*56,58*). In addition, layer IVC in monkey area 17 contains extremely few NPY somata (*56,115*), much as layer IV throughout

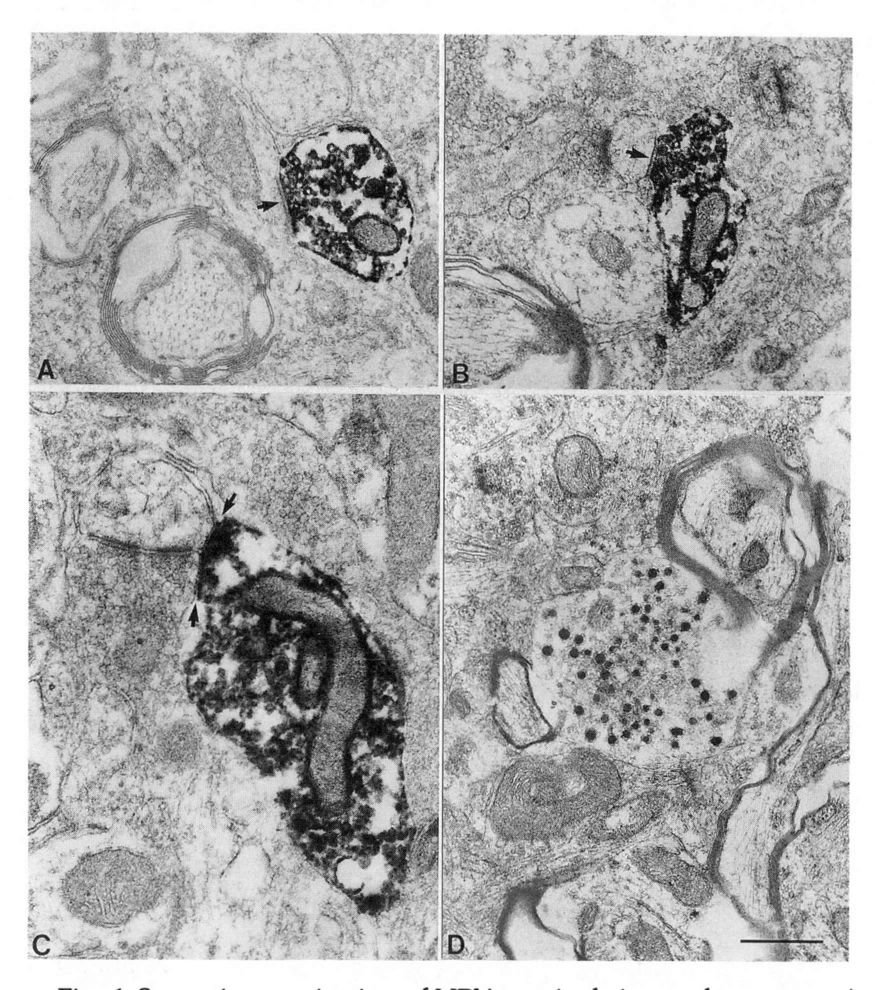

Fig. 6. Synaptic organization of NPY terminals in monkey cortex. A. and B. Electron micrographs of synaptic terminals, immunostained for NPY, forming contacts (arrows) with unlabeled dendrites. These synaptic contacts have symmetric membrane thickenings, as would be expected from the coexistence of NPY and GABA in cortical interneurons. C. Electron micrograph of a large NPY terminal that is adjacent (arrows indicate cleft) to the unlabeled terminal and an unlabeled dendrite. The two unlabeled processes are jointed by an asymmetric synaptic contact. The NPY terminal's membrane is parallel to the membranes of both unlabeled processes, but it forms no synapse with either process in this or any other section. D. Electron micrograph of a large terminal filled with immunostained dense-core vesicles. Terminals such as these are rare in the monkey cortex. Bar equals 1.0 μm in A and B, and 0.5 μm in C and D.

the cortex of most species is generally reported to contain few NPY somata or processes. What these data suggest is that NPY cortical neurons probably receive extremely few monosynaptic inputs from thalamic relay cells. The extreme paucity of immunoreactive fibers in layer IV also suggests that the NPY cells fail to synapse in a conventional manner on the thalamic-recipient cells. Thus, the first stages at which cortical neurons process the major thalamic inputs occur without much influence from NPY neurons.

2.1.7. NPY Receptors

Because neuropeptides may exert actions at considerable distances from their release sites (e.g., *124*), a real understanding of where NPY cells may exert postsynaptic effects requires detailed knowledge not only of presynaptic elements, but also postsynaptic receptor sites. Such knowledge is particularly important for the cerebral cortex, where a sizable proportion of vesicle-filled NPY-immunoreactive terminals forms no recognizable synapses *(56,93;* Fig. 6C,D). The laminar patterns of binding produced with ^{3}H- or ^{125}I- NPY closely resemble the distribution of NPY-immunoreactive plexuses: The binding is densest in layers II and III and moderately dense in layers V and VI *(125–129)*. Similar overall patterns in NPY binding are evident across several species of rodents, pigs, and primates *(125,127–130)*. Where the NPY binding sites have been examined most closely, which is in the primary visual cortex of cats and monkeys *(131)*, some indication of mismatch between NPY-immunoreactive fibers and NPY receptors has been detected. Thus, in the cat, binding is highest in layer I and superficial layer II, but is uniformly lower in all deeper layers *(131)*. In the monkey, dense binding is found in two bands in area 17, one of which includes layers I and II and the other interpreted to be layer IVCβ. If the deeper band is actually the junction between layer IVC and V, then the distribution of binding sites in this area *(131)* and the distribution of immunoreactive processes *(56,58,62)* would be in very close agreement. However, the presence of high NPY binding in layer IVCβ, itself, would be a serious mismatch, since the very lowest density of immunoreactive processes is seen in this layer *(56,58)*. The potential sources of transmitter/receptor mismatch are many, as discussed elsewhere *(132)*, but may be attributed in the case of NPY to the existence of multiple receptor types *(128)* or the ability of NPY, released at a site in one layer, to act on neurons in a neighboring layer.

The localization of NPY-binding sites in the monkey visual cortex is also significant for the dramatic differences in laminar distribution evident between the first visual area (area 17) and the second visual area (area 18). Instead of alternating layers of high and low binding, as in area 17, a progressive reduction in binding is seen in traversing the thickness of area 18 *(131)*. Such interareal differences are also very apparent in comparing patterns of immunoreactivity for NPY between areas 17 and 18 (*see* Section 2.1.1.), and these reflect the very dramatic differences in cell number, density, organization and chemistry that exist at the border between the first and second visual areas of the monkey cerebral cortex.

2.1.8. Plasticity of NPY Expression

Regulation of neuronal activity has been shown to control the immunoreactive levels of several cortical neurotransmitters, enzymes, receptors, and neuropeptides (e.g., *79,133–137*). Levels of NPY are also affected by changes in neuronal activity. Limbic seizures, induced by kainic acid injections *(138,139)* or by lesions of the hilus of the dentate gyrus *(140)*, lead to rapid and dramatic increases in the cortical levels of NPY, as determined by RIA, and in the levels of preproNPY messenger RNA, as determined by *in situ* hybridization *(140)*. In fact, the increased mRNA levels are seen as an increase in the number of neurons labeled by *in situ* hybridization, indicating that the seizures induced novel or increased expression in neurons that normally express levels of NPY message too low to detect.

With more subtle manipulations of neuronal activity, no changes in NPY immunoreactivity have been found. Whereas monocular deprivation of adult monkeys leads to reductions in immunoreactivity for GABA, GAD, $GABA_A$ receptors, and tachykinins in the primary visual cortex *(78,79,134,141)*, and to increased immunoreactivity for Type II calmodulin-dependent protein kinase *(133)* and histochemical staining for acetylcholinesterase *(72,142)*, deprivation of neonates *(62)* and adults (S. Hendry, unpublished observations) produces no change in NPY immunoreactivity in monkey area 17. The failure of NPY to change may be the result of the laminar distribution of the immunoreactive cells, since it is the somata and processes within layer IVC that are most affected by deprivation, and NPY neurons and processes tend to avoid this layer (*see* Section 2.1.1.).

Whereas reduction in thalamocortical afferent activity does not appear to affect grossly NPY immunoreactivity, complete elimination of the afferents arising from cholinergic neurons in the basal forebrain does produce changes in NPY and SRIF. Both peptides show increased immunoreactive levels, including what appear to be higher densities of immunostained somata and fibers following destruction of the cholinergic forebrain cells *(143)*. The greater immunostaining has been interpreted as sprouting of novel processes, but may represent novel expression of NPY instead.

2.1.9. Cortical Development

Cortical development is so extreme in its complexity that, in the absence of markers to follow specific neuronal populations, even the most basic trends are difficult to identify. Through the use of ^{3}H-thymidine autoradiography to label neurons according to their birth dates (i.e., when they leave the mitotic cycle), the fundamental "inside-out" development of cortical layers has been demonstrated *(144–146)*. In addition, it is now well-understood that the earliest neurons migrate to the cortex and form a primordial layer, which is subsequently split by the arrival of the cells that make up the cortical plate and form layers II–VI *(147–150)*. Those earliest cells that wind up deep to the cortical plate are designated the subplate. Following neuronal migration, the cortical neurons develop morphologically and establish appropriate connections over a rather lengthy period (e.g., *151*).

Several studies have described the development of morphological features and laminar distribution of NPY neurons. The cells of the rat cortex were reported to acquire slowly adult characteristics during the first three postnatal weeks *(47)*. Their distribution both prenatally *(152)* and postnatally *(47)* was reported to follow the inside-out development of the cerebral cortex, with immunoreactive somata appearing first in the deeper layers and subcortical white matter, and subsequently in more superficial layers. However, a much different pattern for rat cortex has been described. A transient population of NPY-immunoreactive cells was found to occupy the base of the cortical plate from embryonic day 17 or 19 to postnatal day 4 *(153,154)*. The cells are intensely immunoreactive during that period, but either die out or lose their immunoreactivity after day 4. Such a pattern has not been detected immunocytochemically in other species *(52,155)* but has been identified by *in situ* hybridization in the fetal monkey visual cortex (Fig. 7). Cells containing preproNPY message in area 17 of fetal monkeys are densely packed in presumptive layer III, where they form a uniform band of labeling. This band of cells is restricted to area 17. As with the rat cortex, NPY-expressing cells of monkey visual cortex make a sudden appearance (during the middle of the third trimester) and then quickly lose their chemical signature (within 2 wk of birth). Most observations of this kind, which indicate transient expression of a substance during development, are usually interpreted as evidence for a role of that substance in some maturational process. It would be tempting to do the same for NPY, but there are no data currently available to indicate that the peptide has developmental significance as a trophic, tropic, or other kind of factor.

Instead of addressing the question of what NPY might be doing during cortical development, several groups have used NPY immunoreactivity as a marker for a specific neuronal population in the cerebral cortex and have discovered general mechanisms of cortical growth and maturation:

1. Populations of chemically defined neurons underwent waves of development in which morphologically distinct classes were generated and then apparently eliminated by programed cell death *(52)*;
2. Cells of the subplate were positively identified as neurons based on their immunoreactivity for NPY and other neuronal markers *(53,156)*, and their role in the development of cortical connectivity has been investigated. The subplate cells receive functional synaptic contacts from thalamocortical neurons, apparently as those axons pass through a "waiting period" before invading the cortex proper *(157)*. The effects of chemical lesions that eliminate these cells prior to the invasion of thalamocortical axons indicate the subplate neurons are required for the thalamocortical axons arising in a specific thalamic nucleus to recognize and terminate in the appropriate cortical area *(158)*.
3. Two populations of subplate cells were described based on related chemical and connectional features *(159)*. A major population establishes the earliest axonal projections to the thalamus and the contralateral hemisphere, and may, thus, serve as pioneers for the connections formed by neurons developing from the cortical plate *(160)*. None of these is

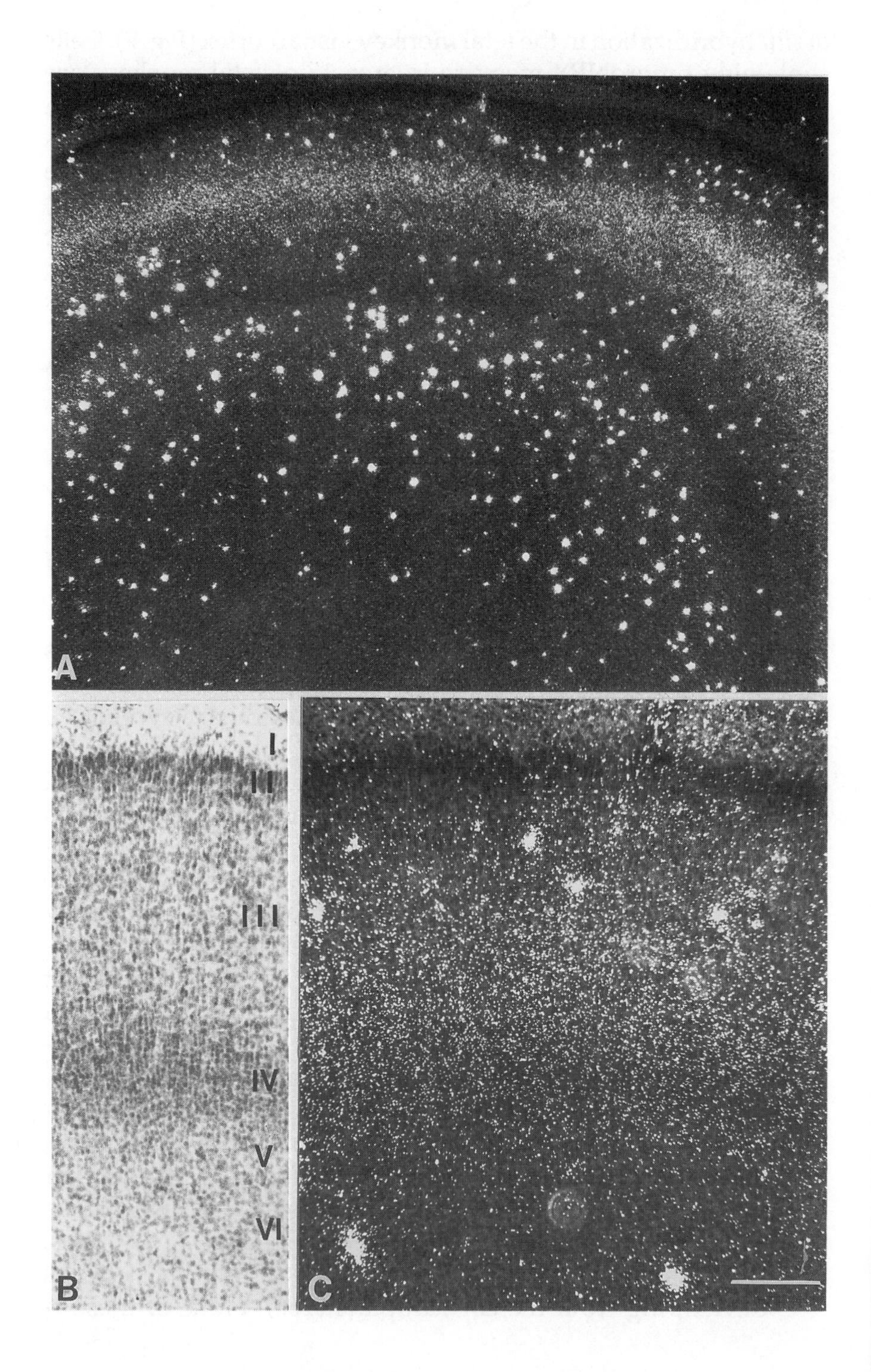
A
B
C
I
II
III
IV
V
VI

immunoreactive for NPY, SRIF, or a calcium-binding protein, calbindin, but they are, instead, labeled by the retrograde transport of ^{3}H-D-aspartate, indicating that they may employ glutamate or aspartate as a neurotransmitter. A second subplate population forms exclusively local connections within the subplate and overlying cortex. These are the NPY-immunoreactive neurons. Thus, NPY immunostaining allows one of two distinct populations of subplate neurons to be conclusively identified. Most subplate neurons underlying the kitten cortex, including those immunoreactive for NPY, are eliminated by a wave (or waves) of programed cell death *(146,150)*;

4. Studies from several species indicate that the NPY cells are among the earliest to acquire adult characteristics of position and morphology *(47,52,155)*. Other chemically distinct classes follow at later dates.

There are two limitations to the studies of cortical NPY cells in development: (1) They do not address what NPY, itself, may be doing as a chemical agent in the development of the cerebral cortex. It is important to recognize that was neither the intention of these studies nor a possible outcome of studies in which NPY

Fig. 7. *(opposite page)* Micrographs of preproNPY localization of fetal monkey visual cortex (area 17). A. Low-magnification dark field autoradiogram of a seciton through area 17 of an E121 monkey fetus (gestation in macaque monkeys is approx 165 d). Localization of mRNAs encoding for preproNPY was done using a ^{35}S-labeled complementary RNA probe. Grain clusters indicative of labeled somata are present in a thin and very superficial band, and a much wider and deep zone that includes layers V and VI and the subjacent white matter. The white-matter cells are presumed to be members of subplate population, indentified by NPY immunoreactivity in kitten visual cortex *(see text)*. In addition to the grain clusters, a diffuse labeling is seen in the middle of area 17. B and C. and Bright-field (B) and dark-field (C) photomicrographs of the cortex proper from the E121 monkey fetus. Comparison of Nissl staining and autoradiographic labeling in the same section indicates that the diffuse band of labeling occupies presumptive layer III, beneath the cortical plate (CP) in area 17. At the same age, no band is evident in any other area of monkey cortex, including the adjacent second visual area (area 18). Within 2 wk, the diffuse band is no longer present in area 17. From previously unpublished experiments by S. Hendry and C. M. Gall. Bar equals 400 μm in A, and 150 μm in B and C.

localization and cell characterization were the goals; and (2) labeling for NPY, either by immunocytochemistry or *in situ* hybridization, largely excludes the neurons present within the cortical plate, so the development of the vast majority of cortical neurons must be addressed by studies of markers other than NPY (e.g., *159*). However, labeling for NPY and other peptides has allowed specific groups of cortical neurons to be identified unambiguously, and from the properties of these cells, a critical role for subplate neurons in the development of the entire cortex has been postulated *(161)*.

2.1.10. NPY in Cortical Neuropathology

Alzheimer's disease and senile dementia of the Alzheimer's type (SDAT) are marked by significant degeneration of the cerebral cortex, the loss of a large proportion of pyramidal neurons, and the presence of amyloid plaques within the neuropil and neurofibrillary tangles within individual neurons *(162)*. Following reports of dramatic reductions in the levels of the acetylcholine-synthesizing enzyme, choline acetyltransferase, within the cortex of Alzheimer's patients *(163)*, a concerted effort was made to identify other transmitter-specific systems that might be at risk. Apparently contradictory results have been obtained from radioimmunoassay studies of NPY in neurologically normal and Alzheimer's patients. Three studies report no change in the levels of cortical NPY, measured by RIA *(41,164)*, whereas one study reports 60–80% reductions in NPY from the temporal, frontal, and occipital cortex and hippocampus of Alzheimer's patients as compared with age-matched controls *(42)*. The NPY in the latter study was identified as the native peptide when HPLC analysis determined it was not an altered form of the prepropeptide *(42)*. It is not clear what the source of the apparent contradiction among studies of Alzheimer's patients might be, although differences in postmortem delays and average age of the patients have been noted *(42)*. What is clear from immunocytochemical studies is that the staining for NPY is far from normal in Alzheimer's patients: Immunoreactive somata display unusual morphologies, their processes appear truncated, and the staining of fiber plexuses is greatly diminished *(65,108,165)*. In addition, the binding of ^{3}H-NPY is also reduced in the temporal cortex of Alzheimer's patients, apparently because of a reduction in the number of receptors rather than a change in their affinity *(129)*. The combination of RIA, immunocytochemistry, and ligand-binding studies strongly suggests that NPY is affected in Alzheimer's disease and SDAT, and follows the widely reported reduction in SRIF levels within the cortex of Alzheimer's patients *(166–169)*. A comparison of methods suggests that the reduction in NPY and SRIF levels is owing not to the degeneration of these neurons, but to the dramatic pruning of their processes *(108)*.

Unlike the parallel changes in SRIF and NPY within cortical neurons of Alzheimer's patients, the two peptides are reportedly very different in Parkinson's disease. Throughout the cerebral cortex, including the hippocampal formation of Parkinsonian patients, SRIF levels are greatly reduced, but NPY levels are left unchanged *(170)*. Although originally interpreted to indicate a basic similarity between Alzheimer's and Parkinson's diseases, these findings of no NPY changes in Parkinsonian patients and more recent findings of NPY loss in Alzheimer's patients indicate, instead, an essential difference in the two diseases.

A key component of Alzheimer's disease is the correlation of increased density of amyloid plaques in the cerebral cortex with the reduced mental status of the patient *(171)*. The nature of the plaques is, thus, of great interest to many investigators. Immunocytochemical studies suggest that NPY- *(172)*, SRIF- *(173,174)*, and tachykinin-immunoreactive fibers *(175)* have some preference for the peripheries of the amyloid plaques. Processes stained for any of these peptides are found within the neural portion of 10–15% of the cortical plaques. Yet many other proteins and peptides related to neurotransmission are present within plaques *(176,177)*, and for processes containing two peptides that are colocalized with NPY (SRIF and tachykinins), the association with plaques is seemingly no greater than chance *(175)*. It would appear, then, that any relationship between NPY fibers and neuritic plaques is closer to accidental than causal.

The significance of findings that neurons immunoreactive for NPY and the colocalized SRIF undergo rather extensive changes in Alzheimer's disease and SDAT is open to many interpretations. Not all cortical neuropeptides are equally affected in these diseases, since NPY and SRIF are disproportionately depleted compared with substance P, cholecystokinin, and vasoactive intestinal polypeptide *(108)*. Obviously, then, some selectivity exists as to which classes of cortical neurons are affected in these disease states. However, despite an initial report that SRIF-immunoreactive neurons possess neurofibrillary tangles *(176)*, which would indicate these cells are selectively vulnerable in Alzheimer's disease, a more recent study found that no SRIF- or NPY-immunoreactive cortical cells in Alzheimer's patients contain the microtubule-associated, tau protein *(108)*, the major protein constituent of paired helical filaments within the neurofibrillary tangles *(177)*. In addition, it is a longstanding observation that populations of pyramidal cells of the cerebral cortex are targeted preferentially in Alzheimer's disease *(162)*. These findings suggest that the changes in NPY/SRIF-immunoreactive neurons are only a small part of a massive degenerative process within the cortex of Alzheimer's patients and may be significant only in that NPY cells are *not* spared from the neuronal death brought on by Alzheimer's disease.

2.1.11. Conclusions

The life history of NPY neurons tells us much about the cerebral cortex. They begin as an exuberant population, like many specific populations of cortical cells, and progressively adopt adult features that include morphological heterogeneity, laminar and areal differences in distribution, and for one area, a columnar pattern of somata and processes. Changes in neuronal activity can alter the expression of NPY in cortical cells, and they appear to be as sensitive as, but no more sensitive than, the general population of cortical neurons to neuropathological changes. As a microcosm of the cortical world, the NPY cells serve nicely. One key question remains, however, and that is: How does NPY function as a transmitter or other kind of signaling agent in the cerebral cortex? The presence of NPY in vesicle-filled processes, its release under appropriately depolarizing conditions, and the presence of specific populations of receptor sites suggest that NPY functions as a chemical that transmits messages from one cell to another (reviewed in *5,178,179*). What are those messages? Although few answers have been found for neurons of the cerebral neocortex, studies of another cortical region, the hippocampus, have brought the field close to an answer (*see* Section 2.2.).

2.2. Hippocampal Formation

The hippocampal formation is a part of the cerebral cortex that includes the CA fields (Ammon's horn), the dentate gyrus, and the subicular complex. A closely related area of the cerebral cortex is the entorhinal cortex. As might be expected from the fact that the hippocampal formation is a part of the cerebral cortex, many of the features described in the preceding section on the neocortex can be attributed specifically to the hippocampus.

2.2.1. Cell Types

NPY immunoreactive neurons of the hippocampal formation and entorhinal cortex are nonpyramidal cells *(55,180,181)*. Within a single region, such as subfield CA1 of the rat hippocampus, cells of different sizes and morphologies are NPY-immunoreactive (*180*; Fig. 8C). Perhaps most marked is the variety of immunoreactive neurons in hilus of the dentate gyrus, where NPY is found within types of dentate pyramidal basket cells, polymorphic cells, and spheroid cells (*180*; Fig 8D), each of which was correlated with the morphology of Golgi-stained neurons *(182)*. In the CA fields, some immunoreactive neurons closely resemble Golgi-stained basket cells, whereas others appear to be members of other classes. When all the regions within the hippocampal formation are considered, a very large number of morphological types of NPY cells have been described in the human *(181)*, monkey *(55,180)*, and rat *(180,183)*.

The existence of NPY immunoreactivity exclusively within nonpyramidal cells of the hippocampal formation does not imply that the peptide is present only in interneurons. Several studies have described nonpyramidal efferent cells in this region *(184,185)*, including GABA-immunoreactive projection neurons *(186–189)*. Whether these include any cells in which NPY might also serve as a transmitter agent is unknown.

As in other areas of the cerebral cortex, many of the NPY-immunoreactive neurons also display SRIF-like immunoreactivity *(183,190)*. Extensive coexistence was found throughout CA1, CA3, the dentate gyrus, subicular complex, and entorhinal cortex, but the proportion of NPY-immunoreactive neurons in which SRIF immunoreactivity was localized varied across the hippocampal formation of rats, from 39% in dorsal CA1 to 75% in ventral CA3 *(183)*. Similar variability has been reported for NPY/SRIF coexistence in neurons of the human hippocampal formation. These data on coexistence of SRIF and NPY immunoreactivities in rat and human hippocampal neurons, and the discovery of GABA in all SRIF neurons of the cat hippocampus *(104)* suggest that NPY and GABA may coexist in neurons of the hippocampus.

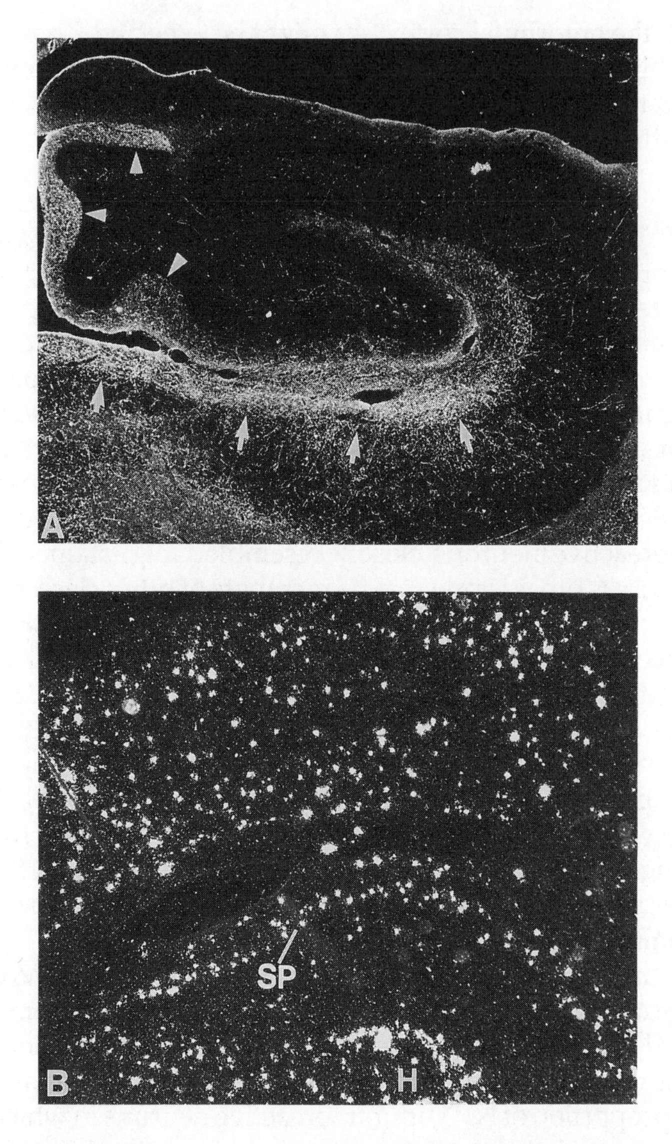

Fig. 8. NPY localization in the hippocampal formation. A. Dark-field photomicrograph of NPY-immunoreactive fibers, that are dense in the molecular layer of the dentate gyrus (arrowheads) and stratum moleculare of the CA fields (arrows). Fibers are present in the dentate hilus, and many are present in the subicular complex. B. Dark-field autoradiogram of preproNPY mRNA in the rat hippocampus, detected by *in situ* hybridization histochemistry with a ^{35}S-labeled complementary RNA

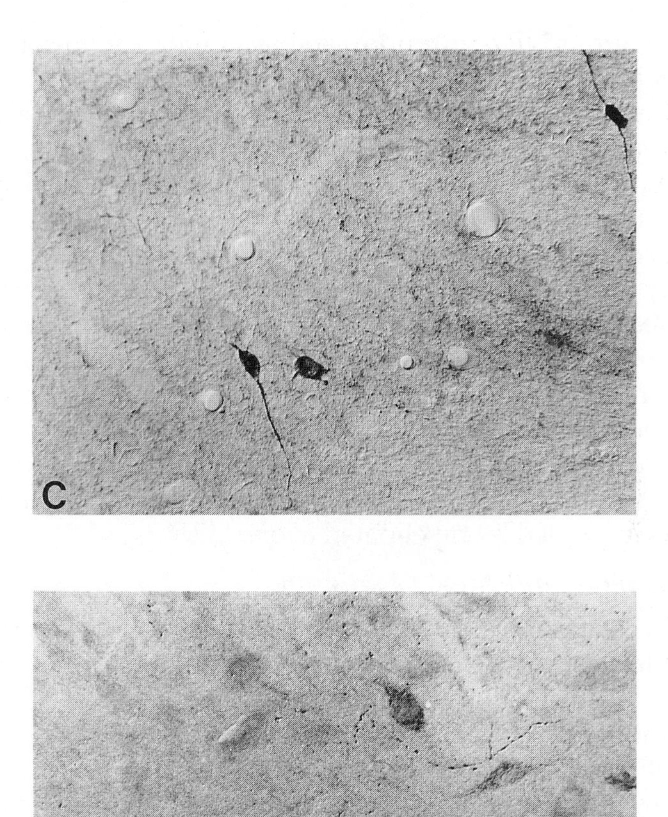

probe. Labeled somata in the CA fields are in and around stratum pyramidale (SP) and in the dentate hilus (H). Numerous somata are also present in the overlying neocortex. C. NPY-immunoreactive neurons at the border of stratum oriens of CA1 and the overlying fibers of the alveus in the monkey hippocampus. The neurons are small and nonpyramidal. D. Immunoreactive somata and processes in the hilus of the monkey dentate gyrus. This region contains many NPY cells, but in the monkey, they are weakly immunoreactive. Bar equals 600 µm in A, 400 µm in B, 40 µm in C, and 25 µm in D.

Numerous NPY-immunoreactive somata are present in the fiber tracts associated with the hippocampal formation, including the fimbria, alveus, and angular bundle of rats and monkeys *(180)* and humans *(181,190,191)*. The greatest number is evident in the white matter of the human hippocampal formation. In all species, the NPY-immunoreactive white matter cells give rise to processes that either remain within the fiber tracts or turn to invade the adjacent gray matter. These features of NPY-immunoreactive neurons in the subcortical white matter of the hippocampal formation closely resemble the characteristics of similarly placed neurons in the cerebral neocortex. Unlike the situation in one area of neocortex, the cat primary visual area, the total number of NPY-immunoreactive neurons in the hippocampal white matter does not appear to decline with maturity, but becomes diluted by the expansion in the tissue occupied by myelinated axons *(191)*.

2.2.2. Laminar and Areal Distribution

NPY-immunoreactive somata and processes are unevenly distributed in each part of the hippocampal formation. In the CA fields of the rat, somata labeled by immunocytochemistry *(180)* and *in situ* hybridization (*192*; Fig. 8B) are present within the layer of pyramidal cell bodies (stratum pyramidale) and in the layers of the apical and basal dendrites of the pyramidal cells (stratum radiatum and stratum oriens). The greatest density of labeled neurons in the hippocampal formation is found in the hilus of the dentate gyrus of rats (*180,192*; Fig. 8B), monkeys *(55,69,180)*, and humans *(69,181)*. In the rat and monkey, a prominent plexus of NPY fibers is present in the contiguous stratum moleculare of the CA fields and the outer third of the molecular layer of the dentate gyrus (*180*; Fig. 8A). These are layers of the termination of the temporo-ammonic/perforant path axons, which arise from neurons in the entorhinal cortex *(193,194)*. However, this correlation should not be taken to mean that the perforant path axons are NPY-immunoreactive *(180)*. Instead, the fibers in stratum moleculare and the outer third of the dentate molecular layer, and the remaining NPY-immunoreactive fibers that are scattered throughout the other laminae of the CA fields and dentate gyrus appear to arise from neurons intrinsic to the hippocampal formation.

The binding of radioactive NPY in the rat hippocampal formation is greatest throughout stratum radiatum and stratum oriens of the CA fields *(125,126,183,195)*. The same is true for binding of labeled peptide YY (PYY), which recognizes a high-affinity receptor site *(125)*. Along the septo-temporal axis in the rat hippocampus, greater binding is evident in more temporal regions, but the laminar difference remains *(195)*. These data indicate that a dramatic mismatch between peptide and binding site exists for the hippocampal formation, since it is these laminae that have some of the lowest densities of NPY-immunoreactive fibers. In addition, there are no indications of greater NPY or PYY binding to stratum moleculare or the outer third of the dentate molecular layer *(125,183,195)* that might correspond to the dense distribution of immunoreactive fibers in these layers. These findings indicate that immunoreactive elements and ligand-binding sites in the hippocampal formation may match well by area, but are mismatched by layer.

The location of the NPY- and PYY-binding sites indicates they are present on either the dendrites of hippocampal pyramidal cells or the axons that innervate the dendrites. Because the binding sites survive destruction of the pyramidal cells following injections of quinolinic acid, they are interpreted to be presynaptic *(125)*, possibly mediating the presynaptic effects of NPY application on these neurons *(122,196)*.

Systems of NPY-immunoreactive somata and fibers generally do not adopt tight laminar distributions in the subicular complex and entorhinal cortex. In the subiculum, itself, the greatest density of immunoreactive fibers is found in the molecular layer, but in the pre- and *para*-subiculum and in the lateral and medial entorhinal cortex, a more homogeneous pattern of fibers is evident *(197)*. The areas of the entorhinal cortex are also notable for the much greater density of fibers than that seen in the hippocampal formation. In the entorhinal cortex, the plexuses of NPY fibers include axons arising from immunoreactive interneurons and axons of afferent systems originating in the piriform cortex, lateral nucleus of the amygdala, and locus coeruleus *(197)*.

2.2.3. Plasticity of NPY Expression

Limbic seizures, induced by systemic injections of kainic acid or pentylenetetrazol *(198,199)* or by unilateral lesions of the dentate hilus *(140)*, lead to dramatic increases in NPY within the hippocampal formation. This increase is certainly pronounced when measured by neurochemical and molecular biological methods

(138,139,198,200,201) and can be replicated by more general manipulations, such as electroconvulsive shock *(202)* or adrenalectomy *(203)*. The increase in NPY expression is even more extraordinary when viewed with immunocytochemistry and *in situ* hybridization histochemistry. In normal rats, neurons immunoreactive for NPY *(48)* or expressing NPY mRNAs *(140)* are not found among the granule cells of the dentate gyrus or in the mossy fiber terminations of their axons in the dentate hilus and stratum lucidum of CA3. However, within days of the onset of limbic seizures, very dense immunoreactivity begins to appear in the mossy fiber termination zones and continues to build for 2 mo after seizure onset *(139)*. Despite the immunoreactivity in mossy fiber terminals, no NPY immunostaining is evident in the somata of the dentate gyrus, which give rise to the mossy fibers. However, *in situ* hybridization histochemistry clearly shows intense labeling for mRNAs to preproNPY in these neurons *(199)*. The increased labeling for preproNPY mRNA follows very quickly after a seizure-producing hilus lesion, with the obvious increases in hybridization levels detected within 10 h *(140)*. These results demonstrate that the granule cell somata, which give rise to the mossy fiber system, display a novel expression of preproNPY, and that the mossy fibers, themselves, display a novel immunoreactivity for NPY under limbic seizure conditions in rats. Different mechanisms involving loss of NPY neurons may be at work in human temporal lobe epilepsy *(204)*.

In addition to the novel expression of preproNPY in the granule cells of the dentate gyrus, cells in stratum pyramidale of CA1 and in the superficial layers of the entorhinal cortex are also induced to express NPY message following limbic seizures *(140)*. The levels of mRNA in these cells and in the cells of the dentate gyrus do not remain high: 2 d following the onset of seizures, the dentate granule cells again display no detectable signal for preproNPY, and by 4 d, the levels throughout the hippocampal formation and entorhinal cortex return to normal *(140)*.

The changes in the mossy fiber system of rats undergoing limbic seizures are not restricted to NPY expression. Preproenkephalin mRNA and enkephalin peptide are also elevated in the mossy fibers, whereas preprodynorphin, dynorphin peptide, and CCK are reduced in the same system (*see 205* for review). Thus, the seizures do not specifically affect NPY, but appear to induce changes in several neuroactive peptides.

Mossy fiber axons are capable of changes not only in their chemical characteristics, but also in their morphological features. Immunocytochemical findings suggest that some of the NPY-immunoreactive mossy fibers give off new collaterals following limbic seizures, which pass through the granule cell layer of the dentate gyrus and terminate in the supragranular molecular layer *(198)*. These data corroborate other findings that indicate mossy fibers sprout in the hippocampus of animals undergoing seizures *(206,207)* and provide a clearer image of the plastic responses of hippocampal neurons.

Does the novel expression of NPY within the granule cell/ mossy fiber system and other neuronal populations contribute to the seizure sensitivity of these animals, or is it a reaction to the seizures and, possibly, an attempt to control them? Physiological and pharmacological evidence indicates that NPY reduces excitatory synaptic effects onto hippocampal pyramidal neurons *(122,123,196)*, which would be one powerful way of bringing unrestrained excitatory circuits under control. Comparisons of NPY levels with seizure behavior also suggests that the induction of high levels of peptide is a response rather than a cause of seizures, since the induction of NPY expression begins after electrographically recorded seizures have reached their peak *(140)*. Thus, NPY is not a causative agent for the seizures, but may instead be part of a compensatory mechanism that serves to control seizure activity.

2.2.4. Neuropathology and NPY Neurons

The hippocampal formation and entorhinal cortex are among the most severely affected regions in Alzheimer's disease (e.g., *208*). As expected, then, NPY immunoreactivity in the hippocampal formation is greatly altered in Alzheimer's patients *(209)*, but the severity of changes varies across this region. Immunoreactive fibers are most dramatically reduced in the hilus of the dentate gyrus, CA1, the subiculum, and the entorhinal cortex *(209)*. Two of these, CA1 and the entorhinal cortex, are also the regions containing the greatest density of neuritic plaques and neurofibrillary tangles *(209)*. By contrast, such regions as the subiculum and CA3 exhibit few changes in NPY-immunoreactive elements and few plaques or tangles. The correlation between loss of NPY elements and pathological changes might indicate a causative role for the peptide in the progression of Alzheimer's disease or might simply reflect the greater loss of all neuronal elements, including NPY-immunoreactive ones, in certain hippocampal regions of Alzheimer's patients. The large number of immunoreactive neurons that remain in the hippocampal formation of these patients has been taken to suggest that NPY cells in this region of the cerebral cortex are uniquely resistant to degeneration and may keep essential neuronal circuits functioning at a minimal level *(209)*.

2.2.5. Conclusions

Many aspects of neuronal organization and plasticity that are evident for NPY expression in neurons of the neocortex are also pronounced in cells of the hippocampal formation. These include similar lamina- and area-specific distributions of neurons and receptors, similar responses to manipulations of neuronal activity, and similar changes in cases of neuropathology. Thus, comparison of NPY neurons and receptors in these regions of the cerebral cortex reinforces the idea that the two share many fundamental features of functional organization. For those interested in neocortical physiology, such a conclusion brings a great deal of hope for understanding the effect of NPY neurotransmission on the responses of cortical neurons, since the powerful inhibition that NPY exerts presynaptically on excitatory synapses *(122,123,196*; Bleakman et al., this vol.) may be another feature of NPY that is shared by the hippocampus and the cerebral neocortex.

2.3. Olfactory Bulb

The main olfactory bulb is a cortical telencephalic region whose features closely resemble those of the cerebral cortex. As in the cerebral cortex, a conspicuous population of NPY-immunoreactive neurons is present in the deepest layer of the main olfactory (the deep granule cell layer) and in the subjacent white matter *(210–213)*. There is also the strong indication in comparing reports of rodents *(210,211,214,215)* with those of cats and laboratory primates *(213)* and humans *(212)* that a much greater proportion of NPY cells are present in the white matter of the primates and cats than in rodents. Additional NPY somata are present at the border between the glomerular layer and external plexiform layer.

Studies of the rodent main olfactory bulb have placed the NPY cells into two well-recognized classes of neurons *(210,211,214,215)* described in Golgi and electron microscopic studies *(216,217)*. These are the superficial short axon cell and the deep short axon cell. Processes of the deep cells were found to contribute to a dense fiber plexus that innervates the granule cell layer *(211)*, although numerous fibers are also present in the white matter and were suggested to arise from an extrinsic source of innervation *(210)*. These studies *(210,211,214)* concluded that the NPY-immunoreactive neurons fit neatly within the known morphological framework of the main olfactory bulb and suggested that NPY may serve as a marker for a proportion of previously classified neurons.

A much different interpretation has been offered from examination of the main olfactory bulb and related structures (accessory olfactory bulb and anterior olfactory nucleus), principally in cats and kittens *(213)*. These authors point out that previous descriptions of NPY cells in the main olfactory bulb and accounts of Golgi-stained neurons do not agree in their details (e.g., the absence of spines on "superficial short axon cells" in the former reports and their presence in the latter). Instead, they see the NPY neurons of the olfactory bulb as an extensive system of somata, present throughout the white matter of the olfactory bulb and peduncle. Based on the morphology of their local axonal plexuses, all NPY neurons of this region, including those with somata near the glomerular layer of the main olfactory bulb, are interpreted to be of a single class of "axonal loop" cells. In the white matter, this class possesses ipsilateral long axonal projections to the granule cell layer, thus accounting for the extensive immunoreactive fiber plexus. During early postnatal development in kittens, individual fibers from the white matter cells reportedly can be followed as part of a second long axonal system, which projects to the contralateral hemisphere through the olfactory limb of the anterior commissure. The commissural system is lost at later ages (i.e., the third postnatal month).

In many respects, the scheme of development by NPY neurons described in the kitten olfactory bulb *(213)* is similar to that described for the cerebral cortex *(52,161)*, including the retraction of axon collateral systems that is critical for the development of connections in the cerebral cortex *(218)*. In fact, the presence of NPY-immunoreactive neurons in the white matter throughout the cerebral cortex, hippocampus, olfactory bulb, and olfactory peduncle may be indicative of a class of interstitial cell *(219)* common to many areas of the telencephalon that may serve common functions in these areas during development and in adulthood. Studies of the chemical nature of cells in the olfactory bulb support the suggestion that NPY neurons are very similar across telencephalic regions, for, as with the cells of the cerebral cortex, the NPY neurons in the olfactory bulb display somatostatin-like immunoreactivity *(220)* and appear to contain the enzyme NADPH-diaphorase *(211)*.

2.4. Striatum

The caudate nucleus, putamen, and nucleus accumbens septi contain some of the highest levels of NPY-like immunoreactivity in the brain of rats *(38,48)*, monkeys *(221)*, and humans *(40)*, and the greatest density of neurons containing NPY message in the human brain *(43)*. These findings are so consistent that early reports of very little detectable NPY in microdissections of rat striatum *(222)* can be attributed to methodological problems. In the human striatum, there is some question as to whether the concentration of NPY-like immunoreactivity is greater in the nucleus accumbens *(39)* or in the caudate nucleus/putamen *(14)*, but in either case, chemical differences between these two functionally and connectionally different regions of the striatum are apparent. Some reports indicate that the numerical density of immunoreactive somata does not vary to any great degree between the caudate nucleus and nucleus accumbens *(14,223)*, and so the difference in NPY concentration between these regions is said to arise from the presence of NPY within a system of afferent axons that selectively innervates one of them (i.e., the projection from the amygdala or ventral tegmental area to the nucleus accumbens; *39*). However, other studies document a much greater density of NPY immunoreactive somata and fibers in the human nucleus accumbens and closely adjacent bed nucleus of the stria terminalis *(224)*, which would easily account for the greater peptide concentration in the ventral striatum.

2.4.1. Distribution Within the Caudate and Putamen

The striatum is a chemically heterogeneous structure, in which neuronal elements expressing various enzymes, peptides, and receptors are concentrated into compartments. The result is a pattern of patches or striosomes displaying one set of chemical characteristics, surrounded by a distinct matrix displaying a different set of characteristics *(225–227)*. This pattern is correlated with the distribution of striatal afferent terminals and efferent neurons

(227,228), which indicates that the chemically distinct compartments are fundamental units of striatal functional organization. When assayed by RIA, the microdissected human caudate nucleus and putamen reveal no internal differences that might reflect a patch/ matrix distribution *(39)*, which is consistent with the results of immunocytochemical studies showing no clustering of NPY-immunoreactive somata or fibers in the human *(14)*. Clustering of NPY somata is also absent from the striatum of the monkey *(55,229)* and rat *(113)*. By contrast, the distribution of NPY cells in the cat striatum is uneven: Clusters of intensely immunoreactive somata and fibers alternate with regions containing extremely few NPY elements *(229)*. The pattern of NPY immunoreactivity, itself, has not been directly correlated with striosome distribution, but it would appear from the documented patterns of peptide and enzyme coexistence in striatal NPY neurons that NPY somata preferentially occupy the matrix. This might be an obvious conclusion, since in the striatum, the great majority of NPY neurons of all species contain both SRIF-like immunoreactivity and NADPH-diaphorase histochemical activity *(113,223)*, and both of these coexisting substances have been localized to the acetylcholinesterase-rich matrix of the cat striatum *(230)*. However, careful analysis of SRIF localization in the rat striatum demonstrates that only the immunostained fibers occupy the matrix; the somata are present in both compartments *(231)*. It is likely that a similar situation exists for NPY neurons in the cat, where the fiber staining is robust and gives the impression of an uneven distribution of the peptide. In the monkey and human, where fibers are not so well-immunostained, a more homogeneous pattern is observed. For all species, then, the NPY/SRIF neurons may be distributed as a bridge between the striosomes and the matrix, with somata in both regions and fibers concentrated in one.

The NPY cells and fibers in the human nucleus accumbens are also distributed unevenly *(224)*. The clusters of NPY-immunostained fibers are reportedly much less distinct than the clusters immunostained for SRIF, but the two apparently overlap. Preliminary evidence suggests the NPY/SRIF-immunostained clusters are complementary to clusters of dopaminergic terminals, immunostained for tyrosine hydroxylase *(224)*. These data indicate that the NPY innervation of the human nucleus accumbens is also selective for the matrix.

In addition to their compartmental organization of striosome and matrix, the caudate nucleus and putamen also display more global heterogeneities. Perhaps most dramatic is the division of the striatum into cortical-recipient zones, either along a rostro-caudal gradient (i.e., frontal cortex to head of caudate, occipital, and temporal lobes to tail; 232) or as elongated strips that form medio-lateral domains *(233)*. At least two reports indicate that the density of NPY neurons in the primate caudate nucleus follows the rostro-caudal gradient. In the caudate nucleus of humans and of Tamarins, a species of Old World primates, many more neurons are present in the head and body than in the tail *(69)*, but in the striatum of squirrel monkeys, a species of New World monkeys, a greater numerical density of somata is found in the posterior striatum than the anterior *(229)*. Since neither of these studies employed quantitative methods, the perceived variations within the caudate nuclei of different species may be either illusory or tied to a more complex innervation pattern such as that documented by Selemon and Goldman-Rakic *(233)*.

2.4.2. NPY Neuronal Class in the Striatum

The consensus of many studies in several species is that the NPY-immunoreactive striatal neurons make up a single class of medium aspiny neurons, a type of striatal interneuron *(234,235;* Fig. 9A). Failure to label NPY-immunoreactive neurons retrogradely in the cat striatum from injection sites in the substantia nigra and globus pallidus *(229)* and the very low concentrations of NPY-immunoreactive fibers in these main striatal targets confirms that the NPY cells of the striatum are interneurons. Several other lines of evidence, including little variability in NPY cell size and dendritic shape, and the absolute *(113,223)* or predominant coexistence *(224,229)* of NPY with SRIF and NADPH-diaphorase, are consistent with a single morphological class of NPY striatal neuron. However, it is not clear whether the striatal NPY cells may be divided into two subpopulations based on the presence of GABA in one and its absence in another (Fig. 9C,D). Although some NPY cells are also GABA immunoreactive, some NPY somata and terminals clearly are not *(93,236)*. However, even following the intraventricular administration of colchicine to enhance cell-body immunoreactivity, the number of neurons immunostained for GABA *(93,236)* was lower than that immunostained previously for GAD *(237,238)*. These findings suggest that the absence of GABA

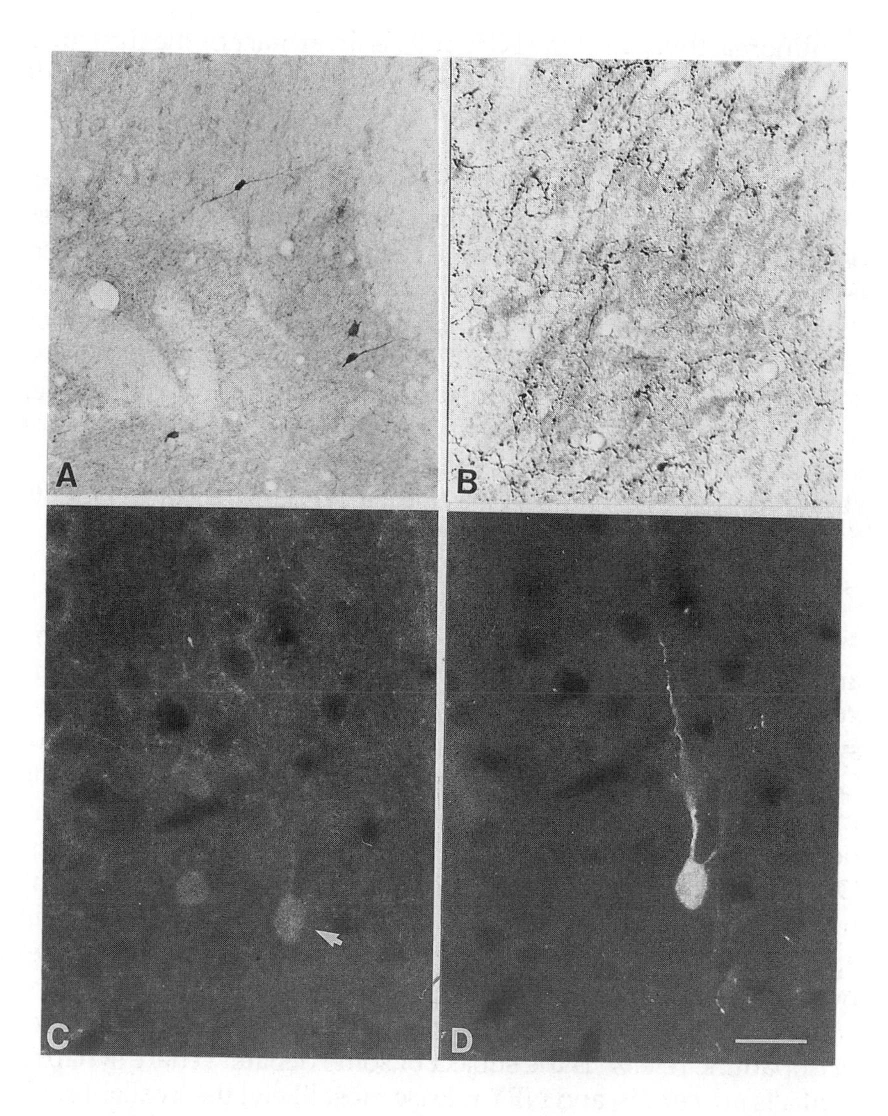

Fig. 9. NPY immunoreactivity in the monkey striatum. A. Immunostained somata in the medial, caudal part of the putamen. Relatively small somata are scattered throughout the macaque monkey striatum. B. NPY-immunoreactive fibers in the caudate nucleus. The NPY fibers in the monkey striatum are thin and highly beaded. C and D. Immunofluorescent photomicrographs showing GAD (C) and NPY (D) immunostaining in the same section through the monkey caudate nucleus. Of the two cells that are GAD-positive, one (arrow) is also NPY immunoreactive. Similar findings in this and other species indicate that NPY exists within a small subpopulation of striatal GABA neurons. Bar equals 80 µm in A, 20 µm in B, and 15 µm in C and D.

immunoreactivity in some NPY cells is an artifact of the difficulty in localizing GABA in striatal cells. The presence of GABA in all NPY neurons would make it more likely that the striatal NPY cells are a homogeneous population.

On a more general level of striatal organization, immunoreactivity for NPY may help to distinguish between two broad classes of GABA neurons in the caudate nucleus and putamen. It has been estimated that approx 90% of the striatal GABA neurons project to the globus pallidus and substantia nigra, whereas the remainder are interneurons *(237,239,240)*. NPY immunostaining marks at least some of the interneuronal population. One other intrinsic molecule, the calcium-binding protein parvalbumin, is present within presumed GABA interneurons *(241)* and may provide a signal for striatal GABA interneurons that either overlaps or complements that seen for NPY immunostaining.

2.4.3. Synaptic Circuits of NPY Cells

Major striatal afferents arise from neurons in the cerebral cortex and the substantia nigra. The former appears to be a glutamatergic system (e.g., *242; see 243* for review), and the latter is a dopaminergic system *(244)*. Intrinsic circuitry is dominated by the population of GABA neurons, which includes not only interneurons, but also the main projection cells of the striatum *(240,245)*. NPY medium spiny neurons receive convergent inputs from each of these sources. Dopaminergic axons, immunoreactive for the enzyme tyrosine hydroxylase, are unmyelinated processes that are frequently in apposition to NPY-immunoreactive neurons, but much less frequently form conventional synaptic contacts with these neurons *(120,246)*. Whether the sites of nonsynaptic apposition are also sites of dopamine release is the subject of some debate. Where synaptic contacts are certain and NPY release most likely, the available evidence demonstrates a monosynaptic input from at least some dopaminergic axons onto a population of striatal NPY cells. That these afferent axons arise from neurons in the substantia nigra has been directly demonstrated in the rat *(247,248)*. In addition, a prominent monosynaptic input from corticostriatal neurons to NPY cells has also been documented for the rat striatum *(247,248)*. Thus, the two afferent systems that provide the major drive for striatal cells both form monosynaptic contacts with NPY cells.

Synaptic interactions between GABA neurons and NPY cells have been well documented *(93,236,249)*. The termination of GABA axons onto NPY somata and dendrites of the nucleus accumbens *(249)* and the caudate nucleus and putamen *(93,236)* suggests that the plexuses formed by GABA interneurons or the recurrent collaterals of GABA projection neurons terminate with some frequency onto NPY interneurons, which, themselves, are GABAergic *(see above)*. In return, axons of the NPY cells terminate on GABA-immunoreactive neurons, principally onto their dendrites *(93,236)*, and selectively target a type of GABA neuron that receives very sparse synaptic inputs overall. These NPY-recipient cells are unlike the majority of striatal GABA neurons, which are densely innervated and receive no NPY inputs *(93)*. So a simple striatal circuit involving NPY neurons includes cortical and midbrain dopaminergic afferents driving NPY cells, GABA terminals from interneurons or projection neurons inhibiting the cells, and the resulting influence of the NPY neurons falling, in part, onto a distinct subpopulation of GABA neurons.

As in the cerebral cortex, many of the NPY axons in the striatum do not form conventional synapses *(93,250)*, and thus may or may not release NPY. However, regions of membrane apposition, in which the plasma membranes of an immunoreactive axon and an adjacent neuronal profile are parallel to one another and are separated by a cleft of 100 nm or less, are very commonly encountered. These sites of apposition include axo-axonic contacts between NPY-positive and unlabeled axonal profiles *(93,236)*. Although less common than in the cerebral cortex, the striatal axo-axonic appositions may be sites at which NPY effects are exerted presynaptically. All conventional synapses are of the symmetric variety *(93,248,250)*, as would be predicted from the NPY/GABA coexistence. From these data, one is left with an extraordinary number of possibilities of how and where the NPY neurons of the striatum may exert their effects. Much of the uncertainty (e.g., do the sites of membrane apposition represent sites of NPY function?) cannot be addressed with methods of peptide localization. Even the greatly improving methods of receptor localization will not adequately answer such questions, so a great deal of uncertainty remains as to how widespread and powerful the influence of NPY neurons is in the normal striatum.

2.4.4. Regulation of NPY Expression

Manipulations of inputs from the substantia nigra and the cerebral cortex alter the expression of NPY in the striatum. Interactions between dopaminergic afferents and NPY neurons were first detected biochemically as reduced NPY release following administration of amphetamine *(251)*. Subsequent studies have demonstrated that destruction of dopaminergic neurons in the substantia nigra increases the numerical density and staining intensity of NPY-immunoreactive neurons in the ipsilateral caudate-putamen of the rat *(252,253)*. A similar increase occurs in the number of striatal cells expressing preproNPY mRNA and preproSRIF mRNA following destruction of the nigral dopaminergic neurons *(254)*. The changes in immunoreactivity and message levels are seen 3 wk following the placement of lesions. Thus, the change in immunoreactivity is apparently a change in peptide levels, regulated at gene transcription; under normal conditions, the transcription of NPY and SRIF genes and the levels of these peptides are partially suppressed by dopaminergic neurotransmission *(252–254)*. Pharmacological studies suggest that the normal suppression is exerted through D1 receptors and that enhancement of NPY immunoreactivity may occur by dopamine effects on D2 receptors *(253)*.

From this basic finding of NPY regulated by dopaminergic inputs, the following has been added:

1. Inputs from corticostriatal neurons also regulate NPY immunoreactivity. Thermocoagulation of the frontal and parietal lobes of the rat cerebral cortex leads to increased NPY immunoreactivity in the striatum *(255)*. This increase in striatal immunoreactivity is seen on both sides of the brain, but is more robust ipsilateral to the cortical lesion *(255)*.
2. The increased NPY staining produced by cortical lesions occurs in only the dorsal and lateral parts of the striatum, whereas that which follows substantia nigra lesions occurs mainly in ventral and medial parts of the striatum *(255)*.
3. Destruction of the cerebral cortex prevents the change in immunoreactivity induced by substantia nigra lesions *(255)*.
4. Lesions of the intralaminar thalamic nuclei, the principal source of thalamostriate projections, lead to some increased NPY immunoreactivity in the medial part of the striatum *(255)*.

5. Unlike the cells of the caudate-putamen, the NPY neurons of the rat nucleus accumbens show *reduced* NPY immunoreactivity following destruction of the substantia nigra *(256)*. The reduction is bilateral, but is more dramatic contralateral to the lesion of the substantia nigra. Furthermore, the effects of these lesions are reversed by apomorphine treatment, but only in the anterior half of the nucleus accumbens, ipsilateral to the lesion.

These data indicate that transmission through each of the major afferents to the striatum regulates NPY expression, but that the regulatory events are complex: They can occur in both hemispheres, even though afferents to only one are removed, each afferent appears to control one part of the striatum, but the afferents appear to interact in such a way that suggests the cerebral cortex ultimately is in control of striatal NPY plasticity, and the effects in the more dorsal striatal regions (caudate-putamen) and the nucleus accumbens are mirror images of one another.

2.4.5. NPY and Striatal Neuropathology

Huntington's disease is a hereditary condition marked by dementia and choreiform movements. Neuronal loss in Huntington's disease is largely confined to the striatum, where as many as 80% of the neurons degenerate *(257)*. Many of the neurons that survive are NPY-immunoreactive *(258)*. By RIA, the concentration of NPY and of the coexisting peptide, SRIF *(259)*, is found to increased in the striatum of Huntington's patients as compared with normal controls *(258)*. In the same material, the density of immunoreactive somata and/or processes increases dramatically *(258,260)*. These increases are interpreted not as sprouting of new fibers or generation of new neurons, but as a compacting of neuronal survivors into a much-depleted neuropil. The key element to this selective survival, which is unique among currently studied neuropathologies *(261)*, may be the coexistence of NADPH-diaphorase in the NPY/SRIF striatal neurons (*see* Section 2.4.2.). The original description of NADPH-diaphorase-positive neurons described them as "solitary active cells," and found them to be resistant to numerous chemical and surgical insults *(117)*. More recent studies have confirmed that the NADPH-diaphorase cells survive following certain excitotoxic lesions of the striatum *(262,263)* and experimentally induced ischemia *(264)*. Although compelling as a correlation between neuronal phenotype and survival in pathologies, the persistence of the NPY/SRIF/NADPH-diaphorase neurons in Huntington's disease and in experimental conditions is at present an unexplained phenomenon. The mechanism by which these neurons are able to maintain themselves in the face of massive degeneration throughout the striatum is simply not known.

2.5. Septum, Basal Forebrain, and Amygdala

In many respects, the NPY immunoreactivity in these subcortical telencephalic structures is similar to the pattern seen for the cerebral cortex and striatum:

1. Most NPY-immunoreactive cells are small interneurons;
2. SRIF-like immunoreactivity coexists in many of these cells; and
3. The distribution of the somata and of immunoreactive fibers varies among the subdivisions of each of these complex regions.

2.5.1. Septum

Most reports of NPY immunoreactivity in the septum indicate that few somata and fibers are present in this region *(55,69,222)*. However, a detailed study of the human septal complex reports the presence of numerous NPY somata and fibers in the medial septal area, and fewer somata and fibers in the lateral area *(224)*. Patches of immunoreactive fibers, averaging several hundred square microns in area, are distributed throughout the medial septum, but individual fibers do not have the characteristic "woolly" appearance of SRIF-immunoreactive fibers. Although more SRIF-immunoreactive than NPY-immunoreactive somata are seen in the medial septum, the distribution of fibers and the morphology of somata are the same for the two peptides, suggesting considerable coexistence. Additionally, in certain divisions of the lateral septal area (e.g., the laterodorsal division), about half of the NPY cells are found to display SRIF-like immunoreactivity *(224)*.

2.5.2. Basal Forebrain

The nucleus of the diagonal band of Broca is closely related topographically and cytoarchitectonically to the medial septal area, and in these two nuclei, the NPY innervation is very similar *(224)*.

Patches of NPY fibers are present in both of these nuclei and in the olfactory tubercle. In the human, the regions of the patches are said to differ in cell size and structure from the islands of Calleja *(224)*. In the monkey, the patches of NPY terminals are reported to occupy the plexiform layer of the olfactory tubercle and to surround the islands of Calleja *(55)*.

In the human brain, a virtually continuous system of NPY-immunoreactive fibers extends from the subcallosal gyrus through the medial septal area, into the nucleus of the diagonal band of Broca and olfactory tubercle, and finally in the basal nucleus of Meynert *(224)*. One report suggests that the basal nucleus of monkeys may contain few NPY cells *(55)*, but both radioimmunoassay *(39)* and immunocytochemistry *(69)* have accented the similar high level of NPY innervation in the basal forebrain of humans and nonhuman primates. It is not known whether this global system of NPY fibers arises from an afferent system common to all of the basal telencephalon or from similar groups of local circuit neurons present in each nucleus.

2.5.3. Amygdala

The many nuclei of the amygdala can be included in a rostral mass, densely interconnected with the forebrain, and a caudal mass, densely interconnected with the hypothalamus and brainstem *(265)*. Although some studies have indicated that the density of NPY immunoreactive somata and fibers is rather homogeneous across the amygdala *(44,48)*, most have found the distribution of cells and processes to vary with nuclear and subnuclear borders (*55,69,266;* Fig. 10A,B). In general, the nuclei and subnuclei rostral to the origin of the stria terminalis are relatively poor in NPY somata and fibers, whereas the more caudal nuclei are relatively rich *(266)*. In the rat, most NPY-immunoreactive somata are present in the medial nucleus, where they are clustered at the ventral and lateral borders, and at the base of the stria terminalis *(266)*. Other nuclei have somata scattered more or less diffusely. Although the nomenclature of the nuclei differs, a similar pattern is seen for neurons in the rat amygdala containing preproNPY mRNA *(192)*. In humans and in nonhuman primates, dense collections of NPY somata are also seen in the medial nucleus, as well as in the basal (designated basolateral by Smith et al. *[55]*, but more appropriately termed basal, following the nomenclature of Crosby and Humphrey *[267]*)

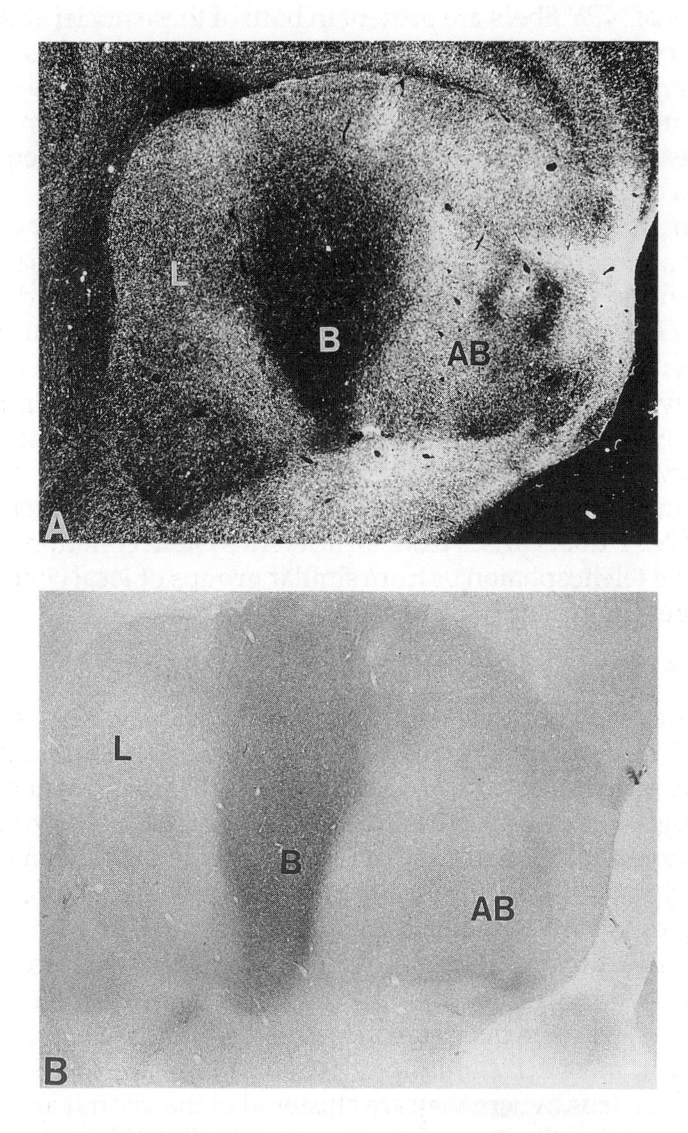

Fig. 10. NPY immunoreactivity in monkey amygdala. A and B. Sections 120 µm apart, stained immunocytochemically for NPY (A) and histochemically for acetylcholinesterase (B). By using the cholinesterase for NPY (A) and histochemically for acetylcholinesterase (B). By using the cholinesterase as a guide to the organization of the amygdala, the densest distribution of NPY fibers was found in the accessory basal nucleus (AB), particularly the superficial and magnocellular divisions. The lat-

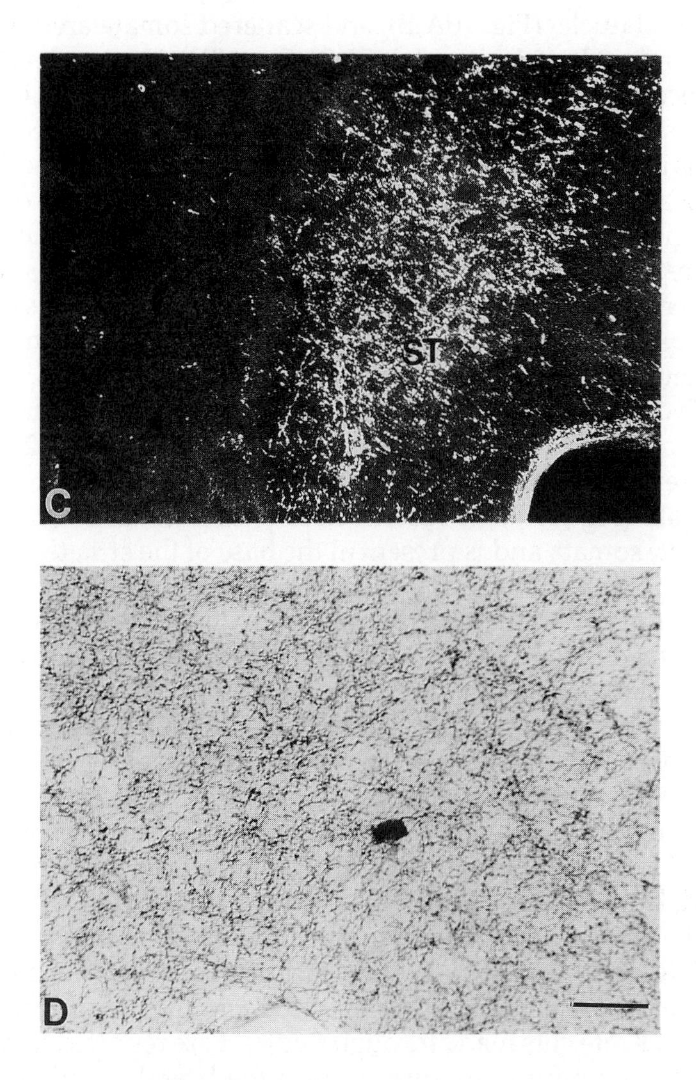

eral nucleus (L) has fewer immunoreactive fibers, and the basal nucleus (B) has virtually none. C. Dark-field photomicrograph of NPY-immunoreactive fibers in the stria terminalis (ST). The presence of fibers in this tract indicates that part of the output of the amygdala is carried by NPY-immunoreactive axons. D. High-magnification photomicrograph of a field typical of the accessory basal nucleus of the amygdala. A small immunostained soma is surrounded by a dense field of fibers. Bar equals 350 μm in A and B, and 20 μm in C and D.

and lateral nuclei (Fig. 10A,B), and scattered somata are found in the central and cortical nuclei (*55,69*; Fig. 10D). Dense plexuses of fibers occupy the central nucleus, particularly its medial third, and the dorsal half of the medial nucleus in rats *(266)*. The central nucleus is also densely innervated in monkeys and humans *(55,69)*. The NPY plexus is dense and coextensive with the plexus of noradrenergic fibers arising from brainstem nuclei. However, no loss in the density of the NPY fibers and a virtually complete loss of the noradrenergic fibers are seen following injections of the neurotoxin, 6-hydroxydopamine, suggesting that the NPY axons do not arise from neurons in the noradrenergic brainstem nuclei *(266)*.

Two or three morphological types of cells are NPY-immunoreactive in the amygdala. Most are described as small (12 μm in diameter) and bipolar *(266)*; they are present throughout the amygdala (Fig. 10D). A second class possesses large (22 μm in diameter) somata and is present at the base of the stria terminalis. Even larger multipolar cells, 40 μm in diameter, are either extreme examples of the second class or make up a distinct third class *(266)*. Most of the bipolar type of NPY neurons and possibly most of the multipolar type(s) in the rat amygdala *(267)* and all of the NPY neurons in the human amygdala *(69)* are also immunoreactive for SRIF. These NPY/SRIF neurons appear to be neither selectively preserved nor selectively targeted in Alzheimer's disease *(268)*. As in the cerebral cortex, other peptides, such as cholecystokinin and vasoactive intestinal polypeptide, are present in neurons of the amygdala that are not NPY-immunoreactive *(267)*.

The presence of NPY-immunoreactive fibers in the stria terminalis (Fig. 10C), a major efferent tract of the amygdala, suggests that at least some of the NPY cells in this complex are output neurons *(38)*. Knife cuts made through this tract are reported to reduce the levels of immunoreactive fibers in the lateral septal area, the suprachiasmatic nucleus of the hypothalamus, and the bed nucleus of the stria terminalis *(44)*. These data do support the existence of an NPY amygdalofugal projection, but the difficulties in interpreting lesion data, particularly when subtractive analysis is required, are legendary. In fact, the basic finding of reduced forebrain or hypothalamic immunostaining following destruction of the stria terminalis has not been replicated *(266)*. More direct methods, such as simultaneous retrograde labeling and immunostaining of neurons in the amygdala, would undoubtedly lead to a conclusive answer concerning NPY-output neurons, but results of such studies have not been reported. It appears safest, then, to conclude that most of the NPY somata in the amygdala are interneurons, with axons restricted either to the nucleus of origin or to adjacent amygdaloid nuclei.

2.5.4. Summary

NPY neurons of the mammalian telencephalon are remarkably similar from region to region and may, thus, be viewed as a single group that varies in only subtle ways from one place to the next. They are among the earliest neurons to develop adult characteristics of position and morphology in each part of the developing telencephalon. In the adult brain, the NPY cells of the telencephalon are interneurons that contain the peptide, SRIF, and the enzyme, NADPH-diaphorase. The presence of NPY immunoreactivity serves to label distinctly one class of GABAergic, inhibitory neuron in some regions, most notably the cerebral cortex. From region to region, the expression of NPY is highly plastic, and is regulated by both global and localized manipulations. Perhaps because of the presence of NADPH-diaphorase within them, the NPY somata of the forebrain are able to survive trauma and chemical insult to a much greater degree than other neurons and may, thus, selectively survive in some neuropathological diseases (e.g., Huntington's disease), but they appear to be only part of a general degenerative process for most regions in Alzheimer's patients.

3. Diencephalon

3.1. Thalamus

The thalamus proper is divided into the dorsal and ventral thalamus. The two have different developmental origins, and are distinguished in the adult mammal by the presence of cortical relay neurons in dorsal thalamus and their absence in the ventral thalamus *(105)*. Two other regions, the epithalamus and subthalamus, are related parts of the diencephalon. The dorsal thalamus, itself, is divided into six groups of nuclei (anterior, medial, lateral, ventral, intralaminar, and midline) based on the physiology, structure, and connections of their neurons *(268)*.

The concentration of NPY measured in the dorsal and ventral thalamus is among the lowest in the CNS *(14,222)*. Immunocytochemical studies in most species are consistent with these data, and find very few somata and scattered fibers in the dorsal thalamus. NPY fibers are present within certain midline thalamic nuclei and in the anterior thalamic nuclei of the rat *(48)* and monkey *(55,269)*, as well as in the medial and ventral geniculate nuclei and pulvinar of the cat *(270)*. Some of the fibers that reach that monkey dorsal thalamus ascend vertically from the region of the periventricular hypothalamic nucleus (*269;* Fig. 11B). By contrast with the distribution of immunoreactive fibers, putative receptor sites for NPY are relatively high in the lateral posterior, mediodorsal, and medial geniculate nuclei of the rat thalamus *(125)*. Extremely few somata are detected in the dorsal thalamus, and these are concentrated in the midline nuclei of monkeys *(55)*, but are more widely scattered in the lateral posterior, mediodorsal, and parafascicular nuclei of cats *(270)*.

The immunocytochemical and ligand-binding data indicate that a marked species variability exists in the NPY innervation of the dorsal thalamus, and that the distributions of immunoreactive fibers and receptors are mismatched. Examination of the ventral thalamus, particularly the reticular nucleus, adds to the confusion. In the reticular nucleus, very few fibers and no somata are reported to be immunoreactive for NPY *(55,222,269,270)* but by *in situ* hybridization histochemistry, most somata are found to express preproNPY mRNA (*192;* Fig. 11A). Since the neurons of the reticular nucleus innervate the entire dorsal thalamus *(100)*, a dense NPY fiber plexus would be expected in every thalamic nucleus if the NPY message were transcribed and the peptide transported into the axon of these neurons. The absence of such a plexus raises several possibilities from the technical (the concentration of NPY is too low in neurons and fibers of the dorsal and ventral nucleus to

Fig. 11. *(opposite page)* NPY in the diencephalon. A. Low-magnification autoradiogram of neurons labeled for preproNPY mRNA. Cell bodies in the arcuate (Ar) and dorsomedial (DM) nuclei of the hypothalamus are labeled very intensely. In addition, the densely packed somata of the thalamic reticular nucleus (R) are labeled, even though this nucleus consistently displays no neurons immunoreactive for NPY. Numerous NPY somata are present in the cerebral cortex. B. Dark-field photomicrograph of NPY-immunoreactive fibers in the monkey thalamus. A major plexus of fibers occupies the paraventricular nucleus (Pv) of the thalamus, as well as rhomboid and centromedial nucleus, each of them along the mid-

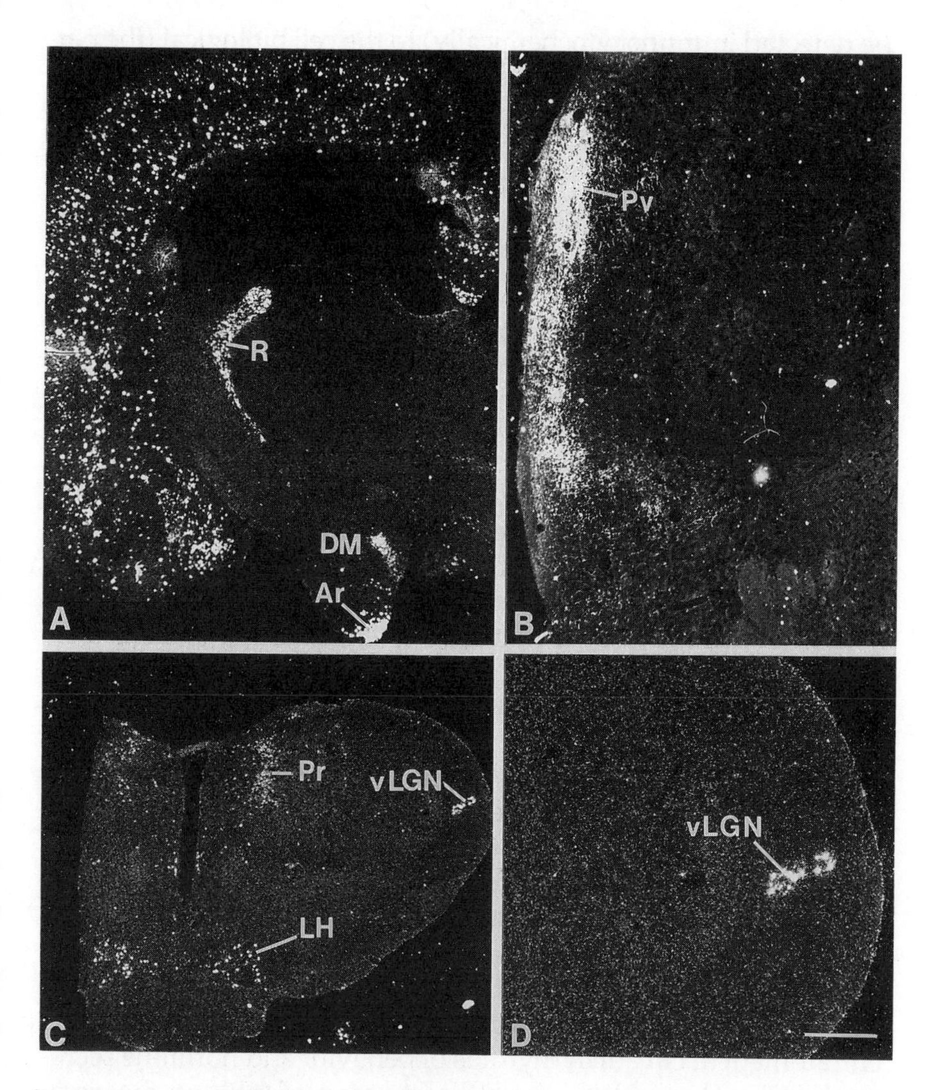

line. C. Low-magnification autoradiogram of neurons containing preproNPY message in a section posterior to the one in A. Loose aggregates of labeled somata are apparent in the region of the lateral hypothalamus (LH). A very thin zone of intense labeling is also apparent laterally, in the ventral lateral geniculate nucleus (vLGN). Labeled somata are also apparent in the pretectal complex (Pr). D. The region of the vLGN in the same section as C. The thin zone of labeling is made up of several closely packed somata. These neurons have been found in other species to give rise to axons that terminate in the suprachiasmatic nucleus of the hypothalamus. Bar equals 1.2 mm in A and C, 45 μm in B, and 270 μm in D.

be detected immunocytochemically) to the cell biological (the patterns of mRNA and receptor localization are remnants of a robust NPY synthesis that occurred during the development of thalamic neurons, but was subsequently suppressed). The latter is supported by findings of transient NPY immunoreactivity in neurons of the dorsal thalamus during late prenatal development *(153)*. It is intriguing that the expression of NPY in the reticular nucleus would extend the coexistence of NPY and SRIF into neurons of the diencephalon, since the somata of this nucleus are reported to be SRIF-immunoreactive in cats and monkeys *(269,271)*.

A second ventral thalamic nucleus, the intergeniculate leaflet, contains numerous somata that are NPY-immunoreactive *(48,55,270,272;* Fig. 12C,D). The report that NPY cells are present in the intergeniculate leaflet of rodents and cats, but not monkeys *(272)* appears in error, since numerous immunoreactive somata are evident in the primate equivalent of this nucleus, the pregeniculate nucleus *(269,273)*. The NPY neurons make up a subpopulation of long-projection cells in the intergeniculate leaflet that send their axons to the suprachiasmatic nucleus of the hypothalamus *(274–277)*. Yet the intergeniculate leaflet is not a simple, one-peptide nucleus. Two additional subpopulations of neurons, in which NPY immunoreactivity is conspicuously absent, also project to the suprachiasmatic nucleus or to the leaflet of the contralateral side *(275)*. The commissural neurons are immunoreactive for met-enkephalin, but the neurochemical characteristics of the nonNPY neurons projecting to the hypothalamus are not known *(275)*. Whether predominantly from the NPY neurons or the nonimmunoreactive neurons, studies of the geniculo-hypothalamic projection show it to be "part of the complex system of visual circuits involved in the entrainment of circadian rhythms of behavior and hormone secretion" *(275)*. The evidence for a direct action of NPY on the circadian functions of the suprachiasmatic nucleus (*see* Section 3.2.2.) strongly suggests at least the NPY population of geniculo-hypothalamic neurons is important for normal circadian rhythms.

3.2. Hypothalamus

The hypothalamus is a complex neural structure composed of many functionally and structurally distinct nuclei. Swanson *(278)* points out that the hypothalamus can be divided into 12 compart-

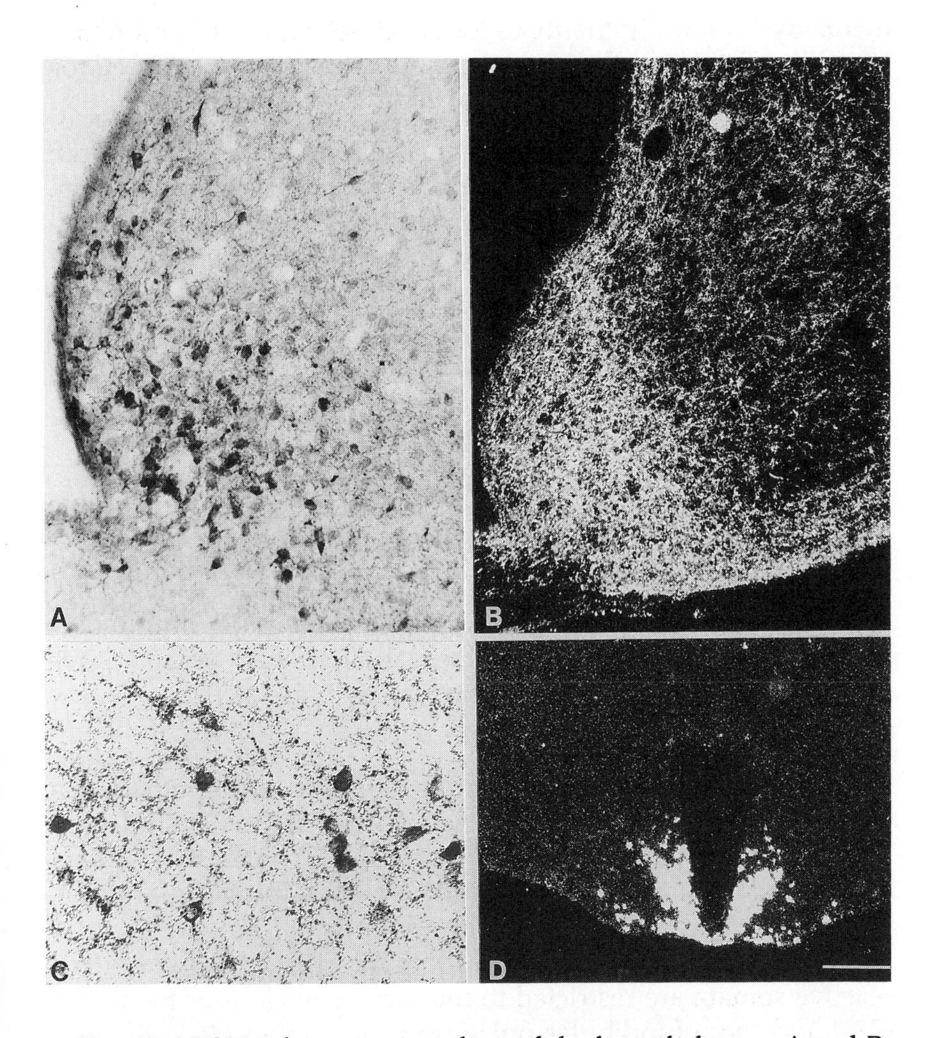

Fig. 12. NPY in the arcuate nucleus of the hypothalamus. A and B. Bright-field and dark-field photomicrographs of NPY immunoreactivity in the arcuate nucleus of the rat. The nucleus is packed with many small somata (A) and a very high density of immunostained fibers (B). Similar types and densities of immunostained neurons and fibers are also present in the arcuate nucleus of the monkey hypothalamus (C). In agreement with the immunostaining pattern, labeling by *in situ* hybridization histochemistry demonstrates the presence of tightly packed somata in the rat arcuate nucleus containing preproNPY mRNA. From previously unpublished experiments by C. M. Gall. Bar equals 60 μm in A and B, 45 μm in C, and 120 μm in D.

ments by combining the three longitudinal zones of Crosby and Woodburne *(279)* with the four rostro-caudal regions of LeGros Clark *(280)*. The three longitudinal zones are:

1. Periventricular, in which most neuroendocrine cells are found;
2. Medial, composed of nuclei innervated by the telencephalon, particularly the limbic system; and
3. Lateral, made up of loose collections of neurons innervated by the fibers of the medial forebrain bundle *(278)*.

The four rostro-caudal regions are the preoptic, supraoptic, tuberal, and mammillary. For the most part, investigators of NPY innervation of the hypothalamus have described their results by referring to these standard levels.

The hypothalamus contains some of the highest concentrations of NPY-like immunoreactivity in the brains of rats *(38)*, monkeys *(221)*, and humans *(40)*. Levels of NPY are higher than those of SRIF, the peptide traditionally defined as "hypothalamic" *(40)*. As discussed in some detail below, hypothalamic NPY immunoreactivity is present principally in fibers and terminals (*see also* chapters by McDonald and Koenig, and Stanley, this vol.).

3.2.1. Immunoreactive Somata

Reports of NPY immunoreactive somata in the hypothalamus vary to some degree among species. The consensus of studies in rats *(20,281)* and most other rodents *(282,283)* holds that somata displaying NPY immunoreactivity or containing preproNPY mRNA *(49,50)* are present predominantly within the arcuate nucleus and lateral hypothalamus (Fig. 12; cf Fig. 11A). Immunoreactive somata are restricted to the arcuate nucleus of the rabbit *(284)* and the infundibular nucleus of the human *(250)*. By contrast, in some rodents and in other species, additional groups of immunoreactive somata are found in the median eminence (golden hamster; *282*), the periventricular nucleus, infundibular nucleus, ventromedial nucleus (cats: *285*; monkeys: Fig. 13A,B), and supraoptic nucleus (monkey: *55*; Fig. 13C,D). How many of these differ-

Fig. 13. *(opposite page)* NPY in the monkey hypothalamus. A. Low-magnification photomicrograph of immunostained somata and fibers in the paventricular nucleus of the monkey hypothalamus. B. Higher magnification of NPY fiber density in the magnocellular paraventricular nucleus, including some fibers that surround the somata of the nucleus

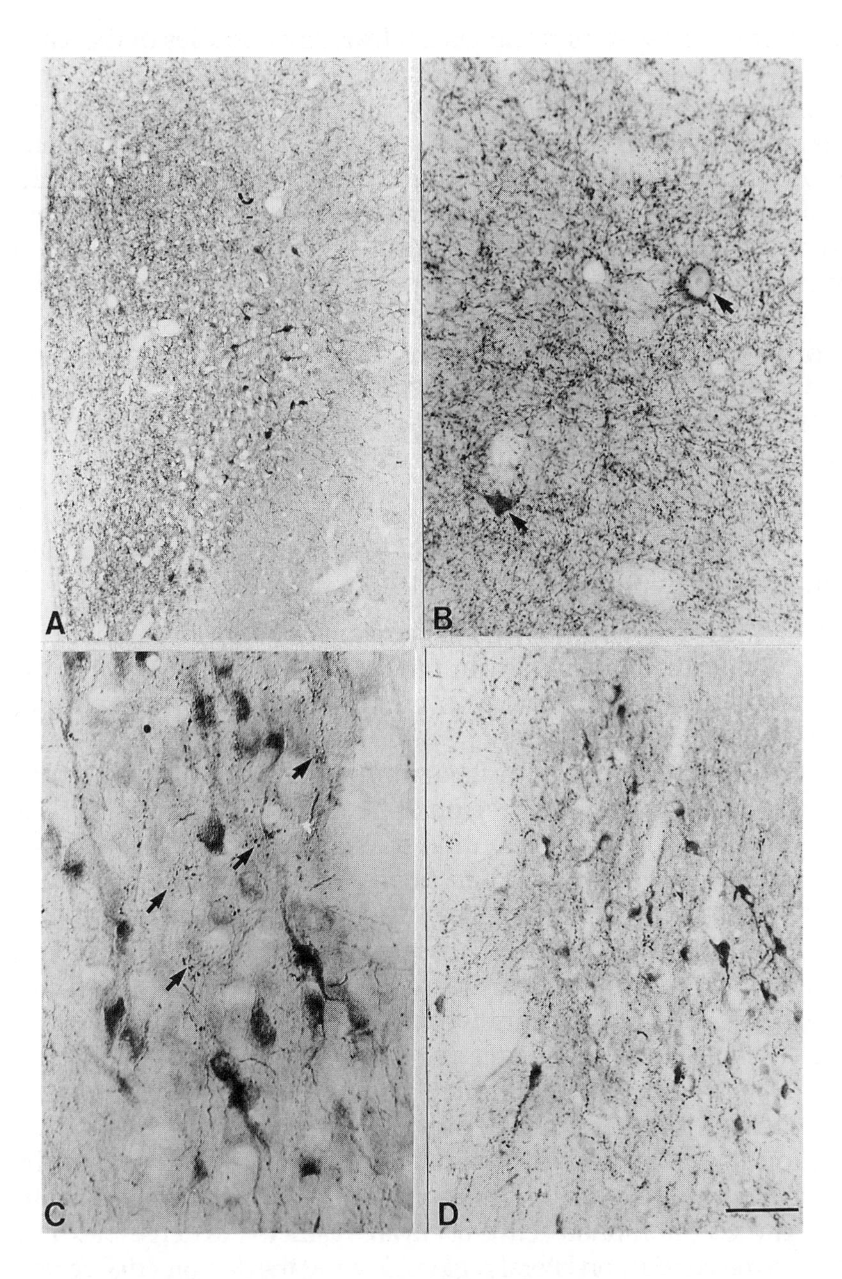

(arrows). C and D. Photomicrographs of NPY-immunostained somata and fibers in the supraoptic nucleus of the monkey hypothalamus. The somata are relatively large and give rise to long processes. In addition, clusters of punctate processes are immunostained (arrows). Bar equals 150 μm in A, 30 μm in B, 20 μm in C, and 80 μm in D.

ences are truly species related is unclear, since studies of the same species, such as domestic cats *(285,286)*, or of closely related species, such as two types of ground squirrels *(283,287)*, report different patterns of cell-body immunoreactivity. These differences may be attributable to diurnal or seasonal cycles, or to the levels of circulating hormones among individual subjects *(221,288–292)*, but are more likely the result of technical features, namely the use of colchicine in most studies of the hypothalamus to arrest axoplasmic transport and, thus, enhance cell-body immunostaining. The effect of differences in dose and application of colchicine are not known, and the ability of drugs, such as colchicine, not only to enhance NPY staining in a stable neuronal population, but also induce NPY expression in novel populations remains barely explored.

3.2.2. Immunoreactive Fibers

The following general features are found for the NPY innervation of the hypothalamus:

1. *The distribution of NPY fibers is widespread.* Most hypothalamic nuclei are innervated by immunoreactive fibers, although the density of innervation varies greatly across the hypothalamus. In the human *(250)*, monkey *(55)*, and cat *(285,286)*, the densest innervation is found along the midline, in what corresponds to the periventricular region of Crosby and Woodburne (*279*; Fig. 12A,B). The distribution of NPY fibers in the rat is very similar, but can best be described as densest in one of the four rostro-caudal zones, namely the tuberal hypothalamus *(20)*. The lowest density of NPY fibers is in the mammillary bodies, which are surrounded by an immunoreactive capsule of fibers *(286)*, but are themselves virtually devoid of fibers *(55,286)*.
2. *Within a single nucleus, the distribution of immunoreactive fibers is often inhomogeneous.* Particularly among some of the more densely innervated nuclei, such as the suprachiasmatic, supraoptic, and paraventricular nuclei of the rat hypothalamus, the density of NPY fibers is uneven *(20,294)*. In the suprachiasmatic nucleus, immunoreactive fibers are restricted to dense plexuses in the ventral and lateral regions, leaving the dorsomedial region completely without an NPY innervation *(294)*. In the paraventricular nucleus, the medial division is more densely innervated than the lateral division *(282)*. By contrast, the distribution of NPY fibers in the periventricular nucleus and retrochias-

matic area is dense, but uniform *(20)*. The intranuclear variations in NPY fiber density may be related to the uneven distribution of specific target neurons, such as those displaying corticotropin-releasing factor in the paraventricular nucleus *(295)*.

3. *Multiple sources of NPY innervation exist for the hypothalamus.* Experimental studies have shown that NPY axons innervating the rat suprachiasmatic nucleus arise from the ventral lateral geniculate (intrageniculate leaflet) nucleus of the thalamus. Thus, lesions or knife cuts that disrupt connections between the two nuclei lead to reduced numbers of immunoreactive fibers in the suprachiasmatic nucleus *(274,276)*, and injections of retrogradely transported tracers in the suprachiasmatic nucleus label NPY-immunoreactive cells in the ventral lateral geniculate nucleus *(277)*. The innervation appears to be bilateral and exclusive, so that bilateral destruction of the ventral lateral geniculate nuclei leads to complete loss of NPY fibers in both suprachiasmatic nuclei *(274)*. Other studies have confirmed that the suprachiasmatic nucleus is not innervated by NPY-immunoreactive axons arising from the brainstem *(296)*, although the presence of two types of NPY-immunoreactive terminals in the suprachiasmatic nucleus *(297,298)* might be one indication of two sources of NPY innervation. For both types of terminals, those in which immunoreactivity is diffusely distributed and those in which the immunoreactivity is confined to dense-core vesicles, conventional symmetric synapses are formed *(297,298)*, principally onto the dendrites of neurons in the suprachiasmatic nucleus *(298)*. Many of the recipient neurons display vasoactive intestinal polypeptide-like immunoreactivity *(297)*, which suggests a close functional interaction between two populations of peptidergic cells.

The NPY fibers innervating the paraventricular nucleus of the hypothalamus arise from catecholamine-containing cell bodies in the brainstem *(299)* and from a shorter projection system that originates in the arcuate nucleus of the hypothalamus *(281)*. The results of Sawchenko and coworkers *(299)* demonstrate that both the parvicellular division and the magnocellular division of the paraventricular nucleus are innervated by NPY-immunoreactive axons, and that these axons arise principally from adrenergic cell groups in the medulla and from noradrenergic somata in the A1 cell group. The NPY innervation of the parvicellular division is densest among neurons that are known to be immunoreactive for corticotropin-releasing factor and thyrotropin-releasing factor and to project to the neurohemal zone of the median eminence *(see 299)*. Innervation of the magnocellular division appears to be equally dense for the vasopressin and oxytocin neurons. The hypothalamic innervation by NPY axons originating in the brainstem has been implicated in feeding behavior *(296)*, and in the control of fluid balance and blood pressure *(300)*.

4. *The density of NPY fibers in a single nucleus varies across species.* Species-specific patterns are apparent when the total NPY pattern for the hypothalamus is examined, since individual nuclei may contain widely different densities of immunoreactive fibers. Such differences are perhaps most dramatic for the suprachiasmatic nucleus (SCN), which receives a prominent NPY innervation in rats *(20)*, moderate innervation in cats (*301*; cf. *286*), squirrel monkeys *(55)*, and macaque monkeys *(273)*, and a sparse innervation in humans *(273)*. This reduction in NPY innervation in the SCN of some species is intriguing, since studies in rats show that application of APP (presumably acting as NPY) to the SCN *(302)* or the destruction of the NPY neurons in the rat ventral lateral geniculate nucleus that project to the SCN *(303–305)* affects circadian function. The differences between monkeys and humans have been attributed to the direct innervation by retinal axons of NPY-immunoreactive somata in the human SCN, thus short-circuiting the geniculo-hypothalamic system while still preserving the function of NPY in circadian rhythms *(306)*.

Species differences in NPY-neuronal distribution are also apparent in the supraoptic nucleus. Whereas only immunostained fibers are found in the rodent nucleus *(282,283)*, immunostained somata and fibers are evident in the monkeys (*269*; Fig. 13). However, no NPY somata have been localized to the human supraoptic nucleus *(250)*, and the concentration of NPY-like immunoreactive material in the monkey supraoptic nucleus is not very high and could be accounted for by the presence of NPY fibers *(221)*. These data may indicate that the monkey is unique in the presence of NPY neurons within the supraoptic nucleus or that the immunostaining is the result of a crossreacting substance in monkey hypothalamic neurons.

5. *The levels of hypothalamic NPY immunoreactivity vary with the state of the animal.* For several homeostatic functions, in which the hypothalamus plays a major role, fluctuations of NPY concentration within specific populations of neurons have been implicated as a controlling factor. Rapid changes in NPY levels have been reported for the paraventricular *(296,307)* and arcuate nuclei *(307)* following relatively short periods (48 h) of food deprivation. Feeding returns the NPY levels in these nuclei to normal, but produces an increase in NPY immunoreactivity in the lateral hypothalamus *(307)*. These data are consistent with the experimentally determined role of NPY in feeding behavior. They are supported by an extensive body of data that demonstrates that the injection of NPY into hypothalamic sites, including the paraventricular nucleus and the perifornical area, elicits feeding and drinking behavior, and leads to hyperphagia and obesity if chronically administered *(308–314)*. The specific source(s) of the extremely dense NPY innervation in these regions, whether intrahypothalamic or extrinsic, is not currently known.

There is considerable interest in the antagonistic actions on feeding behavior of NPY and a second neuropeptide, cholecystokinin (CCK). Where application of NPY into specific hypothalamic sites induces feeding behavior and where starvation raises the levels of NPY immunoreactivity, the opposite is true of CCK: application of that neuropeptide produces satiety *(315,316)* and counteracts the effects of NPY application *(317)*. That ip injections of CCK also appear to reduce the hypothalamic concentration of NPY *(318)* suggests that the phasic regulation of one hypothalamic peptide, NPY, may be controlled by a second peptide, CCK. An anatomical basis for the functional interaction between NPY- and CCK-containing systems may be the dense innervation of both the paraventricular nucleus of the hypothalamus and the median eminence by CCK-immunoreactive (e.g., *319*) and NPY-immunoreactive axons.

Studies of the effects of hypothalamic application of NPY indicate that it is not only feeding behavior, but also drinking behavior that is controlled by this peptide *(309)*. Such control, produced by exogenous application of NPY, is likely to be part of a natural mechanism, as indicated by studies in salt-loaded rats. Where drinking behavior has been induced by the presentation of hypertonic saline, concentrations of NPY within the median eminence, preoptic area, mediobasal hypothalamus, and neurointermediate lobe of the pituitary increased by 50–100% *(291)*. A concomitant and dramatic reduction in arginine vasopressin within the pituitary *(291)* suggests this, too, is a site where interactions between NPY and a second peptide influence the general behavior of an animal *(see 284)*.

As outlined above (Section 3.6.), an NPY-immunoreactive axonal system originating in the ventral lateral geniculate nucleus and terminating in the suprachiasmatic nucleus may be a key component in the regulation of circadian rhythms. Of considerable significance then, are reports that the levels of NPY-like immunoreactivity vary within the rat suprachiasmatic nucleus during the daily light–dark cycle. Data from quantitative immunocytochemical methods suggest that gradually increasing NPY levels during a 12-h light phase are countered by gradually decreasing levels during dark phase *(290)*, and that peaks in immunoreactivity occur at the times when the lights are turned on and off. Despite the interpretive difficulties that come from quantifying immunocytochemical signals and the very broad interindividual results for single time-points *(290)*, these findings are suggestive of functionally significant shifts in NPY levels that may help to entrain circadian rhythms.

Each of these examples demonstrates two points of significance to nervous system function: (1) Global functions mediated by any neural circuit can be attributed to neurons that are chemically and anatomically specific. Such a case has been made previously and powerfully for groups of neurosecretory neurons in the hypothalamus (*see 278* for review) and would seem to be intuitively obvious for the CNS, in general. Yet specific examples of chemically defined neurons mediating well-defined functions are rare. The involvement of NPY neurons in feeding, drinking, and circadian functions are among the best of the available examples. (2) As in other parts of the body, peptides in the CNS often work through mutual interactions. Examples of peptide interactions in the hypothalamo-hypophyseal portal system and in the gut abound *(1)*. It is becoming more apparent, particularly with NPY in the hypothalamus, that similar mechanisms operate among cells in the brain and spinal cord.

3.2.3. NPY Receptors

Despite the dramatic concentration of NPY-immunoreactive fibers in nuclei of the hypothalamus, the level of [^{125}I]-NPY binding sites is extremely low, except in the suprachiasmatic nucleus *(125,126,321)*. This mismatch between peptide and putative receptor in the hypothalamus has been well noted by Martel and coworkers *(322)*. They appear to favor the presence of multiple classes of NPY receptors, including a mainly hypothalamic one that is not localized under presently used incubation conditions, to account for the mismatch. Certainly, more than one class of NPY receptor has been identified and referred to as Y1 and Y2 subtypes *(323,324; see also* Grundemar et al., this vol.). That ligands other than NPY, such as the related peptide YY (PYY), produce significantly greater binding in nuclei of the lateral hypothalamus *(128,321,324)* lends some support to the suggestion that the appropriate type(s) of NPY receptors will eventually be localized at high density in the hypothalamus.

3.2.4. Summary

It is within the diencephalon that the greatest confusion exists for NPY localization and the greatest certainty exists for its function. Different approaches for localizing NPY neurons (immunocytochemistry and *in situ* hybridization histochemistry) and the localization of NPY receptors give very different answers to where NPY might be functioning, particularly the reticular nucleus of the thalamus and in the hypothalamus. How to account for the presence of preproNPY mRNA in neurons of the reticular nucleus and the absence of NPY-like immunoreactivity in the same cells is a matter for speculation. The explanation for high levels of peptide, but low levels of receptors in the hypothalamus is equally elusive. Yet in the hypothalamus are sites at which the release and reception of NPY appear to mediate homeostatic functions, such as eating, drinking, and circadian rhythms. Since similar problems of interpretation have been encountered for other neuropeptides, their resolution for NPY, and the application of the findings to studies of physiology are likely to have a significant impact on understanding not only NPY in the diencephalon, but also peptides throughout the CNS.

4. Brainstem and Spinal Cord

4.1. Midbrain

The nuclei of the midbrain are conspicuous by the relatively sparse distribution of NPY fibers and somata. Both immunocytochemistry *(30)* and *in situ* hybridization histochemistry *(50)* indicate there are NPY somata in the central gray, interpeduncular nucleus and inferior colliculus of the rat. The somata are most numerous in the inferior colliculus, where they adopt a superficial distribution, and appear to vary in size from small, medially, to large, laterally *(30)*. It is presumed that the NPY cells are a class of interneuron in this midbrain auditory relay nucleus. Other immunocytochemical studies have been rather unhelpful, since they report either the presence of no NPY cells in the midbrain *(222)* or the existence of many, widely distributed NPY cells in a pattern that cannot otherwise be confirmed *(324)*. NPY fibers are present in each of the three regions that contain immunoreactive somata. Fibers in the caudal, medial part of the inferior colliculus appear to be continuous with a relatively dense plexus in the periaqueductual gray *(30)*.

4.2. Pons and Medulla

NPY-immunoreactive neurons are present in two general types of nuclei in the lower brainstem: cranial nerve nuclei and monoamine-containing nuclei. The presence of NPY in the former may be related directly to the role of NPY in visceral function, whereas its presence in the latter serves to distinguish among groups of long-projection neurons that send their axons to the forebrain and spinal cord.

4.2.1. Cranial Nerve Nuclei

In the respiratory brainstem nuclei of the cat, both somata and fibers are found to be NPY immunoreactive. Somata are present in the dorsal respiratory nucleus (ventrolateral solitary nucleus), the ventral respiratory nucleus (nucleus ambiguus and Botzinger complex), and the pneumotaxic center (nucleus parabrachialis and Kolliker-Fuse nucleus) of cats that have received local injections of colchicine *(325)*. The greatest density of somata is in the Botzinger complex, and the lowest density in the ventrolateral solitary nucleus. NPY fibers are also unevenly distributed, with the greatest density in the Kolliker-Fuse nucleus *(325)*. When the injections of colchicine are made not directly into the parenchyma of the brain but into the ventricles, only the somata in the solitary nucleus remain NPY immunoreactive in the cat *(325)*, rat *(222,299,324)*, and rabbit *(326)*. They are also the only immunostained cells in the human *(327)*. These findings may indicate that NPY cells make up a large proportion of the respiratory brainstem, but that normally the cell bodies contain low concentrations of the peptide. Alternatively, there remains the likelihood that only the neurons of the solitary nucleus normally contain NPY or an NPY-like molecule, and the direct injections of colchicine induce NPY expression in novel groups of somata.

Other populations of NPY-immunoreactive neurons are present in the dorsal motor nucleus of the vagus of rabbits *(326)*, and the spinal trigeminal nucleus of rats *(30)* and humans *(327)*. At least some of the NPY somata in the dorsal motor nucleus are preganglionic neurons, identified by the retrograde transport of Fast Blue from injections into the nerve trunk, and are not members of a catecholamine cell group *(326)*. Local injections of colchicine are required to visualize the NPY-like immunoreactivity in these cells. The small NPY cells in the spinal trigeminal nucleus are superficially located and probably give rise to the dense plexus of fibers in the dorsal laminae of this nucleus (Fig. 14 B,D). They are, thus, likely to be interneurons, similar to the cells in the dorsal horn of the spinal cord *(see* Section 4.3.; *30)*.

4.2.2. Monoaminergic Neurons

NPY immunoreactivity has been localized within noradrenergic, adrenergic, and serotonergic neurons of the mammalian brainstem. The existence of NPY within some neurons and not others serves to subdivide the central monoaminergic cells into chemically distinct groups *(228)* and to distinguish among connectionally specific subpopulations within certain groups *(329)*.

The monoaminergic neurons containing NPY include the *noradrenergic cells* of the A1 group in the ventrolateral medulla, a subpopulation of neurons in the A2 group of the dorsal medulla *(18,21,328,330)* and the locus coeruleus *(18,328,331)*, and the A4 group of the pons *(328)*, the *adrenergic cells* of the C1 and C2 groups *(326–328,332)*, and solitary nucleus *(18,328,332)* and some of the *serotonergic cells* of the nucleus raphe pallidus of rabbits *(326)* and

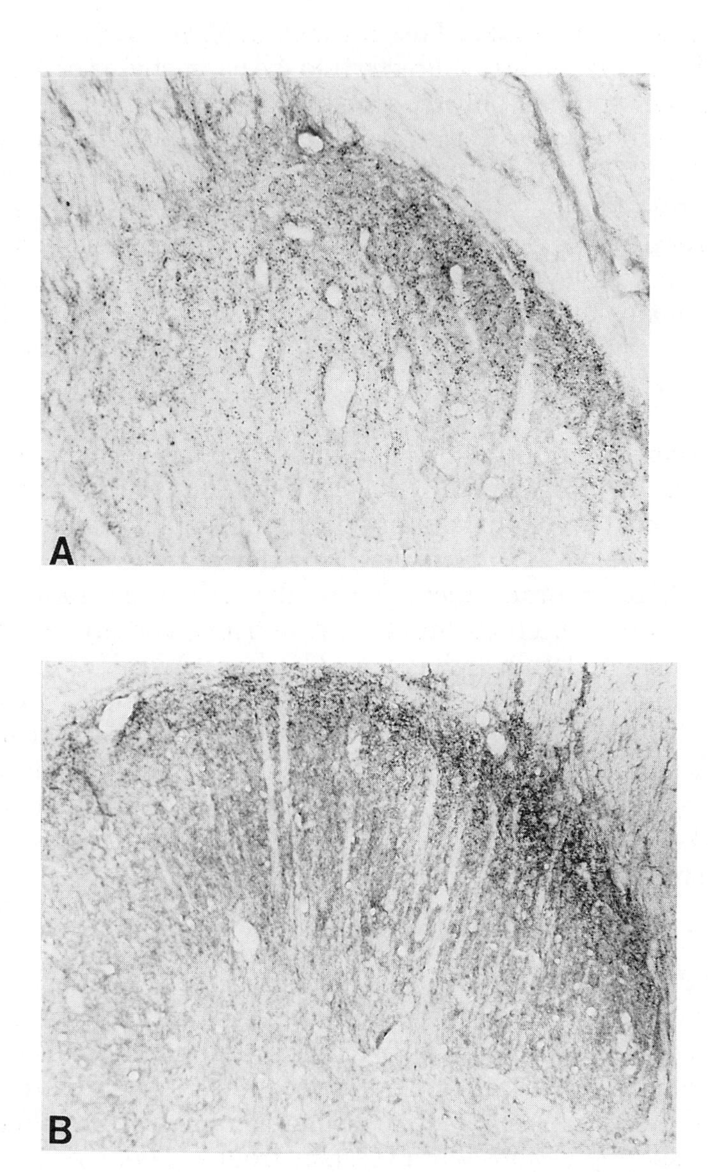

Fig. 14. NPY-immunoreactive fibers in somatic sensory nuclei of the spinal cord and medulla. A and B) Low-magnification photomicrographs of NPY fibers in the dorsal horn of the spinal cord and in the spinal trigeminal nucleus (B). The fibers are densest in the substantia gelatinosa. At higher magnification, the immunoreactivity is seen as

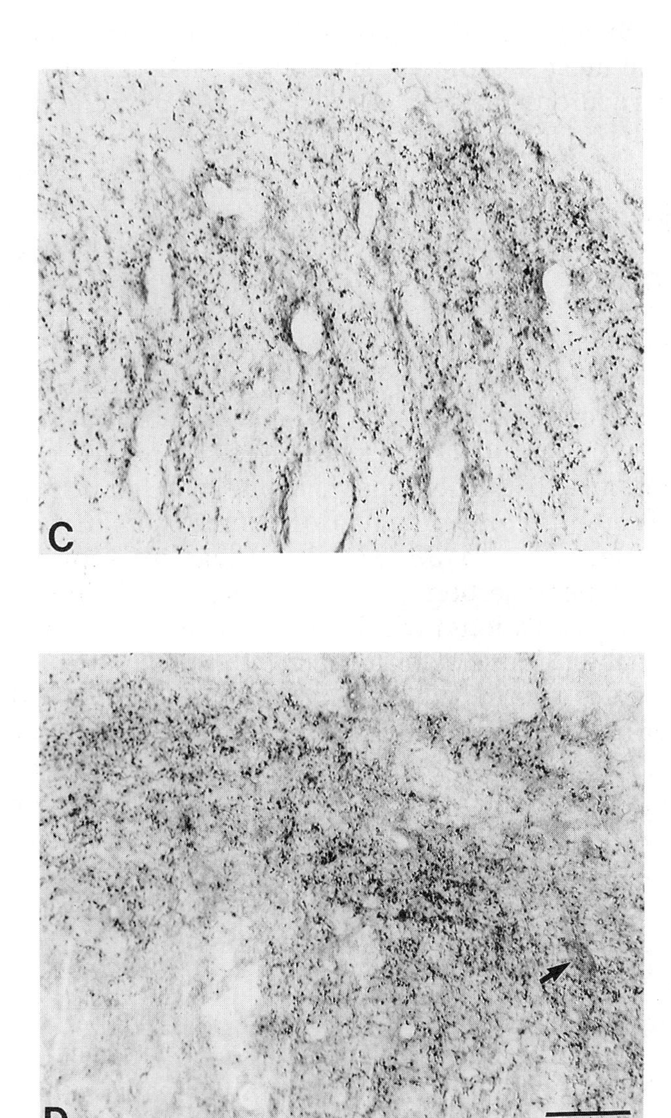

numerous punctate profiles. A cell body that appears to be lightly immunoreactive is seen in the spinal trigeminal nucleus (arrow). The patterns of spinal cord immunoreactivity are reported to vary dramatically across cord segments and across species. Bar equals 120 μm in A and B, and 30 μm in C and D.

presumed B2-3 group of humans *(327)*. Devoid of NPY immunoreactivity are most noradrenergic neurons in A2 and all the noradrenergic neurons in the A5 and A7 groups of the pons, as well as all of the dopaminergic neurons of the brainstem, hypothalamus, and forebrain *(328)*. These findings, which cut across several species, suggest that most adrenergic neurons are NPY immunoreactive, all dopaminergic cells, are not and noradrenergic and serotonergic neurons vary both among and within nuclei.

In the locus coeruleus, NPY coexistence divides noradrenergic neurons into distinct subpopulations. Detailed studies of the locus coeruleus find that less than a quarter of the tyrosine hydroxylase-immunoreactive (noradrenergic) neurons also display NPY-like immunoreactivity *(331)*. Long recognized as a source of projections to disparate areas of the CNS, including the spinal cord, hypothalamus, and cerebral cortex, the neurons of the locus coeruleus are a heterogeneous group *(329,331)*. Yet it was not apparent until recently that the projections to different targets arise from distinct neuronal subpopulations in the locus coeruleus. By examining the connectional and immunohistochemical properties of the same neurons, it was determined that a large proportion of NPY-immunoreactive noradrenergic neurons of the locus coeruleus project to the hypothalamus *(299,329)* and possibly to the entorhinal cortex *(335)*. Conversely, very few of the NPY/noradrenergic neurons send their axons either to the spinal cord or the cerebral neocortex *(329)*. These data, which were gained when NPY immunoreactivity was used an independent label for a subpopulation of locus coeruleus neurons, demonstrate that, as seen in the forebrain, chemically distinct neurons are often connectionally distinct.

4.3. Spinal Cord

The NPY innervation of the spinal cord arises from some intraspinal neurons, located in the superficial dorsal horn *(45,336,337)*, and from supraspinal neurons, found principally in the ventrolateral medulla *(329,338)*. The intraspinal neurons are diffusely scattered in laminae II and III in rodents *(336,337)*, cats, marmosets, horses *(337)*, and humans *(44)*. Some somata have been reported in layer I of rats *(336)*. The cells are described as large (20 μm in diameter) with long dendrites oriented rostrocaudally or obliquely *(336)*. The supraspinal NPY neurons are members of the adrenergic C1 group *(329,338)*. Experimental observations of no NPY-immunoreactive neurons in the dorsal root ganglia *(37)* and no effects on immunoreactivity following dorsal rhizotomy *(336,337)*, sciatic nerve section, or application of capsaicin to the sciatic nerve *(337)* indicate that intraspinal and supraspinal neurons are the only sources of NPY innervation in the mammalian spinal cord.

NPY-immunoreactive fibers and, thus, the concentration of NPY are much greater at sacral spinal levels than at more rostral levels *(45,337,339)*. This finding is consistent across species. Immunoreactive fibers are densest in the dorsal horn at all levels (Fig. 14A,C), and at thoracic levels are particularly dense around the preganglionic sympathetic neurons of the intermediolateral cell column *(336,337,340)*. The increase in NPY fibers is striking in the ventral horn at sacral levels, where a dense plexus surrounds the motoneurons of the perineal striated muscle (Onuf's nucleus; *337*). In addition, NPY innervation is relatively dense in the sexually dimorphic nuclei of the rat lumbar cord *(339,341)* and appears to be more intense in males than females *(341)*. The suggestion has been made that the NPY innervation of one sexually dimorphic nucleus, the cremaster nucleus, may serve to control the blood flow and the intrascrotal position, and thus the temperature, of the testes *(341)*.

The NPY innervation of the intermediolateral cell column in the thoracic cord is a route whereby neurons of the ventrolateral medulla may control blood pressure in an NPY-dependent manner *(see 342,343)*. Electron microscopic examinations demonstrate that a minority of the synaptic contacts in the intermediolateral cell column of the rat spinal cord are formed by NPY-immunoreactive fibers. In addition, only 30% of the NPY terminals located in single thin sections form conventional synaptic contacts, principally onto small dendritic shafts *(340)*. These data have been interpreted to indicate that NPY afferents from the ventrolateral medulla provide a modulatory input to the preganglionic sympathetic neurons of the intermediolateral cell column and that other neurotransmitter-specific afferents provide the main afferent drive to these cells *(340)*.

4.4. Summary

In the brainstem and spinal cord, NPY neurons include a mixture of interneurons in some regions (inferior colliculus, spinal trigeminal nucleus, and spinal cord) and long-projection neurons in others (the catecholamine cell groups of the pons and medulla).

The greatest attention has been paid to the catecholamine cell groups, where NPY immunoreactivity can be used to distinguish one nucleus from another and, within a single nucleus, one type of projection neuron from another. The long projections have been implicated in some specific CNS functions, including the control of blood pressure.

5. Concluding Remarks

The past decade has seen the publication of several dozens of studies, employing NPY immunoreactivity to characterize specific neuronal elements in the CNS. Most of the straightforward observations about the distribution of NPY neurons would appear to have been made, and many "second-generation" studies, employing experimental methods to characterize these neurons, have also been published. We are left with the picture of a widely distributed molecule, contained within very similar neurons throughout the cerebral cortex and nuclei of the forebrain, and in very different neurons in the hypothalamus, brainstem, and spinal cord. The next generation of studies is likely to take two paths. As the role of NPY in neuronal physiology becomes clearer, then the presence of the peptide at specific sites of contact between neurons will take on greater meaning. Thus, increased attention will probably be paid to the electron microscopic localization of NPY and, ultimately of its receptors, within neuronal circuits. Conversely, as the global roles of NPY in CNS functions are better understood, then the regulation of peptide expression during development and in adulthood is likely to be a very closely examined topic. It is here that the use of *in situ* hybridization histochemistry will probably pay great dividends.

In a sense, the field has reached the end of an era, not only for NPY, but also for neuropeptides in general. Since we have come to expect neuropeptide existence and coexistence in diverse collections of neurons, the fact that a particular neuropeptide is widely distributed no longer appears especially significant. How a chemically specific population is organized, how it participates in the development of a region of the CNS, and how its regulation may control the function of the CNS are now deemed far more important. The field has moved steadily away from phenomenology and toward understanding of how specific groups of peptide-containing neurons contribute to the function of areas and nuclei. In many ways, this setting of a new agenda for neuropeptide research has grown out of and owes its rapid progress to the intensive study of neuropeptide Y.

Acknowledgments

My gratitude to C. M. Gall and E. G. Jones, who allowed me unlimited access to published and unpublished material they have produced, and to J. P. Card for his many very helpful suggestions on the manuscript.

References

1. Krieger, D. T. (1983) Brain peptides: what, where and why? *Science* **222,** 975–985.
2. Tatemoto, K., Carlquist, M., and Mutt, V. (1982) Neuropeptide Y—a novel brain peptide with structural similarities for peptide YY and pancreatic polypeptide. *Nature* **296,** 659–666.
3. Emson, P. C. and De Quidt, M. E. (1984) NPY—a new member of the pancreatic polypeptide family. *TINS* **4,** 31–35.
4. O'Donohue, T. L., Chronwall, B. M., Pruss, R. M., Mezey, E., Kiss, J. Z., Eiden, L. E., Massari, V. J., Tessel, R. E., Pickel, V. M., DiMaggio, D. A., Hotchkiss, A. J., Crowley, W. R., and Zukowska-Grojec, Z. (1985) Neuropeptide Y and peptide YY neuronal and endocrine systems. *Peptides* **6,** 755–768.
5. Allen, J. M. and Bloom, S. R. (1986) Neuropeptide Y: a putative neurotransmitter. *Neurochem. Int.* **8,** 1–8.
6. Gray, T. S. and Morley, J. S. (1986) Neuropeptide Y: anatomical distribution and possible function in mammalian nervous system. *Life Sci.* **38,** 389–401.
7. Barnstable, C. J. (1980) Monoclonal antibodies which recognize different cell types in the rat retina. *Nature* **286,** 231–235.
8. Hockfield, S., McKay, R. D., Hendry, S. H. C., and Jones, E. G. (1983) A surface antigen that identifies ocular dominance columns in the visual cortex and laminar features of the lateral geniculate nucleus. Cold Spring Harbor Symp. *Quant. Biol.* **35,** 877–889.

9a. Zipursky, S. L., Venkatesh, T. R., Teplow, D. B., and Bener (1984) Neuronal development in the drosophila retina monoclonal antibodies as molecular probes. *Cell* **36,** 15–26.

9b. Katz. L. C. (1987) Local circuitry of identified projection neurons in cat visual cortex brain slices. *J. Neurosci.* **7,** 1223–1249.

10. Ritchie, T. C., Thomas, M. A., and Coulter, J. D. (1989) A nerve terminal protein with a selective distribution in spinal cord and brain. *J. Neurosci.* **9,** 2697–2709.
11. Zacco, A., Cooper, V., Chantler, P. D., Fisher-Hyland, S., Horton, H. L., and Levitt, P. (1990) Isolation, biochemical characterization and ultrastuctural

analysis of the limbic system-associated membrane protein (LAMP), a protein expressed by neurons comprising functional neural circuits. *J. Neurosci.* **10,** 73–90.

12. Walaas, S. I., Nairn, A. C., and Greengard, P. (1983) Regional distribution of calcium- and cyclic adenosine 3':5'-monophosphate-regulated protein phosphorylation systems in mammalian brain. I. Particulate systems. *J. Neurosci.* **3,** 291–301.
13. Walaas, S. I., Nairn, A. C., and Greengard, P. (1983) Regional distribution of calcium- and cyclic adenosine 3':5'-monophosphate-regulated protein phosphorylation systems in mammalian brain. II. Soluble systems. *J. Neurosci.* **3,** 302–311.
14. Dawbarn, D., Hunt, S. P., and Emson, P. C. (1984) Neuropeptide Y: regional distribution, chromatic characterization and immunohistochemical demonstration in post-mortem human brain. *Brain Res.* **296,** 168–173.
15. DiMaggio, D. A., Chronwall, B. M., Buchanan, K., and O'Donohue, T. L. (1985) Pancreatic polypeptide immunoreactivity in rat brain is actually neuropeptide Y. *Neuroscience* **15,** 1149–1157.
16. De Quidt, M. E. and Emson, P. C. (1986) Distribution of neuropeptide Y-like immunoreactivity in the rat central nervous system- I. Radioimmunoassay and chromatographic characterisation. *Neuroscience* **18,** 527–543.
17. Loren, I., Alumets, J., Hakanson, R., and Sundler, F. (1979) Immunoreactive pancreatic polypeptide (PP) occurs in the central and peripheral nervous system: preliminary immunocytochemical observations. *Cell Tissue Res.* **200,** 179–186.
18. Lundberg, J. M., Hokfelt, T., Anggard, A., Kimmel, J., Goldstein, M. & Markey, K. (1980) Coexistence of an avian pancreatic polypeptide (APP) immunoreactive substance and catecholamines in some peripheral and central neurons. *Acta Physiol. Scand.* **110,** 107–109.
19. Lundberg, J. M., Terenius, L., Hokfelt, T., and Tatemoto, K. (1984) Comparative immunohistochemical and biochemical analysis of pancreatic polypeptide-like peptides with special reference to presence of neuropeptide Y in central and peripheral neurons. *J. Neurosci.* **4,** 2376–2386.
20. Card, J. P., Brecha, N., and Moore, R. Y. (1983) Immunohistochemical localization of avian pancreatic polypeptide-like immunoreactivity in the rat hypothalamus. *J. Comp. Neurol.* **217,** 123–136.
21. Jacobowitz, D. M. and Olschowka, J. A. (1982) Bovine pancreatic polypeptide-like immunoreactivity in brain and peripheral nervous system: coexistence with catecholaminergic nerves. *Peptides* **3,** 569–590.
22. Olschowka, J. A., O'Donohue, T. L., and Jacobowitz, D. M. (1981) The distribution of bovine pancreatic polypeptide-like immunoreactive neurons in rat brain. *Peptides* **2,** 309–311.
23. Moore, R. Y., Gustafson, E. L., and Card, J. P. (1984) Identical immunoreactivity of afferents to the rat suprachiasmatic nucleus with antisera against avian pancreatic polypeptide, molluscan cardioexcitatory peptide and neuropeptide Y. *Cell Tissue Res.* **236,** 41–46.
24. Inui, A., Mizuno, N., Ooya, M., Suenaga, K., Morioka, H., Ogawa, T., Ishida, M., and Baba, S. (1985) Cross-reactivities of neuropeptide Y and peptide YY with pancreatic polypeptide antisera: evidence for the existence of pancreatic polypeptide in the brain. *Brain Res.* **330,** 386–389.

25. Minth, C. D., Bloom, S. R., Polak, J. M., and Dixon, J. E. (1984) Cloning, characterization and DNA sequence of a human cDNA encoding neuropeptide tyrosine. *Proc. Natl. Acad. Sci. USA* **81,** 4577–4581.
26. Larhammar, D., Ericsson, A., and Persson, H. (1987) Structure and expression of the rat neuropeptide Y gene. *Proc. Natl. Acad. Sci. USA* **84,** 2068–2072.
27a. Covenas, R., Aguirre, J. A., Alonso, J. R., Dios, M., Lara, J., and Aijon, J. (1990) Distribution of neuropeptide Y-like immunoreactive fibers in the cat thalamus. *Peptides* **11,** 45–50.
27b. Higuchi, H., Yang, H.-Y. T., and Sabol, S. L. (1988) Rat neuropeptide Y precursor gene expression. *J. Biol. Chem.* **263,** 6288–6295.
28. Chan-Palay, V. (1988) Techniques for the simultaneous demonstrations of neuropeptide Y gene expression and peptide storage in single neurons of the human brain. *Histochemistry* **90,** 123–127.
29. Chronwall, B. M., Chase, T. N., and O'Donohue, T. L. (1984) Coexistence of neuropeptide Y and somatostatin in rat and human cortical and rat hypothalamic neurons. *Neurosci. Lett.* **52,** 213–217.
30. De Quidt, M. E. and Emson, P. C. (1986b) Distribution of neuropeptide Y-like immunoreactivity in the rat central nervous system. II. Immunohistochemical analysis. *Neuroscience* **18,** 545–618.
31. Bons, N., Mestrie, N., Petter, A., Danger, J. M., Pelletier, G., and Vaudry, H. (1990) Localization and characterization of neuropeptide Y in the brain of Microcebus murinus (Primate, Lemurian). *J. Comp. Neurol.* **298,** 343–361.
32. Rakic, P. (1988) Specification of cerebral cortical areas. *Science* **241,** 170–176.
33. Wise, S. P. and Herkenham, M. (1982) Opiate receptor distribution in the cerebral cortex of the rhesus monkey. *Science* **218,** 387–389.
34. Hendry, S.H.C., Schwark, H. D., Jones, E. G., and Yan, J. (1987) Numbers and proportions of GABA immunoreactive neurons in different areas of monkey cerebral cortex. *J. Neurosci.* **7,** 1503–1520.
35. Hendry, S. H. C., Jones, E. G., Hockfield, S., and McKay, R. D. G. (1988) Neuronal populations stained with the monoclonal antibody, Cat-301, in the mammalian cerebral cortex and thalamus. *J. Neurosci.* **8,** 518–542.
36. Lewis, D. A., Campbell, M. J., and Morrison, J. H. (1986) An immunohistochemical characterization of somatostatin-28 and and somatostatin 28 (1-12) in monkey prefrontal cortex. *J. Comp. Neurol.* **248,** 1–18.
37. Lewis, D. A., Campbell, M. J., Foote, S. L., Goldstein, M., and Morrison, J. H. (1987) The distribution of tyrosine hydoxylase-immunoreactive fibers in primate neocortex is widespread but regionally specific. *J. Neurosci.* **7,** 279–290.
38. Allen, Y. S., Adrian, T. E., Allen, J. M., Tatemoto, K., Crow, T. J., Bloom, S. R., and Polak, J. M. (1983) Neuropeptide Y distribution in rat brain. *Science* **221,** 877–879.
39. Beal, M. F., Mazurek, M. F., and Martin, J. B. (1987) A comparison of somatostatin and neuropeptide Y distribution in monkey brain. *Brain Res.* **405,** 213–219.
40. Adrian, T. E., Allen, J. M., Bloom, S. R., Ghatei, M. A., Rossor, M., Roberts, G. W., Crow, T. J., Tatemoto, K., and Polak, J. M. (1983) Neuropeptide Y distribution in human brain. *Nature* **306,** 584–586.
41. Dawbarn, D., Rossor, M. N., Mountjoy, C. Q., Roth, M., and Emson, P. C. (1986) Decreased somatostatin but not neuropeptide Y immunoreactivity in cerebral cortex in senile dementia of Alzheimer type. *Neurosci. Lett.* **70,** 154–159.

42. Beal, M. F., Mazurek, M. F., Chattha, G. K., Svendsen, C. N., Bird, E. D., and Martin, J. B. (1986a) Neuropeptide Y immunoreactivity is reduced in cerebral cortex in Alzheimer's disease. *Ann. Neurol.* **20,** 282–288.
43. Brene, S., Lindefors, N., Kopp, J., Sedvall, G., and Persson, H. (1989) Regional distribution of neuropeptide Y mRNA in postmortem human brain. *Mol. Brain Res.* **6,** 241–249.
44. Allen, J. M., Yeats, J. C., Adrian, T. E., and Bloom, S. R. (1984) Radioimmunoassay of neuropeptide Y. *Reg. Pept.* **8,** 61–70.
45. Allen, J. M., Gibson, S. J., Adrian, T. E., Polak, J. M., and Bloom, S. R. (1984) Neuropeptide Y in human spinal cord. *Brain Res.* **308,** 145–148.
46. Beal, M. F., Mazurek, M. F., Lorenz, L. J., Chattha, G. K., Ellison, D. W., and Martin, J. B. (1986c) An examination of neuropeptide Y postmortem stability in an animal model simulating human autopsy conditions. *Neurosci. Lett.* **64,** 69–74.
47. McDonald, J. K., Parnavelas, J. G., Karamanlidis, A. N., and Brecha, N. (1982) The morphology and distribution of peptide-containing neurons in the adult and developing visual cortex of the rat. IV. Avian pancreatic polypeptide. *J. Neurocytol.* **11,** 985–995.
48. Nakagawa, Y., Shiosaka, S., Emson, P. C., and Tohyama, M. (1985) Distribution of neuropeptide Y in the forebrain and diencephalon: an immunohistochemical analysis. *Brain Res.* **361,** 52–60.
49. Gehlert, D. R., Chronwall, B. M., Schafer, M. P., and O'Donohue, T. L. (1987) Localization of neuropeptide Y messenger ribonucleic acid in rat and mouse brian [sic] by in situ hybridization. *Synapse* **1,** 25–31.
50. Morris, B. J. (1989) Neuronal localisation of neuropeptide Y gene expression in rat brain. *J. Comp. Neurol.* **290,** 358–368.
51. Wahle, P., Meyer, G., and Albus, K. (1986) Localization of NPY-immunoreactivity in the cat's visual cortex. *Exp. Brain Res.* **61,** 364–374.
52. Wahle, P. and Meyer, G. (1987) Morphology and quantitative changes of transient NPY-ir neuronal populations during early postnatal development of the cat visual cortex. *J. Comp. Neurol.* **261,** 165–192.
53. Chun, J. J. M., Nakamura, M. J., and Shatz, C. J. (1987) Transient cells of the developing mammalian telencephalon are peptide-immunoreactive neurons. *Nature* **325,** 617–620.
54. Demeulemeester, H., Vandesande, F., Orban, G. A., Brandon, C., and Vanderhaeghen J.J. (1988) Heterogeneity of GABAergic cells in cat visual cortex. *J. Neurosci.* **8,** 988–1000.
55. Smith, Y., Parent, A., Kerkerian, L., and Pelletier, G. (1985) Distribution of neuropeptide Y immunoreactivity in the basal forebrain and upper brainstem of the squirrel monkey (Samiri sciureus). *J. Comp. Neurol.* **236,** 71–89.
56. Hendry, S. H. C., Jones, E. G., and Emson, P. C. (1984) Morphology, distribution and synaptic relations of somatostatin and neuropeptide Y immunoreactive neurons in rat and monkey neocortex. *J. Neurosci.* **4,** 2497–2517.
57. Hendry, S. H. C., Jones, E. G., DeFelipe, J., Schmechel, D., Brandon, C., and Emson, P. C. (1984b) Neuropeptide containing neurons of the cerebral cortex are also GABAergic. *Proc. Natl. Acad. Sci. USA* **81,** 6526–6530.
58. Kuljis, R. and Rakic, P. (1989) Distribution of neuropeptide Y-containing perikarya and axons in various neocortical areas in the macaque monkey. *J. Comp. Neurol.* **280,** 383–392.

59. Kuljis, R. and Rakic, P. (1989) Multiple types of neuropeptide Y-containing neurons in primate neocortex. *J. Comp. Neurol.* **280,** 393–409.
60. Kuljis, R. and Rakic, P. (1989) Neuropeptide Y-containing neurons are situated predominantly outside cytochrome oxidase puffs in macaque visual cortex. *Vis. Neurosci.* **2,** 57–62.
61. Kuljis, R. and Rakic, P. (1990) Hypercolumns in primate visual cortex can develop in the absence of cues from photoreceptors. *Proc. Natl. Acad. Sci. USA* **87,** 5303–5306.
62. Tigges, M., Tigges, J., McDonald, J. K., Slattery, M., and Fernandes, A. (1989) Postnatal development of neuropeptide Y-like immunoreactivity in area 17 of normal and visually deprived rhesus monkeys. *Vis. Neurosci.* **2,** 315–328.
63. Vincent, S. R., Johansson, O., Hokfelt, T., Meyerson, R., Sachs, C., Elde, R. P., Terenius, L., and Kimmel, J. (1982) Neuropeptide coexistence in human cortical neurones. *Nature* **298,** 65–67.
64. Chan-Palay, V., Allen, Y. S., Lang, W., Haesler, U., and Polak, J. M. (1985) Cytology and distribution in normal human cerebral cortex of neurons immunoreactive with antisera against neuropeptide Y. *J. Comp. Neurol.* **238,** 382–389.
65. Chan-Palay, V., Lang, W., Allen, Y. S., Haesler, U., and Polak, J. M. (1985b) Cortical neurons immunoreactive with antisera against neuropeptide Y are altered in Alzheimer's-type dementia. *J. Comp. Neurol.* **238,** 390–400.
66. Chan-Palay, V. and Yasargil, G. (1986) Immunocytochemistry of human brain tissue with a polyclonal antiserum against neuropeptide Y. *Anat. Embryol.* **174,** 27–33.
67. Van Reeth, O., Goldman, S., Schiffman, S., Verstappen, A., Pelletier, G., Vaudry, H., and Vanderhagen, J. J. (1987) Distribution of neuropeptide Y immunoreactivity in human visual cortex and underlying white matter. *Peptides* **8,** 1107–1117.
68. Blinkenberg, M., Kruse-Larsen, C., and Mikkelsen, J. D. (1990) An immunohistochemical localization of neuropeptide Y (NPY) in its amidated form in human frontal cortex. *Peptides* **11,** 129–137.
69. Schwartzberg, M., Unger, J., Weindl, A., and Lange, W. (1990) Distribution of neuropeptide Y in the prosencephalon of man and Cotton-head Tamarin (*Saguinus oedipus*): colocalization with somatostatin in neurons of striatum and amygdala. *Anat. Embryol.* **181,** 157–166.
70. Horton, J. C. and Hubel, D. H. (1981) Regular patchy distribution of cytochrome oxidase staining in primary visual cortex of macaque monkey. *Nature* **292,** 762–764.
71. Humphrey, A. L. and Hendrickson, A. E. (1983) Background and stimulus-induced patterns of high metabolic activity in the visual cortex (area 17) of the squirrel and macaque monkey. *J. Neurosci.* **3,** 345–358.
72. Horton, J. C. (1984) Cytochrome oxidase patches: a new cytoarchitectonic feature of monkey visual visual cortex. *Phil. Trans. Roy. Soc. Lond. B.* **304,** 199–253.
73. Wong-Riley, M. and Carroll, E. W. (1984) The effect of impulse blockage on cytochrome oxidase activity in monkey visual system. *Nature* **307,** 262–264.
74. Hendrickson, A. E. (1985) Dots, stripes, and columns in monkey visual cortex. *Trends Neurosci.* **8,** 406–410.

75. Livingstone, M. and Hubel, D. H. (1984) Anatomy and physiology of a color system in the primate visual cortex. *J. Neurosci.* **4,** 309–356.
76. Hendrickson, A. E., Hunt, S. P., and Wu, J.-Y. (1981) Immunocytochemical localization of glutamic acid decarboxylase in monkey striate cortex. *Nature* **292,** 605–607.
77. Celio, M. R., Scharer, L., Morrison, J. H., Norman, A. W., and Bloom, F. E. (1986) Calbindin immunoreactivity alternates with cytochrome c-oxidase-rich zones in some layers of the primate visual cortex. *Nature* **323,** 715–717.
78. Hendry, S. H. C., Jones, E. G., and Burstein, N. (1988) Activity-dependent regulation of tachykinin-like immunoreactivity in neurons of monkey primary visual cortex. *J. Neurosci.* **8,** 1225–1238.
79. Hendry, S. H. C., Fuchs, J., deBlas, A., and Jones, E. G. (1990) Distribution and plasticity of immunocytochemically localized $GABA_A$ receptors in adult monkey visual cortex. *J. Neurosci.* **10,** 2438–2450.
80. Mountcastle, V. B. (1957) Modality and topographic properties of single neurons of cat's somatic sensory cortex. *J. Neurophysiol.* **20,** 408–434.
81. Jones, E. G., Burton, H., and Porter, R. (1975) Commissural and cortico-cortical 'columns' in the somatic sensory cortex of primates. *Science* **190,** 572–574.
82. Goldman, P. S. and Nauta, W. J. H. (1977) Columnar distribution of cortico-cortical fibers in the frontal association, limbic and motor cortex of the developing rhesus monkey. *Brain Res.* **122,** 393–413.
83. Hubel, D. H. and Wiesel, T. N. (1962) Receptive fields, binocular interaction, and functional architecture in the cat's visual cortex. *J. Physiol. (Lond.)* **160,** 106–154.
84. Shatz, C. J. and Stryker, M. P. (1978) Ocular dominance in layer IV of the cat's visual cortex and the effects of monocular deprivation. *J. Physiol.* **281,** 267–283.
85. Rakic, P., Goldman-Rakic, P. C., and Gallagher, D. (1988) Quantitative autoradiography of major neurotransmitter receptors in the monkey striate and extrastriate cortex. *J. Neurosci.* **8,** 3670–3690.
86. Feldman, M. (1984) Morphology of the neocortical pyramidal cell, in *Cerebral Cortex,* vol.1. Peters, A. and Jones, E. G., eds., Plenum, New York, pp. 123–200.
87. Fairen, A., DeFelipe, J., and Regidor, J. (1984) in *Cerebral Cortex,* vol. 1. Peters, A. and Jones, E. G., eds., Plenum, New York, pp. 201–254.
88. Hendry, S. H. C. (1987) Recent advances in understanding the intrinsic circuitry of the cerebral cortex, in *Higher Brain Functions: Recent Explorations of the Brain's Emergent Properties.* Wise, S. P., ed., Wiley, New York, pp. 241–283.
89. Code, R. A. and Winer, J. A. (1985) Commisurral neurons in layer III of cat primary auditory cortex (AI): pyramidal and non-pyramidal cell input. *J. Comp. Neurol.* **242,** 485–510.
90. Citation withdrawn.
91. Lund, J. S., Lund, R. D., Hendrickson, A. E. , Bunt, A. H., and Fuchs, A. F. (1975) The origin of efferent pathways from the primary visual cortex, area 17, of the macaque monkey as shown by retorgrade transport of horseradish peroxidase. *J. Comp. Neurol.* **164,** 287–304.
92. Jones, E. G. and Wise, S. P. (1977) Size, laminar and columnar distribution of efferent cells in the sensory-motor cortex of monkeys. *J. Comp. Neurol.* **175,** 391–438.

93. Aoki, C. and Pickel, V. M. (1989) Neuropeptide Y in the cerebral cortex and the caudate-putamen nuclei: ultrastructural basis for interactions with GABAergic and non-GABAergic neurons. *J. Neurosci.* **9,** 4333–4354.
94. Escobar, M. I., Pimienta, H., Caviness, V. S., Jr., Jacobson, M., Crandall, J. E., and Kosik, K. S. (1986) Architecture of apical dendrites in the murine neocortex: dual apical dendritic systems. *Neuroscience* **17,** 975–989.
95. Campbell, M. J. and Morrison, J. H. (1989) Monoclonal antibody to neurofilament protein (SMI-32) labels a subpopulation of pyramidal neurons in the human and monkey neocortex. *J. Comp. Neurol.* **282,** 191–205.
96. Jones, E. G. and Hendry, S. H. C. (1984) Basket cells, in *Cerebral Cortex*, vol.1. Peters, A. and Jones, E. G., eds., Plenum, New York, pp. 309–336.
97. Peters, A. (1984) Chandelier cells, in *Cerebral Cortex*, vol.1. Peters, A. and Jones, E. G., eds., Plenum, New York, pp. 361–380.
98. Lewis, D. A. and Lund, J. S. (1990) Heterogeneity of chandelier neurons in monkey neocortex: corticotropin-releasing factor- and parvalbumin-immunoreactive populations. *J. Comp. Neurol.* **293,** 599–615.
99. DeFelipe, J., Hendry, S. H. C., Hashikawa, T., Molinari, M., and Jones, E. G. (1990) A microcolumnar structure of monkey cerebral cortex revealed by immunocytochemical studies of double bouquet cell axons. *Neuroscience* **37,** 655–673.
100. Jones, E. G. (1975) Varieties and distribution of non-pyramidal cells in the somatic sensory cortex of the squirrel monkey. *J. Comp. Neurol.* **160,** 205–268.
101. Peters, A. and Kimerer, L. M. (1981) Bipolar neurons in the rat visual cortex: a combined Golgi-electron microscope study. *J. Neurocytol.* **10,** 921–946.
102. Hendry, S. H. C., Jones, E. G., and Beinfeld, M. C. (1983) Cholecystokinin-immunoreactive neurons in rat and monkey cerebral cortex make symmetric synapses and have intimate associations with blood vessels. *Proc. Natl. Acad. Sci. USA* **80,** 2400–2404.
103. Connor, J. R. and Peters, A. (1984) Vasoactive intestinal polypeptide-immunoreactive neurons in rat visual cortex. *Neuroscience* **12,** 1027–1044.
104. Somogyi, P., Hodgson, A. J., Smith, A. D., Nunzi, M. G., Gorio, A., and Wu, J.-Y. (1984) Different populations of GABAergic neurons in the visual cortex and hippocampus of the cat contain somatostatin- or cholecystokinin-immunoreactive material. *J. Neurosci.* **4,** 2590–2603.
105. Jones, E. G. and Hendry, S. H. C. (1986) Peptide-containing neurons of the primate cerebral cortex. In: *Neuropeptides in Neurologic and Pyschiatric Disease*, J. B. Martin and J. D. Barchas, eds., Raven Press, New York, pp.163-178.
106. Hokfelt, T., Johansson, O., Ljungdahl, A., Lundberg, J. M., and Schultzberg, M. (1980) Peptidergic neurones. *Nature* **284,** 515–521.
107. Papadopoulos, G. C., Parnavelas, J. G., and Cavanagh, M. E. (1987) Extensive co-existence of neuropeptides inthe rat visual cortex. *Brain Res.* **420,** 95–99.
108. Kowall, N. W. and Beal, M. F. (1988) Cortical somatostatin, neuropeptide Y, and NADPH diaphorase neurons: normal anatomy and alterations in Alzheimer's disease. *Ann. Neurol.* **23,** 105–114.
109. Benoit, R., Ling, N., Alford, B., and Guillemin, R. (1982) Seven peptides derived from pro-somatostatin in rat brain. *Biochem. Biophys. Res. Commun.* **107,** 944–950.

110. de Lima, A. D. and Morrison, J. H. (1989) Ultrastructural analysis of somatostatin-immunoreactive neurons and synapses in the temporal and occipital cortex of the macaque monkey. *J. Comp. Neurol.* **283,** 212–227.
111. Szabo, S. and Reichlin, S. (1981) Somatostatin in rat tissues is depleted by cysteamine administration. *Endocrinology* **109,** 255–257.
112. Chattha, G. K. and Beal, M. F. (1987) Effect of cysteamine on somatostatin and neuropeptide Y in rat striatum and cortical synaptosomes. *Brain Res.* **401,** 359–364.
113. Vincent, S. R., Johansson, O., Hokfelt, T., Skirboll, L., Elde, R. P., Terenius, L., Kimmel, J., and Goldstein, M. (1983) NADPH-diaphorase: a selective histochemical marker for striatal neurons containing both somatostatin and avian pancreatic polypeptide (APP)-like immunoreactivities. *J. Comp. Neurol.* **217,** 252–263.
114. Sharp, F. R., Gonzalez, M. F., and Sagar, S. M. (1987) Fetal frontal cortex transplanted to injured motor/sensory cortex of adult rats. II. VIP-, somatostatin and NPY-immunoreactive neurons. *J. Neurosci.* **7,** 3002–3015.
115. Sandell, J. H. (1986) NADPH diaphorase histochemistry in the macaque striate cortex. *J. Comp. Neurol.* **251,** 388–397.
116. Duckett, S. and Pearse, A. G. E. (1964) The nature of the solitary active cells of the central nervous system. *Experientia* **20,** 259–260.
117. Thomas, E. and Pearse, A. G. E. (1964) The solitary active cells: histochemical demonstration of damage restistant nerve cells with a TPN diaphorase reaction. *Acta. Neuropathol.* **3,** 238–249.
118. Schmechel, D. E., Vickery, B. G., Fitzpatrick, D., and Elde, R. P. (1984) GABAergic neurons of mammalian cerebral cortex: widespread subclass defined by somatostatin content. *Neurosci. Lett.* **47,** 227–232.
119. Jones, E.G., DeFelipe, J., Hendry, S. H. C., and Maggio, J. (1988) A study of tachykinin-immunoreactive neurons in monkey cerebral cortex. *J. Neurosci.* **8,** 1206–1244.
120. Citation withdrawn.
121. Aoki, C. and Pickel, V. M. (1990) Neuropeptide Y in cortex and striatum. *Ann. NY Acad. Sci.* **611,** 186–205.
122. Colmers, W. F., Lukowiak, K., and Pittman, Q. J. (1987) Presynaptic action of neuropeptide Y in area CA1 of the rat hippocampal slice. *J. Physiol.* **383,** 285–299.
123. Colmers, W. F., Lukowiak, K., and Pittman, Q. J. (1988) Neuropeptide Y action in the rat hippocampal slice: site and mechanism of presynaptic inhibition. *J. Neurosci.* **8,** 3827–3837.
124. Jan, L. Y. and Jan, Y. N. (1982) Peptidergic transmission in sympathetic ganglia of the frog. *J. Physiol.* **327,** 219–246.
125. Lynch, D. R., Walker, M. W., Miller, R. J., and Snyder, S. H. (1989) Neuropeptide Y receptor binding sites in rat brain: differential autoradiographic localizations with 125I-peptide YY and ^{125}I-neuropeptide Y imply receptor heterogeneity. *J. Neurosci.* **9,** 2607–2619.
126. Martel, J.-C., St-Pierre, S., and Quirion, R. (1986) Neuropeptide Y receptors in rat brain: autoradiographic localization. *Peptides* **7,** 55–60.
127. Martel, J.-C., St-Pierre, S., Bedard, P. J., and Quirion, R. (1987) Comparison of [^{125}I] Bolton-Hunter neuropeptide Y binding sites in the forebrain of various mammalian species. *Brain Res.* **419,** 403–407.

128. Martel, J.-C., Fournier, A., St. Pierre, S., and Quirion, R. (1990a) Quantitative autoradiographic distribution of [^{125}I] Bolton-Hunter neuropeptide Y receptor binding sites in rat brain. Comparison with [^{125}I] peptide YY receptor sites. *Neuroscience* **36,** 25–283.
129. Martel, J.-C., Alagar, R., Robitaille, Y., and Quirion, R. (1990b) Neuropeptide Y receptor binding sites in human brain. Possible alteration in Alzheimer's disease. *Brain Res.* **519,** 228–235.
130. Inui A., Oya, M., Okita, M., Inoue, T., Sakatani, N., Moroka, H., Shii, K., Yokono, K., Mizuno, N., and Baba, S. (1988) Peptide YY receptors in the brain. *Biochem. Biophys. Res. Commun.* **150,** 25–32.
131. Rosier, A. M., Orban, G. A., and Vandesande, F. (1990) Regional distribution of binding sites for neuropeptide Y in cat and monkey visual cortex determined by in vitro receptor autoradiography. *J. Comp. Neurol.* **293,** 486–498.
132. Herkenham, M. (1987) Mismatches between neurotransmitter and receptor localizations in brain: observations and implications. *Neuroscience* **23,** 1–38.
133. Hendry, S. H. C. and Kennedy, M. B. (1986) Immunoreactivity for a calmodulin-dependent protein kinase is selelctively increased in macaque striate cortex after monocular deprivation. *Proc. Natl. Acad. Sci. USA* **83,** 1536–1540.
134. Hendry, S. H. C. and Jones, E. G. (1986) Reduction in number of GABA immunostained neurons in deprived-eye dominance columns of monkey area 17. *Nature* **320,** 750–753.
135. Warren, R., Tremblay, N., and Dykes, R. W. (1989) Quantitative study of glutamic acid decarboxylase immunoreactive neurons and cytochrome oxidase activity in normal and partially deafferented rat hindlimb somatosensory cortex. *J. Comp. Neurol.* **288,** 583–592.
136. Welker, W. I., Soriano, E., and Van der Loos, H. (1989) Plasticity in the barrel cortex of adult mouse: effects of peripheral deprivation on GAD-immunoreactivity. *Exp. Brain Res.* **74,** 412–452.
137. Welker, W. I., Soriano, E., Dorfli, J., and Van der Loos, H. (1989) Plasticity in the barrel cortex of adult mouse: transient increase of GAD-immunoreactivity following sensory stimulation. *Exp. Brain Res.* **78,** 659–664.
138. Marksteiner, J. and Sperk, G. (1988) Concomitant increase of somatostatin, neuropeptide Y and glutamate decarboxylase in the frontal cortex of rats with decreased seizure threshold. *Neuroscience* **26,** 379–385.
139. Marksteiner, J., Sperk, G., and Maas, D. (1989) Differential increases in bran levels of neuropeptide Y and vasoactive intestinal polypeptide after kainic acid-induced seizures in the rat. *Naunyn-Schmiedeberg's Arch. Pharmacol.* **339,** 173–177.
140. Gall, C., Lauterborn, J., Isackson, P., and White, J. (1990) Seizures, neuropeptide regulation, and mRNA expression in the hippocampus. *Prog. Brain Res.* **83,** 371–390.
141. Hendry, S. H. C. and Jones, E. G. (1988) Activity-dependent regulation of GABA expression in the visual cortex of adult monkeys. *Neuron* **1,** 701–712.
142. Graybiel, A. M. and Ragsdale, C. W., Jr. (1982) Psuedocholinesterase staining in the primary visual pathway of the macaque monkey. *Nature* **299,** 439–442.

143. Gaykema, R. P. A., Compaan, J. C., Nyakas, C., Horvath, E., and Luiten, P. G. M. (1989) Long-term effects of cholinergic basal forebrain lesions on neuropeptide Y and somatostatin immunoreactivity in rat neocortex. *Brain Res.* **489,** 392–396.
144. Angevine, J. B., Jr. and Sidman, R. L. (1961) Autoradiographic study of cell migration during histogenesis of cerebral cortex in the mouse. *Nature* **192,** 766–768.
145. Rakic, P. (1974) Neurons in the rhesus monkey visual cortex: systematic relation between time of origin and eventual disposition. *Science* **183,** 425–427.
146. Luskin, M. B. and Shatz, C. J. (1985b) Neurogenesis of the cat's primary visual cortex. *J. Comp. Neurol.* **242,** 611–631.
147. Marin-Padilla, M. (1971) Early prenatal ontogenesis of the cerebral cortex (neocortex) of the cat *(Felis domesticus)*: A Golgi study. I. The primordial neocortical organization. Z. *Anat. Entwickl-Gesch.* **134,** 117–145.
148. Marin-Padilla, M. (1990) Origin, formation and prenatal maturation of the human cerebral cortex: an overview. *J. Craniofacial Gen. Dev. Biol.* **10,** 137–146.
149. Rickmann, M., Chronwall, B. M., and Wolff, J. R. (1977) On the development of non-pyramidal neurons and axons outside the cortical plate: The early marginal zone as a pallial anlage. *Anat. Embryol.* **151,** 285–307.
150. Luskin, M. B. and Shatz, C. J. (1985a) Studies of the earliest generated cells of the cat's visual cortex: cogeneration of subplate and marginal zones. *J. Neurosci.* **5,** 1062–1075.
151. McConnell, S. K. (1988) Development and decision-making in the mammalian cerebral cortex. *Brain Res. Rev.* **13,** 1–23.
152. Cavanagh, M. and Parnavelas, J. G. (1990) Development of neuropeptide Y (NPY) immunoreactive neurons in the rat occipital cortex: a combined immunohistochemical-autoradiographic study. *J. Comp. Neurol.* **297,** 553–563.
153. Foster, G. A. and Schulzberg, M. (1984) Immunohistochemical analysis of the ontogeny of neuropeptide Y immunoreactive neurons in the foetal rat brain. *Int. J. Dev. Neurosci.* **2,** 387–407.
154. Woodham, P. L., Allen, Y. S., McGovern, J., Allen, J. M., Bloom, S. R., Balazs, R., and Polak, J. M. (1985) Immunohistochemical analysis of the early ontogeny of the neuropeptide Y system in rat brain. *Neuroscience* **15,** 173–202.
155. Huntley, G. W., Hendry, S. H. C., Killackey, H. P., Chalupa, L. M., and Jones, E. G. (1988) Temporal sequence of neurotransmitter expression by developing neurons of fetal monkey visual cortex. *Dev. Brain Res.* **43,** 69–96.
156. Chun, J. J. M. and Shatz, C. J. (1989) The earliest-generated neurons of the cat cerebral cortex: characterization by MAP2 and neurotransmitter immunohistochemistry during fetal life. *J. Neurosci.* **9,** 1648–1667.
157. Friauf, E., McConnell, S. K., and Shatz, C. J. (1990) Functional synaptic circuits in the subplate during fetal and early postnatal development of cat visual cortex. *J. Neurosci.* **10,** 2601–2613.
158. Ghosh, A., Antonini, A., McConnell, S. K., and Shatz, C. J. (1990) Requirement for subplate neurons in the formation of thalamocortical connections. *Nature* **347,** 179–181.
159. Antonini, A. and Shatz, C. J. (1990) Relation beween putative transmitter phenotypes and connectivity of subplate neurons during cerebral cortical development. *Eur. J. Neurosci.* **2,** 744–761.

160. McConnell, S. K., Ghosh, A., and Shatz, C. J. (1989) Subplate neurons pioneer the first axon pathway from the cerebral cortex. *Science* **245,** 978–982.
161. Shatz, C. J., Chun, J. J. M., and Luskin, M. B. (1988) The role of the subplate in the development of the mammalian telencephalon, in Peters, A. and Jones, E. G., eds., *Cerebral Cortex, vol. 7, Development and Maturation of Cerebral Cortex.* Plenum, New York, pp. 35–58.
162. Terry, R. D. and Davies, P. (1980) Dementia of the Alzheimer type. *Ann. Rev. Neurosci.* **3,** 77–95.
163. Bowen, D. M., Smith, C. B., White, P., and Davison, A. N. (1976) Neurotransmitter related enzymes and indices of hypoxia in senile dementia and other abiotrophies. *Brain* **99,** 459–496.
164. Foster, N. L., Tamminga, C. A., and O'Donohue, T. L. (1986) Brain choline acetyltransferase and neuropeptide Y concentrations in Alzheimer's disease. *Neurosci. Lett.* **63,** 71–75.
165. Nakamura, S. and Vincent, S. R. (1986) Somatostatin- and neuropeptide Y-like immunoreactive neurons in the neocortex in senile dementia of Alzheimer's type. *Brain Res.* **370,** 11–20.
166. Davies, P., Katzman, R., and Terry, R. D. (1980) Reduced somatostatin-like immunoreactivity in cerebral cortex from cases of Alzheimer's disease and Alzheimer senile dementia. *Nature* **288,** 279–280.
167. Rossor, M. N., Emson, P. C., Mountjoy, C. Q., Roth, M., and Iversen, L. L. (1980) Reduced amounts of immunoreactive somatostatin in the temporal cortex in temporal cortex in senile dementia. *Neurosci. Lett.* **20,** 373–377.
168. Beal, M. F. and Martin, J. B. (1986) Neuropeptides in neurological disease. *Ann. Neurol.* **20,** 547–565.
169. Quirion, R., Martel, J. C., Robataille, Y., Etienne, P., Wood, P., Nair, N. P. V., and Gauthier, S. (1986) Neurotransmitter and receptor deficits in senile dementia of the Alzheimer's type. *Can. J. Neurol. Sci.* **13,** 503–510.
170. Allen, J. M., Cross, A. J., Crow, T. J., Javoy-Agid, F., Agid, Y., and Bloom, S. R. (1985) Dissociation of neuropeptide Y and somatostatin in Parkinson's disease. *Brain Res.* **337,** 197–200.
171. Blessed, G., Tomlinson, B. E., and Roth, M. (1968) The association between quantitative measures of dementia and of senile changes in the cerebral grey matter of elderly subjects. *Brit. J. Psychiat.* **114,** 797–811.
172. Dawbarn, D. and Emson, P. C. (1985) Neuropeptide Y-like immunoreactivity in neuritic plaques of Alzheimer's disease. *Biochem. Biophys. Res. Commun.* **126,** 289–294.
173. Armstrong, D. M., LeRoy, S., Shields, D. and Terry, R. D. (1985) Somatostatin-like immunoreactivity within neuritic plaques. *Brain Res.* **338,** 71–79.
174. Morrison, J. H., Rogers, J., Scherr, S., Benoit, R., and Bloom, F. E. (1985) Somatostatin immunoreactivity in neuritic plaques of Alzheimer's patients. *Nature* **314,** 90–92.
175. Armstrong, D. M., Benzing, W. C., Evans, J., Terry, R. D., Shields, D., and Hansen, L. A. (1989) Substance P and somatostatin coexist within neuritic placques: implications for the pathogenesis of Alzheimer's disease. *Neuroscience* **31,** 663–671.

176a. Perry, E. K., Tomlinson, B. E., Blessed, G., Perry, R. H., Cross, A. J., and Crow, A. J. (1981) Neuropathological and biochemical observations on the noradrenergic systems in aging and in Alzheimer's disease. *J. Neurol. Sci.* **51,** 278–287.

176b. Roberts, G. W., Crow, T. J., and Polak, J. M. (1985) Location of neuronal tangles in somatostatin neurones in Alzheimer's disease. *Nature* **314,** 92–94.
177a. Kosik, K. S., Joachim, C. L., and Selkoe, D. J. (1986) The microtubule-associates protein, tau, is a major antigenic component of paired helical filaments in Alzheimer's disease. *Proc. Natl. Acad. Sci. USA* **82,** 4531–4534.
177b. Armstrong, D. M., Bruce, G., Hersh, L. B., and Terry, R. D. (1986) Choline acetyltransferase immunoreactivity in neuritic placques of Alzheimer's brain. *Neurosci. Lett.* **71,** 229–234.
178. Dockray, G. J. (1986) Neuropeptide Y: in search of a function. *Neurochem. Int.* **1,** 9–11.
179. Maccarrone, C. and Jarrott, B. (1986) Neuropeptide Y: a putative neurotransmitter. *Neurochem. Int.* **8,** 13–22.
180. Kohler, C., Eriksson, L., Davies, S., and Chan-Palay, V. (1986a) Neuropeptide Y innervation of the hippocampal region in the rat and monkey brain. *J. Comp. Neurol.* **244,** 384–400.
181. Chan-Palay, V., Kohler, C., Haesler, U., Lang, W., and Yasargil, G. (1986) Distribution of neurons and axons immunoreactive with antisera against neuropeptide Y in the normal human hippocampus. *J. Comp. Neurol.* **248,** 360–375.
182. Amaral, D. G. (1978) A Golgi study of cell types in the hilar region of the hippocampus of the rat. *J. Comp. Neurol.* **182,** 851–914.
183. Kohler, C., Eriksson, L. G., Davies, S., and Chan-Palay, V. (1987a) Co-localization of neuropeptide tyrosine and somatostatin immunoreactivity in neurons of individual subfields of the rat hippocampal region. *Neurosci. Lett.* **78,** 1–6.
184. Alonso, A. and Kohler, C. (1982) Evidence for separate projectinos of hippocampal pyramidal and non-pyramidal neurons to different parts of the septum in the rat brain. *Neurosci. Lett.* **31,** 209–214.
185. Schwerdtfeger, W. K. and Buhl, E. H. (1986) Various types of non-pyramidal hippocampal neurons project to the septum and contralateral hippocampus. *Brain Res.* **386,** 145–154.
186. Seress, L. and Ribak, C. E. (1983) GABAergic cells in the dentate gyrus appear to be local circuit and projection neurons. *Exp. Brain Res.* **50,** 173–182.
187. Ribak, C. E., Seress, L., Peterson, G. M., Seroogy, K. B., Fallon, J. H., and Schmued, L. C. (1986) A GABAergic inhibitory component within the hippocampal commissural pathway. *J. Neurosci.* **6,** 3492–3498.
188. Germroth, P., Schwerdtfeger, W. K., and Buhl, E. H. (1989) GABAergic neurons in the entorhinal cortex project to the hippocampus. *Brain Res.* **494,** 187–192.
189. Citation withdrawn.
190. Chan-Palay, V. (1987) Somatostatin immunoreactive neurons in the human hippocampus and cortex shown by immunogold/silver intensification on Vibratome sections: coexistence with neuropeptide Y neurons, and effects in Alzheimer's dementia. *J. Comp. Neurol.* **260,** 201–223.
191. Lotstra, F., Schiffmann, S. N., and Vanderhaeghen, J.-J. (1989) Neuropeptide Y-containing neurons in the human infant hippocampus. *Brain Res.* **478,** 211–226.
193. Hjorth-Simonsen, A. and Jeune, B. (1972) Origin and termination of the hippocampal perforant path in the rat studied by silver impregnation. *J. Comp. Neurol.* **144,** 215–232.

194. Rosene, D. L. and Van Hoesen, G. W. (1987) The hippocampal formation of the primate brain: a review of some comparative aspects of cytoarchitecture and connections in cerebral cortex, vol. 6, Jones, E. G. and Peters, A., eds., Plenum, New York, pp. 345–456.
195. Kohler, C., Schultzberg, M., and Radesater, A.-C. (1987b) Distribution of neuropeptide Y receptors in the rat hippocampal region. *Neurosci. Lett.* **75,** 141–146.
196. Colmers, W. F., Lukowiak, K., and Pittman, Q. K. (1985) Neuropeptide Y reduces orthodromically evoked population spike in rat hippocampal CA1 by a possibly presynaptic mechanism. *Brain Res.* **346,** 404–408.
197. Kohler, C., Smialowska, M., Eriksson, L. G., and Chan-Palay, V. (1986b) Origin of the neuropeptide Y innervation of the rat retrohippocampal region. *Neurosci. Lett.* **65,** 287–292.
198. Marksteiner, J., Lassmann, H., Saria, A., Humpel, C., Meyer, D. K., and Sperk, G. (1990a) Neuropeptide levels after pentylenetetrazol. *Eur. J. Neurosci.* **2,** 98–103.
199. Marksteiner, J., Ortler, M., Bellman, R., and Sperk, G. (1990) Neuropeptide Y biosynthesis is markedly induced in mossy fibers during temporal lobe epilepsy of the rat. *Neurosci. Lett.* **112,** 143–148.
200. Meyer, D. K., Olenik, C., and Sperk, G. (1988) Chronic effects of systemic applications of kainic acid on mRNA levels of neuropeptides in rat brain. *Soc. Neurosci. Abs.* **14,** 6.
201. Yount, G. L., Gall, C. M., and White, J. D. (1989) Stimulation of hippocampal preproneuropeptide Y expression following recurrent seizure. *Soc. Neurosci. Abs.* **15,** 1274.
202. Wahlestedt, C., Blendy, J. A., Kellar, K. J., Heilig, M., Widerlov, E., and Ekman, R. (1990) Electroconvulsive shocks increase the concentration of neocortical and hippocampal neuropeptide Y (NPY)-like immunoreactivity in the rat. *Brain Res.* **507,** 65–68.
203. Dean, R. G. and White, B. D. (1990) Neuropeptide Y expression in rat brain: effects of adrenalectomy. *Neurosci. Lett.* **114,** 339–334.
204. de Lanerolle, N. C., Kim, J. H., Robbins, R. J., and Spencer, D. D. (1989) Hippocampal interneuron loss and plasticity in human temporal lobe epilepsy. *Brain Res.* **495,** 387–395.
205. Gall, C. (1990) Comparative anatomy of the hippocampus with special reference to differences in the distributions of neuroactive peptides, in *Cerebral Cortex,* vol. 8. Jones, E. G. and Peters, A., eds., pp.163–209.
206. Cronin, J. and Dudek, F. E. (1988) Chronic seizures and collateral sprouting of dentate mossy fibers after kainic acid treatment in rats. *Brain Res.* **474,** 181–184.
207. Sutula, T., Xiao-Xian, H., Cavacos, J., and Scott, G. (1988) Synaptic reorganization in the hippocampus induced by abnormal functional activity. *Science* **239,** 1147–1150.
208. Hyman, B. T., van Hoesen, G. W., Damasio, A. R., and Barnes, C. L. (1984) Alzheimer's disease: Cell specific pathology isolates the hippocampal formation. *Science* **225,** 1168–1170.
209. Chan-Palay, V., Lang, W., Haesler, U., Kohler, C., and Yasargil, G. (1986) Distribution of altered hippocampal neurons and axons immunoreactive

with antisera against neuropeptide Y in Alzheimer's-type dementia. *J. Comp. Neurol.* **248,** 376–394.
210. Gall, C., Seroogy, K. B., and Brecha, N. (1986) Distribution of VIP- and NPY-like immunoreactivities in rat main olfactory bulb. *Brain Res.* **374,** 389–394.
211. Scott, J. W., McDonald, J. K., and Pemberton, J. L. (1987) Short axon cells of the rat olfactory bulb display NADPH-diaphorase activity, neuropeptide Y-like immunoreactivity and somatostatin-like immunoreactivity. *J. Comp. Neurol.* **260,** 378–391.
212. Ohm, T. G., Braak, E., Probst, A., and Weindl, A. (1988) Neuropeptide Y-like immunoreactive neurons in the human olfactory bulb. *Brain Res.* **451,** 295–300.
213. Sanides-Kohlrausch, C. and Wahle, P. (1990) Morphology of neuropeptide Y-immunoreactive neurons in the cat olfactory bulb and olfactory peduncle: postnatal development and species comparison. *J. Comp. Neurol.* **291,** 468–489.
214. Matsutani, S., Senba, E., and Tohyama, M. (1988) Neuropeptide- and neurotransmitter-related immunoreactivities in the developing rat olfactory bulb. *J. Comp. Neurol.* **272,** 33–342.
215. Matsutani, S., Senba, E., and Tohyama, M. (1989) Distribution of neuropeptidelike immunoreactivities in the guinea pig olfactory bulb. *J. Comp. Neurol.* **280,** 577–586.
216. Price, J. L. and Powell, T. P. S. (1970) The mitral and short axon cells of the olfactory bulb. *J. Cell Sci.* **9,** 379–409.
217. Schneider, S. P. and Macrides, F. (1978) Laminar distributions of interneurons in the main olfactory bulb of the adult hamster. *Brain Res. Bull.* **3,** 73–82.
218. O'Leary, D. D. M. (1989) Do cortical areas emerge from a protocortex? *Trends in Neurosci.* **12,** 400–406.
219. Kostovic, I. and Rakic, P. (1980) Cytology and time of origin of interstitial neurons in the white matter in infant and adult human and monkey telencephalon. *J. Neurocytol.* **9,** 19–242.
220. Seroogy, K., Hokfelt, T., Buchan, A., Brown, J. C., Terenius, L., Norman, A. W., and Goldstein, M. (1989) Somatostatin-like immunoreactivity in rat main olfactory bulb: extent of coexistence with neuropeptide Y-, tyrosine hydroxylase- and vitamin D-dependent calcium binding protein-like immunoreactivities. *Brain Res.* **496,** 389–396.
221. Khorram, O., Roselli, C. E., Ellinwood, W. E., and Spies, H. G. (1987) The measurement of neuropeptide Y in discrete hypothalamic and limbic regions of male rhesus macaques with a human NPY-directed antiserum. *Peptides* **8,** 159–163.
222. Chronwall, B. M., DiMaggio, D. A., Massari, V. J., Pickel, V. M., Ruggiero, D. A., and O'Donohue, T. L. (1985) The anatomy of neuropeptide Y-containing neurons in rat brain. *Neuroscience* **15,** 1159–1181.
223. Kowall, N. W., Ferrante, R. J., Beal, M. F., and Martin, J. B. (1985) Characteristics, distribution and interrelationships of somatostatin, neuropeptide Y and NADPH-diaphorase in human caudate nucleus. *Soc. Neurosci. Abs.* **11,** 209.
224. Gaspar, P., Berger, B., Lesur, A., Borsotti, J. P., and Febvret, A. (1987) Somatostatin 28 and neuropeptide Y innervation in the septal area and related cortical and subcortical structures of the human brain. Distribution, relationships and evidence for differential coexistence. *Neuroscience* **22,** 49–73.

225. Graybiel, A. M. and Ragsdale, C. W., Jr. (1978) Histochemically distinct compartments in the striatum of human, monkey and cat demonstrated by acetylcholinesterase staining. *Proc. Natl. Acad. Sci. USA* **75,** 5723–5726.
226. Graybiel, A. M. (1983) Compartmental organization of the mammalian striatum. *Prog. Brain Res.* **58,** 239–245.
227. Gerfen, C. R. (1984) The neostriatal mosaic: Relationships among striatal input, output and peptidergic systems. *Nature* **311,** 461–464.
228. Graybiel, A. M. (1982) Correlative studies of histochemistry and fiber connections in the central nervous system, in *Cytochemical Methods in Neuroanatomy*. Palay, S. L. and Chan-Palay, V., eds., Alan Liss, New York, pp.45–67.
229. Smith, Y. and Parent, A. (1986) Neuropeptide Y-immunoreactive neurons in the striatum of cat and monkey: morphological characteristics, intrinsic organization and co-localization with somatostatin. *Brain Res.* **372,** 241–252.
230. Sandell, J. H., Graybiel, A. M., and Chesselet, M.-F. (1986) A new enzyme marker for striatal compartmentalization: NADPH diaphorase activity in the caudate nucleus and putamen of the cat. *J. Comp. Neurol.* **243,** 326–334.
231. Gerfen, C. R. (1985) The neostriatal mosaic. I. Compartmental organization of projections from the striatum to the substantia nigra in the rat. *J. Comp. Neurol.* **236,** 454–476.
232. Kemp, J. M. and Powell, T. P. S. (1970) The cortico-striate projection in the monkey. *Brain* **93,** 525–546.
233. Selemon, L. D. and Goldman-Rakic, P. S. (1985) Longitudinal topography and interdigitation of corticostriatal projections in the rhesus monkey. *J. Neurosci.* **5,** 776–794.
234. DiFiglia, M., Pasik, P., and Pasik, T. (1976) A Golgi study of neuronal types in the neostriatum of monkeys. *Brain Res.* **114,** 245–256.
235. Dimova, R., Vuillet, J., and Seite, R. (1980) Study of the rat neostriatum using a combined Golgi-electron microscope technique and serial sections. *Neuroscience* **5,** 1581–1596.
236. Vuillet, J., Kerkerian-Le Goff, L., Kachidian, P., Dusticier, G., Bosler, O., and Nieoullon, A. (1990) Striatal NPY-containing neurons receive GABAergic afferents and may also contain GABA: an electron microscopic study in the rat. *Eur. J. Neurosci.* **2,** 672–681.
237. Bolam, J. P., Powell, J. F., Wu, J. Y., and Smith, A. D. (1985) Glutamate deacrboxylase-immunoreactive structures in the rat neostriatum: a correlated light and electron microscopic study including a combination of Golgi impregnation with immunocytochemistry. *J. Comp. Neurol.* **237,** 1–20.
238. Penny, G. R., Afsharpour, S., and Kitai, S. T. (1986) The glutamate decarboxylase-, leucine enkephalin-, methionine enkephalin-, and substance P-immunoreactive neurons in the neostriatum of the rat and cat: evidence for partial population overlap. *Neuroscience* **17,** 1011–1045.
239. Oertel, W. H. and Mugnaini, E. (1984) Immunocytochemical studies of GABAergic neurons in rat basal ganglia and their relations to other neuronal systems. *Neurosci. Lett.* **47,** 233–238.
240. Kita, H. and Kitai, S. T. (1988) Glutamate decarboxylase immunoreactive neurons in rat neostriatum: their morphological types and populations. *Brain Res.* **447,** 346–352.

241. Kita, H., Kosaka, T., and Heizmann, C. W. (1990) Prvalbumin-immunoreactive neurons in the rat neostriatum: a light and electron microscopic study. *Brain Res.* **536,** 1–15.
242. McGeer, P. L., McGeer, E. G., Scherer, U., and Singh, K. (1977) A glutamergic corticostriatal path? *Brain Res.* **128,** 369–373.
243. Streit, P. (1984) Gutamate and aspartate as transmitter candidates for systems of the cerebral cortex, in *Cerebral Cortex*, vol. 2. Jones, E. G. and Peters, A., eds., Plenum, New York, pp. 119–143.
244. Lindvall, O. and Bjorklund, A. (1974) The organization of the ascending catecholamine neuron systems in the rat brain. *Acta. Physiol. Scand. (Suppl.)* **412,** 1–48.
245. Ribak, C. E., Vaughn, J. E., and Roberts, E. (1979) The GABA neurons and their axon terminals in rat corpus striatum as demonstrated by GAD immunocytochemistry. *J. Comp. Neurol.* **187,** 261–284.
246. Kubota, Y., Inagaki, S., Kito, S., Shimada, S., Okayama, T., Hatanaka, H., Pelletier, G., Takagi, H., and Tohyama, M. (1988) Neuropeptide Y-immunoreactive neurons receive synaptic inputs from dopaminergic axon terminals in the rat neostriatum. *Brain Res.* **458,** 389–393.
247. Vuillet, J., Kerkerian, L., Kachidian, P., Bosler, O., and Nieoullon, A. (1989a) Ultrastructural correlates of functional relationships between nigral dopaminergic or cortical afferent fibres and neuropeptide Y-containing neurons in the rat striatum. *Neurosci. Lett.* **100,** 99–104.
248. Vuillet, J., Kerkerian, L., Salin, P., and Nieoullon, A. (1989b) Ultrastructural features of NPY-containing neurons in the rat striatum. *Brain Res.* **477,** 241–251.
249. Massari, V. J., Chan, J., Chronwall, B. M., O'Donohue, T. L., Oertel, W. H., and Pickel, V. M. (1988) Neuropeptide Y in the rat nucleus accumbens: ultrastructural localization in aspiny neurons receiving synaptic input from GABAergic terminals. *J. Neurosci. Res.* **19,** 171–186.
250. Pelletier, G., Desy, L., Kerkerian, L., and Cote, J. (1984) Immunocytochemical localization of neuropeptide Y (NPY) in the human hypothalamus. *Cell Tiss Res.* **238,** 203–205.
251. Tatsuoka, Y., Riskind, P. N., Beal, M. F., and Martin, J. B. (1987) The effect of amphetamine on the in vivo release of dopamine, somatostatin neuropeptide Y from rat caudate nucleus. *Brain Res.* **411,** 200–203.
252. Kerkerian, L., Bosler, O., Pelletier, G., and Nieoullon, A. (1986) Striatal neuropeptide Y neurons are under the influence of the nigrostriatal dopaminergic pathway: immunohistochemical evidence. *Neurosci. Lett.* **66,** 106–112.
253. Kerkerian, L., Salin, P., and Nieoullon, A. (1988) Pharmacological characterization of dopaminergic influence on expression of neuropeptide Y immunoreactivity by rat striatal neurons. *Neuroscience* **26,** 809–817.
254. Lindefors, N., Brene, S., Herrera-Marschitz, M., and Persson, H. (1990) Neuropeptide gene expression in brain is differentially regulated by midbrain dopamine neurons. *Exp. Brain Res.* **80,** 489–500.
255. Kerkerian, L., Salin, P., and Nieoullon, A. (1990) Cortical regulation of striatal neuropeptide Y (NPY)- containing neurons in the rat. *Eur. J. Neurosci.* **2,** 181–189.

256. Salin, P., Kerkerian, L., and Nieoullon, A. (1990) Expression of neuropeptide Y immunoreactivity in the rat nucleus accumbens is under the influence of the dopaminergic mesencephalic pathway. *Brain Res.* **81,** 363–371.
257. Bruyn, G. W. (1968) Huntington's chorea: historical, clinical and laboratory synopsis, in *Handbook of Clinical Neurology, vol. 6, Diseases of the Basal Ganglia.* Vinken, P. J. and Bruyn, G. W., eds., Elsevier/North-Holland, Amsterdam, pp. 298–378.
258. Dawbarn, D., De Quidt, M. E., and Emson, P. C. (1985) Survival of basal ganglia neuropeptide Y-somatostatin neurones in Huntington's disease. *Brain Res.* **340,** 251–260.
259. Aronin, N., Cooper, P. E., Lorenz, L. J., Bird, E. D., Sagar, S. M., Leeman, S. E., and Martin, J. B. (1983) Somatostatin is increased in the basal ganglia on Huntington's disease. *Ann. Neurol.* **13,** 519–526.
260. Marshall, P. E. and Landis, D. M. M. (1985) Huntington's disease is accompanied by changes in the distribution of somatostatin-containing neuronal processes. *Brain Res.* **329,** 71–82.
261. Allen, Y. S., Bloom, S. R., and Polak, J. M. (1986) The neuropeptide Y-immunoreactive neuronal system: discovery, anatomy and involvement in neurodegenerative disease. *Human Neurobiol.* **5,** 227–234.
262. Beal, M. F., Kowall, N. W., and Ellison, D. W. (1986) Replication of the neurochemical characteristics of Huntington's disease by quinolinic acid. *Nature* **321,** 168–171.
263. Beal, M. F., Kowall, N. W., and Swartz, K. J. (1989) Differential sparing of somatostatin-neuropeptide Y and cholinergic neurons following striatal excitotoxic lesions. *Synapse* **3,** 38–47.
264. Uemura, Y., Kowall, N. W., and Beal, M. F. (1990) Selective sparing of NADPH-diaphorase-somatostatin-neuropeptide Y neurons in ischemic gerbil striatum. *Ann. Neurol.* **27,** 620–625.
265. Price, J. L., Russchen, F. T., and Amaral, D. G. (1987) The limbic region. II: The amygdaloid complex, in *The Handbook of Chemical Neuroanatomy, vol. 5. Integrated Systems of the CNS, Part 1. Hypothalamus, Hippocampus, Amygdala, Retina.* Bjorklund, A., Hokfelt, T., and Swanson, L. W., eds., Elsevier, New York, pp. 279–388.
266. Gustafson, E. L., Card, J. P., and Moore, R. Y. (1986) Neuropeptide Y localization in the rat amygdaloid complex. *J. Comp. Neurol.* **251,** 349–362.
267a. Crosby, E. C. and Humphrey, T. (1941) Studies of the vertebrate telencephalon: II. The nuclear pattern of the anterior olfactory nucleus, tuberculum olfactorium and the amygdaloid complex in adult man. *J. Comp. Neurol.* **74,** 309–352.
267b. McDonald, A. J. (1989) Coexistence of somatostatin with neuropeptide Y, but not with cholecystokinin or vasoactive intestinal polypeptide, in neurons of the rat amygdala. *Brain Res.* **500,** 37–45.
268a. Unger, J. W., McNeill, T. H., Lapham, L. L., and Hamill, R. W. (1988) Neuropeptides and neuropathology in the amygdala in Alzheimer's disease: relationship between somatostatin, neuropeptide Y and subregional distribution of neuritic plaques. *Brain Res.* **452,** 293–302.
268b. Jones, E. G. (1985) *The Thalamus.* Plenum, New York, p. 935.

269. Molinari, M., Hendry, S. H. C., and Jones, E. G. (1987) Distributions of certain neuropeptides in the primate thalamus. *Brain Res.* **426,** 270–289.
270. Citation withdrawn.
271. Oertel, W. H., Graybiel, A. M., Mugnaini, E., Elde, R. P., Schmechel, D. E., and Kopin, I. (1983) Coexistence of glutamic acid decarboxylase- and somatostatin-like immunoreactivity in neurons of the feline nucleus reticularis thalami. *J. Neurosci.* **3,** 1322–1332.
272. Ueda, S., Kawata, M., and Sano, Y. (1986) Identification of neuropeptide Y immunoreactivity in the suprachiasmatic nucleus and the lateral geniculate nucleus of some mammals. *Neurosci. Lett.* **68,** 7–10.
273. Moore, R. Y. (1989) The geniculohypothalamic tract in monkey and man. *Brain Res.* **486,** 190–194.
274. Card, J. P. and Moore, R. Y. (1982) Ventral lateral geniculate nucleus efferents to the rat suprachiasmatic nucleus exhibit avian pancreatic polypeptide-like immunoreactivity. *J. Comp. Neurol.* **206,** 390–396.
275. Card, J. P. and Moore, R. Y. (1989) Organization of lateral geniculate-hypothalamic connections in the rat. *J. Comp. Neurol.* **284,** 135–147.
276. Harrington, M. E., Nance, D. M., and Rusak, B. (1985) Neuropeptide Y immunoreactivity in the hamster geniculo-suprachiasmatic tract. *Brain Res. Bull.* **15,** 465–472.
277. Harrington, M. E., Nance, D. M., and Rusak, B. (1987) Double-labeling of neuropeptide Y-immunoreactive neurons which project from the geniculate to the suprachiasmatic nuclei. *Brain Res.* **410,** 275–282.
278. Swanson, L. W. (1987) *The hypothalamus, in Handbook of Chemical Neuroanatomy, vol. 5, Integrated Systems of the CNS, Part 1. Hypothalamus, Hippocampus, Amygdala and Retina.* Bjorklund, A., Hokfelt, T., and Swanson, L. W., eds., Elsevier, Amsterdam pp. 1–124.
279. Crosby, E. C. and Woodburne, R. T. (1940) The comparative anatomy of the preoptic area and the hypothalamus. *Proc. Assoc. Res. Nervous Mental Dis.* **20,** 52–169.
280. Le Gros Clark, W. E. (1938) Morphological aspects of the hypothalamus, in *The Hypothalamus. Morphological, Functional, Clinical and Surgical Aspects.* Le Gros, W. E., Clark, W. E., Beattie, J., Riddoch, G., Dott, N. M., eds., Oliver and Boyd, Edinburgh, pp. 2–68.
281. Bai, F. L., Yamano, M., Shiotani, Y., Emson, P. C., Smith, A. D., Powell, J. F., and Tohyama, M. (1985) An arcuato-paraventricular and dorsomedial hypothalamic neuropeptide Y-containing system which lacks noradrenaline in the rat. *Brain Res.* **331,** 172–175.
282. Sabatino, F. D., Murnane, J. M., Hoffman, R. A., and McDonald, J. K. (1987) Distribution of neuropeptide Y-like immunoreactivity in the hypothalamus of the adult golden hamster. *J. Comp. Neurol.* **257,** 93–104.
283. Reuss, S., Hurlbut, E. C., Speh, J. C., and Moore, R. Y. (1990) Neuropeptide Y localization in telencephalic and diencephalic structures of the ground squirrel brain. *Am. J. Anat.* **188,** 163–174.
284. McDonald, J. K., Koenig, J. I., Gibbs, D. M., Collins, P. and Noe, B. D. (1987) High concentrations of neuyopeptide Y in pituitary portal blood of rats. *Neuroendocrinology* **46,** 538–541.
285. Hu, H., Rao, J. K., Prasad, C., and Jayaraman, A. (1987) Localization of neuropeptide Y-like immunoreactivity in the cat hypothalamus. *Peptides* **8,** 569–573.

286. Leger, L., Charnay, Y., Danger, J.-M., Vaudry, H., Pelletier, G., Dubois, P.-M., and Jouvet, M. (1987) Mapping of neuropeptide Y-like immunoreactivity in the feline hypothalamus and hypophysis. *J. Comp. Neurol.* **255,** 283–292.
287. Muchlinski, A. E. and Johnson, D. J. (1983) Avian pancreatic polypeptide-like immunoreactivity in ground squirrel hypothalamus. *Brain Res. Bull.* **11,** 405–410.
288. Maccarrone, C., Jarrott, B., and Conway, E. L. (1986) Comparison of neuropeptide Y in hypothalamic and brainstem nuclei of young and mature spontaneously hypertensive and normotensive Wistar-Kyoto rats. *Neurosci. Lett.* **68,** 232–238.
289. Calza, L., Giardino, L., Battistini, N., Zanni, M., Galetti, S., Protopapa, F., and Velardo, A. (1989) Increase of neuropeptide Y-like immunoreactivity in the paraventricular nucleus of the fasting rats. *Neurosci. Lett.* **104,** 99–104.
290. Calza, L., Giardino, L. Zanni, M., Velardo, A., Parchi, P., and Marrama, P. (1990) Daily changes of neuropeptide Y-like immunoreactivity in the suprachiasmatic nucleus of the rat. *Reg. Peptides* **27,** 127–137.
291. Hooi, S. C., Richardson, G. S., McDonald, J. K., Allen, J. M., Martin, J. B., and Koenig, J. I. (1989) Neuropeptide Y (NPY) and vasopressin (AVP) in the hypothalamo-neurohypophysial axis of salt-loaded or Brattleboro rats. *Brain Res.* **486,** 214–220.
292. Abe, M., Saito, M., and Shimazu, T. (1990) Neuropeptide Y in the specific hypothalamic nuclei of rats treated neonatally with monosodium glutamate. *Brain Res. Bull.* **24,** 289–291.
293. Card, J. P. and Moore, R. Y. (1984) The suprachiasmatic nucleus of the golden hamster: Immunohistochemical analysis of cell and fiber distribution. *Neuroscience* **13,** 415–431.
294. Card, J. P. and Moore, R. Y. (1988) Neuropeptide Y localization in the rat suprachiasmatic nucleus and periventricular hypothalamus. *Neurosci. Lett.* **88,** 241–246.
295. Wahlestedt, C., Skagerberg, G., Ekman, R., Heilig, M., Sundler, F. and Hakanson, R. (1987) Neuropeptide Y (NPY) in the area of the hypothalamic paraventricular nucleus activates the pituitary-adrenocortical axis in the rat. *Brain Res.* **417,** 33–38.
296. Sahu, A., Kalra, S. P., Crowley, W. R., and Kalra, P. S. (1988) Evidence that NPY-containing neurons in the brainstem project into selected hypothalamic nuclei: implication in feeding behavior. *Brain Res.* **457,** 376–378.
297. Hisano, S., Chikamori-Aoyama, M., Katoh, S., Kagotani, Y., Daikoku, S., and Chihara, K. (1988) Suprachiasmatic nucleus neurons immunoreactive for vasoactive intestinal polypeptide have synaptic contacts with axons immunoreactive for neuropeptide Y: an immunoelectron microscopic study in the rat. *Neurosci. Lett.* **88,** 145–150.
298. Ibata, Y., Takahashi, Y., Okamura, H., Kubo, T., and Kawakami, F. (1988) Fine structure of NPY-containing neurons in the lateral geniculate nucleus and their terminals in the suprachiasmatic nucleus of the rat. *Brain Res.* **439,** 230–235.
299. Sawchenko, P. E., Swanson, L. W., Grzanna, R., Howe, P. R. C., Bloom, S. R., and Polak, J. M. (1985) Colocalization of neuropeptide Y immunoreactivity in brainstem catecholaminergic neurons that project to the paraventricular nucleus of the hypothalamus. *J. Comp. Neurol.* **241,** 138–153.

300. Edwards, G. L., Cunningham, J. T., Beltz, T. J., and Johnson, A. K. (1989) Neuropeptide Y-immunoreactive cells in the caudal medulla project to the median preoptic nucleus. *Neurosci. Lett.* **105,** 19–26.
301. Cassone, V. M., Speh, J. C., Card, J. P., and Moore, R. Y. (1988) Comparative anatomy of the mammalian hypothalamic suprachiasmatic nucleus. *J. Biol. Rhythms* **3,** 71–91.
302. Albers, H. E., Ferris, C. F., Leeman, S. E., and Goldman, B. D. (1984) Avian pancreatic polypeptide phase shifts hamster circadian rhythms when microinjected into the suprachiasmatic region. *Science* **223,** 833–835.
303. Johnson, R. F., Moore, R. Y., and Morin, L. P. (1989) Lateral geniculate lesions later circadian activity rhythms in the hamster. *Brain Res. Bull.* **22,** 411–422.
304. Moore, R. Y. and Card, J. P. (1985) Visual pathways and the entrainment of circadian rhythms, in *The Medical and Biological Effects of Light. Annals of the New York Academy of Science* vol. 453. Wurtman, R. J., Baum, M. J., and Potts, J. T., eds., pp. 123–133.
305. Pickard, G. E., Ralph, M. R., and Menaker, M. (1987) The intergeniculate leaflet partially mediates effects of light on circadian rhythms. *J. Bio. Rhythms* **2,** 35–56.
306. Moore, R. Y. and Card, J. P. (1990) Neuropeptide Y in the circadian timing system. *Ann. NY Acad Sci.* **511,** 247–257.
307. Beck, B., Jhanwar-Uniyal, M., Burlet, A., Chapleur-Chateau, M., Liebowitz, S., and Burlet, C. (1990) Rapid and localized alterations of neuropeptide Y in discrete hypothalamic nuclei with feeding status. *Brain Res.* **528,** 245–249.
308. Stanley, B. G., Chin, A. S., and Leibowitz, S. F. (1985) Feeding and drinking elicited by central action of neurpeptide Y: evidence for a hypothalamic sites(s) of action. *Brain Res. Bull.* **14,** 521–524.
309. Stanley, B. G., Daniel, D. R., Chin, A. S., and Liebowitz, S. F. (1985b) Paraventricular nucleus injections of peptide YY and neuropeptide Y preferentially enhance carbohydrate ingestion. *Peptides* **6,** 1205–1211.
310. Stanley, B. G., Kyrkouli, S. E., Lampert, S., and Liebowitz, S. F. (1986) Neuropeptide Y chronically injected into the hypothalamus: a powerful neurochemical inducer of hyperphagia and obesity. *Peptides* **7,** 1189–1192.
311. Stanley, B. G., Magdalin, W., and Liebowitz, S. F. (1989) A critical site for neuropeptide Y-induced eating lies in the caudolateral paraventricular/perifornical region of the hypothalamus. *Soc. Neurosci. Abstr.* **15,** 894.
312. Stanley, B. G. and Liebowitz, S. F. (1985) Neuropeptide Y injected in the paraventricular hypothalamus: a powerful stimulant of feeding behavior. *Proc. Natl. Acad. Sci. USA* **82,** 3940–3943.
313. Morley, J. E., Levine, A. S., Gosnell, B. A., Kneip, J., and Grace, M. (1987) Effect of neuropeptide Y on ingestive behaviors in the rat. *Am. J. Physiol.* **252,** R599–R609.
314. Pau, M. Y. C., Pau, K. Y. F., and Spies, H. G. (1988) Characterization of central actions of neuropeptide Y on food and water intake in rabbits. *Physiol. Behav.* **44,** 787–802.
315. Gibbs, J., Young, R. C., and Smith, G. P. (1973) Cholecystokinin decreases food intake in rats. *J. Comp. Physiol. Psychol.* **84,** 488–495.

316. Antin, J., Gibbs, J., Holt, J., Young, R. C., and Smith, G. P. (1975) Cholecystokinin elicits the complete behavioral sequence of satiety in rats. *J. Comp. Physiol. Psychol.* **89,** 784–790.
317. Rowland, N. E. (1988) Peripheral and central satiety factors in neuropeptide Y-induced feeding in rats. *Peptides* **9,** 989–992.
318. Pages, N., Gourch, A., Orosco, M., Comoy, E., Bohoun, C., Rodriguez, M., Martinez, J., Jacquot, C., and Cohen, Y. (1990) Changes in brain neuropeptide Y induced by cholecystokinin peptides. *Neuropeptides* **17,** 141–145.
319. Loren, I., Alumets, J., Hakanson, R., and Sundler, F. (1979) Distribution of gastrin and CCK-like peptides in the rat brain. *Histochemistry* **59,** 249–257.
320. Citation withdrawn.
321. Quirion, R., Martel, J.-C., Dumont, Y., Cadieux, A., Jolicoeur, F., St.-Pierre, S., and Fournier, A. (1990) Neuropeptide Y receptors: autoradiographic distribution in the brain and structure-activity relationships. *Ann. NY Acad. Sci.* **611,** 58–72.
322. Martel, J.-C., Fournier, A., St-Pierre, S., and Quirion, R. (1988) Comparative distribution of neuropeptide Y immunoreactivity and receptor autoradiography in rat forebrain. *Peptides* **9 (Suppl. 1),** 15–20.
323. Citation withdrawn.
324. Wahlestedt, C., Grundemar, L., Hakanson, R., Heilig, M., Shen, G. H., Zukowska-Grojec, Z., and Reis, D. J. (1990) Neuropeptide Y receptor subtypes, Y1 and Y2. *Ann. NY Acad. Sci.* **611,** 7–26.
325. Aguirre, J.A., Covenas, R., Dawid-Milner, M. S., Alonso, J. R., Garcia-Herdugo, G., and Gonzalez-Baron, S. (1989) Neuropeptide Y-like immunoreactivity in the brain stem respiratory nuclei of the cat. *Brain Res. Bull.* **23,** 201–207.
326. Blessing, W.W., Howe, P. R. C., Joh, T. H., Oliver, J. R., and Willoughby, J. O. (1986) Distribution of tyrosine hydroxylase and neuropeptide Y-like immunoreactive neurons in rabbit medulla oblongata, with attention to colocalization studies, presumptive adrenaline-synthesizing perikarya and vagal preganglionic cells. *J. Comp. Neurol.* **248,** 285–300.
327. Halliday, G. M., Li, Y. W., Oliver, J. R., Joh, T. H., Cotton, R. G. H., Howe, P. R. C., Geffen, L. B., and Blessing, W. W. (1988) The distribution of neuropeptide Y-like immunoreactive neurons in the human medulla oblongata. *Neuroscience* **26,** 179–191.
328. Everitt, B. J., Hokfelt, T., Terenius, L., Tatemoto, K., Mutt, V., and Goldstein, M. (1984) Differential coexistence of neuropeptide Y (NPY)-like immunoreactivity with catecholamines in the central nervous system of the rat. *Neuroscience* **11,** 443–462.
329. Holets, V. R., Hokfelt, T., Rokaeus, A., Terenius, L., and Goldstein, M. (1988) Locus coeruleus neurons in the rat containing neuropeptide Y, tyrosine hydroxylase or galanin and their efferent projections to the spinal cord, cerebral cortex and hypothalamus. *Neuroscience* **24,** 893–906.
330. Blessing, W. W., Oliver, J. R., Hodgson, A. H., Joh, T. H., and Willoughby, J. O. (1987) Neuropeptide Y-like immunoreactive C1 neurons in the rostral ventrolateral medulla of the rabbit project to sympathetic preganglionic neurons in the spinal cord. *J. Autonom. Nerv. Syst.* **18,** 121–129.

330. Hunt, S. P., Emson, P. C., Gilbert, R., Goldstein, M., and Kimmell, J. R. (1981b) Presence of avian pancreatic polypeptide-like immunoreactivity in catecholamine and methionine-enkephalin containing neurones within the central nervous system. *Neurosci. Lett.* **21,** 125–130.
331. Smialowska, M. (1988) Neuropeptide Y immunoreactivity in the locus coeruleus of the rat brain. *Neuroscience* **25,** 123–131.
333. Hokfelt, T., Lundberg, J. M., Tatemoto, K., Mutt, V., Terenius, L., Polak, J., Bloom, S., Sasek, C., Elde, R., and Goldstein, M. (1983) Neuropeptide y (NPY)—and FMRFamide neuropeptide-like immunoreactivities in catecholamine neurons of the rat medulla oblongata. *Acta Physiol. Scand.* **117,** 315–318.
334. Massari, V. J., Hornby, P. J., Friedman, E. K., Milner, T. A., Gillis, R. A., and Gatti, P. J. (1990) Distribution of neuropeptide Y-like immunoreactive perikarya and processes in the medulla of the cat. *Neurosci. Lett.* **115,** 37–42.
335. Wilcox, B. J. and Unnerstall (1990) Identification of a subpopulation of neuropeptide Y-containing locus coeruleus neurons that project to the entorhinal cortex. *Synapse* **6,** 284–291.
336. Hunt, S. P., Kelly, J. S., Emson, P. C., Kimmel, J. R., Miller, R. J., and Wu, J.-Y. (1981) An immunohistochemical study of neuronal populations containing neuropeptides or gamma-aminobutyrate within the superficial layers of the rat dorsal horn. *Neuroscience* **6,** 1883–1898.
337. Gibson, S. J., Polak, J. M., Allen, J. M., Adrian, T. E., Kelly, J. S., and Bloom, S. R. (1984) The distribution and origin of a novel brain peptide, neurpeptide Y, inthe spinal cord of several mammals. *J. Comp. Neurol.* **227,** 78–91.
339. Sasek, C. A. and Elde, R. P. (1985) Distribution of neuropeptide Y-like immunoreactivity and its relationship to FMRF-amide-like immunoreactivity in the sixth lumbar and first sacral cord segments of the rat. *J. Neurosci.* **5,** 1729–1739.
340. Llewellyn-Smith, I. J., Minson, J. B., Morilak, D. A., Oliver, J. R., and Chalmers, J. P. (1990) Neuropeptide Y-immunoreactive synapses in the intermediolateral cell column of rat and rabbit thoracic spinal cord. *Neurosci. Lett.* **108,** 243–248.
341. Newton, B. W. and Hamill, R. H. (1988) Neuropeptide Y immunoreactivity is preferentially located in rat lumbar sexually dimorphic nuclei. *Neurosci. Lett.* **94,** 10–16.
342. Morris, M. J., Pilowsky, P. M., Minson, J. B., West, M. J., and Chalmers, J. P. (1987) Microinjection of kainic acid into the rostral ventrolateral medulla causes hypertension and release of NPY-like immunoreactivity from rabbit spinal cord. *Clin. Exp. Pharmacol. Physiol.* **14,** 127–132.
343. Pilowsky, P. M., Morris, M. J., Minson, J. B., West, M. J., Chalmers, J. P., Willoughby, J. O., and Blessing, W. W. (1987) Inhibition of vasodepressor neurons in the caudal ventrolateral medulla of the rabbit increases both arterial pressure and the release of neuropeptide Y-like immunoreactivity from the spinal cord. *Brain Res.* **420,** 380–384.

PP, PYY, and NPY

Occurrence and Distribution in the Periphery

F. Sundler, G. Böttcher, E. Ekblad, and R. Håkanson

1. Introduction

Physiological events in peripheral organs are centrally as well as locally regulated. Central control is concerned with major functional adjustments, and is exercised mainly by parasympathetic and sympathetic nerves and by the hypothalamo-pituitary axis. The local regulation is thought to be concerned with precise functional adjustments according to local needs, and is exercised predominantly by endocrine/paracrine cells and local neurons. Endocrine/paracrine cells that secrete regulatory peptides are widely distributed in glands and in organs lined by epithelium, such as the airways, the gastrointestinal tract, and genitourinary tract. Local neurons occur in many organs of the body and are particularly well developed in the gastrointestinal tract, where they form the so-called enteric nervous system. Together, all these communication cells, whether they are concerned with remote or local regulation mechanisms, constitute the neuroendocrine system *(1–4)*. The various members of the neuroendocrine system employ similar modes of communication, and the messengers they use are either small molecules (amines, amino acids) or peptides of varying length.

From: *The Biology of Neuropeptide Y and Related Peptides;*
W. F. Colmers and C. Wahlestedt, Eds. © 1993 Humana Press Inc., Totowa, NJ

PPP Ala — Pro — Leu — Glu — Pro — Val — Tyr — Pro — Gly — Asp — Asn — Ala — Thr —
Pro — Glu — Gln — Met — Ala — Gln — Tyr — Ala — Ala — Glu — Leu — Arg — Arg —
Tyr — Ile — Asn — Met — Leu — Thr — Arg — Pro — Arg — Tyr — NH_2

PYY Tyr — Pro — Ala — Lys — Pro — Glu — Ala — Pro — Gly — Glu — Asp — Ala — Ser —
Pro — Glu — Glu — Leu — Ser — Arg — Tyr — Tyr — Ala — Ser — Leu — Arg — His —
Tyr — Leu — Asn — Leu — Val — Thr — Arg — Glu — Arg — Tyr — NH_2

NPY Tyr — Pro — Ser — Lys — Pro — Asp — Asn — Pro — Gly — Glu — Asp — Ala — Pro —
Ala — Glu — Asp — Leu — Ala — Arg — Tyr — Tyr — Ser — Ala — Leu — Arg — His —
Tyr — Ile — Asn — Leu — Ile — Thr — Arg — Glu — Arg — Tyr — NH_2

Fig. 1. The family of pancreatic polypeptide-related peptides. Amino acid sequences of porcine pancreatic polypeptide (PPP), peptide YY (PYY), and neuropeptide Y (NPY). Boxed amino acids have identical positions in at least two of the peptides.

The neuroendocrine system in the gut and pancreas is well developed; the endocrine/paracrine cells are of many different types, and the nervous control is complex and engages a large number of local neurons (for reviews, *see 2,5–8*). The candidate messengers are numerous; many of the messengers that occur in neurons and endocrine/paracrine cells in the gastroentero-pancreatic region are present also in the brain.

Regulatory peptides often constitute families. Examples of such families are the gastrin/cholecystokinin (CCK)-related peptides, the pancreatic polypeptide (PP)-related peptides (Fig. 1), the vasoactive intestinal peptide (VIP)-related peptides, and the tachykinins. Messengers in the neuroendocrine system often have a dual distribution in nerves and in endocrine/paracrine cells *(3)*. CCK, substance P, and somatostatin may serve as examples. Furthermore, it is usual that several messengers coexist in one and the same cell or neuron. The number of messengers that may coexist in certain endocrine/paracrine cells or neurons may be quite impressive (for reviews, *see 4,5,7,9,10*). The coexisting messengers (sometimes referred to as the chemical coding) in a given population of endocrine cells or neurons often differ from one species to another, between different age groups, and between cells or neurons that display widely differing activity. The functional significance of the chemical coding is as yet poorly understood. The following description of the distribution and properties of PP, neuropeptide Y (NPY), and peptide YY (PYY) illustrates the complexity of the peripheral neuroendocrine system.

2. Pancreatic Polypeptide (PP)

2.1. Historical Background

The first PP to be sequenced was the avian form *(11)*. It was isolated from side fractions in the purification of chicken insulin *(12–14)*. Bovine PP was isolated independently *(15,16)*. Both avian and bovine PP were found to contain 36 amino acid residues with a C-terminal tyrosine amide. Somewhat surprisingly, the two peptides were found to have identical amino acid residues at 16 positions only, with <50% sequence homology between them. Subsequently, the ovine, porcine, canine, and human peptides were isolated and found to differ from bovine PP by substitution of merely one or two residues *(16)*. By now, the PP sequence is known for a wide variety of mammalian and submammalian species, the most recent addition being that of PP in the European common frog *(17)*. The PP sequences of some representative vertebrates are given in Table 1.

Soon after the elucidation of the chemical structure of avian and bovine PP, antibodies became available. The initial immunocytochemical studies revealed avian PP in numerous pancreatic endocrine cells (PP cells) and a few scattered endocrine cells in the gastrointestinal tract *(18)*. The distribution of PP cells was found to differ markedly from that of other endocrine cells in the avian pancreas in that the PP cells occurred scattered in the exocrine parenchyma rather than being accumulated in islets. Studies in humans, on the other hand, showed PP to be localized in islet cells *(19)*.

2.2. The Pancreatic PP Cell

Once PP had been localized to pancreatic endocrine cells, it became of interest to identify the cell type. Using different histochemical staining techniques for the visualization of islet cells and a combination of light and electron microscopic immunocytochemistry, it became clear that the PP-containing cells were distinct from those storing insulin, glucagon, or somatostatin *(18–22)*. It was possible to characterize the PP cell with respect to size and morphology of the secretory granules. In the rat, the PP cell had small, round, electron-dense granules whereas in, e.g., the cat and dog, the PP-cell granules were much larger and less electron dense *(20)*. It was then realized that the PP cells in these latter species were identical to cells previously labeled F cells *(20,23,24)*.

Table 1
Amino Acid Sequence of Pancreatic Polypeptide from European Common Frog (FPP), Chicken (APP), Rat (RPP), and Humans (HPP)

FPP	Ala-Pro-Ser-Glu-Pro-His-His-Pro-Gly-Asp-Gln-Ala-Thr-Gln-Asp-Gln-Leu-Ala-
APP	Gly-Pro-Ser-Gln-Pro-Thr-Tyr-Pro-Gly-Asp-Asp-Ala-Pro-Val-Glu-Asp-Leu-Ile-
RPP	Ala-Pro-Leu-Glu-Pro-Met-Tyr-Pro-Gly-Asp-Asn-Ala-Thr-His-Glu-Gln-Arg-Ala-
HPP	Ala-Pro-Leu-Glu-Pro-Val-Tyr-Pro-Gly-Asp-Asn-Ala-Thr-Pro-Glu-Gln-Met-Ala-
FPP	Gln-Tyr-Tyr-Ser-Asp-Leu-Tyr-Gln-Tyr-Ile-Thr-Phe-Val-Thr-Arg-Pro-Arg-Phe-Nh_2
APP	Arg-Phe-Tyr-Asp-Asn-Leu-Gln-Gln-Tyr-Leu-Asn-Val-Val-Thr-Arg-His-Arg-Tyr-NH_2
RPP	Gln-Tyr-Glu-Thr-Gln-Leu-Arg-Arg-Tyr-Ile-Asn-Thr-Leu-Thr-Arg-Pro-Arg-Tyr-NH_2
HPP	Gln-Tyr-Ala-Ala-Asp-Leu-Arg-Arg-Tyr-Ile-Asn-Met-Leu-Thr-Arg-Pro-Arg-Tyr-NH_2

It may be mentioned in this context that the human pancreas seems to harbor two ultrastructurally distinguishable types of PP cells, one having fairly small electron-dense granules (as in, e.g., the rat) and the other having larger, more electron-lucent granules (as in, e.g., the cat and dog) *(25)*. The PP cells are not uniformly distributed within the pancreas. As a rule, they are more numerous in islets of the duodenal (head) portion than in islets of the splenic (tail) portion *(20,26)*. In this respect, the PP-cell frequency was found to be inversely correlated to the frequency of glucagon cells, which predominate in the splenic portion *(26,27)*. The different distribution of the PP and glucagon cells is thought to reflect the fact that the pancreas derives from two different primordia, one ventral and one dorsal, which fuse during embryonic development *(28)*. The ventral primordium gives rise to the PP-cell-rich duodenal portion (Fig. 2A), and the dorsal primordium to the glucagon-cell-rich splenic portion. In some species, notably the chicken, most of the PP cells occur scattered in the exocrine parenchyma (Fig. 2B) and in the ductal epithelium rather than within the islets *(18,20)*. In the rat and most other mammals, the PP cells are more or less confined to the islets where they often occur in the most peripheral cell layers. In, e.g., the human pancreas, it is common to find small islet-like clusters composed almost completely of PP cells.

2.3. PP Cells in the Gut

PP cells occur also in the gastrointestinal tract as demonstrated in the chicken *(18)*, particularly at hatching and a few weeks thereafter *(29)*. Also in mammals, such as the dog, PP cells occur, although in small numbers, in the gastrointestinal tract *(20,30)*. In the feline and canine stomach, PP cells are regularly seen both in the fundus and the antrum. In the rat, PP cells have been demonstrated in the antrum in small numbers during a short period postnatally *(31)*. PP does not seem to occur in neuronal elements, neither in the periphery nor in the brain.

2.4. Functional Aspects

Surprisingly little is known about the functional significance of PP. A role in glucose metabolism has been advocated (for a recent review, *see* 32). Such a role may be exercised via several mechanisms and pathways, including an action on the liver and on food intake. Further, PP is capable of inhibiting stimulated pancreatic

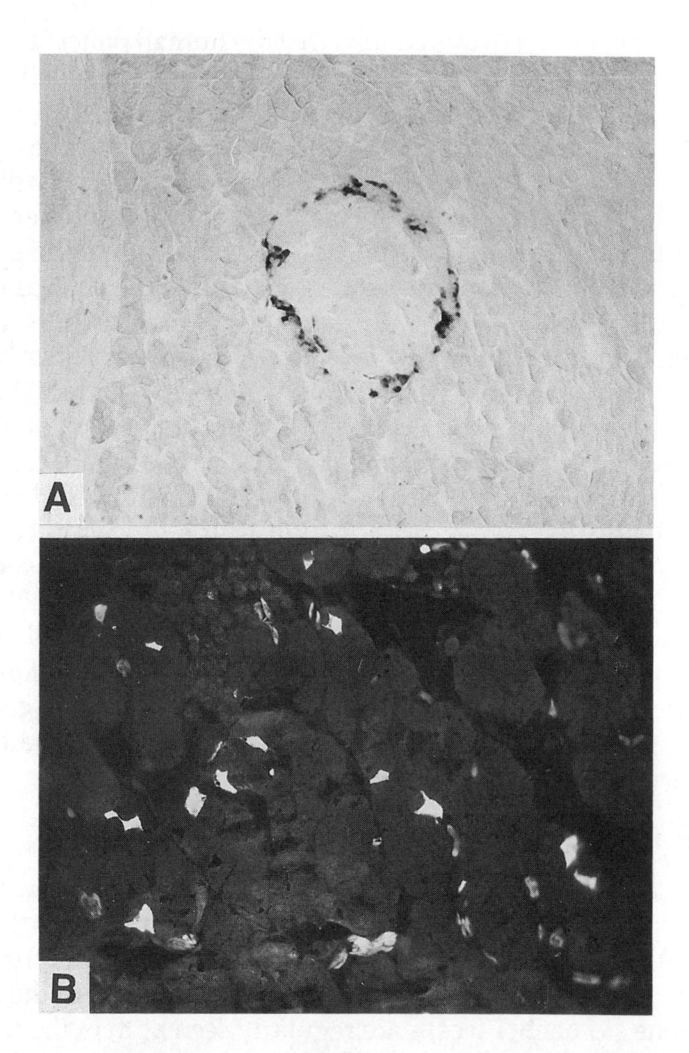

Fig. 2. PP-immunoreactive cells in the rat pancreas (duodenal portion) (A) and in the chicken pancreas (B). In the rat, PP cells are confined to the islet periphery; in the chicken, they occur scattered in the exocrine tissue.

enzyme secretion, reducing gall-bladder pressure, and augmenting the resistance of the choledochal sphincter (*33,34*; for a review, *see 35*). At least in the dog, these effects can be elicited at physiological blood concentrations *(34)*. In addition, PP doses within the physiological range inhibit the motility of the upper digestive tract. As studied in the chicken, PP is suspected of having a trophic effect on the proventriculus, which is the avian counterpart of the oxyntic gland area of the mammalian stomach *(36)*. Greenberg et al. *(37)* suggested that PP might be trophic for the exocrine pancreas.

3. Peptide YY (PYY)

3.1. Historical Background

The isolation of the 36 amino acid residue PYY from porcine upper small intestine was described by Tatemoto and Mutt *(38)* and by Tatemoto *(39)*, who used an assay strategy designed to detect C-terminally α-amidated peptides. Shortly afterward, the same strategy led to the discovery of a structurally related peptide—NPY—in porcine brain *(40,41)*. Displaying close structural resemblance to PP, both PYY and NPY were recognized as members of the "PP family."

Later, the amino acid sequences of murine, canine, and human PYY were established *(42–45)*, and the rat PYY precursor sequence was deduced from its corresponding mRNA/cDNA sequence *(46)*. The precursor organization, with a C-terminal flanking peptide that may be processed and secreted together with PYY, shows great similarity to that of the PP precursor *(47–49)*. In 1985, an additional PYY-like peptide was isolated from anglerfish pancreas by Andrews and coworkers. Anglerfish peptide Y (aPY) contains 37 amino acid residues; it is nonamidated with glycine as the C-terminal residue *(50a)*. More recently PYY homologues have been isolated from daddy sculpin islets *(50b)* and from the intestine of the European common frog *(50c)*.

3.2. Localization in Intestinal Endocrine Cells

PYY is found in endocrine cells in the intestines of a wide range of mammalian and submammalian species *(50–59)* (Fig. 3). PYY cells are particularly numerous in the distal ileum and in the colon/rectum. They are of the open type, flask-shaped, and with an accumulation of secretory granules at the base of the cell. At the electron microscopic level, the granules are relatively large, rounded, and homogeneously electron dense *(57,60,61)*.

In the colon, the PYY cells at times issue long, slender, basal processes, suggesting a paracrine role. PYY in the intestines is costored and probably cosecreted with glicentin (enteroglucagon) (Fig. 4). Nothing is known about the functional significance of the coexistence of PYY and glicentin.

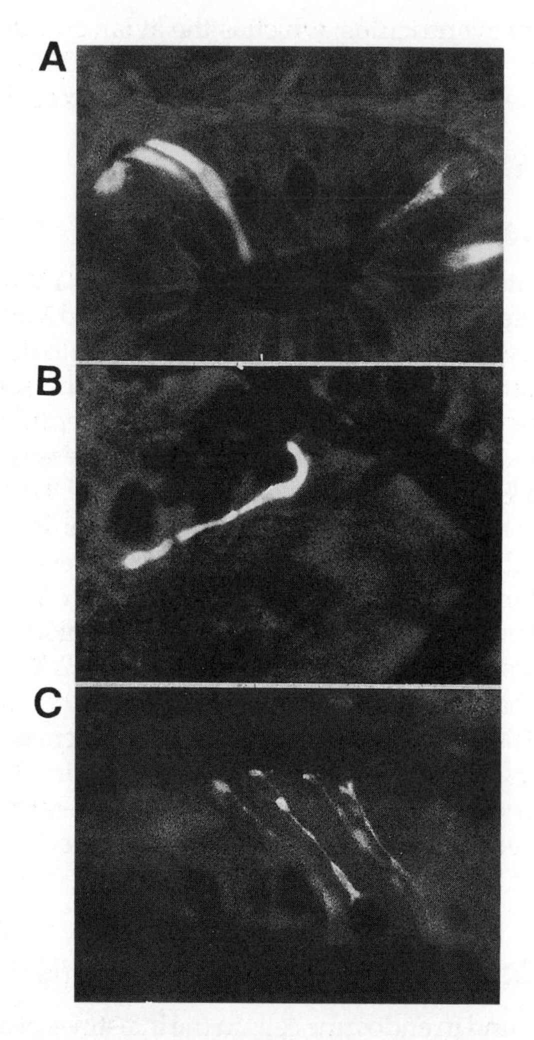

Fig. 3. PYY-immunoreactive endocrine cells in large intestine of humans (A), rat (B), and a bony fish, *Pleuronectes platessa* (C). Note elongated cell shape indicating contact both with the lumen and with the basal membrane.

3.3. Localization in Gastric Endocrine Cells

PYY-immunoreactive endocrine cells are regularly found in the gastric mucosa of mammals *(2,4,25,54,62)* and submammalian vertebrates *(59)*. In the antral mucosa of mammals, the PYY-immunoreactive cells are situated in the lower half (rodents) or the midportion (carnivores and humans) of the glands. The cells are

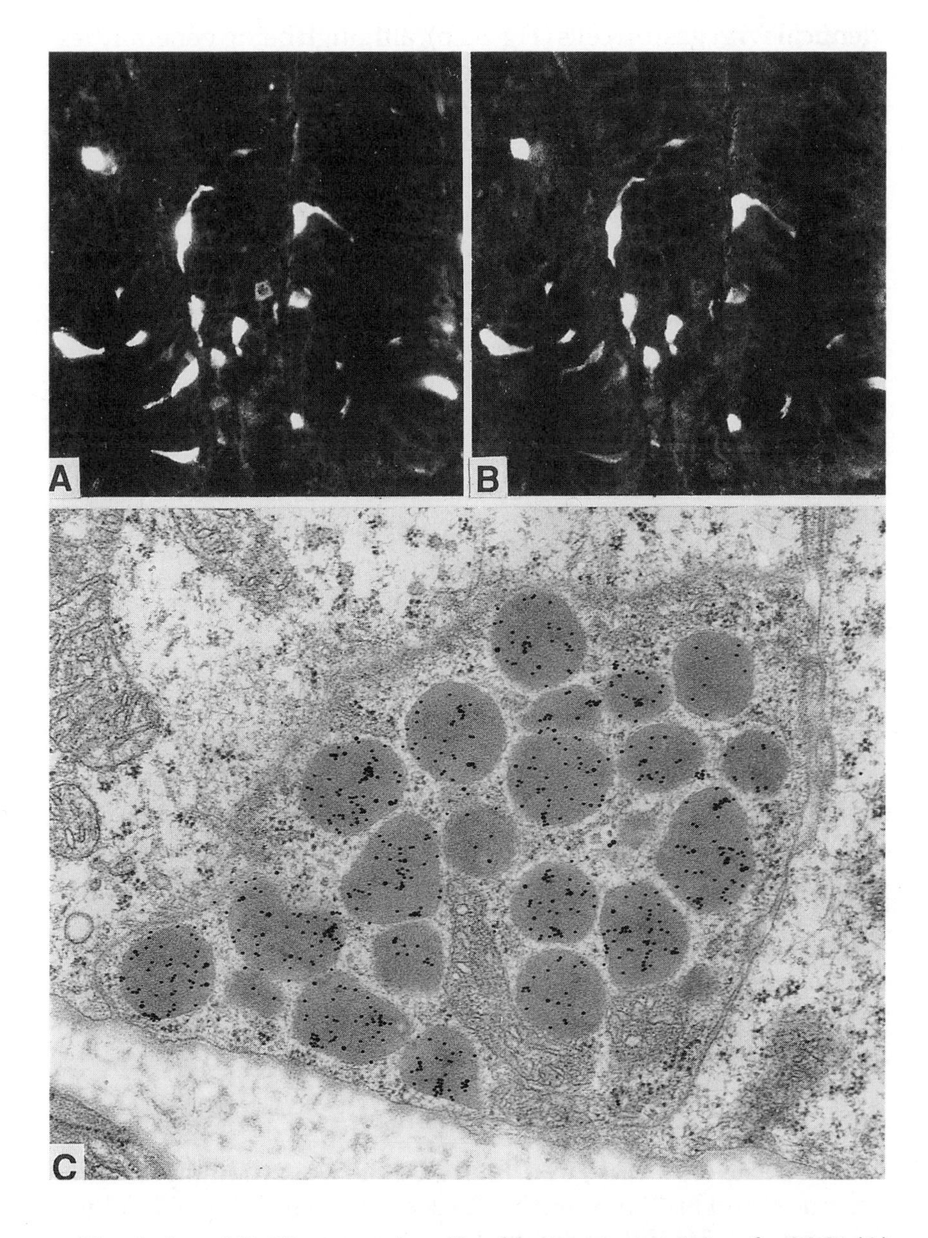

Fig. 4. A and B: Human colon. Double immunostaining for PYY (A) and glicentin (gut glucagon) (B). PYY-containing cells are identical with those storing glicentin. C. Immunogold staining for PYY on ultrathin section of cat colon. In the endocrine cell process, gold particles are accumulated over the secretory granules, which are comparatively large, rounded, and homogenously electron dense.

identical with gastrin cells (Fig. 5a,b), although being generally less numerous *(4,62,63)*. A very minor proportion of the PYY-immunoreactive cells in the antrum display somatostatin immunoreactivity. In the rat, the majority of PYY/gastrin-immunoreactive cells also contain immunoreactive GABA *(4)*.

In the acid-producing part of the stomach, PYY-immunoreactive cells are few, except in the mucosa of the minor curvature in the rat, where such cells are quite numerous. Most of the PYY-immunoreactive cells in the oxyntic mucosa display somatostatin immunoreactivity, as studied in the mouse, rat, and guinea pig (Fig. 5 c,d).

The presence of a PYY-like peptide in the gastric mucosa is supported by data from immunochemical analysis *(63–66)*. The elution profiles from HPLC separation suggest that only a proportion of the PYY-immunoreactive material in the stomach may be identical to authentic PYY *(63)*. However, additional support for the presence of "true" PYY is lended by the finding of PYY mRNA in extracts of rat antrum *(46)*.

3.4. Localization in Pancreatic Endocrine Cells

PYY (or a closely related peptide) occurs in the pancreas of mammalian as well as of submammalian species *(25,59,60,67–71)*. There is, however, some species variation in the cellular localization. Thus, PYY coexists with glucagon in rats and mice, whereas in many other species, PYY seems to be colocalized with PP (Fig. 6).

Available chemical data support the concept of pancreatic PYY *(52,71,72)*. Recently Krasinski and coworkers *(73)* measured PYY mRNA in the rat pancreas and intestine during ontogeny, and found peak levels in the pancreas shortly after birth, whereas in the intestine, PYY mRNA levels progressively increased up to adulthood.

3.5. Occurrence in Peripheral Neurons

Since its discovery, PYY (and PP) has been thought to be the hormonal and NPY the neuronal representative of the PP family of peptides *(74–76)*. Interestingly, however, a growing body of evidence points toward a neuronal role also for PYY. Thus, a PYY-like peptide is present in neuronal elements in the upper gut of mice, rats, cats, ferrets, and pigs. In the cat and ferret, PYY-immunoreactive nerve fibers are quite numerous in the myenteric ganglia of

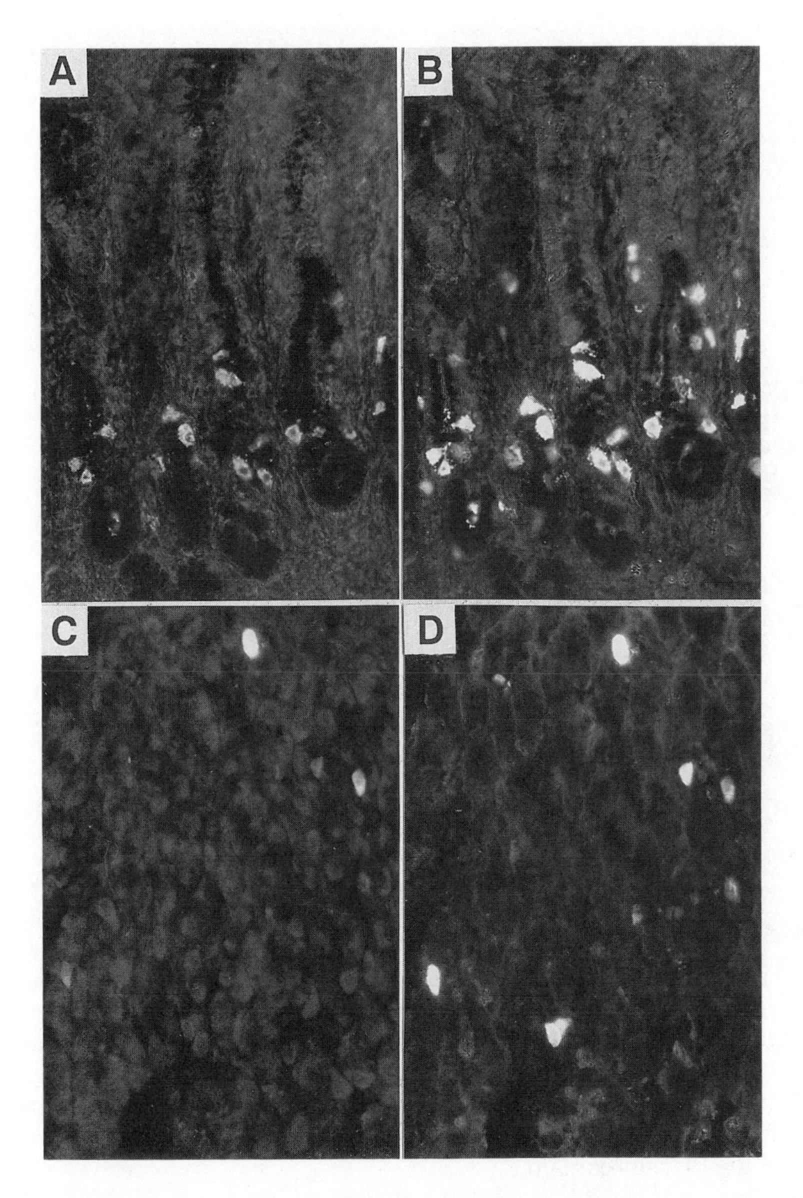

Fig. 5. Sections from the stomach of cat and guinea pig. A and B: Cat antrum mucosa. Double immunostaining for PYY (A) and gastrin (B). C and D: Guinea pig oxyntic mucosa. Double immunostaining for PYY (C) and somatostatin (D). Note PYY immunoreactivity in a major subpopulation of the gastrin cells in the antrum and in a minor subpopulation of the somatostatin cells in the oxyntic mucosa.

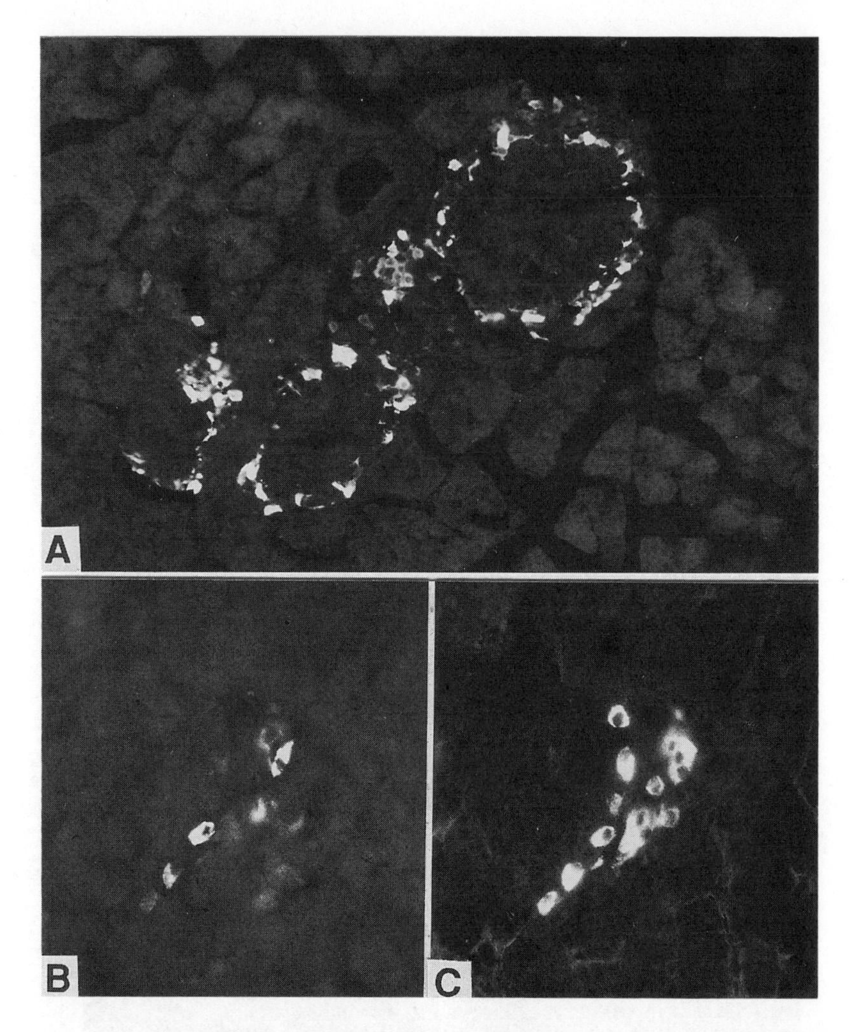

Fig. 6. A: Rat pancreas. PYY immunofluorescence in numerous cells in the periphery of islets. Most of these cells are identical with glucagon cells (not shown). B and C: Section from pig pancreas immunostained for PP (B) and, after elution of the antibodies, for PYY (C). A subpopulation of the PP cells contain PYY.

the stomach and upper small intestine *(63)* (Fig. 7). A moderate number of such fibers occur in the smooth muscle of these regions. In the mouse, rat, and pig, PYY-immunoreactive fibers seem to be restricted to a smooth-muscle bundle lining the minor side of the stomach close to the esophagogastric junction (Sundler et al., to be published). This muscle is supplied with ganglionic formations

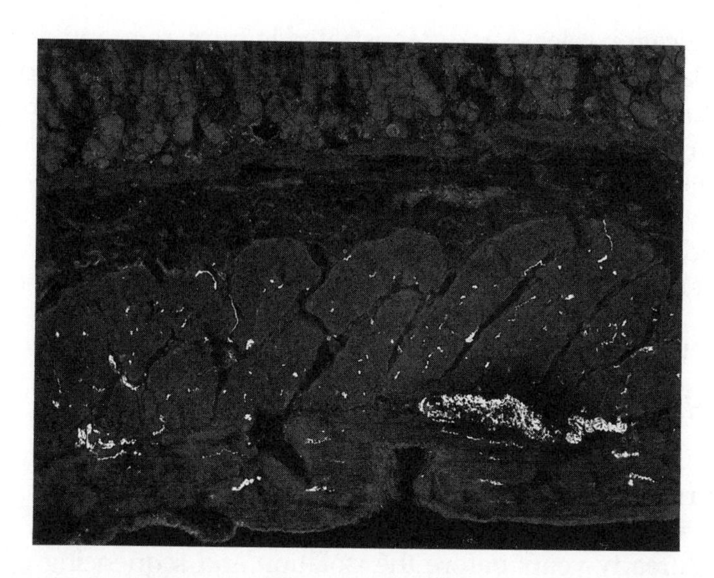

Fig. 7. Ferret stomach, oxyntic region. PYY immunofluorescent nerve fibers are densely accumulated in the myenteric ganglia (lower part of fig.) and occur scattered in the smooth muscle (middle part of fig.); they are absent from the mucosa (upper part of fig.).

on the serosal surface (and lacks traditional myenteric ganglia embedded in the muscle). PYY-containing nerve cell bodies are regularly seen in the serosal ganglia together with cell bodies containing other neuropeptides, such as VIP and gastrin-releasing peptide (GRP) *(77)*. A recent study has also shown PYY-like material in the rat superior cervical ganglia and in the sympathetic nerves in the submandibular salivary glands *(78)*.

3.6. Functional Aspects

PYY is released in response to a meal, and after intestinal perfusion with oleic acid or with glucose *(79–86)*. Increased plasma PYY levels are seen after vagal stimulation in the dog *(64)*, but not in the pig *(86)*, and after administration of GRP (gastrin-releasing peptide) *(86)* or cholecystokinin (CCK) *(87)*. PYY inhibits stimulated gastric and pancreatic secretion (*64,81,84,88–90*; cf. *91,92*). PYY also has inhibitory effects on gastric and intestinal motility *(52,82,93–95)*, and inhibits intestinal myoelectric activity *(96,97)*. A study in the cat has indicated a stimulatory action of PYY on rectal tone and anal canal pressure *(98)*. PYY exerts inhibitory effects on stimulated insulin and glucagon release as demonstrated in mice, rats, and dogs *(99–103)*. Basal insulin secretion is unaffected in humans *(104)* and stimulated in dogs *(100)*. Interestingly, plasma PYY concentrations are raised within 30 min after a meal, long before the luminal contents have reached the distal gut (cf *91*). Elevated plasma PYY levels have also been registered after intestinal resections, and a role for PYY in intestinal adaptation has been suggested. A role for PYY as a stimulant of proliferation in the intestinal epithelium has been proposed *(105)*.

4. Neuropeptide Y (NPY)

4.1. Historical Background

NPY is a tyrosine-rich peptide of 36 amino acid residues with a C-terminal α-amide group *(40)*. It was first isolated from porcine brain, and found to display 50% homology with PP and 70% with PYY. Already years before the isolation and sequencing of NPY, the presence of a PP-like peptide in neuronal elements in the brain and in the periphery had been demonstrated by immunocytochemistry *(106)*. When NPY antibodies became available, it became clear that the PP-like peptide in both the brain and the periphery was identical to NPY, which was found to have a widespread distribution throughout the body. Based on the very large number of immunocytochemical studies published since then, it can be concluded that NPY occurs in a subpopulation of sympathetic neurons and in subpopulations of parasympathetic or local neurons, notably in the enteric nervous system *(3,7,107)*. NPY-immunoreactive fibers are numerous around blood vessels. However, not all blood vessels are surrounded by NPY fibers, nor are NPY fibers confined to blood vessels. Such fibers occur also in nonvascular smooth muscle, around acini of exocrine glands and beneath surface epithelia.

In addition, NPY is a constituent of the adrenal medulla *(108)*. In some species, such as cat and cow, the NPY-containing cells seem to be identical with those storing noradrenaline *(109,110)*. In other species, notably the rat, NPY seems to coexist with adrenaline (*111–113*: for a review, *see 114*).

4.2. NPY in Sympathetic Neurons

Noradrenaline (NA) is the classical transmitter of postganglionic sympathetic nerves. However, immunocytochemistry has shown a number of peptides to coexist with NA in subpopulations of sympathetic neurons *(115–118)*. The most prominent neuropeptide in the sympathetic nervous system is NPY. Immunocytochemical investigations have demonstrated NPY in perivascular nerve fibers throughout the body *(114)* (Fig. 8A,B). Most perivascular NPY-immunoreactive fibers seem to have a sympathetic origin. Simultaneous immunostaining for NPY and tyrosine hydroxylase (TH) or dopamine-β-hydroxylase (DBH), enzymes present in sympathetic neurons, has provided evidence that NA and NPY coexist in many nerve cell bodies in sympathetic ganglia and in numerous perivascular sympathetic nerve fibers *(67,107,118–120)*. In fact, NPY has been detected in about 40–50% of the superior cervical and stellate ganglion cells and in nearly 80% of the coeliac ganglion cells *(67,118)*, indicating that NPY occurs in a major subpopulation of the NA-containing sympathetic ganglion cells (Fig. 8C). Extirpation of sympathetic ganglia or chemical symphathectomy (treatment with 6-hydroxydopamine) causes a reduction in the number of NPY-immunoreactive nerve fibers around cerebral blood vessels and a reduction of the NPY concentration by 70% in cerebral and cranial blood vessels *(121)*. Interestingly, it has been shown that although sympathetic fibers to blood vessels usually contain NPY, sympathetic fibers to exocrine glands and fat cells do not *(117,118,122)*. Consequently, they represent two different populations of sympathetic neurons that differ with respect to their chemical coding (Fig. 9). Indeed, for practical purposes, NPY may serve as a marker for those sympathetic fibers that innervate the cardiovascular system. However, there are exceptions to this rule. In the guinea pig uterine artery, for instance, NPY has been found to coexist with NA in one population of nerve fibers and with VIP in another *(123)*. The NPY/VIP-containing axons are distinct from the sympathetic fibers.

Clearly, NPY is costored with NA in many sympathetic nerve fibers, especially those that surround blood vessels. This has been demonstrated in many species, including humans *(67,114,117–120,124)*. Perivascular NPY-containing fibers are particularly numerous around arteries and arterioles. Large elastic arteries have a sparser innervation than large muscular arteries. As the diameter of the muscular arteries decreases, the density of the NPY innervation increases.

On the whole, arteries are more densely innervated than veins *(114,119)*. NPY-immunoreactive nerve fibers are notably numerous around cerebral arteries *(121,124)*. The rostal part of the circle of

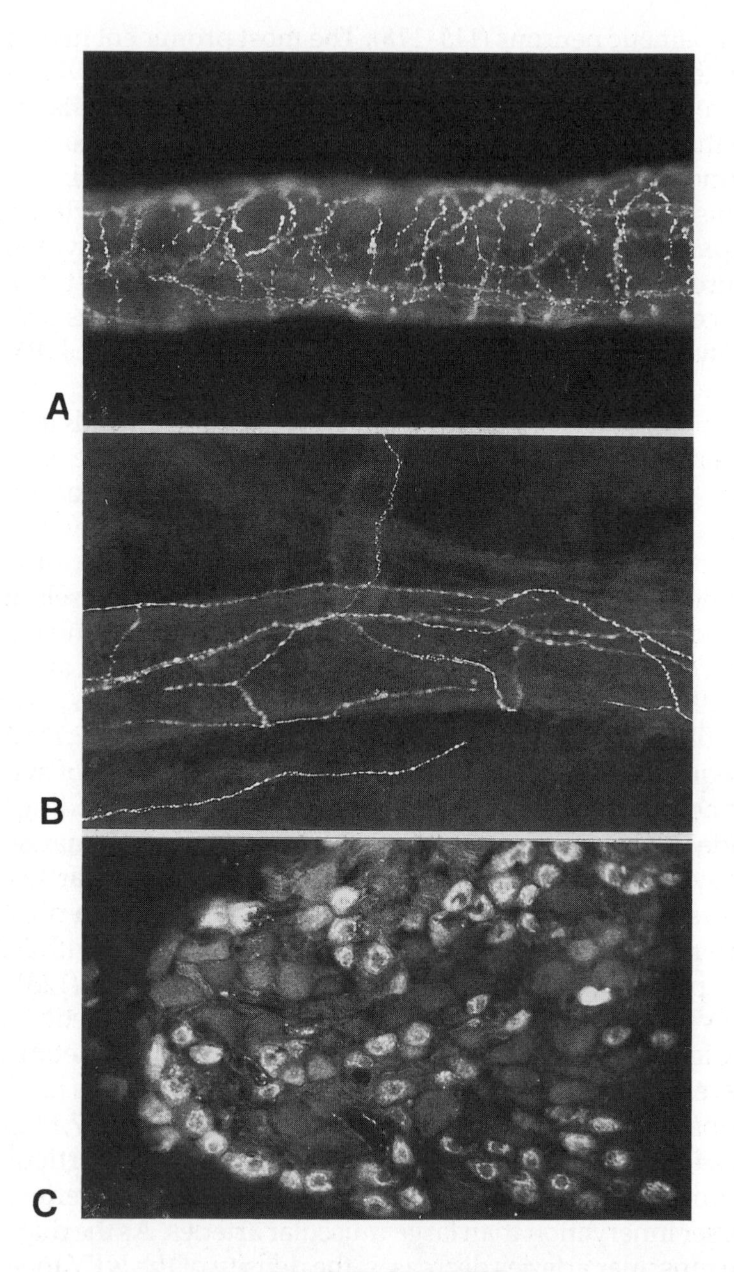

Fig. 8. A and B: Perivascular NPY-containing nerve fibers in basilar artery (A) and testicular artery (B) of guinea pig. Whole mount preparations. C: NPY-containing nerve cell bodies in a sympathetic ganglion (superior cervical ganglion of rat). NPY is contained in a major subpopulation of the nerve cell bodies, which are known to produce noradrenaline.

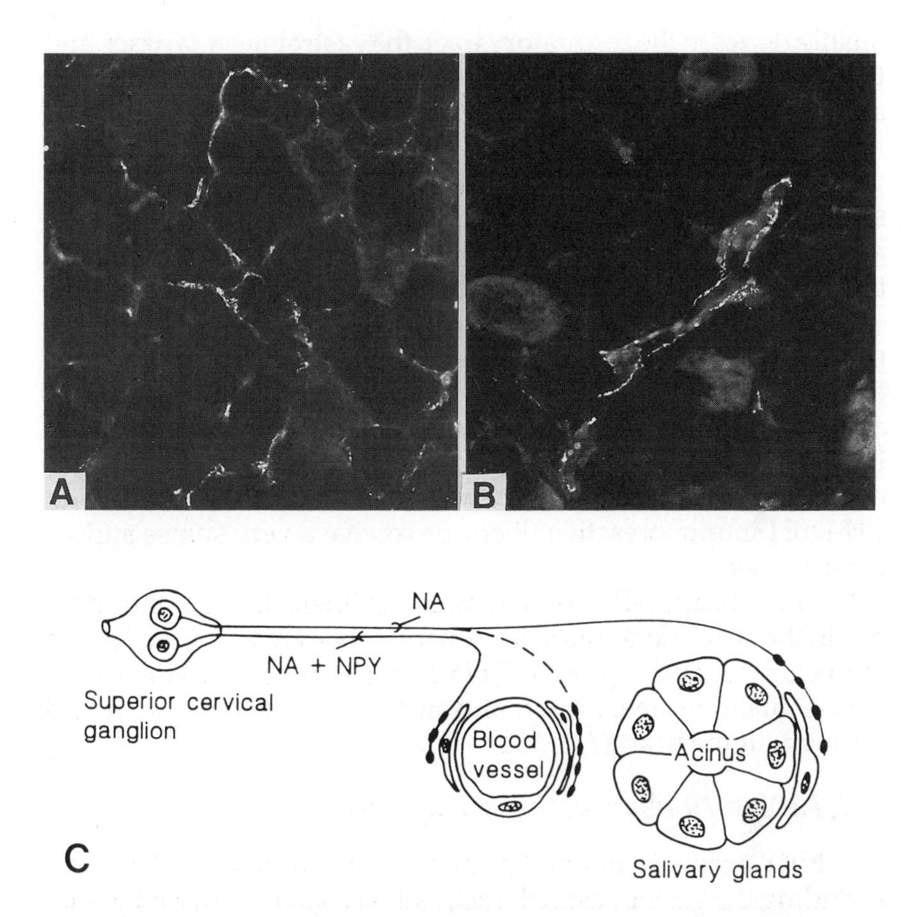

Fig. 9. Example of pathway-specific chemical coding. A and B: Submandibular gland of ferret. Sections immunostained for noradrenergic nerve fibers using dopamine-β-hydroxylase antiserum (A) and for NPY, which is also contained in noradrenergic fibers (B). C: Schematic drawing to illustrate the sympathetic innervation. The submandibular gland of the ferret (and the cat) receives noradrenergic nerve fibers from the superior cervical ganglion. Those that innervate blood vessels also contain NPY; those that innervate acini do not. (For a different situation, *see* Fig. 12).

Willis appears to be more densely supplied with NPY-containing fibers than the caudal part and the basilar and vertebral arteries *(121)*. NPY-containing fibers are also numerous around coronary arteries *(119,120,125)*. Beside the cerebral arteries and coronary arteries, the thyroid artery and the uterine and spermatic arteries are among those that have a conspicuously rich NPY innervation.

Small arteries in the respiratory tract, the gastrointestinal tract, and the genitourinary tract also receive a rich supply. In the liver, spleen, and kidney, NPY-containing perivascular fibers are few. Large veins have a sparse innervation of NPY fibers, whereas medium-sized veins receive a somewhat richer supply. NPY fibers are few around small veins and venules. The findings described above are reminiscent of earlier observations on the adrenergic innervation of blood vessels *(126–128)*.

Immunocytochemical studies have demonstrated NPY in perivascular nerve fibers also in skeletal muscle *(129)*. Such fibers are particularly numerous around small arteries and arterioles, and there is an almost complete overlap between NPY- and TH/DBH-immunoreactive fibers, confirming the anticipated coexistence of NPY and NA *(129)*. The veins and venules lack NPY- and TH-DBH-immunoreactive fibers, or receive a very sparse supply of such fibers.

In the heart, NPY-containing sympathetic fibers are numerous in the atria and the auricles, whereas they are less numerous in the ventricles and septum *(67,118,120,130)*. Also nonsympathetic NPY-containing fibers are found in the conduction system and throughout the heart *(131–134)*.

4.3. NPY in Nonsympathetic Neurons

NPY occurs in nonsympathetic neurons in several organs, including the gastointestinal tract, salivary glands, thyroid gland, pancreas, urogenital system, and airways (for a review, *see* 107).

4.3.1. Gastrointestinal Tract

NPY-immunoreactive nerve fibers are numerous throughout the gastrointestinal tract in all layers of the wall (Fig. 10). The esophagus receives a dense network of NPY fibers *(135–138)*. In the stomach, NPY-containing fibers occur mainly in the smooth muscle and myenteric ganglia, although they are present also in the mucosa and submucosa (around blood vessels and at the basal portion of the glands) *(136,139–142)*. Immunoreactive nerve cell bodies are found in the myenteric ganglia of the esophagus and stomach *(7,136,142)*. In the small intestine, NPY-containing nerve fibers are numerous in all layers and around blood vessels. The dense mucosal innervation in this part of the gut is particularly notable, and immunoreactive nerve cell bodies are numerous in

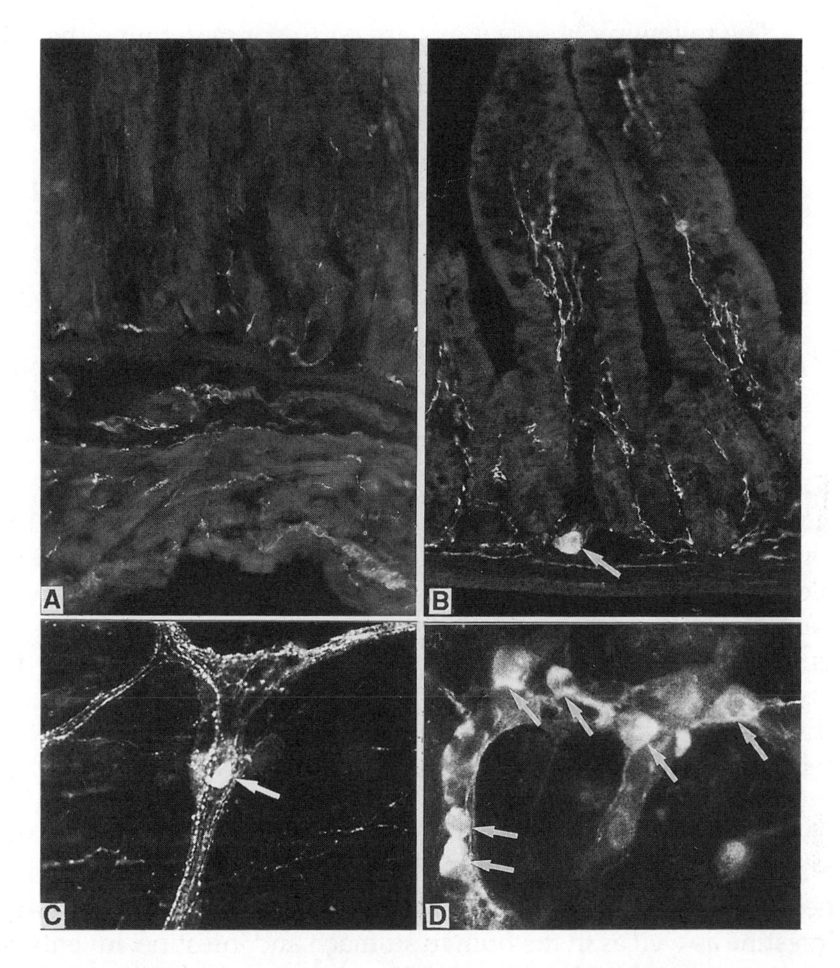

Fig. 10. NPY-immunoreactive nerve fibers in rat stomach (A) and small intestine (B–D). In the stomach, NPY fibers are numerous in the smooth muscle and myenteric ganglia, whereas they are fewer in the mucosa. In the small intestine, numerous NPY fibers are found in the smooth muscle, mucosa, and intramural ganglia. NPY-immunoreactive nerve cell bodies are found in both submucous (arrows in B and D) and in myenteric ganglia (arrow in C).

both submucous and myenteric ganglia *(7,139,140,143–145)*. The topography of the NPY innervation of the large intestine resembles that of the small intestine, except that the fibers are less numerous *(7,137,139,140,146)*. The perivascular NPY innervation of the gut is generally dense, and the distribution resembles that of the sympathetic nerve fibers.

The following observations are pertinent in defining the origin of NPY-containing fibers in the gut wall:

1. Intramural ganglia are rich in NPY-immunoreactive nerve cell bodies;
2. Extrinsic denervation causes a loss of perivascular NPY-immunoreactive fibers, whereas the bulk of enteric NPY fibers remains *(139,144,146,147)*;
3. Perivascular NPY fibers contain TH and DBH *(139,148,149)*; and
4. Chemical sympathectomy (6-hydroxydopamine) eliminates the NPY-containing perivascular nerves as well as a few NPY-immunoreactive fibers in the myenteric ganglia *(139,143,147)*.

Together these observations suggest that NPY-containing nerve fibers in the gut derive from at least two different sources (Fig. 10). One population is the sympathetic nerve fibers that supply mainly blood vessels. Another population is intrinsic to the gut (enteric neurons), and supplies mainly the mucosa and smooth muscle. In guinea pig small intestine, the submucous as well as some of the myenteric NPY-immunoreactive neurons also contain immunoreactive choline acetyl transferase (ChAT), CCK, calcitonin gene-related peptide (CGRP), and somatostatin *(150a)*, whereas other myenteric NPY-storing nerve cell bodies contain dynorphin, enkephalin, and VIP *(5)*. In the pig, rat, and mouse small intestine, the whole population of myenteric and submucous NPY-containing neurons stores VIP *(150b)* (Fig. 11). In the rat stomach and large intestine as well as in the human stomach and intestine, myenteric neurons contain both NPY and VIP, whereas submucous neurons contain either NPY or VIP *(137,142,146,151,152)*. Microsurgical techniques have been used to outline the projections of NPY-containing enteric neurons (*139; see also* 5 and 7). In the guinea pig small intestine, most myenteric NPY-containing neurons project anally for approx 2 mm to the circular muscle and other myenteric neurons. Some myenteric NPY-containing neurons (those storing NPY/ChAT/CCK/CGRP/somatostatin), as well as the submucous NPY-containing neurons, project to the mucosa *(150)*. In the guinea-pig large intestine, myenteric NPY-containing neurons give off ascending oral projections of about 24 mm in length *(153)*. In the rat intestine, NPY- (and VIP-) containing myenteric neurons project anally to other myenteric neurons and to smooth muscle. These projec-

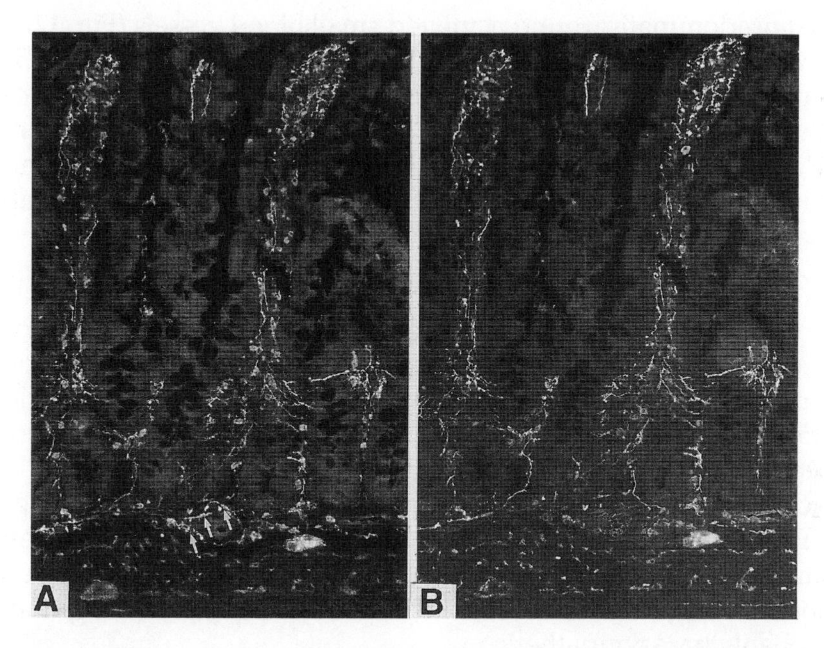

Fig. 11. Cryostat section from rat small intestine double immunostained for NPY (A) and VIP (B). Note complete coexistence of NPY and VIP in mucosa, smooth muscle, and intramural ganglia except that perivascular NPY fibers (arrows) do not contain VIP.

tions are approx 2 mm in length in the small intestine *(144)* and approx 4 mm in the large intestine *(146)*. The submucous NPY- (and VIP-) containing neurons in rat small intestine issue ascending projections, approx 4 mm in length *(146)*, whereas no oroanal projections could be detected in the large intestine *(146)*. In both locations, the submucous neurons supply the mucosa/submucosa and other submucous neurons. The canine small intestine harbors NPY-immunoreactive myenteric neurons that issue orally directed, 30 mm long fibers to reach other myenteric neurons and circular muscle *(154)*.

4.3.2. Salivary Glands

NPY-containing nerve fibers are numerous in the salivary glands of the rat. One population of such fibers is coarse and intensely immunoreactive, and predominates around blood vessels and ducts. These fibers contain TH and are thus probably noradrenergic. Another population consists of delicate, fine-beaded fibers that predominate around acini and small blood vessels (Fig. 12A). These fibers lack TH and are thus probably nonadrenergic (*107,155*; Sundler et al., unpublished observations). Immunocytochemical double staining has revealed the latter fibers to store also VIP and SP, and to originate in parasympathetic ganglia, i.e., the otic ganglion and (Fig. 12B.C) small ganglia in the hilus region of the submandibular gland *(3,107,155–157)*.

4.3.3. Thyroid Gland

Apart from a rich and intensely NPY-immunoreactive nerve fiber network around blood vessels, there is a population of more delicate NPY-containing fibers predominating around the follicles *(158,159)*. The perivascular fibers contain DBH indicating a noradrenergic nature, whereas some of those distributed around the follicles lack DBH and are thus probably nonadrenergic. The nonadrenergic NPY-containing fibers also store VIP and seem to emanate from small ganglia located close to the thyroid capsule *(160)*. These nonadrenergic NPY/VIP-containing neurons of local origin are probably parasympathetic.

4.3.4. Respiratory Tract

NPY-containing nerve fibers are widely distributed in the respiratory tract, including the nasal mucosa and the middle ear (*67,161–163*; for reviews, *see 164–165*). In the nasal mucosa and in the wall of the Eustachian tube, NPY-containing fibers are numerous around blood vessels and moderate in number around acini of seromucous glands. In the tracheobronchial wall, NPY-containing fibers are fairly numerous around small blood vessels and glandular acini, and within smooth muscle. Most of the NPY-containing nerve fibers contain the adrenergic markers TH and DBH, and are thus classified as noradrenergic. As studied in some detail in the rat *(156)*, ferret *(166)*, and pig *(167,168)*, there is evidence for a population of nonadrenergic NPY-containing neurons supplying the nasal mucosa and the tracheobronchial mucosa and smooth muscle. In the nasal mucosa, the NPY-containing nonadrenergic fibers predominate in the seromucous glands and seem to emanate from the sphenopalatine ganglion *(157)*, whereas those in the tracheobronchial wall probably originate in small ganglia located within the tracheal wall. Frequent coexistence with VIP has been noticed in the nonadrenergic NPY-containing fibers in the airways, and a

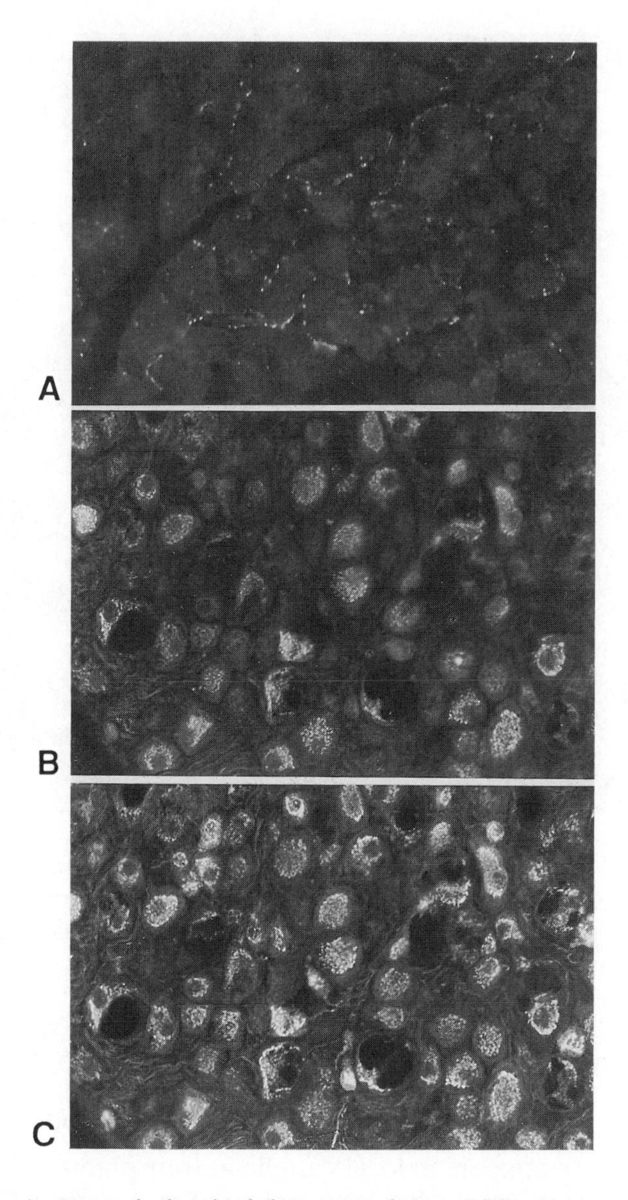

Fig. 12. A. Parotid gland of the rat. Delicate, NPY-containing nerve fibers are distributed around many of the glandular acini. Most of these fibers are nonadrenergic, and contain also VIP and SP (not shown). B and C. Otic ganglion, which is the source of these nonadrenergic NPY-containing fibers. Double immunostaining for NPY (B) and VIP (C). Many of the cell bodies contain both NPY and VIP, some contain VIP but lack NPY.

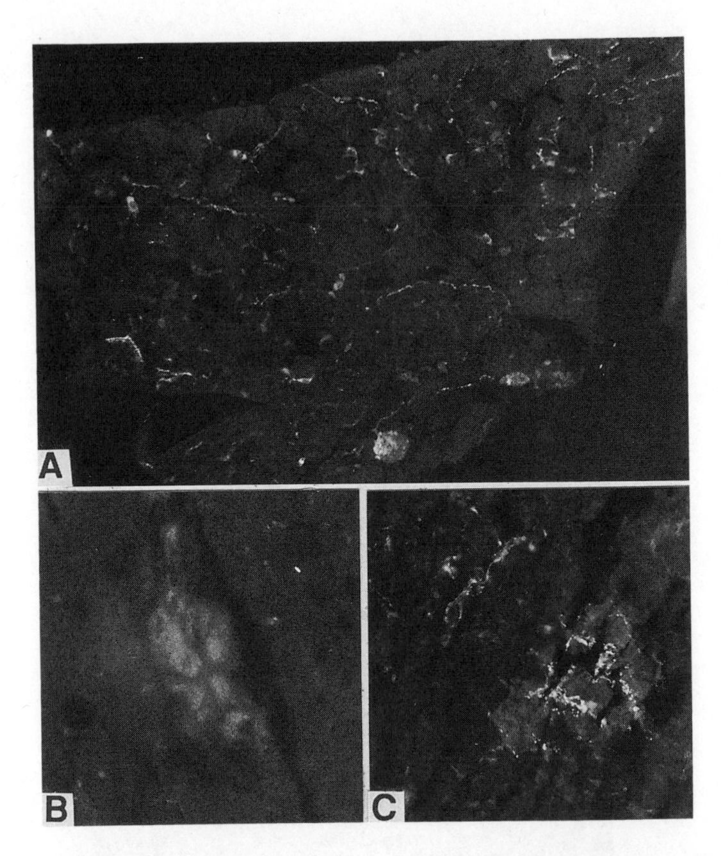

Fig. 13. A and B. Guinea-pig pancreas. A. NPY-containing nerve fibers occur scattered in the exocrine parenchyma. B. Small intrapancreatic ganglion harboring NPY-containing nerve cell bodies. These nerve cells are nonadrenergic and indicated the presence of nonadrenergic NPY-containing fibers also in the pancreas. C. Dog pancreas. Numerous NPY-containing nerve fibers within an islet. These fibers are noradrenergic and contain also galanin (for details, *see 8*).

parasympathetic nature of these neurons is highly likely. The fact that NPY appears in local ganglia in transplanted lungs indicates that these ganglia have the potential to express NPY also in humans *(169,170)*.

4.3.5. Pancreas

NPY-immunoreactive nerve fibers have been demonstrated in the pancreas in a number of species *(8,148)* (Fig. 13A). In addition to a rich supply of NPY-immunoreactive nerve fibers around blood vessels, such fibers were found around both ducts and acini in the exocrine parenchyma as well as within the islets. Intrapancreatic ganglionic formations were found to contain NPY-immunoreactive cell bodies (Fig. 13B). As studied in the rat, sympathectomy (surgical or 6-hydroxydopamine treatment) caused the disappearance of NPY-fibers around blood vessels and within the pancreatic islets *(119,147,148)*. In canine pancreatic islets, NPY-immunoreactive nerve fibers are numerous (Fig. 13C), and most of these fibers also contain TH and galanin *(171)*. Thus, pancreatic NPY-immunoreactive fibers are of two kinds: sympathetic and nonsympathetic (parasympathetic).

4.3.6. Urogenital Tract

NPY-containing nerve fibers occur in the kidneys *(172)* as well as in the urinary bladder *(173)*. In the kidney, the NPY-containing nerve fibers are mainly adrenergic, whereas in the urinary bladder, nonadrenergic NPY fibers prevail. The detrusor muscle is particularly rich in nonadrenergic NPY-containing nerve fibers. The pelvic ganglia harbor numerous NPY-immunoreactive nerve cell bodies. Since postganglionic denervation eliminates all NPY-containing nerve fibers in the smooth muscle of the bladder wall, whereas 6-hydroxydopamine treatment is virtually ineffective, the pelvic ganglia are the potential source of the NPY fibers in the urinary bladder *(173)*.

In the ovary and oviduct of several species, a large number of NPY-containing nerve fibers can be demonstrated. Sequential staining for DBH and NPY together with results of 6-hydroxydopamine treatment revealed that NPY is stored in noradrenergic nerves *(174)*. In the guinea-pig uterus, a subpopulation of the adrenergic nerves contains NPY. During pregnancy, these NPY-containing nerve fibers degenerate. This is followed by a restoration after parturition *(175)*. However, in the rat uterine cervix, only perivascular NPY-containing fibers seem to be noradrenergic, whereas NPY fibers in the myometrium are not *(176)*. A possible origin for the nonadrenergic NPY fibers is the paracervical ganglia, which contain neurons immunoreactive to both NPY and VIP *(177)*.

Generally speaking, perivascular NPY-containing nerve fibers seem to be sympathetic. One exception is, however, the uterine artery, which in a number of species, including humans, harbors

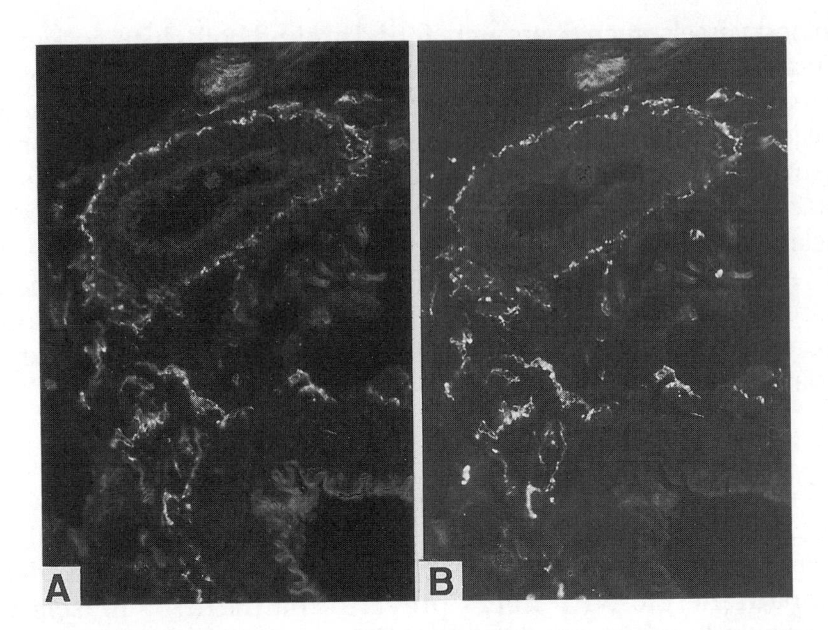

Fig. 14. Guinea-pig uterine artery. Cross-section double immunostained for NPY (A) and VIP (B). All perivascular NPY fibers contain also VIP.

nonadrenergic NPY-containing nerve fibers *(123,178,179)*. In addition to NPY, other peptides such as VIP, dynorphin, and somatostatin, have been found to be colocalized in these nerve fibers (Fig. 14).

4.4. Functional Aspects

The effects of NPY that are best studied are those concerned with vasoconstriction and interaction with NA. NPY has important pre- and postjunctional effects at the sympatho-effector junction *(180)*. Presynaptically, it inhibits transmitter release from sympathetic nerve endings. Postsynaptically, it has a rather poor direct contractile effect on most isolated peripheral blood vessel preparation. However, it potentiates the response to nerve stimulation and to other vasopressor agents *(119,180a,180b)*. Interestingly, the vasopressor effect of NPY in vivo is quite strong. Intravital microscopy revealed that NPY was highly potent in contracting the arterioles in the vascular bed of skeletal muscle. NPY was in fact ten times more potent than NA *(181)*. The high vasoconstrictor potency of NPY in this vascular bed is interesting in view of the fact that the arterioles of the skeletal muscle are likely to contribute greatly to the total peripheral vascular resistance.

In the guinea-pig ileum, NPY is thought to inhibit motor activity through a neuromodulatory effect, which may reflect a suppressed release of acetylcholine *(182,183)*. The effect of NPY on electrogenic ion transport has been studied in the rat intestinal epithelia *(184)*. NPY causes a reduction in the short-circuit current. This effect is insensitive to the neurotoxin tetrodotoxin or to α-adrenoreceptor antagonists, and is therefore thought to reflect a direct effect on the epithelial cells. Net chloride absorption is increased, whereas no effect on sodium movements has been detected.

5. Concluding Remarks

It is at present possible, using highly specific antibodies, to demonstrate selectively the cellular localization of PP, PYY, and NPY, by immunocytochemistry. The functional properties of the three peptides are, however, much less distinguishable. Particularly, there is a prominent overlap in the pharmacological actions of PYY and NPY in, e.g., the gut (for a review, *see 185*). Thus, many of the data given in this chapter under Section 3.6. could equally well have been placed under the corresponding NPY section (Section 4.4.). Additional studies are needed to evaluate whether NPY or PYY is the physiologically relevant ligand for the Y receptors at a given place.

Acknowledgments

Grant support was from Swedish MRC (Proj. No. 4499 and 1007), Swedish Diabetes Association, Nordic Insulin Foundation, Hoechst Diabetes Foundation, Crafoord's Foundation, Wiberg's Foundation, Påhlsson's Foundation, Bergvall's Foundation, and the Medical Faculty, University of Lund.

References

1. Håkanson, R. and Sundler, F. (1983) The design of the neuroendocrine system: a unifying concept and its consequences. *Trends Pharmacol. Sci.* **4,** 41–44.

2. Sundler, F. and Håkanson, R. (1988) Peptide hormone-producing endocrine/-paracrine cells in the gastro-entero-pancreatic region, in *Handbook of Chemical Neuroanatomy, vol. 6. The Peripheral Nervous System*. Björklund, A., Hökfelt, T., and Owman, C., eds. Elsevier, Amsterdam. pp. 219–295.
3. Sundler, F., Böttcher, G., Ekblad, E., and Håkanson, R. (1989) The neuroendocrine system of the gut. *Acta Oncol.* **28,** 303–314.
4. Sundler, F., Ekblad, E., and Håkanson, R. (1991) The neuroendocrine system of the gut—an update. *Acta Oncol.* **30,** 419–427.
5. Furness, J. B. and Costa, M. (1987) *The Enteric Nervous System.* Churchill Livingstone, Edinburgh.
6. Solcia, E., Capella, C., Buffa, K., Usellini, L., Fiocca, R., and Sessa, F. (1987) Endocrine cells of the digestive system, in *Physiology of the Gastrointestinal Tract.* 2nd ed. Johnsson, L. R., ed., Raven, New York, pp. 111–130.
7. Ekblad, E., Håkanson, R., and Sundler, F. (1991) Microanatomy and chemical coding of peptide-containing neurons in the digestive tract, in *Neuropeptide Function in the Gastrointestinal Tract.* Daniel, E. E., ed., CRC Press, Boca Raton, FL, pp. 131–179.
8. Sundler, F. and Böttcher, G. (1991) Islet innervation, with special reference to neuropeptides, in *The Endocrine Pancreas.* Samols, E., ed., Raven, New York, pp. 29–52.
9. Lundberg, J. M. and Hökfelt, T. (1983) Coexistence of peptides and classical neurotransmitters. *Trends Neurosci.* **6,** 325–333.
10. Håkanson, R., Böttcher, G., Ekblad, E., Grunditz, T., and Sundler, F. (1990) Functional implications of messenger coexpression in neurons and endocrine cells, in *Neuropeptides and Their Receptors* Schwartz, T. W., Hilsted, L. M., and Rehfeld, J. F., eds. Alfred Benzon Symp. **29,** Munksgaard, Copenhagen, pp. 211–232.
11. Kimmel, J. R. and Pollock, H. G. (1971) A new pancreatic polypeptide hormone. *Fed. Proc.* **30,** 1318 (abs.).
12. Kimmel, J. R., Pollock, H. G., and Hazelwood, R. L. (1968) Isolation and characterization of chicken insulin. *Endocrinology* **83,** 1323–1330.
13. Langslow, D. R., Kimmel, J. R., and Pollock, H. G. (1973) Studies on the distribution of a new avian pancreatic polypeptide and insulin among birds, reptiles, amphibians and mammals. *Endocrinology* **93,** 558–565.
14. Kimmel, J. R., Hayden, L. J., and Pollock, H. G. (1975) Isolation and characterization of a new pancreatic polypeptide hormone. *J. Biol. Chem.* **250,** 9369–9376.
15. Lin, T. M. and Chance, R. E. (1972) Spectrum of gastrointestinal actions of a new bovine pancreas polypeptide (BPP) *Gastroenterology* **62,** 852 (abs.).
16. Lin, T. M. and Chance, R. E. (1974) Pancreatic polypeptide in mammalian tissues, in *Endocrinology of the Gut.* Chey, W. Y. and Brooks, F.P., eds., Charles B. Slack Inc., Thorofare, NJ, pp. 143–145.
17. McKay, D. M., Shaw, C., Thim, L., Johnston, C. F., Halton, D. W., Fairweather, I., and Buchanan, K. D. (1990) The complete primary structure of pancreatic polypeptide from the European common frog *Rana temporaria. Regul. Peptides* **31,** 187–198.
18. Larsson, L.-I., Sundler, F., Håkanson, R., Pollock, H. G., and Kimmel, J. R. (1974) Localization of APP, a postulated new hormone, to a pancreatic endocrine cell type. *Histochemistry* **42,** 377–382.

19. Larsson, L.-I., Sundler, F., and Håkanson, R. (1975) Immunohistochemical localization of human pancreatic polypeptide (HPP) to a population of islet cells. *Cell Tissue Res.* **156,** 167–171.
20. Larsson, L.-I., Sundler, F., and Håkanson, R. (1976) Pancreatic polypeptide—a postulated new hormone: identification of its cellular storage site by light and electron microscopic immunocytochemistry. *Diabetologia* **12,** 211–226.
21. Baetens, D., DeMey, J., and Gepts, W. (1977) Immunohistochemical and ultrastructural identification of the pancreatic polypeptide-producing cell (PP-cell) in the human pancreas. *Cell Tissue Res.* **185,** 239–246.
22. Schweisthal, M. R., Schweisthal, J. V., and Frost, C. C. (1978) Localization of human pancreatic polypeptide in an argyrophil fourth cell type in islets of the rat pancreas. *Am. J. Anat.* **152,** 257–282.
23. Forssmann, W. G., Helmstädter, V., Metz, J., Greenberg, J., and Chance, R. E. (1977) The identification of the F-cell in the dog pancreas as the pancreatic polypeptide producing cell. *Histochemistry* **50,** 281–290.
24. Greider, M. H., Gersell, D. J., and Gingerich, R. L. (1978) Ultrastructural localization of pancreatic polypeptide in the F cell of the dog pancreas. *J. Histochem. Cytochem.* **26,** 1103–1108.
25. Solcia, E., Fiocca, R., Capella, C., Usellini, L., Sessa, F., Rindi, G., Schwartz, T. W., and Yanaihara, N. (1985) Glucagon- and PP-related peptides of intestinal L cells and pancreatic/gastric A or PP cells. Possible interrelationships of peptides and cells during evolution, fetal development and tumor growth. *Peptides* **6 Suppl. 3,** 223–229.
26. Orci, L., Baetens, D., Ravazzola, M., Stefan, Y., and Malaisse-Lagae, F. (1976) Ilots á polypeptide pancréatique (PP) et ilots á glucagon: distribution topographique distincte dans le pancreas de Rat. *CR Acad. Sc. Paris* **283,** 1213–1216.
27. Baetens, D., Malaisse-Lagae, F., Perrelet, A., and Orci, L. (1979) Endocrine pancreas: three-dimensional reconstruction shows two types of islets of Langerhans. *Science* **206,** 1323–1325.
28. Orci, L. (1982) Macro- and micro-domains in the endocrine pancreas. Banting lecture 1981. *Diabetes* **31,** 538–565.
29. Alumets, J., Håkanson, R., and Sundler, F. (1978) Distribution, ontogeny and ultrastructure of pancreatic polypeptide (PP) cells in the pancreas and gut of the chicken. *Cell Tissue Res.* **194,** 377–386.
30. Sundler, F., Böttcher, G., Håkanson, R., and Schwartz, T. W. (1984) Immunocytochemical localization of the icosapeptide fragment of the PP precursor: a marker for "true" PP cells? *Regul. Peptides* **8,** 217–224.
31. Citation withdrawn.
32. Andersen, D. K. (1990) The role of pancreatic polypeptide in glucose metabolism, in *Gastrointestinal Endocrinology*, Thompson, J. C., ed., Academic, New York, pp. 333–358.
33. Greenberg, G. R., Mc Cloy, R. F., Adrian, T. E., Chadwick, V. S., Baron, J. H., and Bloom, S. R. (1978) Inhibition of pancreas and gallbladder by pancreatic polypeptide. *Lancet* **II,** 1280–1282.
34. Taylor, I. L., Solomon, T. E., Walsh, J. H., and Grossman, M. I. (1979) Pancreatic polypeptide: Metabolism and effect on pancreatic secretion in dogs. *Gastroenterology* **76,** 524–528.

35. Lin, T. M. (1980) Pancreatic polypeptide: isolation, chemistry and biological function, in *Gastrointestinal Hormones.* Jerzy Glass, G. B., ed., Raven, NY, pp. 275–306.
36. Laurentz, D. A. and Hazelwood, R. L. (1979) Does the third pancreatic hormone (APP) play a trophic role in the growth of the embryonic chick proventiculus? *Proc. Soc. Exp. Biol. Med.* **160,** 144–149.
37. Greenberg, G. R., Mitznegg, P., and Bloom, S. R. (1977) Effect of pancreatic polypeptide on DNA-synthesis in the pancreas. *Experientia* **33,** 1332,1333.
38. Tatemoto, K. and Mutt, V. (1980) Isolation of two novel candidate hormones using a chemical method for finding naturally occurring polypeptides. *Nature* **285,** 417,418.
39. Tatemoto, K. (1982) Isolation and characterization of peptide YY (PYY) a candidate gut hormone that inhibits pancreatic exocrine secretion. *Proc. Natl. Acad. Sci. USA* **I79,** 2514–2518.
40. Tatemoto, K., Carlquist, M., and Mutt, V. (1982) Neuropeptide Y—a novel brain peptide with structural similarities to peptide YY and pancreatic polypeptide. *Nature* **296,** 659–660.
41. Tatemoto, K. (1982) Neuropeptide Y: Complete amino acid sequence of the brain peptide. *Proc. Natl. Acad. Sci. USA* **79,** 5485–5489.
42. Corder, R., Gaillard, R. C., and Böhlen, P. (1988) Isolation and sequence of rat peptide YY and neuropeptide Y. *Regul. Peptides* **21,** 253–261.
43. Eberlein, G. A., Eysselein, V. E., Reeve, J. R., Jr., Shively, J. E., Schaeffer, M., Niebel, W., and Goebell, H. (1988) Isolation and structure of human PYY. *Biomed. Res.* **9 (Suppl. 1)** 38. (Abs.)
44. Tatemoto, K. Nakano, I., Makk, G., Angwin, P., Mann, M., Schilling, J., and Go, V. L. W. (1988) Isolation and primary structure of human peptide YY. *Biochem. Biophys. Res. Commun.* **157,** 713–717.
45. Eysselein, V. E., Eberlein, G. A., Grandt, D., Schaeffer, M., Zehres, B., Behn, U., Schaefer, D., Goebell, H., Davis, M., and Lee, T. D. (1990) Structural characterization of canine PYY. *Peptides* **11,** 111–116.
46. Leiter, A. B., Toder, A., Wolfe, H. J., Taylor, I. L., Cooperman, S., Mandel, G., and Goodman, R. H. (1987) Peptide YY. Structure of the precursor and expression in exocrine pancreas. *J. Biol. Chem.* **262,** 12,984–12,988.
47. Schwartz, T. W., Gingerich, R. L., and Tager, H. S. (1980) Biosynthesis of pancreatic polypeptide: Identification of a precursor and a cosynthesized product. *J. Biol. Chem.* **225,** 494–498.
48. Schwartz, T. W., Hansen, H. F., Håkanson, R., Sundler, F., and Tager, H. S. (1984) Human pancreatic icosapeptide: Isolation, sequence and immunocytochemical localization of the COOH-terminal fragment of the pancreatic polypeptide precursor. *Proc. Natl. Acad. Sci. USA* **81,** 708–712.
49. Boel, E., Schwartz, T. W., Norris, K. E., and Fiil, N. P. (1984) A cDNA encoding a small common precursor for human pancreatic polypeptide and pancreatic icosapeptide. *EMBO J.* **3,** 909–912.
50a. Andrews, P. C., Hawke, D., Shively, J. E., and Dixon, J. E. (1985) A nonamidated peptide homologous to porcine peptide YY and neuropeptide Y. *Endocrinology* **116,** 2677–2681.
50b. Cutfield, S. M., Carne, A., and Cutfield, J. F. (1987) The amino-acid sequences of sculpin islet somatostatin-28 and peptide YY. *FEBS Lett.* , **214,** 57–61.

50c. Conlon, J. M., Chartrel, N., and Vaudry, H. (1992) Primary structure of frog PYY: Implications for the molecular evolution of the pancreatic polypeptide family. *Peptides* **13,** 145–149.
51a. El-Salhy, M., Wilander, E., Grimelius, L., Terenius, L., Lundberg, J. M., and Tatemoto, K. (1982) The distribution of polypeptide YY (PYY)—and pancreatic polypeptide (PP)—immunoreactive cells in the domestic fowl. *Histochemistry* **75,** 25–30.
51b. El-Salhy, M., Grimelius, L., Lundberg, J. M., Tatemoto, K., and Terenius, L. (1982) Immunocytochemical evidence for the occurrence of PYY, a newly isolated gut polypeptide in endocrine cells in the gut of amphibians and reptiles. *Biomed. Res.* **3,** 303–306.
52. Lundberg, J. M., Tatemoto, K., Terenius, L., Hellström, P. M. Mutt, V., Hökfelt, T., and Hamberger, B. (1982) Localization of peptide YY (PYY) in gastrointestinal endocrine cells and effects on intestinal blood flow and motility. *Proc. Natl. Acad. Sci. USA* **79,** 4471–4475.
53. El-Salhy, M., Grimelius, L., Wilander, E., Ryberg, B., Terenius, L., Lundberg, J. M., and Tatemoto, K. (1983) Immunocytochemical identification of polypeptide YY (PYY) cells in the human gastrointestinal tract. *Histochemistry* **77,** 15–23.
54. El-Salhy, M., Wilander, E., Juntti-Berggren, L., and Grimelius, L. (1983) The distribution and ontogeny of polypeptide YY (PYY)—and pancreatic polypeptide (PP)—immunoreactive cells in the gastrointestinal tract of the rat. *Histochemistry* **78,** 53–60.
55. Leduque, P., Pauline, C., and Dubois, P. M. (1983) Immunocytochemical evidence for a substance related to the bovine pancreatic polypeptide-peptide YY group of peptides in the human fetal gastrointestinal tract. *Regul. Peptides* **6,** 219–230.
56. Böttcher, G., Sjölund, K., Ekblad, E., Håkanson, R., Schwartz, T. W., and Sundler, F. (1984) Co-existence of peptide YY and glicentin immunoreactivity in endocrine cells of the gut. *Regul. Peptides* **8,** 261–266.
57. Böttcher, G., Alumets, J., Håkanson, R., and Sundler, F. (1986) Co-existence of glicentin and peptide YY in colorectal L-cells in cat and man. An electron microscopic study. *Regul. Peptides* **13,** 283–291.
58. Lluis, F. and Thompson, J. C. (1988) Neuroendocrine potential of the colon and rectum. *Gastroenterology* **94,** 832–852.
59. Perez-Tomas, R., Ballesta, J., Pastor, L. M., Madrid, J. F., and Polak, J. M. (1989) Comparative immunohistochemical study of the gastroenteropancreatic endocrine system of three reptiles. *Gen. Comp. Endocrinol.* **76,** 171–191.
60. Ali-Rachedi, A., Varndell, I. M., Adrian, T. E., Gapp, D. A., van Noorden, S., Bloom, S. R., and Polak, J. M. (1984) Peptide YY (PYY) immunoreactivity is co-stored with glucagon-related immunoreactants in endocrine cells of the gut and pancreas. *Histochemistry* **80,** 487–491.
61. Nilsson, O., Bilchik, A. J., Goldenring, J. R., Ballantyne, G. H., Adrian, T. E., and Modlin, I. M. (1991) Distribution and immunocytochemical colocalization of peptide YY and enteroglucagon in endocrine cells of the rabbit colon. *Endocrinology* **129,** 139–148.
62. Ondolfo, J. P., Lehy, T., Labeilee, D., and Grés, L. (1989) Growth pattern of the polypeptide-YY cell population in the upper digestive tract of the

rat during the perinatal period and after weaning. *Cell Tissue Res.* **258,** 569–576.

63. Böttcher, G., Ekblad, E., Ekman, R., Håkanson, R., and Sundler, F. (1992) Peptide YY as a neuropeptide in the gut: Immunocytochemical and immunochemical evidence. *Neuroscience* (in press).
64. Taylor, I. L. (1985) Distribution and release of peptide YY in dog measured by specific radioimmunoassay. *Gastroenterology* **88,** 731–737.
65. Miyachi, Y., Jitsuishi, W., Miyoshi, A., Fujita, S., Mizuchi, A., and Tatemoto, K. (1986) The distribution of polypeptide YY-like immunoreactivity in rat tissues. *Endocrinology* **118,** 3163–2167.
66. Roddy, D. R., Koch, T. R., Reilly, W. M., Carney, J. A., and Go, V. L. W. (1987) Identification and distribution of immunoreactive peptide YY in the human, canine, and murine gastrointestinal tracts: species-related antibody recognition differences. *Regul. Peptides* **18,** 201–212.
67. Lundberg, J. M., Terenius, L., Hökfelt, T., and Goldstein, M. (1983) High levels of neuropeptide Y in peripheral noradrenergic neurons in various mammals including man. *Neurosci. Lett.* **42,** 167–172.
68. El-Salhy, M., Grimelius, L., Emson, P. C., and Falkmer, S. (1987) Polypeptide YY- and neuropeptide Y-immunoreactive cells and nerves in the endocrine and exocrine pancreas of some vertebrates: an onto- and phylogenetic study. *Histochemical J.* **19,** 111–117.
69. Ding, W.-G., Koyama, S., Fujimura, M., Miyazaki, M., and Kimura, H. (1990) CGRP and PYY immunoreactive structures in the pancreas of various vertebrates. *Digestion* **46 Suppl. 1,** 23,24.
70. Cheung, R., Andrews, P. C., Plisetskaya, E. M., and Youson, J. H. (1991) Immunoreactivity to peptides belonging to the pancreatic polypeptide family (NPY, aPY, PP, PYY) and to glucagon-like peptide in the endocrine pancreas and anterior intestine of adult lampreys, *Petromyzon marinus*: An immunohistochemical study. *Gen. Comp. Endocrinol.* **81,** 51–63.
71. Böttcher, G., Sjöberg, J., Ekman, R., Håkanson, R., and Sundler, F. (1992) Immunoreactive peptide YY in the mammalian pancreas; Immunocytochemical localization and immunochemical characterization. *Regul. Peptides* (in press).
72. Kreymann, B., Ghatei, M. A., Domin, J., Kanse, S., and Bloom, S. R. (1991) Developmental patterns of glucagon-like peptide-1-(7-36) amide and peptide-YY in rat pancreas and gut. *Endocrinology* **129,** 1001–1005.
73. Krasinski, S. D., Wheeler, M. B., and Leiter, A. B. (1991) Isolation, characterization and developmental expression of the rat peptide-YY gene. *Mol. Endocrinol.* **5,** 433–440.
74. Solomon, T. E. (1985) Pancreatic polypeptide, peptide YY and neuropeptide Y family of regulatory peptides. *Gastroenterology* **88,** 838–841.
75. Hill, F. L. C., Zhang, T., Gomez, G., and Greeley, G. H., Jr. (1991) Peptide YY, a new gut hormone (a mini-review). *Steroids* **56,** 77–82.
76. Laburthe, M. (1990) Peptide YY and neuropeptide Y in the gut. Availability, biological actions and receptors. *Trends Endocrinol. Metabol.* **1,** 168–174.
77. Skak-Nielsen, T., Poulsen, S. S., and Holst, J. J. (1987) Immunohistochemical detection of ganglia in the rat stomach serosa, containing neurons immunoreactive for gastrin-releasing peptide and vasoactive intestinal peptide. *Histochemistry* **87,** 47–52.

78. Häppölä, O., Wahlestedt, C., Ekman, R., Soinila, S., Panula, P., and Håkanson, R. (1990) Peptide YY-like immunoreactivity in sympathetic neurons of the rat. *Neuroscience* **39,** 225–230.
79. Adrian, T. E. and Ferri, G. L. (1985) Human distribution and release of a putative new gut hormone, peptide YY. *Gastroenterology* **89,** 1070–1077.
80. Aponte, G. W., Fink, A. S., Meyer, J. H., Tatemoto, K., and Taylor, I. L. (1985) Regional distribution and release of peptide YY with fatty acids of different chain length. *Am. J. Physiol.* **249,** G745–G750.
81a. Pappas, T. N., Debas, H. T., Goto, Y., and Taylor, I. L. (1985) Peptide YY inhibits meal-stimulated pancreatic and gastric secretion. *Am. J. Physiol.* **248,** G118–G123.
81b. Pappas, T. N., Debas, H. T., and Taylor, I. L. (1985) Peptide YY: Metabolism and effect on pancreatic secretion in dogs. *Gastroenterology* **89,** 1387–1392.
82. Pappas, T. N., Debas, H. T., Chang, A. M., and Taylor, I. L. (1986) Peptide YY release by fatty acids is sufficient to inhibit gastric emptying in dogs. *Gastroenterology* **91,** 1386–1389.
83. Adrian, T. E., Bacarese-Hamilton, A. J., Smith, H. A., Chohan, P., Manolas, K. J., and Bloom, S. R., (1987) Distribution and postprandial release of porcine peptide YY. *J. Endocrinol.* **113,** 11–14.
84a. Adrian, T. E., Savage, A. P., Sagor, G. R., Allen, J. M., Bacarese-Hamilton, A. J., Tatemoto, K., Polak, J. M., and Bloom, S. R. (1985) Effect of peptide YY on gastric, pancreatic and biliary function in humans. *Gastroenterology* **89,** 494–499.
84b. Aponte, G. W., Park, K., Hess, R., Garcia, R., and Taylor, I. L. (1989) Meal-induced peptide tyrosine inhibition of pancreatic secretion in the rat. *FASEB J.* **3,** 1949–1955.
85. Greeley, G. H., Jr., Hashimoto, T., Izukura, M., Gomez, G., and Jeng, J. (1989) A comparison of intraduodenally and intracolonically administered nutrients on the release of peptide-YY in the dog. *Endocrinology* **125,** 1761–1765.
86. Sheikh, S. P., Holst, J. J., Orskov, C., Ekman, R., and Schwartz, T. W. (1989) Release of PYY from pig intestinal mucosa; luminal and neural regulation. *Regul. Peptides* **26,** 253–266.
87. Greeley, G. H., Jr., Jeng, J., Gomez, G., Hashimoto, T., and Hill, F. L. (1989) Evidence for regulation of peptide-YY release by the proximal gut. *Endocrinology* **124,** 1438–1443.
88. Lluis, F., Gomez, G., Fujimura, M., Greeley, G. H., Jr., and Thompson, J. C. (1987) Peptide YY inhibits nutrient-, hormonal-, and vagally-stimulated pancreatic exocrine secretion. *Pancreas* **2,** 454–462.
89. Hosotani, R., Inoue, K., Kogire, M., Tatemoto, K., Mutt, V., and Suzuki, K. (1989) Effect of natural peptide YY on pancreatic secretion and cholecystokinin release in conscious dogs. *Dig. Dis. Sci.* **34,** 468–473.
90. Putnam, W. S., Liddle, R. A., and Williams, J. A. (1989) Inhibitory regulation of rat exocrine pancreas by peptide YY and pancreatic polypeptide. *Am. J. Physiol.* **256,** G698–G703.
91. Jeng, Y.-J., Hill, F. L. C., Lluis, F., Gomez, G., Izukura, M., Kern, K., Chuo, S., Ferrar, S., Greeley, G. H., Jr. (1990) Peptide YY release and actions, in *Gastrointestinal Endocrinology: Receptors and Post-receptor Mechanisms.* Thompson, J. C., ed., Academic, San Diego, pp. 371–386.

92. DeMar, A. R., Taylor, I. L., and Fink, A. S. (1991) Pancreatic polypeptide and peptide YY inhibit the denervated canine pancreas. *Pancreas* **6,** 419–426.
93. Suzuki, T., Nakaya, M., Itoh, Z., Tatemoto, K., and Mutt, V. (1983) Inhibition of interdigestive contractile activity in the stomach by peptide YY in Heidenhain pouch dogs. *Gastroenterology* **85,** 114–121.
94. Allen, J. M., Fitzpatrick, M. L., Yeats, J. C., Darcy, K., Adrian, T. E., and Bloom, S. R. (1984) Effects of peptide YY and neuropeptide Y on gastric emptying in man. *Digestion* **30,** 255–262.
95. Soper, N. J., Chapman, N. J., Kelly, K. A., Brown, M. L., Phillips, S. F., and Go, V. L. W., (1990) The "Ileal brake" after ileal pouch-anal anastomosis. *Gastroenterology* **98,** 111–116.
96. Al-Saffar, A., Hellström, P. M., and Nylander, G. (1985) Correlation between peptide YY-induced myoelectric activity and transit of small-intestinal contents in rats. *Scand. J. Gastroenterol.* **20,** 577–582.
97. Krantis, A., Potvin, W., and Harding, R. K. (1988) Peptide YY (PYY) stimulates intrinsic enteric motor neurones in the rat small intestine. *Naunyn Schmiedebergs Arch. Pharmacol.* **338,** 287–292.
98. Hellström, P. M., Lundberg, J. M., Hökfelt, T., and Goldstein, M. (1989) Neuropeptide Y, peptide YY, and sympathetic control of rectal tone and anal canal pressure in the cat. *Scand. J. Gastroenterol.* **24,** 231–243.
99. Szecówka, J., Tatemoto, K., Rajamäki, G., and Efendic, S. (1983) Effects of PYY and PP on endocrine pancreas. *Acta Physiol. Scand.* **119,** 123–126.
100. Inui, A., Inoue, T., Sakatani, N., Oya, M., Morioka, H., Mizuno, N., and Baba, S. (1987) Biological actions of peptide YY: Effects on endocrine pancreas, pituitary-adrenal axis, and plasma catecholamine concentrations in the dog. *Horm. Metabol. Res.* **19,** 353–357.
101. Greeley, G. H., Jr., Lluis, F., Gomez, G., Ishizuka, J., Holland, B., and Thompson, J. C. (1988) Peptide YY antagonizes beta-adrenergic-stimulated release of insulin in dogs. *Am. J. Physiol.* **254,** E513–517.
102. Guo, Y. S., Singh, P., DeBuono, J. F., and Thompson, J. C. (1988) Effect of peptide YY on insulin release stimulated by 2-deoxyglucose and neuropeptides in dogs. *Pancreas* **3,** 128–134.
103. Guo, Y. S., Singh, P., Draviam, E., Greeley, G. H., Jr., and Thompson, J. C. (1989) Peptide YY inhibits the insulinotropic action of gastric inhibitory polypeptide. *Gastroenterology* **96,** 690–694.
104. Adrian, T. E., Sagor, G. R., Savage, A. P., Bacarese-Hamilton, A. J., Hall, G. M., and Bloom, S. R. (1986) Peptide YY kinetics and effects on blood pressure and circulating pancreatic and gastrointestinal hormones and metabolites in man. *J. Clin. Endocrino. Metab.* **63,** 803–807.
105. Goodlad, R. A., Ghatei, M. A., Domin, J., Bloom, S. R., and Gregory, H. (1989) Plasma enteroglucagon, peptide YY and gastrin in rats deprived of luminal nutrition, and after urogastrone-EGF administration. A proliferative role for PYY in the intestinal epithelium. *Experientia* **45,** 168–169.
106. Lorén, I., Alumets, J., Håkanson, R., and Sundler, F. (1979) Immunoreactive pancreatic polypeptide (PP) occurs in the central and peripheral nervous system: preliminary immunocytochemical observations. *Cell Tissue Res.* **200,** 179–186.
107. Sundler, F., Ekblad, E., Grunditz, T., Håkanson, R., Luts, A., and Uddman, R. (1989) NPY in peripheral, non-adrenergic neurons, in *Neuropeptide Y.*

Mutt, V., Fuxe, K., Hökfelt, T., and Lundberg, J. M., eds., Raven, New York. pp. 93–102.

108. Allen, J. M., Adrian, T. E., Polak, J. M., and Bloom, S. R. (1983) Neuropeptide Y (NPY) in the adrenal gland. *J. Autonomic Nervous System* **9,** 559–563.
109. Varndell, I. M., Polak, J. M., Allen, J. M., Teremghi, G., Bloom, S. R. (1984) Neuropeptide tyrosine (NPY) immunoreactivity in norepinephrine-containing cells and nerves of the mammalian adrenal gland. *Endocrinology* **114,** 1460–1462.
110. Majane, E. A., Alho, H., Kataoka, Y., Lee, C. H., and Yang, H.-Y. T. (1985) Neuropeptide Y in bovine adrenal glands: distribution and characterization. *Endocrinology* **117,** 1162–1168.
111. de Quidt, M. E. and Emson, P. C. (1986) Neuropeptide Y in the adrenal gland: characterization, distribution and drug effects. *Neuroscience* **19,** 1011–1022.
112. Kuramoto, H. Kondo, H., and Fujita, T. (1986) Neuropeptide tyrosine (NPY) - like immunoreactivity in adrenal chromaffin cells and intraadrenal nerve fibers of rats. *Anat. Rec.* **214,** 321–328.
113. Lundberg, J. M., Hökfelt, T., Hemsén, A., Theodorsson-Norheim, E., Pernow, J., Hamberger, B., and Goldstein, M. (1986) Neuropeptide Y-like immunoreactivity in adrenaline cells of adrenal medulla and in tumors and plasma of pheochromocytoma patients. *Regul. Peptides* **13,** 169–182.
114. Sundler, F., Håkanson, R., Ekblad, E., Uddman, R., and Wahlestedt, C. (1986) Neuropeptide Y in the peripheral adrenergic and enteric nervous systems. *Int. Rev. Cytol.* **102,** 243–269.
115. Hökfelt, T., Elfvin, L. G., Elde, R., Schultzberg, M., Goldstein, M., and Luft, R. (1977) Occurrence of somatostatin-like immunoreactivity in some peripheral sympathetic noradrenergic neurons. *Proc. Natl. Acad. Sci. USA* **74,** 3587–3591.
116. Schultzberg, M., Hökfelt, T., Terenius, L., Elfvin, L. G., Lundberg, J. M., Brandt, J., Elde, R. P., and Goldstein, M. (1979) Enkephalin immunoreactive nerve fibers and cell bodies in sympathetic ganglia of the guinea pig and rat. *Neuroscience* **4,** 249–270.
117. Lundberg, J. M., Hökfelt, T., Änggård, A., Terenius, L., Elde, K., Markey, K., Goldstein, M., and Kimmel, J. (1982) Organizational principles in the peripheral sympathetic nervous system: subdivision by coexisting peptides (somatostatin—avian pancreatic polypeptide—and vasoactive intestinal polypeptide-like immunoreactive materials). *Proc. Natl. Acad. Sci. USA* **79,** 1303–1307.
118. Lundberg, J. M., Terenius, L., Hökfelt, T., Martling, C.-R., Tatemoto, K., Mutt, V., Polak, J., Bloom, S., and Goldstein, M. (1982) Neuropeptide Y (NPY)-like immunoreactivity in peripheral noradrenergic neurons and effects of NPY on sympathetic functions. *Acta Physiol. Scand.* **116,** 477–480.
119. Ekblad, E., Edvinsson, L., Wahlestedt, C., Uddman, R., Håkanson, R., and Sundler, F. (1984) Neuropeptide Y co-exists and co-operates with noradrenaline neurons in perivascular nerve fibers. *Regul. Peptides* **8,** 225–235.
120. Uddman, K., Ekblad, E., Edvinsson, L., Håkanson, R., and Sundler, F. (1985) Neuropeptide Y-like immunoreactivity in perivascular nerve fibers of the guinea pig. *Regul. Peptides* **10,** 243–257.
121. Edvinsson, L. Copeland, J. R., Emson, P. C., McCulloch, J., and Uddman, R. (1987) Nerve fibers containing neuropeptide Y in the cerebrovascular bed.

Immunocytochemistry, radioimmunoassay and vasomotor effects. *Cerebral Blood Flow Metab.* **7,** 45–57.

122. Cannon, B., Nedergaard, J., Lundberg, J. M., Hökfelt, T., Terenius, L., and Goldstein, M. (1986) Neuropeptide tyrosine (NPY) is co-stored with noradrenaline in vascular but not in parenchymal sympathetic nerves of brown adipose tissue. *Exp. Cell Res.* **164,** 546–550.
123. Morris, J. L., Gibbins, I. L., Furness, J. B., Costa, M., and Murphy, R. (1985) Colocalization of neuropeptide Y, vasoactive intestinal polypeptide and dynorphin in non-adrenergic axons of the guinea-pig uterine artery. *Neurosci. Lett.* **62,** 31–37.
124. Edvinsson, L., Emson, R., McCulloch, J., Tatemoto, K., and Uddman, R. (1983) Neuropeptide Y: cerebrovascular innervation and vasomotor effects in the cat. *Neurosci. Lett.* **43,** 79–84.
125. Allen, J. M. and Bloom, S. R. (1986) Neuropeptide Y: a putative neurotransmitter. *Neurochem. Int.* **8,** 1–8.
126. Ehinger, B., Falck, B., and Sporrong, B. (1966) Adrenergic fibers to the heart and to peripheral vessels. *Biol. Anat.* **8,** 35–45.
127. Schenk, E. A. and El Baldawi, A. (1968) Dual innervation of arteries and arterioles. A histochemical study. *Z. Zellforsch.* **91,** 170–177.
128. Burnstock, G. (1975) Innervation of vascular smooth muscle: histochemistry and electron microscopy. *Clin. Exp. Pharmacol. Physiol.* **Suppl 2,** 7–20.
129. Pernow, I., Öhlén, A., Hökfelt, T., Nilsson, O., and Lundberg, J. M. (1987) Neuropeptide Y: presence in perivascular noradrenergic neurons and vasoconstrictor effects on skeletal muscle blood vessels in experimental annals and man. *Regul. Peptides* **19,** 313–324.
130. Gu, J., Polak, J. M., Adrian, T. E., Allen, J. M., Tatemoto, K., and Bloom, S. R. (1983) Neuropeptide tyrosine (NPY)—a major cardiac neuropeptide. *Lancet* **I,** 1008–1010.
131. Hassall, C. J. S. and Burnstock, G. (1984) Neuropeptide Y-like immunoreactivity in cultured intrinsic neurones of the heart. *Neurosci. Lett.* **52,** 111–115.
132. Dalsgaard, C.-J., Franco-Cereceda, A., Saria, A., Lundberg, J.-M., Theodorsson-Norheim, E., and Hökfelt, T. (1986) Distribution and origin of substance P- and neuropeptide Y-immunoreactive nerves in the guinea-pig heart. *Cell Tissue Res.* **243,** 477–485.
133. Allen, J. M., Martin, J. B., and Heinrich, G. (1988) Neuropeptide Y is intrinsic to the heart, in *Advances in Atrial Peptide Research.* Brenner, B. M. and Laragh, J. H., eds., Raven, New York, pp. 155–160.
134. Forsgren, S. (1989) Neuropeptide Y-like immunoreactivity in relation to the distribution of sympathetic nerve fibers in the heart conduction system. *J. Mol. Cell. Cardiol.* **21,** 279–290.
135. Aggestrup, S., Emson, P., Uddman, R., Sundler, F., Landkaer Jensen, S., and Rahbek Sörensen, H. (1987) Distribution and content of neuropeptide Y in the human lower esophageal sphincter. *Digestion* **36,** 68–73.
136. Wattchow, D. A., Furness, J. B., Costa, M., O'Brien, P. E., and Peacock, M. (1987) Distribution of neuropeptides in the human esophagus. *Gastroenterology* **93,** 1363–1371.
137. Wattchow, D. A., Furness, J. B., and Costa, M. (1988) Distribution and coexistence of peptides in nerve fibers of the external muscle of the human gastrointestinal tract. *Gastroenterology* **95,** 32–41.

138. Keast, J. R., Furness, J. B., and Costa, M. (1987) Distribution of peptide-containing neurons and endocrine cells in the rabbit gastrointestinal tract, with particular reference to the mucosa. *Cell Tissue Res.* **248,** 565–577.
139. Furness, J. B., Costa, M., Emson, P. C., Håkanson, R., Moghimzadeh, E., Sundler, F., Taylor, I. L., and Chance, R. E. (1983) Distribution, pathways, and reactions to drug treatment of nerves with neuropeptide Y- and pancreatic polypeptide-like immunoreactivity in the guinea-pig digestive tract. *Cell Tissue Res.* **234,** 71–92.
139a. Ekblad, E., Ekelund, M., Graffner, H., Håkanson, R., and Sundler, F. (1985) Peptide-containing nerve fibers in the stomach wall of rat and mouse. *Gastroenterology* **89,** 73–85.
140. Keast, H. R., Furness, J. B., and Costa, M. (1987) Distribution of peptide-containing nerve fibres and endocrine cells in the gastrointestinal mucosa in five mammalian species. *J. Comp. Neurol.* **236,** 403–422.
141. Lee, Y., Shiosaka, S., Emson, P. C., Powell, K. F., Smith, A. D., and Tohyama, M. (1985) Neuropeptide Y-like immunoreactive structures in the rat stomach with special references to the noradrenaline neuron system. *Gastroenterology* **89,** 118–126.
142. Ekblad, E., Håkanson, R., and Sundler, F. (1991) Innervation of the stomach of rat and man with special reference to the endocrine cells, in *The Stomach as an Endocrine Organ.* Håkanson, R. and Sundler, F., eds. Elsevier Science Publishers, Amsterdam, The Netherlands, pp. 79–95.
143. Sundler, F., Moghimzadeh, E., Håkanson, R., Ekelund, M., and Emson, P. C. (1983) Nerve fibers in the gut and pancreas of the rat displaying neuropeptide Y immunoreactivity. Intrinsic and extrinsic origin. *Cell Tissue Res.* **230,** 487–493.
144. Ekblad, E., Winther, C., Ekman, R., Håkanson, R., and Sundler, F. (1987) Projections of peptide-containing neurons in rat small intestine. *Neuroscience* **20,** 169–188.
145. Timmermans, J.-P., Schevermann, D. W., Stach, W., Adraesen, D., and De Groodt-Lasseel, M. H. A. (1990) Distinct distribution of CGRP-enkephalin-, galanin-, neuromedin U-, neuropeptide Y-, somatostatin-, substance P-, VIP- and serotonin-containing neurons in the two submucosal ganglionic neural networks of the porcine small intestine. *Cell Tissue Res.* **260,** 369–379.
146. Ekblad, E., Ekman, R., Håkanson, R., and Sundler, F. (1988) Projections of peptide-containing neurons in rat colon. *Neuroscience* **27,** 655–674.
147. Su, H. C., Bishop, A. E., Power, R. F., Hamada, Y., and Polak, J. M. (1987) Dual intrinsic and extrinsic origin of CGRP- and NPY-immunoreactive nerves of rat gut and pancreas. *J. Neurosci.* **7,** 2674–2687.
148. Ekblad, E., Wahlestedt, C., Ekelund, M., Håkanson, R., and Sundler, F. (1984) Neuropeptide Y in the gut and pancreas: Distribution and possible vasomotor function. *Front Horm. Res.* **12,** 85–90.
149. Larsson, L. T., Malmfors, G., Ekblad, E., Ekman, R., and Sundler, F. (1991) NPY hyperinnervation in Hirschsprung's disease: Both adrenergic and nonadrenergic fibers contribute. *J. Pediat. Surg.* **26,** 1207–1214.
150a. Furness, J. B., Costa, M., Gibbins, I. L., Llewellyn-Smith, I. J., and Oliver, J. R. (1985) Neurochemically similar myenteric and submucous neurons directly traced to the mucosa of the small intestine. *Cell Tissue Res.* **241,** 155–163.

150b. Ekblad, E., Håkanson, R., and Sundler, F. (1984) VIP and PHI coexist with a NPY-like peptide in intramural neurones of the small intestine. *Regul. Peptides* **10,** 47–55.
151. Ekblad, E., Arnbjörnsson, E., Ekman, R., Håkanson, R., and Sundler, F. (1989) Neuropeptides in the human appendix: Distribution and motor effects. *Dig. Dis. Sci.* **34,** 1217–1230.
152. Larsson, L. T., Malmfors, G., Ekblad, E., Ekman, R., and Sundler, F. (1991) NPY hyperinnervation in Hirschsprung's disease: both adrenergic and nonadrenergic fibers contribute. *J. Pediat. Surg.* **26,** 1207–1214.
153. Messenger, J. P. and Furness, J. B. (1990) Projections of chemically specified neurons in the guinea-pig colon. *Arch. Histol. Cytol.* **53,** 467–495.
154. Daniel, E. E., Furness, J. B., Costa, M., and Belbeck, L. (1987) The projections of chemically identified nerve fibers in canine ileum. *Cell Tissue Res.* **247,** 377–384.
155. Leblanc, G. G. and Landis, S. C. (1988) Target specificity of neuropeptide Y-immunoreactive cranial parasympathetic neurons. *J. Neurosci.* **8,** 146–155.
156. Leblanc, G., Trimmer, B. A., and Landis, S. C. (1987) Neuropeptide Y-like immunoreactivity in rat cranial parasympathetic neurons: Coexistence with vasoactive intestinal peptide and choline acetyltransferase. *Proc. Natl. Acad. Sci. USA* **84,** 3511–3515.
157. Hardebo, J. E., Suzuki, N., Ekblad, E., Owman, Ch. (1992) Vasoactive intestinal polypeptide and acetylcholine coexist with neuropeptide Y, dopamine-β-hydroxylase, tyrosine hydroxylase, substance P or calcitonin gene-related peptide in neuronal subpopulations in cranial parasympathetic ganglia of rat. *Cell Tissue Res.* **267,** 291–300.
158. Grunditz, T., Håkanson, R., Rerup, C., Sundler, F., and Uddman, R. (1984) Neuropeptide Y in the thyroid gland: neuronal localization and enhancement of stimulated thyroid hormone secretion. *Endocrinology* **115,** 1537–1542.
159. Grunditz, T., Ekman, R., Håkanson, R., Sundler, F., and Uddman, R. (1988) Neuropeptide Y and vasoactive intestinal peptide coexist in rat thyroid nerve fibers emanating from the thyroid ganglion. *Regul. Peptides* **23,** 193–208.
160. Grunditz, T., Håkanson, R., Sundler, F., and Uddman, R. (1988) Neuronal pathways to the rat thyroid gland revealed by retrograde tracing and immunocytochemistry. *Neuroscience* **24,** 321–335.
161. Håkanson, R., Sundler, F., Mogimzadeh, E., and Leander, S. (1983) Peptide containing nerve fibers in the airways: distribution and functional implications. *Eur. J. Respir. Dis.* **(Suppl.) 64,** 115–140.
162. Sheppard, M. N., Polak, J. M., Allen, J. M., and Bloom, S. R. (1984) Neuropeptide tyrosine (NPY): a newly discovered peptide is present in the mammalian respiratory tract. *Thorax* **39,** 326–330.
163. Uddman, R., Sundler, F., and Emson, P. (1984) Occurrence and distribution of neuropeptide Y-immunoreactive nerves in the respiratory tract and middle ear. *Cell Tissue Res.* **237,** 321–327.
164. Lundberg, J. M., Martling, C.-R., and Hökfelt, T. (1986) Airways, oral cavity and salivary glands: classical transmitters and peptides in sensory and autonomic motor neurons, in *Handbook of Chemical Neuroanatomy*, vol. 6. The Peripheral Nervous System. (Björklund, A., Hökfelt, T., and Owman, C., eds., Elsevier, Amsterdam, pp. 391–444.

165. Lacroix, J.-S. (1989) Adrenergic and nonadrenergic mechanisms in sympathetic vascular control of the nasal mucosa. *Acta Physiol. Scand.* **136, (Suppl. 581),** 1–63.
166. Luts, A. and Sundler, F. (1989) Peptide-containing nerve fibers in the respiratory tract of the ferret. *Cell Tissue Res.* **258,** 259–267.
167. Lacroix, J.-S., Änggård, A., Hökfelt, T., O'Hare, M. M. T., Fahrenkrug, J., and Lundberg, J. M. (1990) Neuropeptide Y: presence in sympathetic and parasympathetic innervation of the nasal mucosa. *Cell Tissue Res.* **259,** 119–128.
168. Martling, C.-R., Matran, R., Alving, K., Hökfelt, T., and Lundberg, J. M. (1990) Innervation of lower airways and neuropeptide effects on bronchial and vascular tone in the pig. *Cell Tissue Res.* **260,** 223–233.
169. Springall, D. R., Polak, J. M., Howard, L., Power, R. F., Krausz, T., Manickam, S., Banner, N. R., Khagani, A., Rose, M., and Yacoub, M. H. (1990) Persistence of intrinsic neurones and possible phenotypic changes after extrinsic denervation of human respiratory tract by heart-lung transplantation. *Am. Rev. Respir. Dis.* **141,** 1538–1546.
170. Springall, D. R., Bloom, S. R., and Polak, J. M. (1991) Neural, endocrine and endothelial regulatory peptides, in *The Lung: Scientific Foundations,* vol. 2, Crystal, R. G., Barnes, P. J., Cherniack, N. S., Weibel, E. R., and West, J. B., eds. Raven, New York, pp. 69–90.
171. Ahrén, B., Böttcher, G., Kowalyk, S., Dunning, B., Sundler, F., Taborsky G. J., Jr. (1990) Galanin is co-localized with noradrenaline and neuropeptide Y in dog pancreas and celiac ganglion. *Cell Tissue Res.* **261,** 49–58.
172. Allen, J. M., Raine, A. E. G., Ledingham, J. G. G., and Bloom, S. R. (1985) Neuropeptide Y: a novel renal peptide with vasoconstrictor and natriuretic activity. *Clin. Sci.* **68,** 373–377.
173. Mattiasson, A., Ekblad, E., Sundler, F., and Uvelius, B. (1985) Origin and distribution of neuropeptide Y—,vasoactive intestinal polypeptide—,and substance P—containing nerve fibers in the urinary bladder of the rat. *Cell Tissue Res.* **239,** 141–146.
174. Kannisto, P., Ekblad, E., Holm, G., Owman, Ch. Sjöberg, N.-O., Stjernquist, M., Sundler, F., and Walles, B. (1986) Existence and coexistence of peptides in nerves of the mammalian ovary and oviduct demonstrated by immunocytochemistry. *Histochemistry* **86,** 25–34.
175. Alm, P., Lundberg, L.-M., Wharton, J., and Polak, J. M. (1988) Organization of the guinea-pig uterine innervation. Distribution of immunoreactivities for different neuronal markers. Effects of chemical- and pregnancy-induced sympathectomy. *Histochem. J.* **20,** 290–300.
176. Papka, R. E. and Traurig, H. H. (1988) Distribution of subgroups of neuropeptide Y-immunoreactive and noradrenergic nerves in the female rat uterine cervix. *Cell Tissue Res.* **252,** 533–541.
177. Inayama, C. O., Hacker, G. W., Gu, J., Dahl, D., Bloom, S. R., and Polak, J. M. (1985) Cytochemical relationships in the paracervical ganglion (Frankenhauser) of rat studied by immunocytochemistry. *Neurosci. Lett.* **55,** 311–316.
178. Fallgren B., Ekblad, E., and Edvinsson, L. (1989) Co-existence of neuropeptides and differential inhibition of vasodilator responses by neuropeptide Y in guinea pig uterine arteries. *Neurosci. Lett.* **100,** 71–76.

179. Stjernquist, M., Ekblad, E., Nordstedt, E., and Radzuweit, C. (1991) Neuropeptide Y (NPY) coexists with tyrosine hydroxylase and potentiates the adrenergic contractile response of vascular smooth muscle in the human uterine artery. *Human Reproduct.* **6,** 1034–1038.
180a. Edvinsson, L., Ekblad, E., Håkanson, R., and Wahlestedt, C. (1984) Neuropeptide Y potentiates the effect of various vasoconstrictor agents on rabbit blood vessels. *Br. J. Pharmacol.* **83,** 519–525.
180b. Håkanson, R., Wahlestedt, C., Ekblad, E., Edvinsson, L., and Sundler, F. (1986) Neuropeptide Y: coexistence with noradrenaline. Functional implications, in *Coexistence of Neuronal Messengers: A New Principle in Chemical Transmission.* Hökfelt, T., Fuxe, K., and Pernow, B., eds., *Prog. Brain Res.* vol. 68, pp. 279–287.
181. Pernow, I., Kahan, T., Hjemdahl, P., and Lundberg, J. M. (1988) Possible involvement of neuropeptide Y in sympathetic vascular control of canine skeletal muscle. *Acta. Physiol. Scand.* **132,** 43–50.
182. Garzón, J., Höllt, V., and Sánchez-Blázquez, P. (1986) Neuropeptide Y is an inhibitor of neural function in the myenteric plexus of the guinea-pig ileum. *Peptides* **7,** 623–629.
183. Allen, J. M., Hughes, J., and Bloom, S. R. (1987) Presence, distribution and pharmacological effects of neuropeptide Y in mammalian gastointestinal tract. *Dig. Dis. Sci.* **32,** 506–512.
184. Cox, H. M., Cuthbert, A. W., Håkanson, R., and Wahlestedt, C. (1988) The effect of neuropeptide Y and peptide YY on electrogenic ion transport in the intestinal epithelia. *J. Physiol.* **398,** 656–680.
185. Sheikh, S. P. (1991) Neuropeptide Y and peptide YY: major modulators of gastrointestinal blood flow and function. *Am. J. Physiol.* **261,** 6701–6715.

Characterization of Receptor Types for Neuropeptide Y and Related Peptides

Lars Grundemar, Sören P. Sheikh, and Claes Wahlestedt

1. Introduction

The structurally related peptides neuropeptide Y (NPY), peptide YY (PYY), and pancreatic polypeptide (PP) are neuronal and/or endocrine messengers that are involved in a variety of physiological processes. NPY and PYY are able to evoke potent biological effects as well as modulating responses to other transmitters. In many biological systems, NPY and PYY, but not PP seem to activate the same receptors. Because of the wide distribution of NPY in the central and peripheral nervous system compared to that of PYY, many of these actions have been attributed to NPY.

1.1. Structure

All three peptides have been suggested to form a similar tertiary structure consisting of an N-terminal polyproline helix (residues 1–8) and an amphiphilic α-helix (residues 15–30), connected with a β-turn, creating a hairpin-like loop, which often is referred to as the so-called PP-fold *(1)*. This domain has been identified from the crystal structure of avian PP, and NMR (nuclear magnetic resonance) studies support this three-dimensional configuration. The helices are kept together by hydrophobic interactions. The amidated C-terminal end (residues 30–36) projects away from the hairpin loop *(2–4)*. *See* Fig. 1.

From: *The Biology of Neuropeptide Y and Related Peptides;*
W. F. Colmers and C. Wahlestedt, Eds.

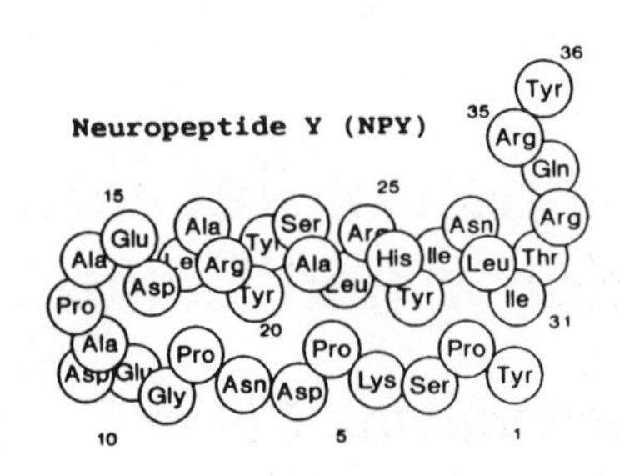

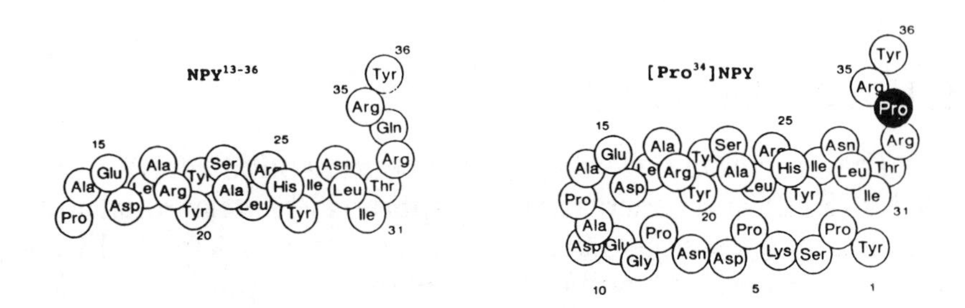

Fig. 1. Pharmacological tools illustrating (above) the proposed secondary structure of NPY, (right) the Y1 receptor agonist [Pro^{34}]NPY, and (left) NPY $_{13-36}$, which is a Y2 receptor agonist.

1.2. Distribution

In the periphery, NPY is present in many sympathetic nerve fibers, especially around blood vessels *(5–8)*. NPY is present also in nonadrenergic perivascular *(9)*, enteric *(6)*, and heart fibers *(10)*. In addition, some NPY immunoreactivity is found in parasympathetic nerves *(11)*. PYY occurs mainly in endocrine cells in the lower gastrointestinal tract *(12)*. However, some PYY immunoreactivity is found also in certain sympathetic fibers *(13)*. PP is predominantly located in endocrine cells of the pancreatic islets *(14)*.

NPY-containing neurons are abundant in the central nervous system *(14,15)*. In contrast, PYY-containing neurons are few, and mainly confined to the brainstem and the cervical spinal cord *(16,17)*. PP, finally does not seem to occur in the central nervous system (CNS) *(18)*.

1.3. Binding Sites

PYY seems to bind to the same binding sites as NPY. In the periphery, specific binding sites for NPY/PYY have been demonstrated in a variety of tissues and cell types, like vascular smooth muscle *(19–23)*, myocardium *(24)*, adrenal medulla *(25)*, kidney *(26,27)*, and in the intestine *(28–30)*.

Binding sites for NPY have been found in a variety of areas in the brain, particularly in the limbic system, hippocampus, the cortex, some thalamic nuclei, and in the brainstem (for a review, *see*, e.g., *31*). PYY binding sites are present in the same areas as NPY binding sites throughout the CNS. Many of these binding sites are thought to be identical to those that bind NPY *(32–34)*.

1.4. Biological Effects

NPY exerts potent biological effects on many targets both in the CNS and in the periphery. The effects of NPY at the sympathetic neuroeffector junction have been studied in some detail. NPY is a well characterized vasoconstrictor and is able to enhance the effects of other pressor agents, such as norepinephrine, but also to suppress the release of transmitters from sympathetic fibers. In the CNS, NPY has, for instance, been associated with modulation of autonomic functions, like regulation of cardiovascular and respiratory activity *(35)*, as well as ingestive behavior *(36)*, and the release of hypothalamic and pituitary hormones *(37)*. NPY seems to fulfill the criteria as a neurotransmitter, since it is stored in synaptic granulae *(38)*, released on electrical nerve stimulation *(8,39)*, and acts at specific receptors (e.g., *1,40*).

2. NPY/PYY Receptor Heterogeneity

The first observation suggesting NPY/PYY receptor heterogeneity in the peripheral nervous system was made in a study in which C-terminal fragments of NPY/PYY were found to evoke effects on prejunctional, but not on postjunctional sites at the sympathetic neuroeffector junction *(41)*. The vas deferens is a much used preparation for the study of prejunctional sympathetic activity. In this assay, the C-terminal fragments NPY_{13-36} and PYY_{13-36} were found to be very effective in suppressing the electrically (sympathetically) evoked twitches. In contrast, although exogenously applied NPY or PYY induced a concentration-dependent contraction of the guinea-pig iliac vein, NPY_{13-36} or PYY_{13-36} seemed to be inactive. Since these fragments showed activity at the prejunctional receptor only, it was suggested that NPY/PYY acted on two different receptor types. The post- and prejunctional receptors were referred to as Y1 and Y2, respectively *(41,42)*. Using radiolabeled NPY or PYY on different tissues, such as neuroendocrine cell lines, two distinct receptor types, which correspond to the Y1 and Y2 receptors, have been characterized *(43,44)*. During the last few years, an increasing body of functional and biochemical evidence has supported the existence of Y1 and Y2 receptor types both in the central and peripheral nervous system *(1,40,45,46)*. The very recent cloning of one of these receptors has revealed that they belong to the G-protein-coupled receptor family of seven transmembrane proteins *(47)*. In addition, very recently, NPY receptors distinct from Y1 and Y2 have been suggested both in the CNS *(48)* and in the periphery *(24,49,50)*. Finally, the existence of specific PP receptors has also been demonstrated *(1,51)*.

The homologous receptors for NPY and related peptides have to date been pharmacologically characterized in the following ways:

1. NPY, PYY, and PP as well as various C-terminal NPY/PYY fragments have different abilities to bind or activate either of the receptor types. PYY and NPY are about equipotent at Y1 and Y2 receptors. Recent data have suggested that NPY, but not PYY, recognizes certain receptors, which are tentatively referred to as Y3 receptors. PP is essentially inactive at either receptor type *(1)*. The C-terminal fragment NPY_{13-36} (shown in Fig. 1) is a Y2 receptor agonist, being about one order of magnitude less potent than NPY, but virtually inactive at the Y1 receptor; cf Fig. 2.
2. The substituted analogs [Pro^{34}]NPY (Fig. 1) or [Leu^{31},Pro^{34}]NPY seem to be full Y1 receptor agonists, while being inactive at Y2 receptors *(56,57)*. In fact, proline in position 34 confers Y1 receptor selectivity; Fig. 2. These receptor-specific ligands have been useful in attempts to identify binding sites and to delineate the physiological significance of Y1 receptors. Figure 3 shows a cartoon of the binding pocket of the Y1 receptor.
3. Some neuroendocrine cell lines seem to express selectively either Y1, Y2, or PP receptor types *(1,26,40,43,44)*. These pure receptor populations have been beneficial in studies of ligand requirements for receptor recognition and on second-messenger systems.
4. Several truncated and/or substituted NPY analogs have been developed in order to identify which parts of the ligand bind or activate the respective receptor type *(56,58–60)*.

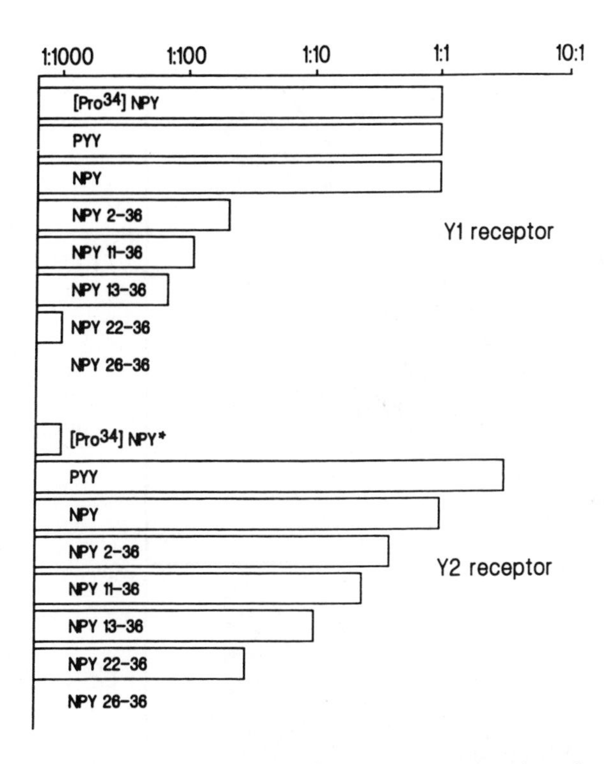

Fig. 2. Log-relative potencies of various NPY-related peptides in comparison with NPY itself at the Y1 (top) and the Y2 receptor (bottom). Data for the Y1 receptor are from concentration-response curves of vascular motor effects, and from radioligand displacement curves using rat aortic smooth muscle cells *(22)* or the neuroblastoma cell line SK-N-MC *(40)*. Data for the Y2 receptor are from bioactivity studies of the NPY-related peptides in rat vas deferens *(53–55)* or from some neuroblastoma cell lines *(40,43,44)*.

5. No antagonists with acceptable potency or receptor selectivity have yet been presented. The irreversible α-adrenoceptor antagonist benextramine seems to evoke a weak antagonism on NPY binding and to attenuate NPY-evoked hypertension. However, the potency seems to be very low *(61)*. An inositol phosphate (Ins[1,2,6]P_3; PP56) has been reported to suppress the amplitude of NPY-evoked vasoconstriction without causing a right shift of the response *(62)*, reflecting noncompetitive antagonism. Instead, and with remarkable selectivity, Ins[1,2,6]P_3 appears to inhibit NPY-evoked elevations of intracellular Ca^{2+} in vascular smooth muscle cells *(63,64)*.

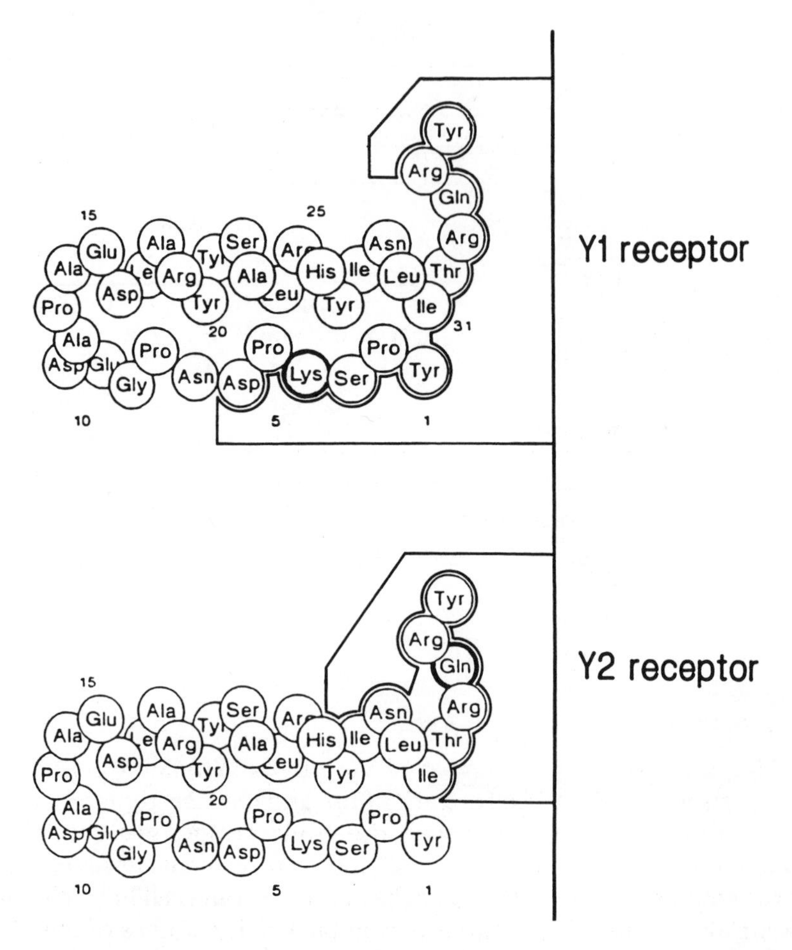

Fig. 3. Proposed secondary structure of the NPY molecule and its receptors, Y1 and Y2. The two receptors recognize different domains of the ligand. The N and C termini, joined together by the hairpin (PP fold), are involved in receptor recognition. Deamidation of the C-terminal end results in loss of Y1 or Y2 receptor affinity. In order to become fully activated, the Y1 receptor requires intact N and C termini of NPY. The deletion of Tyr^1 results in a marked loss of potency (*see* Fig. 2). The Y2 receptor is less stringent in its demand on the N terminus and recognizes a great variety of C-terminal NPY/PYY fragments. In general, longer fragments are more potent than shorter forms. The 12 C-terminal amino acid residues are the minimum length required to activate this receptor. Bold amino acid residues indicate positions where a substitution results in a marked loss of affinity to the receptor *(1,56)*.

3. Second-Messenger Systems

It was suggested quite early that NPY/PYY receptors are coupled to G proteins *(65)*. Since then, a number of studies have indicated that NPY may exert many actions through G-protein associated second-messenger generation (e.g., *40*). The study of second-messenger systems may not be helpful in providing a basis for receptor type classification, since the Y1, Y2, and Y3 receptors seem capable of activating the same intracellular pathways in many systems, resulting in reduced cAMP accumulation and elevated intracellular Ca^{2+} concentrations *(40,50,66)*. Thus, NPY is able to inhibit the stimulated activation of adenylate cyclase in various tissues, such as cerebral vessels *(67)*, ciliary processes *(68)*, vascular smooth muscle cells *(69)*, and spleen *(19)*. This has also been observed in various areas in the CNS *(70–72)*. Studies of human neuroblastoma cell lines have shown that both Y1 and Y2 receptor types inhibit the stimulated formation of cAMP *(40,50,73)*.

In many instances, NPY-elicited elevation of intracellular Ca^{2+} has been demonstrated and/or inferred. Thus, mobilization of intracellular calcium in response to NPY has been observed in central neurons *(74)*, vascular smooth muscle cells *(75)*, and different human cell lines *(73,76)*. Calcium ion influx has been demonstrated following stimulation of Y1 and Y3 receptors *(40,47,50,64)*. Moreover, NPY seems to cause accumulation of inositol phosphates (IP) in some blood vessels *(77)* and in the cerebral cortex *(42,78)*. The release of intracellular calcium ions seems to be activated both by phosphatidyl-inositol-specific phospholipase-C-dependent and independent pathways *(40,66)*.

It is still not known if one and the same G protein (probably G_i or G_o) may mediate the coupling to cAMP and Ca^{2+}. However, through the recent cloning and heterologous expression of the human Y1 receptor, it has become evident that stimulation of this receptor subtype in one and the same cell (CHO and COS-1) can result in changes in both cAMP and Ca^{2+}; however, in 293 cells, no Ca^{2+} coupling was observed *(47,79)*. Thus, there is a possibility, although not definitively shown, that NPY/PYY couples to multiple G proteins. To date, there is no evidence convincingly demonstrating that Y1, Y2, and Y3 receptors differ with respect to second-messenger coupling.

4. Y1 Receptors

A number of physiological processes both in the CNS and in the periphery have been attributed to activation of Y1 receptors. It is well known that iv injections of NPY or PYY equally potently increase arterial blood pressure via receptors distinct from α-adrenoceptors (e.g., *22,80–82*). The Y1 receptor was originally defined negatively by the fact that, unlike the Y2 receptor, it was not activated by C-terminal NPY/PYY fragments, like NPY_{13-36} *(41,42)*; *see also* Figs. 2 and 3. The Y1 receptor has so far mainly been characterized in the periphery, where it is associated with vasoconstriction (e.g., *22,40*).

4.1. Distribution

4.1.1. In the Periphery

From bioactivity studies, the Y1 receptor was suggested early to be postjunctional at the vascular sympathetic neuroeffector junction *(41)*. A large number of binding studies have demonstrated the presence of NPY/PYY receptors in the vascular smooth muscle *(20–23,83)*. In attempts to identify the vascular NPY/PYY receptor type on vascular smooth muscle cells, it was found that the Y1 receptor agonists $[Pro^{34}]$NPY and NPY equally effectively displaced radiolabeled PYY in a monophasic manner, suggesting a homogeneous population of Y1 receptors *(22,84)*. These binding sites have recently been visualized by an autoradiographic approach at the electron-microscopic level. A few Y1 binding sites were also detected on the vascular endothelium *(84)*.

4.1.2. In the Central Nervous System

Autoradiography studies using the Y1 receptor agonist $[Pro^{34}]$NPY on frozen sections of rat brain have indicated that Y1 receptors are abundant in the cerebral cortex and anterior olfactory nucleus. A mixture of Y1 and Y2 types may exist in several other brain areas, such as thalamus, hypothalamus, basal ganglia, and brainstem *(45,85)*. Furthermore, *in situ* hybridization with Y1 receptor cDNA from rat has localized expression of the mRNA encoding this receptor protein in thalamus, cerebral cortex, and parts of hippocampus (*86*, unpublished data).

4.2. Biological Effects

4.2.1. Peripheral Y1 Receptors Evoke Vasoconstriction

The NPY-evoked increase in arterial blood pressure is mediated by Y1 receptors, since Y1 receptor agonists are equally potent and equally effective with NPY/PYY *(22,57,82,87)*. In contrast, NPY 2–36 and shorter C-terminal fragments are markedly less potent or inactive *(22)*. NPY is able to increase the vascular resistance in various vascular beds of many species *(90–93)*. By the use of the Y1 receptor agonist [Pro^{34}]NPY or C-terminal NPY fragments in some bioactivity assays, the presence of Y1 receptors has been suggested in rat, guinea pig, and rat coronary vessels *(22,94)*, guinea-pig inferior caval vein *(22)*, and blood vessels in the rabbit maxillary sinus *(95)*. Removal of the endothelium does not affect the NPY-evoked vasoconstriction, suggesting the vascular rather than the endothelial Y1 receptor evokes vasoconstriction *(96)*. The influx of extracellular Ca^{2+} seems to be required, since several Ca^{2+} channel antagonists suppress NPY-evoked vasoconstriction both in vivo *(97)* and in vitro *(98)*.

NPY is also able to potentiate the vasoconstriction evoked by norepinephrine and other vasoactive agents both in vivo and in vitro *(5,98,99)*. The Y1 receptor seems to mediate the enhancement of norepinephrine-evoked vasoconstriction, since NPY/PYY, but not C-terminal NPY/PYY fragments (e.g., $PYY_{13–36}$ and $NPY_{11–36}$) have this ability *(41,100)*. Whether this response is endothelium-dependent or not is still controversial *(101–103)*. Like the direct NPY-evoked contractile effect, the potentiation by NPY of NE-induced vasoconstriction has been suggested to be Ca^{2+} dependent *(104)*.

Thus, the predominant vascular NPY/PYY receptor seems to be of the Y1 type. However, there is functional and biochemical evidence to suggest a mixture of vascular vasoconstriction-related Y1/Y2 receptors in some vascular beds *(22,105)*. NPY exerts a variety of effects outside the vascular bed and many of these responses have not yet been characterized in terms of receptor subtype classification.

4.2.2. In the Central Nervous System

In the CNS, the Y1 receptor has been linked with different biological actions, for instance, with NPY-induced stimulation of feeding behavior *(36,106)*. Central administration of NPY potently stimulates feeding behavior in rats, even in satiated rats. This effect can be mimicked by the Y1 analog [Leu[31],Pro[34]]NPY, whereas the NPY_{13-36} is much less active. These data suggest that the feeding response is mediated by Y1 receptors *(36)*. Moreover, stimulation of LHRH (luteinizing hormone-releasing hormone) is another central effect of NPY that may be mediated by Y1 receptors *(36)*.

Finally, behavioral studies in the rat have indicated that central administration of NPY, but not NPY_{13-36} evokes a suppression of spontaneous locomotor activity *(107,108)*. This was interpreted as a sedative-anxiolytic effect and may represent yet another Y1-mediated effect in the CNS *(108)*.

4.3. Biochemical Characterization

4.3.1. Ligand Binding Studies

Binding assays have traditionally been used to test the affinity and specificity of neurotransmitter or hormone receptors. The ligand binding characteristics of the Y1 receptor have been investigated using preparations of smooth vascular muscle cells or membranes, and different neuroendocrine cell lines as the source of receptors *(20,22,40,43,44)*. It has been found that several neuroendocrine cell lines selectively express Y1 receptors, which, short of isolation/cloning (*see* Section 9.), has made it possible to study one receptor site only. Furthermore, some of these cell lines have expressed quite high amounts of receptors, e.g., the neuroblastoma cell line SK-N-MC, which has approx 200,000 sites/cell. Analogous with the results from bioactivity studies, a Y1 type of binding site was described. This site bound NPY and PYY with high affinity in the subnanomolar range, whereas the long C-terminal fragments were unable to displace the NPY radiolabel *(43,44)*. This binding profile is present on at least three neuroblastoma cell lines: SK-N-MC, MC-IXC, and CHP-212, and on a rat pheochromocytoma cell line: PC 12 cells. The neuroblastoma cell lines represent easy and reliable systems for testing different NPY analogs for putative agonistic or antagonistic activity. In fact, it was from screening of different NPY analogs on Y1- or Y2-specific cell lines that the specific Y1 agonist, [Leu[31],Pro[34]]NPY was characterized *(57)*. In SK-N-MC cells, the IC_{50} for displacing radiolabeled NPY is 2.1 n*M* and 3.8 n*M* for NPY and [Leu[31],Pro[34]]NPY, respectively, whereas the value for NPY_{13-36} was more than 100 times higher *(57,60)*. The binding profile corresponding to Y1 receptors found on the neuroblastoma cells

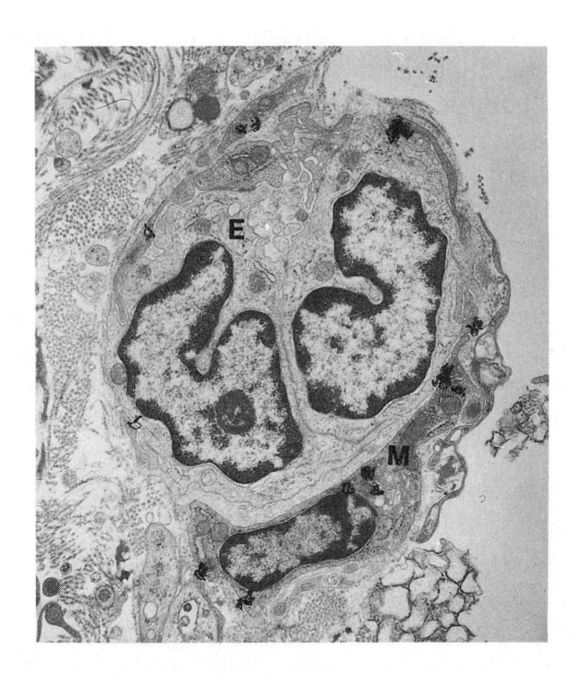

Fig. 4. Electron-microscopic autoradiogram of radiolabeled PYY bound to isolated vascular fragments from rat pancreas; this is likely to to represent Y1-type binding *(see text)*. E; endothelial cell. M; smooth muscle cell.

has been found in more "physiological systems," such as aortic membranes from the rabbit, cultured smooth cells from rat aorta, and intact vascular fragments from rat pancreas *(20,22,84)*. An electron micrograph of Y1-type binding in isolated vascular fragment from rat pancreas is shown as Fig. 4.

4.3.2. Affinity Labeling Studies

Studies in which a radiolabeled ligand is covalently attached to the receptor to enable preliminary structural information on the receptor molecule have been performed for the NPY/PYY receptors subtypes. The Y1 receptor in the neuroblastoma cell line MC-IX-C has been affinity labeled using [^{125}I,Tyr^{36}]PYY and a photoactivated crosslinking reagent ANB-NOS. An M_r = 70,000 protein was the major labeled species *(46)*. The intensity of this labeling could be inhibited by increasing concentrations of unlabeled PYY. Reducing agents did not affect the labeled band, which suggests that the Y1 receptor is not disulfide-linked to other subunits. These results have been confirmed using another neuroblastoma cell line, SK-N-EPl, which also possesses Y1 binding sites *(109)*. In this cell line, an M_r = 68,000 protein was labeled. In a membrane preparation from rat cerebral cortex, which from autoradiographic studies appears to contain predominantly Y1 receptors *(45)*, an M_r = 62,000 protein was labeled using Bolton-Hunter-labeled NPY and disuccinimidyl suberate *(110)*. These authors did not receptor-classify this binding site, which, however, may represent a Y1 binding site. The apparent differences in mol-weight determination between the proteins are likely to be attributed to differences in experimental protocol, type of radiolabel employed, or sample preparation for SDS-gel electrophoresis. Like many hormone receptors, the M_r = 70,000 protein appears to be glycosylated. The sugar residues were initially identified by the adsorption of the crosslinked and solubilized PYY-Y1 receptor complexed to various lectins. These data suggest that carbohydrate moieties of polymeric β1,4-*N*-acetylglucosamine and terminal branched mannose residues probably are linked to the Y1 receptor protein *(46)*. Thus, the Y1 binding site can be clearly distinguished by biochemical studies, and it appears to be a glycosylated protein in the M_r = 62–70,000 range. *See* Fig. 5.

4.4. Structural Ligand Requirements

NPY and its related peptides, in contrast to most other small peptide messengers, retain a distinct tertiary structure in solution *(111)*. There has been a considerable interest in structure–function studies of NPY-related peptides in attempts to develop receptor-specific agonists and antagonists. These experiments have been extensively pursued by using different NPY fragments and analogs both in receptor assays and different biological preparations. In general, PYY is equipotent and equally effective with NPY in all Y1 (and Y2) receptor assays studied *(40)*.

4.4.1. The N-Terminal End of NPY

Characteristic of the Y1 receptor is that a truncation of the first N-terminal residue Tyr^1 (NPY 2-36) results in a marked loss of biological activity or affinity *(22,94)*. Table 1 shows the relative potency of various NPY-related peptides at Y1 receptors. Further N-terminally truncated NPY fragments are even less potent or inactive. For instance, NPY_{13-36} is about 100–400 times less po-

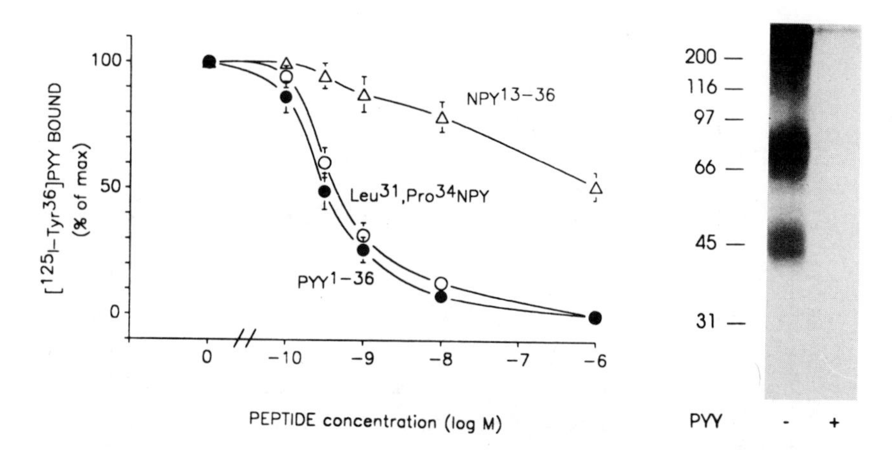

Fig. 5. Left: competition inhibition of [^{125}I,Tyr^{36}]PYY by PYY, [Leu^{31}, Pro^{34}]NPY and NPY$_{13–36}$ in isolated vascular fragments from rat pancreas. The vascular fragments were incubated with 100 μ*M* radiolabeled PYY with or without unlabeled peptides at the indicated concentrations (redrawn from *84*). Right: affinity labeling of Y1 receptors from the human neuroblastoma cell line MC-IXC. Membranes (200 μg/mL) were incubated with 200 p*M* radiolabeled PYY in the absence or presence of 1 μ*M* PYY. After 1 h at 24°C, the membranes were washed and crosslinked with the hetero-bifunctional reagent ANB-NOS (*N*-5-azido-2-nitrobenzoyloxysuccinimide), which is activated by UV-light as described *(46)*.

tent than NPY at this receptor. The requirement for an intact N terminus is also illustrated by the fact that a single substitution of the Y1 receptor agonist [Pro^{34}]NPY from Lys^4 to Glu^4 ([Glu^4,Pro^{34}]NPY) results in a loss of affinity to the Y1 receptor *(1)*. Furthermore, systematic D substitutions in either of the first five amino acid residues in the N-terminal end of NPY result in a marked loss of potency at the Y1 receptor *(59)*. Thus, the Y1 receptor requires an intact N terminus of NPY in order to become fully activated (Figs. 2 and 3).

4.4.2. The C-Terminal End of NPY

The Y1 receptor is slightly less stringent in its demand on this part of the NPY molecule, since substitutions from Ile to Leu in position 31 and from Gly to Pro in position 34 result in analogs ([Pro^{34}]NPY and [Leu^{31},Pro^{34}]NPY), that retain full activity on Y1 receptors, however, being inactive at Y2 receptors (Table 1). It has

Table 1
Relative Potency of NPY-Related Peptides in Comparison with NPY in Tissues Containing Y1 Receptors

Peptide	Relative potency[a]	Test system	Response/ affinity[b]	References
NPY 2–36	1:50	Rat aortic smooth muscle	A	22
NPY 13–36	1:150	Rat aortic smooth muscle	A	22
	1:400	Pig aortic smooth muscle	R/A	*112*
	1:400	Human SK-N-MC cells	R/A	*40*
	1:340	Human HEL cells	A	*113*
NPY 18–36	1:500	Rat aortic smooth muscle	A	22
	1:170	Rabbit aortic smooth muscle	A	20
NPY 26–36	<1:1000	Rat aortic smooth muscle	A	22
Desamido-NPY	0	Guinea-pig iliac vein	R	*41*
[Pro^{34}]NPY	1:1	Rat arterial blood pressure	R	*22,87*
	1:1	Human SK-N-MC cells	R/A	*40*
	1:1	Rat aortic smooth muscle	A	22
	1:1	Guinea-pig caval vein	R	22
[Pro^{20}]NPY	<1:1000	Human SK-N-MC cells	A	*60*
PYY 1–36	2:1	Human SK-N-MC cells	R/A	*40*
	1:1	Rat arterial blood pressure	R	*80,82*
	1:1	Guinea-pig iliac vein	R	*41*
	1:1	Rat aortic smooth muscle	A	21
PP	<1:1000	Human PC 12 cells	A	53
	<1:1000	Human SK-N-MC cells	A	*40,50*

[a]0, no biological effect; [b]R, biological response; A, affinity to binding sites of NPY-related peptides in various experimental setups.

been suggested that deamidated NPY-related peptides no longer form a stable tertiary structure *(111)*. The maintenance of this structure seems to be crucial for receptor recognition, since desamido-NPY is inactive at Y1 and Y2 receptors (Tables 1 and 2).

4.4.3. The Hairpin Loop of NPY (the PP Fold)

The hairpin loop has been suggested to present the N and C termini close together for recognition at the Y1 receptor *(1)*. "Peptide engineering" approaches with substitution of the hairpin loop, with small bridging constructs have been employed in order to delineate which part of the ligand activates the receptor *(58,60)*. Several centrally truncated NPY analogs with intact N and C termini are quite potent at the Y1 receptor, suggesting that the center part of the NPY molecule is not involved in Y1 receptor recognition *(88,120)*. The importance of the close steric arrangement of the N and C termini is illustrated by the fact that substitution to Pro in position 20 in $[Pro^{20}]$NPY breaks the hairpin-like loop, resulting in a loss of affinity to the Y1 receptor (Table 1), retaining, however, full activity at the Y2 receptor (Table 2). The proposed rank order of potency of NPY-related peptides on Y1 receptors is shown in Table 3.

5. Y2 Receptors

In general, Y2 receptors are predominantly known for their association with inhibitory effects in many biological systems. It was reported early on that NPY, in addition to evoking postjunctional effects, was also able to suppress the release of transmitters via prejunctional receptors at sympathetic neuroeffector junctions, such as in the vascular bed *(121,122)* and in the vas deferens *(123)*. The Y2 receptor was originally defined by the observation that NPY_{13-36}/PYY_{13-36}, in contrast to postjunctional Y1 receptors, was almost equipotent with NPY/PYY on prejunctional receptors in the rat vas deferens *(41)*. *See also* Figs. 2 and 3.

5.1. Distribution

5.1.1. In the Periphery

Y2 receptors are generally considered to be localized at prejunctional sites at the sympathetic neuroeffector junction, suppressing the release of transmitters *(41,55,100)*. Y2 receptors may also be located on other nerve fibers, like parasympathetic *(119)* and sensory C fibers

Table 2
Relative Potency of NPY-Related Peptides in Comparison with NPY in Tissues Containing Y2 Receptors

Peptide	Relative potency[a]	Test system	Response/affinity[b]	References
NPY 2–36	1:2-5	Rat vas deferens	R	*115,117*
	1:1	Rat hippocampus	R	*116*
	1:4	Pig spleen	R	*4*
NPY 13–36	1:10-15	Rat vas deferens	R	*55,117*
	1:4	Pig hippocampus	A	*26,43,44*
	1:20	Human SMS-MSN cells	A	*26,43,44*
	1:6	Human SMS-KAN cells	A	*57*
	1:6	Human SK-N-BE2 cells	R/A	*40,50*
	1:3	Guinea-pig sensory C fiber	R	*118*
	1:1	Rat parasymathetic fiber	R	*119*
	1:6-14	Pig spleen	A	*4,19*
	1:10	Pig splenic vessels	R	*19*
NPY 22–36	1:36	Rat vas deferens	R	*55*
	1:10	Human SK-N-BE2 cells	A	*40,50*
NPY 26–36	0	Rat vas deferens	R	*53,55*
NPY 1–24	0	Rat vas deferens	R	*54*
Desamido–NPY	0	Pig spleen	R	*4*
	0	Rat vas deferens	R	*55*
[Pro^{34}]NPY	<1:1000	Pig spleen	A	*56*
	<1:1000	Human SK-N-BE2 cells	R/A	*40,50*

[Pro^{20}]NPY	1:10	Pig hippocampus	A	*60*
PYY 1–36	3:1	Rat vas deferens	R	*55*
	2:1	Human SK-N-BE2 cells	R/A	*40,55*
	1:1	Pig hippocampus	A	*26,43,44*
	1:1	Rat hippocampus	R	*116*
	3:1	Rabbit kidney	A	*26,43,44*
PYY 13–36	1:2-15	Rat vas deferens	R	*41,55*
	1:10	Pig hippocampus	A	*26,43,44*
PP	<1:1000	Pig hippocampus	A	*53*
	<1:1000	Human SK-N-BE2 cells	A	*40,50*

[a]0, no biological response; [b]R, biological response; A, affinity to binding sites of NPY/PYY fragments or NPY analogs in various experimental setups.

Table 3
Proposed Rank Order of Potency of NPY and Related Peptides at the Receptors for NPY and Related Peptides

Receptor	Peptides	Tissue
Y1	[Pro^{34}]NPY = NPY = PYY >> NPY 13–36 >> PP	Blood vessels Cerebral cortex SK-N-MC cells
Y2	PYY ≥ NPY > NPY 13–36 >> [Pro^{34}]NPY, PP	Nerve endings Renal tubular cells Hippocampus SK-N-BE2 cells
Y3	NPY ≥ [Pro^{34}]NPY ≥ NPY 13–36 >> PYY, PP	Brainstem Cardiac membranes Adrenal medulla Hippocampus?
PP	PP >> NPY, PYY	Vas deferens Brainstem PC12 cells

(118). The receptor locations both along the rabbit kidney proximal tubules and on tubule cell have been determined using an autoradiographic approach at the electron-microscopic level *(27)*. A high density of Y2 type of binding sites was found on the basolateral membrane in a particular segment of the proximal tubule *(27)*. Furthermore, the presence of Y2 receptors has also been suggested on rat platelets *(125)*.

5.1.2. In the Central Nervous System

Autoradiographic data from the rat brain using radiolabeled PYY, displaced with Y1- or Y2-selective ligands have suggested a differential localization of Y1 and Y2 receptors *(45,85)*. In general, it appears that the Y2 receptor is the predominant receptor type in the rat brain *(85)*. A particularly dense population of Y2 receptors occurs in hippocampus *(45,85)*. Previous ligand binding data from rat and pig hippocampus suggest that this tissue may contain a homogeneous population of Y2 binding sites *(26,43,44,126)*. More specifically, from electrophysiological studies, the Y2 receptor has been characterized on excitatory terminals forming synapses with rat hippocampal CA1 neurons *(116)*.

5.2. Biological Effects

A variety of physiological processes both in the central nervous system and in the periphery have been attributed to activation of Y2 receptors. Many of these actions are linked to suppression of the release of various transmitters from central *(116,127)* and peripheral nerve fibers *(55,118,119)*.

5.2.1. In the Periphery

The rat vas deferens is a much-used preparation for the study of Y2 receptors. In attempts to characterize this receptor, it has been shown that NPY and PYY and a large number of C-terminal fragments of NPY/PYY (2–36 to 22–36) quite effectively suppress the electrically (sympathetically) paced twitches in a concentration-dependent manner *(53,55)*. Similarly, NPY and PYY as well as several C-terminal NPY fragments have the ability to attenuate stimulated ion secretion in the rat small intestine, suggesting the presence of Y2 receptors *(128)*. Furthermore, C-terminal NPY fragments (NPY_{11-36} to NPY_{16-36}) have been shown to be as effective as the parent molecule in suppressing the release of norepinephrine in the rat mesenteric artery, which is consistent with Y2 receptors *(100)*.

Both NPY and NPY_{13-36} are able to suppress sensory C fiber-mediated contractions of the guinea-pig bronchi *(118)*, and parasympathetic activity supplying the rat heart *(129)* and guinea-pig uterine cervix *(119)*, which suggests a direct action via Y2 receptors on these nerve fibers. Moreover, there is functional and biochemical evidence to suggest the presence of vasoconstriction-related Y2 receptors in the pig spleen *(19)*, guinea-pig caval vein *(22)*, and rat mesenteric artery, respectively *(105)*. Y2 receptors have recently been suggested to inhibit mucociliary activity in the rabbit maxillary sinus *(95)*.

5.2.2. In the Central Nervous System

NPY and PYY, as well as several C-terminal NPY fragments, were able to suppress the release of glutamate from rat hippocampal CA3 neurons via a prejunctional mechanism *(116,127)*. There was a gradual loss of potency with a progressive N-terminal truncation of the molecule, which is characteristic of Y2 receptors *(116)*. There are several lines of evidence suggesting the presence of Y2 receptors in the hippocampus. NPY and NPY_{20-36} have been shown to enhance memory retention, probably reflecting a Y2 receptor-mediated action in the rostral part of the hippocampus *(130)*.

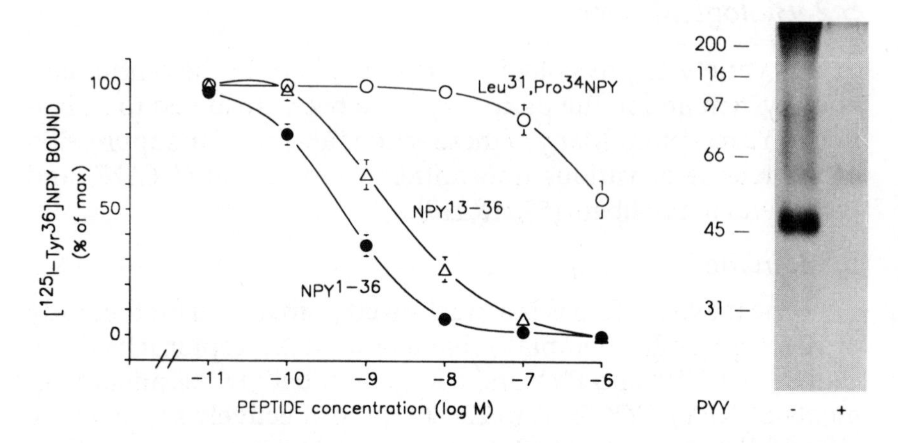

Fig. 6. Left: competition inhibition of [^{125}I,Tyr^{36}]NPY by NPY, [Leu^{31},Pro^{34}]NPY and NPY$_{13-36}$ in hippocampal membranes from the pig. The membranes were incubated with 100 p*M* radiolabeled NPY for 1 h at 24°C. Right: autoradiogram showing affinity labeling of Y2 receptors in rat hippocampal membranes with radiolabeled PYY (200 p*M*) in the absence or presence of 1 μ*M* unlabeled PYY. The crosslinking was performed using the homo-bifunctional reagent EGS (ethylene glycol-*bis*-(succinimidylsuccinate), as described *(131a)*. Positions of marker proteins are indicated.

5.3. Biochemical Characterization

5.3.1. Ligand Binding Studies

NPY receptors were first described by ligand binding studies using preparations from rat brain cortex *(131)*. However, an especially high amount of NPY binding activity was detected in membrane preparations from the hippocampus both in the rat and pig (Fig. 6) *(26,43,44,46,126)*. The hippocampal NPY receptor turned out to be the role model for biochemical studies of the Y2 receptors. The pig hippocampus contains approx 400 fmol of NPY/PYY receptors/mg of tissue, and has an affinity of approx 100 p*M* for NPY *(26,43,44)*. Scatchard transformation of the binding data suggested the presence of one binding site only. The specificity of the receptor is as follows; PYY binds as well as NPY, and the affinity for NPY$_{13-36}$ is only about five times lower, whereas the potency of PP is 10^4 times lower. Furthermore, the radiolabeled (on tyrosine 36) fragment NPY$_{13-36}$ bound with high affinity (K_d = 0.15 n*M*) to a single binding site in the hippocampus membranes, suggesting that the

C-terminal fragment is a relatively good ligand for this NPY receptor *(26,43,44)*. In line with these data, it was also found that several neuroblastoma cell lines selectively expressed Y2 receptors *(26,43,44)*. In the cell lines CHP-234, SMS-KAN, and SMS-MSN, NPY 13–36 was able to displace $[^{125}I,Tyr^{36}]$NPY with a potency four timers lower than that of the intact molecule *(26,43,44)*, whereas the cells did not bind the Y1 receptor analog $[Leu^{31},Pro^{34}]$NPY *(57)*.

Binding sites corresponding to Y2 receptors have also been detected in membrane preparations of secretory epithelial cells from the rat jejunum and rabbit kidney proximal tubules *(26,28,43,44)*. The intestinal receptor binds PYY with slightly higher affinity than NPY. However, the affinity for the long C-terminal fragments is only four to five times lower *(30)*. Similarly, in the rabbit proximal tubules, the NPY_{13-36} binds with a potency only five times less than PYY *(26,43,44)*.

5.3.2. *Affinity-Labeling Studies*

The structure of Y2 receptors from several tissues and species has been probed using affinity-labeling techniques *(46,126,131b)*. In rat and pig hippocampal membranes and in basolateral vesicles from the rabbit kidney, an M_r = 50,000 molecule was the major labeled species *(46,126)*. The labeling of this band was specific for PYY and NPY, and could be displaced by PYY with a half-maximal displacement of 160 p*M*, which is almost identical to the IC_{50} value from membrane binding studies in the rabbit kidney *(46)*. A smaller species of M_r = 38,000 appeared to be a degradation of the larger band, since its labeling intensity was dependent on the presence or absence of various protease inhibitors *(46)*. In addition, studies using the solubilized affinity-labeled Y2 receptor and various lectins suggest the possibility that all the crosslinked molecules are glycosylated with N-terminal *N*-acetylglucosamine residues and that some molecules in addition contain terminal branched mannose residues *(46)*. In the rat enterocyte, an M_r = 52–59,000 was identified. However, in this tissue, the authors also identified a 37–39,000 species. Disregarding the discrepancy regarding the smaller species, it can be concluded that the Y2 receptor appears to be a single glycoprotein of M_r = 50–59,000, whereas the Y1 receptor appears to be a large molecule in the M_r = 62–70,000 range. Thus, analysis of the two NPY receptor types suggests that they are structurally distinct glycoproteins *(46)*.

5.3.3. Solubilization of the Y2 Receptor

NPY receptors have been solubilized in a functional state, i.e., gentle solubilization in a state where the receptor is still able to bind ligand, from cerebral cortex, kidney, and spleen *(132–135)*. All of these reports have employed the zwitterionic detergent CHAPS to solubilize NPY receptors in a functionally active state. In the case of the kidney Y2 receptor, the natural detergent digitonin is even more efficient in solubilizing PYY binding activity than CHAPS *(135)*. Several differences between the solubilized receptors from the different tissues are apparent. First, the Y2 receptors in the rabbit kidney retained a high affinity and appropriate ligand binding characteristics in the crude solubilized state *(134,135)*. By contrast, the solubilized receptors from brain and spleen showed a 10–20-fold decrease in affinity as compared to the membrane-bound receptors *(132,133)*. The second issue concerns receptor coupling to G protein. Both the kidney and brain receptors are sensitive to guanine nucleotides in the form of GTPγS in the membrane-bound state. The solubilized Y2 receptors from the rabbit kidney retain this GTPγS sensitivity, whereas the cortex cerebri and spleen receptors do not. These data suggest that the crude solubilized kidney receptor is still associated with the G protein after the detergent treatment *(135)*. Although it is tempting to speculate that the decrease in binding affinity may be owing to the dissociation of the receptor G-protein complex, this explanation may not apply for the purified Y2 receptor from rabbit kidney. Although it has lost the association with the G protein, it retains the original high binding affinity of ligands *(135)*. Purification of this Y2 receptor to homogeneity was achieved by successive Mono S cation-exchange adsorption, affinity chromatography on wheat germ lectin agarose beads, and affinity chromatography on NPY-AffiGel *(135)*. Electrophoresis and silver staining of the final purified preparation revealed a single protein with an M_r = 60,000. Furthermore, the purified protein can be affinity labeled with radiolabeled PYY, indicating that it contains the ligand binding site of the Y2 receptors *(135)*. Since very few peptide hormone receptors retain their original affinity and specificity in the purified state, the kidney Y2 receptor may be an interesting model to probe for functional activity, for example, reconstituted with purified G proteins in phospholipid vesicles.

5.4. Structural Ligand Requirements

5.4.1. The N-Terminal End of NPY

This part of the molecule is less important for activation of the Y2 receptor than of the Y1 receptor. Hence, NPY_{2-36} is in contrast to the Y1 receptor, about equipotent with NPY, and N-terminally truncated NPY fragments from NPY_{2-36} to $_{22-36}$ are rather potent. Table 2 shows the relative potency of various NPY-related analogs at Y2 receptors. For instance, NPY_{13-36} is about five to ten times less potent than the parent molecule. PYY and its C-terminal fragments appear to be slightly more potent than the corresponding NPY peptide *(55)*. The breaking point of biological activity seems to be between position 25 and 26. Thus, the 12 C-terminal amino acid residues of NPY/PYY are the minimum length required to activate this receptor *(53,55)* (Figs. 2 and 3).

5.4.2. The C-Terminal End of NPY

The Y2 receptor is very stringent in its demand on this part of the NPY molecule, since an intact C-terminal end of NPY is required for activation of the Y2 receptor. This is illustrated by the fact that a substitution from Gly to Pro in position 34 in [Pro^{34}]NPY results in a loss of affinity to the Y2 receptor (Table 2). N-terminal fragments of NPY are inactive *(4,54,136)* (Table 2). From studies where amino acid residues have been substituted systematically, it has been suggested that His^{26} is important for the recognition of the Y2 receptor *(136a)*. Not only is the amino acid sequence of the C-terminal of NPY essential, but also the C-terminal amide group, since desamido-NPY and the C-terminally extended form NPY-Gly-Lys-Arg fail to activate the Y2 receptor (55; Table 2).

5.4.3. The Hairpin Loop

The hairpin loop (PP fold) serves to bring the N and C termini close together *(1)*. Several NPY analogs with the center part of the molecule substituted with links between the N and C termini were shown to be quite potent, suggesting that the hairpin loop of the molecule is not involved in the recognition at the Y2 receptor *(88)*. The importance of an intact C terminus, but not a hairpin loop, is also illustrated by the observation that Pro in position 20 in [Pro^{20}]NPY, which breaks the hairpin loop, is a full Y2 receptor agonist *(60)*. Exchange of the hairpin loop of NPY with that in PP in $PP_{(1-30)}NPY_{(31-36)}$ results in a hybrid peptide that retain very high affinity to the Y2 receptor *(60)*. Conversely, $NPY_{(1-30)}PP_{(31-36)}$ does not recognize this receptor. The proposed rank order of potency of NPY-related peptides on Y2 receptors is shown in Table 3.

6. NPY-Selective Y3 Receptors

There is an increasing number of binding and bioactivity studies from various tissues where the pharmacological order of potency of the NPY-related peptides differs markedly from those of Y1/Y2 receptors. The caveat of these studies seems that these receptors recognize NPY, whereas PYY is several orders of magnitude less potent. There is by now much evidence to suggest the existence of specific NPY receptors, present both in the CNS and periphery. These NPY receptors are referred to as Y3.

6.1. Y3 Receptors in the Central Nervous System

The nucleus tractus solitarius (NTS), the site of termination in the brainstem of primary afferent fibers of arterial baroreceptors, is richly innervated by neurons containing NPY *(137)*. NPY injected into NTS is known to evoke a long-lasting hypotension and bradycardia *(35,48,52,138)*. By injecting NPY-related peptides into NTS, the receptors mediating the NPY-evoked cardiovascular effects have been characterized. Unilateral injections into NTS of NPY, the Y1 receptor agonist $[Pro^{34}]$NPY, or the Y2 receptor agonist NPY 13–36 evoked comparable dose-dependent falls of arterial blood pressure and heart rate of the anesthetized rat. However, the Y1/Y2 receptor agonist PYY was inactive, as were PP and desamido-NPY *(48,52)*.

Glutamate may be the presumed endogenous transmitter of primary baroreceptor afferents *(141)*. Like the cardiovascular effect profile of the NPY-related peptides, local pretreatment of NTS with NPY, $[Pro^{34}]$NPY, or NPY_{13-36} prevented (for hours) the fall in arterial blood pressure and heart rate evoked by subsequent injections of glutamate into the same site in NTS. Pretreatment of NTS with desamido-NPY, PYY, or PP did not affect the cardiovascular responses evoked by glutamate in NTS, suggesting this inhibition to be receptor-specific *(48,52)*. The mechanism behind the NPY-evoked refractoriness to glutamate is still unclear.

Since Y1 and Y2 receptor agonists were equally active, one explanation would be that both Y1 and Y2 receptors mediate the same cardiovascular effects in NTS. However, injection into NTS of PYY, which is about equipotent with NPY on peripheral Y1/Y2 receptors (e.g., *1,40*) was inactive. Also, Tseng et al. *(139)* found PYY to be virtually inactive in NTS. Together these results indicate that in the NTS, the mechanism of action of NPY is distinct from that in the periphery. In support of such a contention, quantitative autoradiography has shown that NPY, but not PYY or PP displaced radiolabeled NPY in the rat brainstem *(142)*. Moreover, functional and biochemical studies have suggested interactions between NPY and α_2-adrenoceptors in NTS *(143–145)*, but not between PYY and α_2-adrenoceptors *(145)*. Together these data indicate the existence of a specific NPY receptor, distinct from the Y1 and Y2 receptor types in the rat brainstem. The cardiovascular effects of NPY in NTS have been linked to G proteins, which are sensitive to pertussis toxin *(144)*. Furthermore, in vitro NPY seems to suppress the stimulated formation of adenylate cyclase in the NTS *(72)*. Thus, the specific NPY receptor in NTS seems to be coupled to G proteins.

Support for the existence of specific NPY receptors has very recently been suggested elsewhere in the brain from electrophysiological studies: In the CA3 neurons of the rat hippocampus NPY, [Leu^{31},Pro^{34}]NPY, and NPY_{13-36}, but not PYY or PP were found to potentiate the neural response to the glutamate receptor agonist *N*-methyl-D-apartate (NMDA) *(146,147)* (*but see* chapter by Bleakman et al., this volume). Further evidence for interaction of NPY with glutamate receptors has been obtained in a study where NPY was shown to enhance NMDA-induced norepinephrine efflux in the rat brain *(148)*.

NPY-containing neurons are abundant in the central nervous system *(15)*, whereas PYY-immunoreactive neurons seem to be rather few *(16,17)*. Therefore, it is not surprising that NPY might act at specific NPY receptor types in the central nervous system. In addition to the presence of Y1/Y2 receptors already demonstrated in the brain *(45,85)*, the present data suggest the existence of a specific NPY receptor in the brain that is distinct from the Y1 and Y2 receptor types; therefore, it could be referred to as a Y3 receptor.

6.2. Y3 Receptors in the Periphery

Nonetheless, NPY-specific binding sites have recently also been suggested in the periphery. On rat cardiac membranes and chromaffin cells of the bovine adrenal medulla, NPY, [Pro^{34}]NPY, and NPY_{13-36} potently displaced radiolabeled NPY, whereas PYY and PP were several orders of magnitude less potent *(24,49,50)*. It is well known that Y1/Y2 receptors equally recognize [$Tyr^{36}I^{125}$-NPY and Tyr^1I^{125}-NPY *(26,43,44)*. By contrast, in cardiac membranes, Tyr^1I^{125}-NPY has a higher specific binding than $Tyr^{36}I^{125}$-NPY, which further supports the existence of binding sites distinct from Y1 and Y2 receptors *(24)*. Radiolabeled PYY did not bind to the adrenal medulla *(50)*. In another study, NPY, but not PYY, was shown to inhibit nicotine-stimulated release of catecholamines from the adrenal medulla *(150)*. Like the NPY receptors in NTS, the NPY receptors in the adrenal medulla and heart seem to be coupled to G proteins *(24,50)*. Conceivably, these properties are those of a common specific NPY receptor subtype, similar in the brain, heart, and adrenal medulla. The proposed rank order of potency of NPY-related peptides on the tentative Y3 receptor is shown in Table 3.

6.3. Structural Ligand Requirements of the Y3 Receptor

The main characteristic of the NPY-specific Y3 receptor is that it, in contrast to Y1 and Y2 receptors, does not recognize PYY. The N terminus of NPY appears to have some importance for activation of this receptor, since NPY_{13-36} has 20–40 times lower affinity than the intact molecule (Table 4). The shape of the hairpin loop (PP fold) of NPY could be important for receptor recognition. The main differences between NPY and PYY are found in the 13–23 segment where PYY differs in 7 of 11 positions. As at the Y1 receptor, substitution to Leu in position 31 and to Pro in position 34 of NPY has virtually no effect on binding affinity to the Y3 receptor. Moreover, like Y1 and Y2 receptors, Y3 receptors require an amidated C-terminal in order to become activated (Table 4).

7. NPY/PYY Receptors on Mast Cells?

The effects of NPY in the peripheral cardiovascular system have mainly concerned vasoconstriction and increase in arterial pressure. However, many studies have, in addition, observed vasodepressor responses to NPY and its C-terminal fragments. Several studies have observed that systemic injection of C-terminal NPY fragments, such as NPY_{17-36} and NPY_{18-36}, elicits a marked hypotension in rats *(113,153–155)*. However, the mechanism behind the depressor response was never revealed. In order to characterize this response, it was shown that high doses of NPY_{18-36},

Table 4
Relative Potency of NPY-Related Peptides in Comparison with NPY in Tissues Containing Y3 Receptors

Peptide	Relative potency[a]	Test system	Response/affinity[b]	References
NPY 13–36	1:2	Rat brainstem	R	*52*
	1:37	Rat heart	A	*24*
	1:25	Rat adrenal medulla	A	*50*
$[Pro^{34}]$NPY	1:1	Rat brainstem	R	*52*
$[Leu^{31}, Pro^{34}]$NPY	1:7	Rat adrenal medulla	A	*50*
Desamido-NPY	0	Rat brainstem	R	*140*
PYY	0	Rat brainstem	R	*48,52,140*
	<1:1000	Rat brainstem	A	*142*
	<1:1000	Rat adrenal medulla	A	*50*
	1:210	Bovine adrenal medulla	A	*150*
	1:100	Rat heart	A	*24*
PP	0	Rat brainstem	R	*52*
	<1:66	Rat adrenal medulla	A	*50*
	<1:1000	Rat heart	A	*24*

[a]0, no biological response; [b]R, biological response; A, affinity to binding sites of NPY-related peptides in various experimental setups.

Table 5
Relative Potency of NPY-Related Peptides in Comparison with NPY at the Mast Cell

Peptide	Relative potency[a]
NPY 22–36	2:1
NPY 15–36	2:1
PYY 1–36	1:1
Desamido-NPY	(++)
[Pro^{34}]NPY	(+)
PP	(+)

[a]The ability of NPY-related peptides to release histamine from rat peritoneal mast cells. (++) active, but no complete concentration–response curves could be constructed; (+) active at very high concentrations only (cf text).

and NPY$_{22-36}$, like NPY itself, evoked a brief and transient pressor response, followed by long-lasting depressor response both in conscious and pithed rats *(156,157)*. The depressor, but not the pressor response to NPY or C-terminal NPY fragments was prevented by pretreatment with histamine H1-receptor antagonists or the histamine liberator compound 48/80, indicating that NPY and C-terminal NPY fragments are capable of releasing histamine from mast cells *(156,157)*.

In further attempts to characterize the mechanism of action, several C-terminal NPY/PYY fragments as well as their respective molecules were shown to evoke a concentration-dependent release of histamine from mast cells *(158,159)*. In contrast to Y1/Y2 receptors, NPY 22–36 was more potent, whereas NPY$_{26-36}$ was equally effective, but less potent than NPY/PYY. Desamido-NPY was active, although less so than the parent molecule, whereas PP and the Y1 receptor agonist [Pro^{34}]NPY were both virtually inactive *(159)*. The relative potency of NPY-related peptides on histamine release from mast cells are shown in Table 5.

The effect profile of the NPY-related peptides on rat peritoneal mast cells suggests that histamine is released by a mechanism that is distinct from the those associated with Y1/Y2/Y3 receptors. Conceivably, positively charged amino acid residues at the C-termini of NPY/PYY activate G proteins in the mast cell membrane by a nonreceptor mechanism, as has been suggested for tachykinins and other basic peptides (for a review, *see 160*). Supporting a nonreceptor mechanism is the very rapid kinetics (<10 s) *(160)* of the

NPY-evoked histamine release *(159)*. Whether the NPY-evoked release of mast-cell histamine is physiologically relevant is unclear. Interestingly, mast cells are particularly numerous around blood vessels and nerves *(161,162)*. Possibly, the NPY-evoked release of mast-cell histamine contributes to the sympathetic control of microcirculation in certain vascular beds.

8. PP Receptors

PP has been shown to bind with a high affinity to binding sites on a pheochromocytoma cell line (PC 12) *(51)*. In contrast, NPY or PYY did not recognize these binding sites, suggesting the presence of specific PP receptors *(1,163)*. In the brain, selective PP binding sites have been found in various areas, which are permeable through the blood–brain barrier, like the area postrema and adjacent to nuclei *(164)*. PP is a pancreatic hormone that inhibits pancreatic exocrine secretion and gall bladder contraction *(165)*. PP has recently been reported to suppress electrically evoked twitches in the rat vas deferens via receptors distinct from Y1 and Y2 *(53)*. Rat PP has proline in position 34 and, interestingly, the Y1-receptor agonist [Pro^{34}]NPY suppresses the twitches, although, much less potently and less effectively than NPY, suggesting that [Pro^{34}]NPY in addition may activate PP receptors in vas deferens *(115)*. The proposed rank order of potency of PP and related peptides is shown in Table 3.

9. Cloning of Receptors

9.1. The Y1 Receptor

Although Y1 receptors were expressed heterologously in *Xenopus* oocytes already, a few years ago *(40)*, this expression system did not allow cloning of the receptor. Instead, its initial cloning was from a rat forebrain cDNA library, and it relied on its homology with already cloned members of the G-protein-coupled superfamily of receptors *(86)*. However, Eva and coworkers were not able to identify their clone, named FC5, and it remained an "orphan" receptor clone when published. The distribution of FC5, as assessed by *in situ* hybridization in rat brain, was rather distinct *(86)*, and the pattern showed certain similarities with the autoradiographic distribution of the Y1 receptor (e.g., *85*). Therefore, Yee et al. *(166)* and

Larhammar et al. *(47)* hypothesized that the "orphan" FC5 clone might correspond to a rat Y1 receptor, and this cDNA was isolated by the polymerase chain reaction (PCR). The latter PCR product was then used to screen a human fetal brain cDNA library, and following identification of a positive clone, hY1-5, its sequencing, and insertion into an expression vector, COS1 cells were transfected; these cells bound ^{125}I-PYY with high affinity and a rank order of potency of competing ligands appropriate for a Y1 receptor. Also since appropriate second-messenger responses to NPY and PYY, i.e., reduction of cAMP accumulation and elevation of intracellular Ca^{2+}, were found, it was concluded that hY1-5 indeed encodes for a human Y1 receptor *(47)*. Independently, a human Y1 receptor was cloned by Herzog et al. *(79)*. The amino acid sequence of this putative receptor is shown in Fig. 7.

9.2. The Y2 Receptor

This receptor type has to the best of our knowledge not been cloned as of May of 1992. However, this probably is only a matter of time, since several research groups are taking various approaches, including biochemical purification and homology hybridization screening (*see also* Section 5.3.3.).

9.3. A Proposed Y3 Receptor

Like the Y1 receptor, a proposed bovine Y3 receptor clone *(152)* was isolated on the basis of its nucleotide homology with other members of the G-protein-coupled superfamily of receptors. Thus, Rimland et al. *(152)* were able to isolate a fragment of the receptor by PCR using a bovine locus coeruleus cDNA library as template; this was followed by the identification and sequencing of a full-length clone, called LC1. After some time as an "orphan" receptor, LC1 was found to confer on transfected cells high-affinity ^{125}I-NPY binding sites *(152)*; the predicted amino acid sequence of LC1 is shown in Fig. 7. More recently, Jazin et al. (unpublished data) isolated a corresponding human clone, but have been unable to establish that this clone confers on transfected cells NPY binding sites or NPY-associated second-messenger responses (Wahlestedt et al., unpublished data). Structurally, the Y1 and the Y3 receptors seem to be only distantly related (cf Fig. 7); for instance, the Y3 receptor appears to be more closely related to the interleukin-8 receptor than to the Y1 receptor. It is therefore at the present time

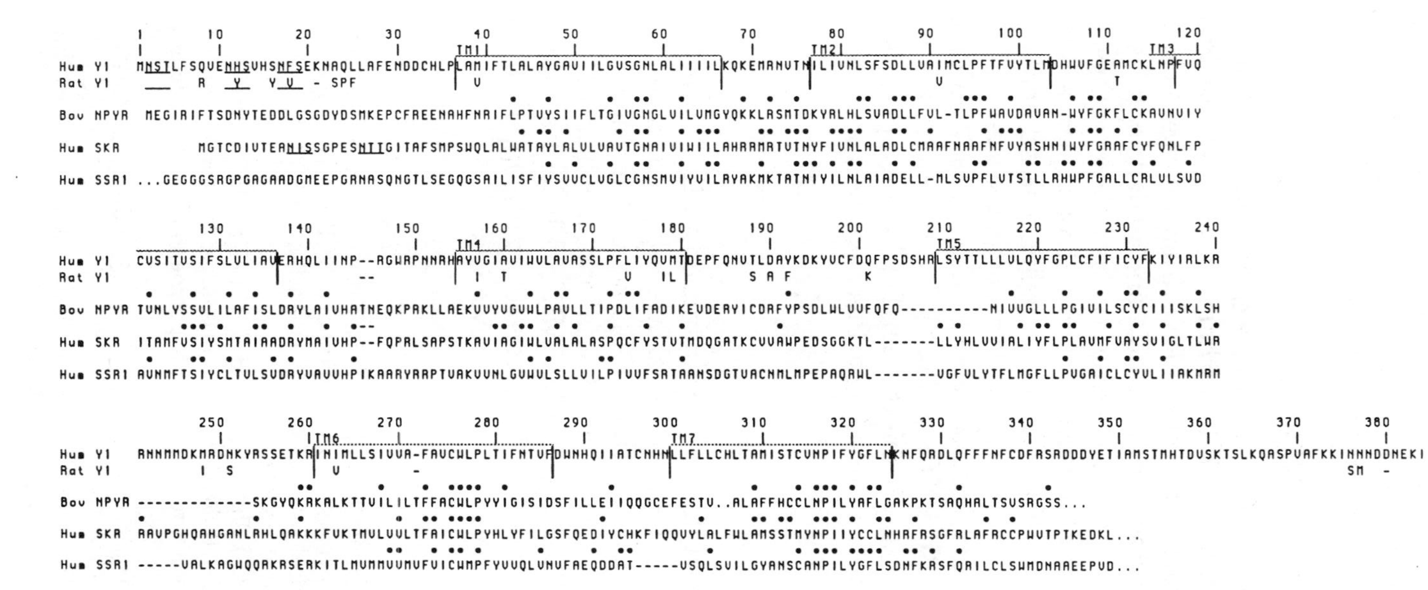

Fig. 7. Amino acid alignment of G-protein-coupled receptors. The human Y1 receptor *(47)* serves as master sequence. In the rat Y1 sequence, only positions that differ from the human sequence are indicated. Dashes represent gaps introduced to optimize alignment. Three underlined tripeptides at the amino terminus conform to the consensus sequence for N-linked glycosylation. The hydrophobic segments likely to be embedded in the cell membrane are bordered with vertical bars and overlined with dotted lines. TM stands for transmembrane. For comparison, the sequence is also shown for a bovine receptor that has been suggested to bind NPY with high affinity (LC1; *152*). Finally, the recently cloned human substance K receptor (SKR) and human somatostatin type 1 receptor (SSR1) have been aligned, demonstrating their structural relationships to the NPY/PYY receptors.

unclear whether the Y3 receptor really is cloned, and the literature awaits confirmatory results using the LCR1 clone and/or the corresponding clone isolated from other species.

9.4. A *Drosophila* NPY Receptor

Very recently, Li et al. *(167)* isolated a cDNA clone, PR4, from *Drosophila melanogaster* by a PCR-based homology approach. Upon microinjection of in vitro transcribed mRNA from this cDNA clone into Xenopus oocytes, it responded to NPY and related peptides in a manner typical of PI-coupled agonists, i.e., by an electrophysiological response reflecting Ca^{2+} mobilization and consequent activation of Ca^{2+}-dependent Cl^- channels. Such coupling makes it unlikely that this *Drosophila* receptor corresponds to any of the mammalian NPY/PYY receptor types discussed above. Also, the rank order of potency, PYY > C2-NPY > NPY > [Pro^{34}]NPY, differs from rank orders discussed previously in this chapter. Finally, attempts to isolate mammalian clones corresponding to PR4 have been unsuccessful (M. Forte, personal communication).

Acknowledgments

During the preparation of this manuscript, L. G. was the recipient of a fellowship from the Royal Swedish Academy of Sciences. S. P. S. wishes to thank Thue W. Schwartz, Lab. Molecular Endocrinology, Copenhagen, Denmark, and John A. Williams, Dept. Physiology, Univ. Michigan, Ann Arbor, MI, for scholarly scientific supervision over the years. C. W. received grant support from NHLBI and NIDA for part of the work reviewed.

References

1. Schwartz, T. W., Fuhlendorff, J., Kjems, L. L., Kristensen, M. S., Vervelde, M., O'Hare, M., Kretenansky, J. L., and Björnholm, B. (1990) Signal epitopes in the three-dimentional structure of NPY interaction with Y1, Y2 and PP receptors. *Ann. NY Acad. Sci.* **611,** 35–47.
2. Glover, I. D., Barlow, D. J., Pitts, J. E., Wood, S. P., Tickle, I. J., Blundell, T. L., Tatemoto, K., Kimmel, J. R., Wollmer, A., Strassburger, W., and Zhang, Y.-S. (1985) Conformal studies of the pancreatic polypeptide hormone family. *Eur. J. Biochem.* **142,** 379–385.
4. MacKerell, A. D., Hemsén, A., Lacroix, J. S., and Lundberg, J. M. (1989) Analysis of structure-function relationships of neuropeptide Y using molecular dynamics simulations and pharmacological activity and binding measurements. *Regul. Pept.* **25,** 295–313.

3. Allen, J., Novotny, J., Martin, J., and Heinrich, G. (1987) Molecular structure of mammalian neuropeptide Y: Analysis by molecular cloning and computer-aided comparison with crystal structure of avian homologue. *Proc. Natl. Acad. Sci. USA* **84**, 2532–2536.
5. Ekblad, E., Edvinsson, L., Wahlestedt, C., Uddman, R., and Håkanson, R. (1984) Neuropeptide Y coexists and co-operates with noradrenaline in perivascular fibres. *Regul. Pept.* **8**, 225–235.
6. Ekblad, E., Håkanson, R., and Sundler, F. (1984) VIP and PHI coexist with an NPY-like peptide in intramural neurones of the small intestine. *Regul. Pept.* **10**, 47–55.
7. Sundler, F., Håkanson, R., Ekblad, E., Uddman, R., Wahlestedt, C. (1986) Neuropeptide Y in peripheral and enteric nervous systems. *Ann. Rev. Cytol.* **102**, 234–269.
8. Sheikh, S. P., Holst, J. J., Skak-Nielsen, T., Knigge, U., Warberg, J., Theodorsson-Norheim, E., Hökfelt, T., Lundberg, J. M., and Schwartz, T. W. (1988) Release of NPY in pig pancreas: dual parasympathetic and sympathetic regulation. *Am. J. Physiol.* 255 **(Gastrointest. Liver Physiol. 18)**, G46–G54.
9. Gibbins, I. L. and Morris, J. L. (1988) Co-existence of immunoreactivity to neuropeptide Y and vasoactive intestinal peptide in non-noradrenergic axons innervating guinea pig cerebral arteries after sympathectomy. *Brain Res.* **444**, 402–406.
10. Forsgren, S. (1989) Neuropeptide Y-like immunoreactivity in relation to the distribution of sympathetic nerve fibers in the heart conduction system. *J. Mol. Cell. Cardiol.* **21**, 279–290.
11. Sundler, F., Ekblad, E., Grunditz, T., Håkanson, R., Luts, A., and Uddman, R. (1989) NPY in peripheral non-adrenergic neurons, in *Neuropeptide Y*. Nobel Conference Series, V. Mutt, T. Hökfelt, K. Fuxe, and J. M. Lundberg, eds. Raven, New York, pp. 92–102.
12. Böttcher, G., Sjölund, K., Ekblad, E., Håkanson, R., Schwartz, T. W., and Sundler, F. (1984) Co existence of peptide YY in glicentin immunoreactivity in endocrine cells of the gut. *Regul. Pept.* **8**, 261–273.
13. Häppölä, O., Wahlestedt, C., Ekman, R., Soinila, S., Panula, P., and Håkanson, R. (1990) Peptide YY- like immunoreactivity in sympathetic neurons of the rat. *Neuroscience* **39**, 225–230.
14. Alumets, J., Håkanson, R., and Sundler, F. (1978) Distribution, ontogeny and ultrastructure of pancreatic polypeptide (PP) cells in pancreas and gut of the chicken. *Cell. Tissue Res.* **194**, 377–386.
14a. Allen, Y. S., Adrian, T. E., and Allen, J. M. (1983) Neuropeptide Y distribution of the rat brain. *Science* **221**, 877–879.
15. De Quidt, M. E. and Emson, P. C. (1986) Distribution of neuropeptide Y-like immunoreactivity in the rat central nervous system. II. Immunohistochemical analysis. *Neuroscience* **18**, 545–618.
16. Brommé, M., Hökfelt, T., and Terenius, L. (1985) Peptide YY (PYY)-immunoreactive neurons in the lower brainstem and spinal cord of the rat. *Acta Physiol. Scand.* **125**, 340–352.
17. Ekman, R., Wahlestedt, C., Böttcher, G., Håkanson, R., and Panula, P. (1986) Peptide YY-like immunoreactivity in the central nervous system of the rat. *Regul. Peptides* **16**, 157–168.

18. Di Maggio, D. A., Chronwall, B. M., Buchman, K., and O'Donohue, T. L. (1985) Pancreatic polypeptide immunoreactivity in rat brain is actually neuropeptide Y. *Neuroscience* **15,** 1149–1157.
19. Lundberg, J. M., Hemsen, A., Larsson, O., Rudehill, A., Saria, A., and Fredholm, B. (1988) Neuropeptide Y receptor in pig spleen: binding characteristics, reduction of cyclic AMP formation and calcium antagonist inhibition of vasoconstriction. *Eur. J. Pharmacol.* **145,** 21–29.
20. Chang, R. and Lotti, V. J. (1988) Specific [^{3}H]proprionyl-neuropeptide Y (NPY) binding in rabbit aortic membranes: comparison with binding in rat brain and biological responses in rat vas deferens. *Biochem. Biophys. Res. Commun.* **151,** 1213–1219.
21. Wahlestedt, C., Håkanson, R., Vaz, C., and Zukowska-Grojec, Z. (1990b) Norepinephrine and neuropeptide Y: vasoconstrictor cooperation in vivo and in vitro. *Am. J. Physiol.* **258,** R736–R742.
22. Grundemar, L., Mörner, J., Högestätt, E., Wahlestedt, C., and Håkanson, R. (1992c) Characterization of vascular receptors for neuropeptide Y. *Br. J. Pharmacol.* **105,** 45–50.
23. Shigeri, Y., Mihara, S.-I., and Fujimoto, M. (1991) Neuropeptide Y receptor in vascular smooth muscle. *J. Neurochem.* **56,** 852–859.
24. Balasubramaniam, A., Sheriff, S., Rigel, D. F., and Fischer, J. E. (1990) Characterization of neuropeptide Y binding sites in rat cardiac ventricular membranses. *Peptides* **11,** 545–550.
25. Cherdchu, C., Deupree, J. D., and Hexum, T. D. (1989) Binding sites for ^{125}I-neuropeptide Y (NPY) on membranes from bovine adrenal medulla. *Eur. J. Pharmacol.* **173,** 115–119.
26. Sheikh, S. P., Sheikh, M. I., and Schwartz, T. W. (1989c) Y2-type receptors for peptide YY on renal proximal tubular cells in the rabbit. *Am. J. Physiol.* **257** *(Renal Fluid Electrolyte Physiol. 26),* F978–F984.
27. Nielsen, S., Sheikh, S. P., and Christensen. E. I. (1991) Axial heterogeneity of peptide YY processing in renal proximal tubules. *Am. J. Physiol.* **260,** F359–F367.
28. Laburthe, M., Chenut, B., Rouyer-Fessard, C., Tatemoto, K., Couvineau, A., Servin, A., and Amiranoff, B. (1986) Interaction of peptide YY with rat intestinal epithelial plasma membranes: binding of the radioiodinated peptide. *Endocrinology* **118,** 1910–1917.
29. Cox, H. M., Cuthbert, A. W., Håkanson, R., and Wahlestedt, C. (1988) The effect of neuropeptide Y and peptide YY on electrogenic ion transport in rat intestinal epithelial. *J. Physiol. Lond.* **398,** 65–80.
30. Servin, A. L., Rouyer-Fessard, C., Balasubramaniam, A., St-Pierre, S., and Laburthe, M. (1989) Peptide YY and neuropeptide Y inhibit vasoactive intestinal peptide-stimulated adenosine 3',5'-monophosphate production in rat small intestine: structural requirements of peptides for interaction with peptide YY prefering receptors. *Endocrinology* **124,** 692–700.
31. Dumont, Y., Martel, J.-C., Fournier, A., St-Pierre, S., and Quirion, R. (1992) Neuropeptide Y and neuropeptide Y receptor subtypes in brain and peripheral tissues. *Progr. Neurobiol.* **38(2),** 125–167.
32. Inui, A., Okita, M., Inoue, T., Sakatani, N., Oya, M., Moriloka, H., Shii, K., Yokono, K., Mizuno, N., and Baba, S. (1988) Characterization of peptide YY receptors in the brain. *Endocrinology* **124,** 402–409.

32a. Busch-Sörensen, M., Sheikh, S. P., O'Hare, M., Tortora, O., and Schwartz, T. W. (1989) Regional distribution of neuropeptide Y and its receptor in the porcine central nervous system. *J. Neurochem.* **52,** 1545–1552.
32b. Martel, J.-C., St-Pierre, S., and Quirion, R. (1986) Neuropeptide Y receptors in rat brain: Autographic localization. *Peptides* **7,** 55–60.
33. Leslie, R. A., McDonald, T. J., and Robertson, H. A. (1988) Autoradiographic localization of peptide YY and neuropeptide Y binding sites in the medulla oblongata. *Peptides* **9,** 1071–1076.
34. Lynch, D. R., Walker, M. W., Miller, R. J., and Snyder, S. H. (1989) Neuropeptide Y receptor binding sites in rat brain: Differential autoradiographic localizations with ^{125}I-peptide YY and ^{125}I-neuropeptide Y imply receptor heterogeneity. *J. Neurosci.* **9(8),** 2607–2619.
35. Barraco, R. A., Ergene, E., Dunbar, J. C., and El-Ridi, M. R. (1990) Cardiorespiratory response patterns elicited by microinjections of neuropeptide Y in the nucleus tractus solitarius. *Brain Res. Bull.* **24,** 465–485.
36. Kalra, S. P., Sahu, A., Kalra, P. S., and Crowley, W. R. (1990) Hypothalamic neuropeptide Y: a circuit in the regulation of gonadotropin secretion and feeding behavior. *Ann. NY Acad. Sci.* **611,** 273–283.
37. McDonald, K. J. (1990) Role of neuropeptide Y in reproductive function. *Ann. NY Acad. Sci.* **611,** 258–272.
38. Fried, G., Terenius, L., Hökfelt, T., and Goldstein, M. (1985b) Evidence for differential localization of noradrenaline and neuropeptide Y (NPY) in neuronal storage vesicles isolated from rat vas deferens. *J. Neurosci.* **5,** 450–458.
39. Lundberg, J. M., Rudehill, A., Sollevi, A., Fried, G., and Wallin, G. (1989) Co-release of neuropeptide Y and noradrenaline from pig spleen in vivo: importance of subcellular storage, nerve impulse frequency and pattern, feedback regulation and resupply by axonal transport. *Neuroscience* **28,** 475–486.
40. Wahlestedt, C., Grundemar, L., Håkanson, R., Heilig, M., Shen, G. H., Zukowska-Grojec, Z., and Reis, D. J. (1990a) Neuropeptide Y receptor subtypes, Y1 and Y2. *Ann. NY Acad. Sci.* **611,** 7–26.
41. Wahlestedt, C., Yanaihara, N., and Håkanson, R. (1986) Evidence for different pre- and postjunctional receptors for neuropeptide Y and related peptides. *Regul. Pept.* **13,** 307–318.
42. Wahlestedt, C., Edvinsson, L., Ekblad, E., and Håkanson, R. (1987) Effects of neuropeptide Y at sympathetic neuroeffector junctions: existence of Y1- and Y2-receptors, in *Neuronal Messengers in Vascular Function.* A. Nobin, C. Owman, and B. Arneklo-Nobin, eds., Elsevier, Amsterdam, The Netherlands, pp. 231–241.
43. Sheikh, S. P., Håkanson, R., and Schwartz, T. W. (1989) Y1 and Y2 receptors for neuropeptide Y. *FEBS Lett.* **245,** 209–214.
44. Sheikh, S. P., O'Hare, M. M. T., Tortora, O., and Schwartz, T. W. (1989b) Binding of monoiodinated neuropeptide Y to hippocampal membranes and human neuroblastoma cell lines. *J. Biol. Chem.* **264,** 6648–6654.
45. Dumont, Y., Fournier, A., St-Pierre, S., Schwartz, T. W., and Quirion, R. (1990) Differential distribution of neuropeptide Y1 and Y2 receptors in the rat brain. *Eur. J. Pharmacol.* **191,** 501–503.
46. Sheikh, S. P. and Williams, J. A. (1990) Structural characterization of Y1 and Y2 receptors for neuropeptide Y and peptide YY by affinity cross-linking. *J. Biol. Chem.* **265,** 8304–8310.

47. Larhammar, D., Blomqvist, A. G., Yee, F., Jazin, E., Yoo, H., and Wahlestedt, C. (1992) Cloning and functional expression of a human neuropeptide Y/peptide YY receptor of the Y1-type. *J. Biol. Chem.* **267,** 10,935–10,938.
48. Grundemar, L., Wahlestedt, C., and Reis, D. J. (1991) Neuropeptide Y acts at an atypical receptor to evoke cardiovascular depression and to inhibit glutamate responsiveness in the brainstem. *J. Pharmacol. Exp. Ther.* **258,** 633–638.
49. Li, W., MacDonald, R. G., and Hexum, T. D. (1990) Characterization of two neuropeptide Y binding proteins in bovine tissues by affinity-labeling. *Soc. Neurosci.* **abstract 220,** 17.
50. Wahlestedt, C., Regunathan, S., and Reis, D. J. (1992) Identification of cultured cells selectively expressing Y1-, Y2-,, or Y3-type receptors for neuropeptide Y/peptide YY. *Life Sci.* **50,** PL7–PL12.
51. Schwartz, T. W., Sheikh, S. P., and O'Hare, M. M. I. (1987) Receptors on pheocromocytoma cells for two members of the PP-fold family—NPY and PP. *FEBS Lett.* **225,** 209–214.
52. Grundemar, L., Wahlestedt, C., and Reis, D. J. (1991) Long-lasting inhibition of the cardiovascular responses to glutamate and the baroreceptor reflex elicited by neuropeptide Y injected into the nucleus tractus solitarius. *Neurosci. Lett.* **122,** 135–139.
53. Jörgensen, JCh., Fuhlendorff, J., and Schwartz,T. W. (1990) Structure-function studies on neuropeptide Y and pancreatic polypeptide—evidence for two PP-fold receptors in vas deferens. *Eur. J. Pharmacol.* **186,** 105–114.
54. Danho, W., Triscari, J., Vincent, G., Nakajima, T., Taylor, J., and Kaiser, E. T. (1988) Synthesis and biological evaluation of pNPY fragments. *Int. J. Peptide Protein Res.* **32,** 496–505.
55. Grundemar, L. and Håkanson, R. (1990) Effects of various neuropeptide Y/peptide YY fragments on electrically-evoked contractions of the rat vas deferens. *Br. J. Pharmacol.* **100,** 190–192.
56. Krstenansky, J. L., Owen, T. J., Payne, M. H., Shatzer, S. A., and Buck, S. H. (1990) C-terminal modifications of neuropeptide Y and its analogs leading to selectivity of the mouse brain receptor over the porcine spleen receptor. *Neuropeptides* **17,** 117–120.
57. Fuhlendorff, J., Gether, U., Aakerlund, L., Langeland-Johansen, N., Thogersen, H., Melberg, S. G., Bang-Olsen, U., Thastrup, O., and Schwartz, T. W. (1990) [Leu^{31},Pro^{34}]Neuropeptide Y—a specific Y1 receptor agonist. *Proc. Natl. Acad. Sci. USA* **187,** 182–186.
58. Krstenansky, J. L., Owen, T. J., Buck, S. H., Hagaman, K. A., and Lean, L. R. (1989) Centrally truncated and stabilized porcine neuropeptide Y analogs: Design, synthesis, and mouse brain receptor binding. *Proc. Natl. Acad. Sci. USA* **86,** 4377–4381.
59. Boublik, J. H., Spicer, M. A., Scott, N. A., Brown, M. R., and Rivier, J. E. (1990) Biologically active neuropeptide Y analogs. *Ann. NY Acad. Sci.* **611,** 27–34.
60. Fuhlendorff, J., Langeland Johansen, N., Melberg, S. G., Thögersen, H., and Schwartz, T. W. (1990b) The antiparallel pancreatic polypeptide fold in the binding of neuropeptide Y to Y1 and Y2 receptors. *J. Biol. Chem.* **265,** 11,706–11,712.

61. Doughty, M. B., Chu, S. S., Miller, D. W., Li, K., and Tessel, R. E. (1990) Benextramine: a long-lasting neuropeptide Y receptor antagonist. *Eur. J. Pharmacol.* **185,** 113,114.
62. Edvinsson, L., Adamsson, M., and Jansen, I. (1990) Neuropeptide Y antagonistic properties of D-myo-inositol-1,2,6 triphosphate in guinea pig basilar arteries. *Neuropeptides* **17,** 99–105.
63. Wahlestedt, C., Reis, D. J., and Edvinsson, L. (1991) A novel inositol phosphate acts as a selective neuropeptideY (NPY) antagonist. *Soc. Neurosci.* 11.6.
64. Wahlestedt, C., Reis, D. J., Yoo, H., Andersson, D., Edvinsson, L. (1992) A novel inositol phosphate selectively inhibits vasoconstriction evoked by the sympathetic co-transmitters neuropeptide Y (NPY) and adenosine triphosphate (ATP). *Neurosci. Lett.* (in press).
65. Undén, A. and Bartfai, T. (1984) Regulation of neuropeptide Y (NPY) binding by guanine nucleotide in the rat cerebral cortex. *FEBS Lett.* **177,** 125–128.
66. Michel, M. C. (1991) Receptors for neuropeptide Y: multiple subtypes and multiple second messengers. *Trends Pharmacol. Sci.* **12,** 389–394.
67. Fredholm, B., Jansen, I., and Edvinsson, L. (1985) Neuropeptide Y is a potent inhibitor of cyclic AMP accumulation in feline cerebral blood vessels. *Acta Physiol. Scand.* **124,** 467–469.
68. Cepelik, J. and Hynie, S. (1990) Inhibitory effects of neuropeptide Y on adenylate cyclase of rabbit ciliary processes. *Curr. Eye Res.* **9,** 121–128.
69. Kassis, S., Olasmaa, M., Terenius, L., and Fishman, P. H. (1987) Neuropeptide Y inhibits cardiac adenylate cyclase through a pertussis toxin-sensitive G protein. *J. Biol. Chem.* **262,** 3429–3431.
70. Petrenko, S., Olianas, M. C., Onali, P., and Gessa, G. L. (1987) Neuropeptide Y inhibits forskolin stimulated adenylate cyclase in rat hippocampus. *Eur. J. Pharmacol.* **136,** 425–428.
71. Westlind-Danielsson, A., Undden, A., Abens, J., Andell, S., and Bartfai, T. (1987) Neuropeptide Y receptors and the inhibition of adenylate cyclase in the human frontal and temporal cortex. *Neurosci. Lett.* **74,** 237–242.
72. Härfstrand, A., Fredholm, B., and Fuxe, K. (1987) Inhibitory effects of neuropeptide Y on cyclic AMP accumulation in slices of the nucleus tractus solitarius region of the rat. *Neurosci. Lett.* **76,** 185–190.
73. Aakerlund, L., Gether, U., Fuhlendorff, J., Schwartz, T. W., and Thastrup, O. (1990) Y1 receptors for neuropeptide Y are coupled to mobilization of intracellular calcium and inhibition of adenylate cyclase. *FEBS Lett.* **260,** 73–78.
74. Perney, T. M. and Miller, R. J. (1989) Two different G-proteins mediate neuropeptide Y and bradykinin-stimulated phospholipid breakdown in cultured sensory neurons. *J. Biol. Chem.* **264,** 7317–7327.
75. Mihara, S., Shigeri, Y., and Fujimoto, M. (1989) Neuropeptide Y-induced intracellular Ca^{2+} increase in vascular smooth muscle cells. *FEBS Lett.* **259,** 79–82.
76. Motulsky, H. J. and Michel, M. C. (1988) Neuropeptide Y mobilizes Ca^{2+} and inhibits adenylate cyclase in human erythroleukemia cells. *Am. J. Physiol.* **255,** E880–E885.
77. Fredholm, B. B., Häggblad, J., Härfstrand, A., and Larsson, O. (1989) Tissue differences in the effects of neuropeptide Y, adenosine and noradrenaline at the second messenger level, in *Neuropeptide Y,* V. Mutt, K. Fuxe, T.

Hökfelt, and J. M. Lundberg, eds. Karolinska Institue Nobel Conference Series, Raven, New York, pp. 181–189.
78. Hinson, J., Rauh, C., and Coupet, J. (1988) Neuropeptide Y stimulates inositol phospholipid hydrolysis in rat brain microprisms. *Brain Res.* **446,** 379–382.
79. Herzog, H., Hort, Y. J., Ball, H. J., Hayes, G., Shine, J., and Selbie, L. A. (1992) Cloned human neuropeptide Y receptor couples to two different second messenger systems. *Proc. Natl. Acad. Sci. USA,* **89,** 5794–5798.
80. Lundberg, J. M. and Tatemoto, K. (1982) Pancreatic poelypeptide family (APP, BPP, NPY and PYY) in relation to α-adrenoceptor-resistant sympathetic vasoconstriction. *Acta Physiol. Scand.* **116,** 393–402.
81. Corder, R., Lowry, P. J., Wilkinson, S. J., and Ramage, A. G. (1985) Comparison of the haemodynamic actions of neuropeptide Y, angiotensin II and noradrenaline in anaesthetised cats. *Eur. J. Pharmacol.* **121,** 25–30.
82. Zukowska-Grojec, Z., Haass, M., and Bayorh, M. (1986) Neuropeptide Y and peptide YY mediate nonadrenergic vasoconstriction and modulate sympathetic responses in rats. *Regul. Pept.* **15,** 99–110.
83. Sheikh, S. P. (1991) Neuropeptide Y and peptide YY: major modulators of gastrointestinal blood flow and function. *Am. J. Physiol.* **261,** G701–G715.
84. Sheikh, S. P., Roach, E., Fuhlendorff, J., and Williams, J. A. (1991b) Localization of Y1 receptors for NPY and PYY on vascular smooth muscle cells in rat pancreas. *Am. J. Physiol.* **260,** G250–G257.
85. Aicher, S. A., Springston, M., Berger, S. B., Reis, D. J., and Wahlestedt, C. (1991) Receptor-selective analogs demonstrate NPY/PYY receptor heterogeneity in rat brain. *Neurosci. Lett.* **130,** 32–36.
86. Eva, C., Keinänen, K., Monyer, H., Seeburg, P., and Sprengel, R. (1990) Molecular cloning of a novel G protein-coupled receptor that may belong to the neuropeptide receptor family. *FEBS Lett.* **271,** 80–84.
87. Potter, E. K., Fuhlendorff, J., and Schwartz, T. W. (1991) [Pro^{34}]neuropeptide Y selectively identifies postjunctional-mediated actions of neuropeptide Y in vivo in rats and dogs. *Eur. J. Pharmacol.* **193,** 15–19.
88. Grundemar, L., Krstenansky, J. L., and Håkanson, R. (1992) Activation of Y_1 and Y_2 receptors by substituted and truncated neuropeptide Y analogs. *Eur. J. Pharmacol.* (accepted).
89. Citation withdrawn.
90. Citation withdrawn.
91. Dacey, R. G., Jr., Bassett, J. E., and Takayasu, M. (1988) Vasomotor responses of rat intracerebral arterioles to vasoactive intestinal peptide, substance P, neuropeptide Y and bradykinin. *J. Cereb. Blood Flow Metab.* **8,** 254–261.
92. Maturi, M. F., Greene, R., Speir, E., Burrus, C., Dorsey, L. M., Markle, D. R., Maxwell, M., Schmidt, W., Goldstein, S. R., and Pettersson, R. E. (1989) Neuropeptide Y, a peptide found in human coronary arteries constricts primarily small coronary arteries to produce myocardial ischemia in dogs. *J. Clin. Invest.* **83,** 1217–1224.
93. Suzuki, Y., Satoh, S., Ikegaki, I., Okada, T., Shibuya, M., Sugita, K., and Asano, T. (1989) Effects of neuropeptide Y and calcitonin gene-related peptide on local cerebral blood flow in rat striatum. *J. Cereb. Blood Flow Metab.* **9,** 268–270.

94. Rioux, F., Bachelard, H., Martel, J.-C., and St-Pierre, S. (1985) The vasoconstrictor effect of neuropeptide Y and related peptides in the guinea pig isolated heart. *Peptides* **7,** 27–31.
95. Cervin, A. (1991) On the sympathetic regulation of mucociliary activity. An experimental study in the rabbit maxillary sinus. *Acad. Thesis. University of Lund.* 1–130.
96. Pernow, J. and Lundberg, J. M. (1988) Neuropeptide Y induces potent contraction of arterial vascular smooth muscle via an endothelium-independent mechanism. *Acta Physiol. Scand.* **134,** 157–158.
97. Franco-Cereceda, A., Lundberg, J. M., and Dahlöf, C. (1985) Neuropeptide Y and sympathetic control of heart contractility and coronary vascular tone. *Acta Physiol. Scand.* **124,** 361–369.
98. Wahlestedt, C., Edvinsson, L., Ekblad, E., and Håkanson, R. (1985) Neuropeptide Y potentiates noradrenaline-evoked vasoconstriction: mode of action. *J. Pharmacol. Exp. Ther.* **234,** 735–741.
99. Abel, P. W. and Han, C. (1989) Effects of neuropeptide Y on contraction, relaxation and membrane potential of rabbit cerebral arteries. *J. Cardiovasc. Pharmacol.* **13,** 52–63.
100. Westfall, T., Chen, W., Ciarlegio, A., Henderson, K., Del Valle, K., Curfman-Falvey, M., and Naes, L. (1990) In vitro effects of neuropeptide Y at the vascular neuroeffector junction. *Ann. NY Acad. Sci.* **611,** 145–155.
101. Hieble, P. J., Ruffolo, R. R., Jr., and Daly, R. N. (1988) Involvement of vascular endothelium in the potentiation of vasoconstrictor responses by neuropeptide Y. *J. Hypertension* **6(suppl. 4),** S239–242.
102. Budai, D., Vu, B. Q., and Duckles, S. P. (1989) Endothelium removal does not affect potentiation by neuropeptide Y in rabbit ear artery. *Eur. J. Pharmacol.* **168,** 97–100.
103. Huidobro-Toro, J. P., Ebel, L., Macho, P., Domenech, R., Fournier, A., St. Pierre, S. (1990) Muscular NPY receptors involved in the potentiation of the noradrenaline-induced vasoconstriction in isolated coronary arteries. *Ann. NY Acad. Sci.* **611,** 362–365.
104. Andriantsitohaina, R. and Stoclet, J. C. (1988) Potentiation by neuropeptide Y of vasoconstriction in rat resistance arteries. *Br. J. Pharmacol.* **95,** 419–428.
105. McAuley, M. A., Howlett, A., and Westfall, T. C. (1991) Antagonism of the actions of [13–36]NPY and [Leu^{31},Pro^{34}]NPY (LP) at Y2 and Y1 receptors in the rat mesenteric arterial bed. *Neuroscience* **abstract 389,** 6.
106. Stanley, B. G. and Leibowitz, S. F. (1985) Neuropeptide Y injected in the paraventricular hypothalamus: a powerful stimulant of feeding behavior. *Proc. Natl. Acad. Sci. USA* **82,** 3940–3943.
107. Heilig, M. and Murison, R. (1987) Intracerebroventricular neuropeptide Y suppresses open field and home cage activity in the rat. *Regul. Pept.* **19,** 221–231.
108. Heilig, M., Wahlestedt, C., and Widerlöv, E. (1988) Neuropeptide Y (NPY) induced activity suppression in the rat: evidence for NPY receptor heterogeneity and for interaction with alpha-adrenergic receptors. *Eur. J. Pharmacol.* **157,** 205–213.
109. Inui, A., Sanok, K., Miura, M., Hirosue, Y., Nakajima, M., Okiba, M., Buba, S., and Kasuga, M. (1991) Differential expression of NPY and PYY recep-

tors in the tumor cell lines derived from the neural crest. 73rd Annual meeting of the Endocrine Society. Wash. DC (175).

110. Mannon, P. J., Taylor, I. L., Kaiser, L. M., and Nguyen, T. D. (1989) Cross linking of neuropeptide Y to its receptor on rat brain membranes. *Am. J. Physiol.* **256,** G.637–G.643.
111. Tonan, K., Kawata, Y., and Hamaguchi (1990) Conformations of isolated fragments of pancreatic polypeptide. *Biochemistry* **29,** 4424–4429.
112. Mihara et al. (1991)
113. Michel, M. C., Schlicker, E., Fink, K., Boublik, J. H., Göthert, M., Willette, R. N., Daly, R. N., Hieble, J. P., Rivier, J. E., and Motulsky, H. J. (1990) Distinction of NPY receptor subtypes in vitro and in vivo. *Am. J. Physiol.* **259,** E131–E139.
114. Lundberg, J. M. and Tatemoto, K. (1982) Pancreatic poelypeptide family (APP, BPP, NPY and PYY) in relation to α-adrenoceptor-resistant sympathetic vasoconstriction. *Acta Physiol. Scand.* **116,** 393–402.
115. Grundemar, L. (1991) Actions of neuropeptide Y on peripheral and central targets: Characterization of receptor subtypes. *Acad. Thesis Univ. of Lund.,* 1–144.
116. Colmers, W. F., Klapstein, G. J., Fournier, A., St-Pierre, S., and Treherne, K. A. (1991) Presynaptic inhibition by neuropeptide Y in rat hippcampal slice in vitro is mediated by a Y2 receptor. *Br. J. Pharmacol.* **102,** 41–44.
117. martel et al. (1990)
118. Grundemar, L., Grundström, N., Johansson, I. G. M., Andersson, R. G. G., and Håkanson, R. (1990) Suppression by neuropeptide Y of capsaicin-sensitive sensory nerve-mediated contraction in guinea-pig airways. *Br. J. Pharmacol.* **99,** 473–476.
119. Stjernquist, M. and Owman, Ch. (1990) Further evidence for a prejunctional action of neuropeptide Y on cholinergic motor neurons in the rat uterine cervix. *Acta Physiol. Scand.* **138,** 95–96.
120. Beck, A., Jung, G., Gaida, W., Köppen, H., Lang, R., and Schnorrenberg, G. (1989) Highly potent and small neuropeptide Y agonist obtained by linking NPY 1-Y via spacer to α-helical NPY 25-36. *FEBS Lett.* **244,** 119–122.
121. Dahlöf, C., Dahlöf, P., Tatemoto, K., and Lundberg, J. M. (1985) Neuropeptide Y (NPY) reduces field stimulation-evoked release of noradrenaline and enhances force concentration in the rat portal vein. *N.S. Arch. Pharmacol.* **328,** 327–330.
122. Haass, M., Cheng, B., Richardt, G., Lang, R. E., and Schömig, A. (1989) Characterization and presynaptic modulation of stimulation-evoked exocytotic co-release of noradrenaline and neuropeptide Y in guinea pig heart. *Naunyn-Schmiedberg's Arch. Pharmacol.* **339,** 71–78.
123. Lundberg, J. M. and Stjärne, L. (1984) Neuropeptide Y (NPY) depresses the secretion of ^{3}H-noradrenaline and the contractile response evoked by field stimulation in the rat vas deferens. *Acta Physiol. Scand.* **120,** 477–479.
124. Leys, K., Schachter, M., and Seven, P. (1987) Autoradiographic localization of NPY receptors in rabbit kidney: comparison with rat, guinea pig and human. *Eur. J. Pharmacol.* **134,** 233–237.
125. Myers, A. K., Farhat, M. Y., Shen, G. H., Debinski, W., Wahlestedt, C., and Zukowska-Grojec, Z. (1990) Platelets as a source and site of action for neuropeptide Y. *Ann. NY Acad. Sci.* **611,** 408–411.

126. Inui, A., Okita, M., Inoue, T., Sakatani, N., Oya, M., Moribka, H., Shii, K., Yokono, K., Mizuno, N., and Baba, S. (1989) Characterization of peptide YY receptors in the brain. *Endocrinology* **124**, 402–409.
127. Colmers, W. F., Lukowiak, K., and Pittman, Q. J. (1987) Presynaptic action of neuropeptide Y in area CA1 of the rat hippocampal slice. *J. Physiol. (Lond)* **383**, 285–299.
128. Cox, H. M. and Cuthbert, A. W. (1990) The effects of neuropeptide Y and its fragments upon basal and electrical stimulated ion secretion in rat jejunum mucosa. *Br. J. Pharmacol.* **101**, 247–252.
129. Potter, E. K., Michell, L., McCloskey, M. J. D., Tseng, A., Goodman, A. E., Shine, J., and McCloskey, D. I. (1989) Pre- and postjunctional actions of neuropeptide Y and related peptides. *Regul. Peptides* **25**, 167–177.
130. Flood, J. F. and Morley, J. E. (1989) Dissociation of the effects of neuropeptide Y in feeding and memory: Evidence for pre- and postsynaptic mediation. *Peptides* **10**, 963–966.
131a. Undén, A., Tatemoto, K., Mutt, V., and Bartfai, T. (1984) Neuropeptide Y receptors in the rat brain. *Eur. J. Biochem.* **145**, 525–530.
131b. Nguyen, T. D., Heintz, G. G., Kaiser, L. M., Staley, C. A., and Taylor, I. L. (1990) Neuropeptide Y differential binding to rat intestinal laterobasal membranes. *J. Biol. Chem.* **265**, 6416–6422.
132. Mannon, P. J., Mervin, S. J., and Taylor, I. L. (1991) Solubilization of the neuropeptide Y receptor from rat brain membranes. *J. Neurochem.* **5**, 1804–1809.
133. Price, J. S. and Brown, M. J. (1990) ^{125}I-neuropeptide Y binding activity of pig spleen cell membranes: effect of solubilisation. *Life Sci.* **47**, 2299–2305.
134. Gimpl, G., Gerstberger, R., Mauss, U., Klotz, K. N., and Lang, R. E. (1990) Solubilization and characterization of active neuropeptide Y receptors from rabbit kidney. *J. Biol. Chem.* **265**, 18142–18147.
135. Sheikh, S. P., Hansen, A. P., and Williams, J. A. (1992) Solubilization and affinity purification of the Y2 receptor for neuropeptide Y and peptide YY from rabbit kidney. *J. Biol. Chem.*, in press.
136. Danger, J. M., Tonon, M. C., Lamacz, M., Martel, J. C., Saint-Pierre, S., Pellitier, G., and Vaudry, H. (1987) Melanotropin release inhibiting activity of neuropeptide Y: structure-activity relationships. *Life Sci.* **40**, 1875–1880.
136a. Minakata, H., Taylor, J. W., Walker, M. W., Miller, R. J., and Kaiser, E. T. (1989) Characterization of amphiphilic secondary structures in neuropeptide Y through the design, synthesis, and study of model peptides. *J. Biol. Chem.* **264**, 7907–7913.
137. Hökfelt, T., Everitt, B. J., Fuxe, K., Kalia, M., Agnati, L. F., Johansson, O., Theodorsson-Norheim, E., and Goldstein, M. (1984) Transmitter and peptide systems in areas involved in the control of blood pressure. *Clin. Exp. Hyper. Theory Practice* **A6 (1 and 2)**, 23–41.
138. Carter, D. A., Vallejo, M., and Lightman, S. L. (1985) Cardiovascular effects of neuropeptide Y in the nucleus tractus solitarius of rats: Relationship with noradrenaline and vasopressin. *Peptides* **6**, 421–425.
139. Tseng, C.-J., Mosqueda-Garcia, R., Appalsamy, M., and Robertson, D. (1988) Cardiovascular effects of neuropeptide Y in rat brainstem mu nuclei. *Circ. Res.* **64**, 55–61.

140. Grundemar et al. (1991)
141. Talman, W. T., Granata, A. R., and Reis, D. J. (1984) Glutaminergic mechanisms in the nucleus tractus solitarius in blood pressure control. *Fed. Proc.* **43,** 39–44.
142. Nakajima, T., Yashima, Y., and Nakamura, K. (1986) Quantitive autoradiographic localization of neuropeptide Y receptors in the rat lower brainstem. *Brain Res.* **380,** 144–150.
143. Agnati, L. F., Fuxe, K., Benfenati, F., Battistini, N., Härfstrand, A., Tatemoto, K., Hökfelt, T., and Mutt, V. (1983) Neuropeptide Y in vitro selectively increases the number of α_2-adrenergic binding sites in membranes of the medulla oblongata of the rat. *Acta Physiol. Scand.* **118,** 293–295.
144. Fuxe, K., Härfstrand, A., Agnati, L. F., von Euler, G., Svensson, T., and Fredholm, B. (1989) On the role of NPY in central cardiovascular regulation, in *Neuropeptide* Y. Karolinska Institue Nobel Conference Series. V. Mutt, K. Fuxe, T. Hökfelt, and J. M. Lundberg, eds., Raven, NY, pp. 201–214.
145. Härfstrand, A., Fuxe, K., Agnati, L., and Fredholm, B. (1989) Reciprocal interactions between alpha 2-adrenoceptor agonist and neuropeptide Y binding sites in the nucleus tractus solitarius of the rat. *J. Neural. Transm.* **75,** 83–99.
146. Citation withdrawn.
147. Citation withdrawn.
148. Roman, F. J., Pascud, X., Duffy, O., and Junien, J. L. (1991) N-methyl-D-aspartate receptor complex modulation by neuropeptide Y and peptide YY in rat hippocampus in vitro. *Eur. J. Pharmacol.* **122,** 202–204.
Saudek, V. and Pelton, J. T. (1990) Sequence-specific ^{1}H NMR assignment and secondary structure of neuropeptide Y in aqueous solution. *Biochem.* **29,** 4509–4515.
149. Citation withdrawn.
150. Higuchi, H., Costa, E., and Yan, H-Y. T. (1988) Neuropeptide Y inhibits the nicotine-mediated release of catecholamines from bovine adrenal chromaffin cells. *J. Pharmacol. Exp. Ther.* **244,** 468–474.
151. Citation withdrawn.
152. Rimland, J. R., Vin, W., Sweetnam, P., Saijoh, K., Nestler, E. J., and Duman, R. S. (1991) Sequence and expression of a neuropeptide Y receptor cDNA. *Mol. Pharmacol.* **40,** 869–875.
153. Boublik, J., Scott, N., Taulane, J., Goodman, M., Brown, M., and Rivier, J. (1989) Neuropeptide Y and neuropeptide Y 18-36. Structural and biological characterization. *Intl. J. Pept. Protein Res.* **33,** 11–15.
154. Brown, M. R., Scott, N. A., Boublik, J., Allen, R. S., Ehlers, R., Landon, M., Crum, R., Ward, D., Bronsther, O., Maisel, A., and Rivier, J. (1989) Neuropeptide Y: Biological and clinical studies, in *Neuropeptide Y*, V. Mutt, K. Fuxe, T. Hökfelt, and J. M. Lundberg, eds. Raven, New York, pp. 201–214.
155. Scott, N. A., Michel, M. C., Boublik, J. H., Rivier, J. E., Motomura, S., Crum, R. L., Landon, M., and Brown, M. R. (1990) Distinction of NPY receptors in vitro and in vivo II. *Am. J. Physiol.* **259,** H174–H180.
156. Grundemar, L., Wahlestedt, C., Shen, S. H., Zukowska-Grojec, Z., and Håkanson, R. (1990) Biphasic blood pressure response to neuropeptide Y in anesthetized rats. *Eur. J. Pharmacol.* **179,** 83–87.

157. Shen, S. H., Grundemar, L., Zukowska-Grojec, Z., Håkanson, R., and Wahlestedt, C. (1991) C-terminal neuropeptide Y fragments are mast cell-dependent vasodepressor agents. *Eur. J. Pharmacol.* **204,** 249–256.
158. Arzubiaga, C., Morrow, J., Roberts, J. II, and Biaggioni (1991) Neuropeptide Y, a putative transmitter in noradrenergic neurons, induces mast cell degranulation but not prostaglandin D_2 release. *J. Allergy Clin. Immun.* **87,** 88–93.
159. Grundemar, L. and Håkanson, R. (1991) Neuropeptide Y, peptide YY and C-terminal fragments release histamine from rat peritoneal mast cells. *Br. J. Pharmacol.* **104,** 776–778.
160. Mousli, M., Bueb, J.-L., Bronner, C., Rouot, B., and Landry, Y. (1990) G protein activation: a receptor-independent mode of action for cationic amphiphilic neuropeptides and venom peptides. *Trends Pharmacol. Sci.* **11,** 358–362.
161. Stead, R. H., Dixon, F., Bramwell, N. H., Riddell, R. H., and Bienenstock, J. (1989) Mast cells are closely apposed to nerves in the human gastrointestinal mucosa. *Gastroenterology* **97,** 575–585.
162. Nilsson, G., Alving, K., Ahlstedt, S., Högfelt, T., and Lundberg, J. M. (1990) Peptidergic innervation of rat lymphoid tissue and lung: Relation to mast cells and sensitivity to capsaicin and immunization. *Cell Tissue Res.* **262,** 125–133.
163. Inui, A., Okita, M., Miura, M., Hirosue, Y., Nakajima, M., Inoue, T., and Baba, S. (1990) Characterization of the receptor for peptide-YY and avian pancreatic polypeptide in chicken and pig brains. *Endocrinology* **127,** 934–941.
164. Whitcomb, D. C., Taylor, I. L., and Vigna, S. R. (1990) Characterization of saturable binding sites for circulating pancreatic polypeptide in rat brain. *Am. J. Physiol.* **259,** G687–G691.
165. Schwartz, T. W. (1983) Pancreatic polypeptide a hormone under vagal control. *Gastroenterology* **85,** 1411–1425.
166. Yee et al. (1991)
167. Li, X.-J., Wu, Y.-N., North, A., and Forte, M. (1992) Cloning, functional expression, and developmental regulation of a neuropeptide Y receptor from *Drosophila melanogaster. J. Biol. Chem.* **267,** 9–12.

Actions of Neuropeptide Y on the Electrophysiological Properties of Nerve Cells

David Bleakman, Richard J. Miller, and William F. Colmers

1. Introduction

Neuropeptide Y (NPY) is a 36 amino acid polypeptide that is found widely distributed throughout the peripheral and central nervous systems *(1,2)*. It is related in structure to the peptide PYY, which has a more limited distribution in the nervous system, but is found in high concentrations in a number of endocrine cells *(3)*. The wide distribution of NPY strongly suggests that it carries out numerous functions in the nervous system, and indeed a very large number of effects of the peptide have been demonstrated following its injection into various regions of the brain (e.g., *4*). In the present chapter, we will review known neurophysiological actions of NPY that may allow us to explain at least some of these actions. We will begin by considering the action of NPY in the periphery and then consider its actions in the central nervous system.

2. The Peripheral Nervous System

NPY is found in high concentrations in a number of peripheral neurons, including those of the sympathetic nervous system and the enteric ganglia, for example *(1–3)*. The actions of NPY at a number of peripheral neuroeffector junctions have been particularly well studied *(5)*. Characteristically, NPY produces both pre-

From: *The Biology of Neuropeptide Y and Related Peptides;*
W. F. Colmers and C. Wahlestedt, Eds.

and postsynaptic effects *(5)*. It is frequently found that the postsynaptic actions of NPY are mediated through Y_1-type NPY receptors, whereas the presynaptic actions are mediated through Y_2 NPY receptors *(6,7)*. In virtually all instances, the presynaptic effects of NPY are manifest as inhibition of neurotransmitter release. For example, NPY is a powerful inhibitor of substance P release from sensory neurons *(8)* and of norepinephrine release from sympathetic neurons *(5)*.

2.1. Sensory Neurons

NPY-containing nerve terminals have been demonstrated to occur in close apposition to the incoming nerve endings of sensory fibers in the dorsal horn of the spinal cord *(9)*. It is thought that substance P acts as the neurotransmitter released from the small nonmyelinated neurons responsible for transmitting nociceptive information into the spinal cord. It has been suggested that NPY modulates the release of transmitters from these sensory nerve endings by acting at presynaptic receptors *(8)*. In confirmation of this hypothesis, it has been shown that the evoked release of the transmitter substance P from dorsal root ganglion neurons (DRG) or of acetylcholine from nodose ganglion neurons grown in vitro *(10)* can be potently reduced by NPY. Further support for this idea has come from the observation that NPY can inhibit the release of transmitter from the peripheral terminals of capsaicin-sensitive sensory neurons in the heart *(11)*. Again, capsaicin specifically stimulates the subpopulation of sensory neurons that transmit nociceptive information. The currently prevailing view is that activation of NPY receptors on these neurons is associated with the reduction in Ca^{2+} influx and that this is the basis of its ability to inhibit neurotransmitter release. The evidence supporting such a view is as follows. When DRG cells are induced to fire a train of action potentials, a stimulus that will evoke the release of neurotransmitter, Ca^{2+} enters the neuron through voltage-sensitive Ca^{2+} channels that open during the spike train. Addition of NPY to cells under these circumstances is not associated with any change in the membrane potential, but is associated with a reduction in Ca^{2+} influx *(12)* (Fig. 1). The most parsimonious explanation for this phenomenon would be a direct block of voltage-sensitive Ca^{2+} channels in these neurons. This hypothesis has been directly tested by using voltage-clamp procedures as indicated in Fig. 2. NPY

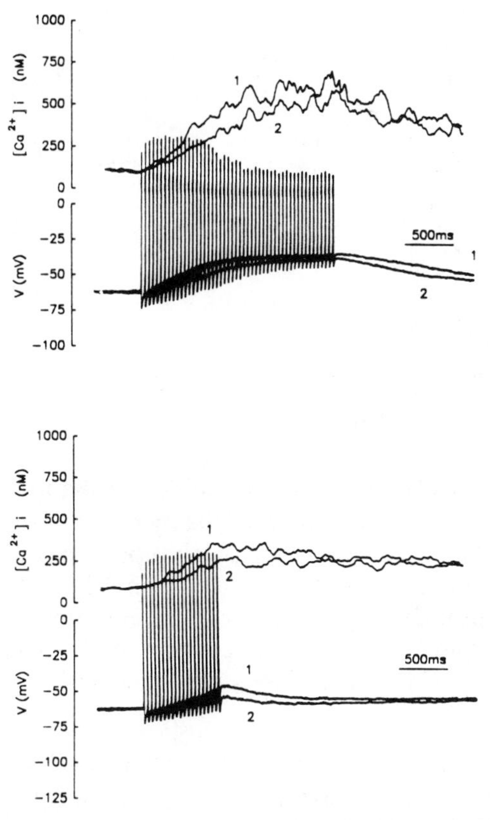

Fig. 1. Action potentials were evoked in single DRG neurons under current-clamp conditions by brief current injection in the absence (1) and presence (2) of NPY (100 n*M*). (i) The $[Ca^{2+}]_i$ signals associated with 50 action potentials in the absence and presence of NPY (100 n*M*). (ii) $[Ca^{2+}]_i$ signals associated with 20 APs in the presence and absence of NPY (100 n*M*).

potently inhibits Ca^{2+} currents in these neurons *(8,12,13)* and indeed similar actions have been observed in myenteric plexus *(14)* and sympathetic neurons *(15)*. This action of NPY is presumably responsible for its ability to inhibit Ca^{2+} influx during a train of action potentials. It will also be noted that peptides, such as NPY_{13-36} are also effective in producing this inhibition, indicating that the effects are mediated by Y_2-type NPY receptors *(12)*.

It is well known that DRG neurons contain more than one kind of voltage-sensitive Ca^{2+} channel, and this is also true of most other types of neurons as well *(16)*. In many instances, triggering

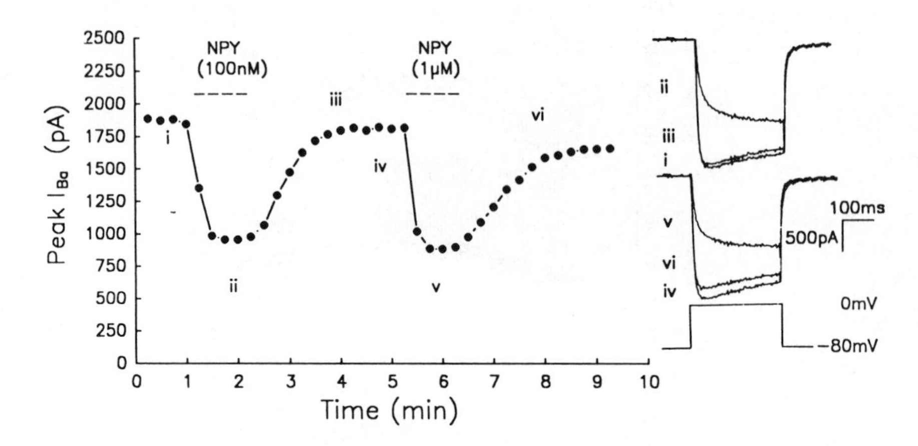

Fig. 2. NPY- and PYY-mediated inhibition of IBa in single DRG neurons. The time-course of inhibition of IBa by two concentrations of NPY is shown with 20 m*M* BAPTA in the patch pipet. The insets show superimposed individual current traces evoked from Vh = –80mV to Vt = 0mV at the time-points indicated. The figure is representative of three similar experiments. (From Bleakman, D., Harrison, N. I., Colmers, W. F., and Miller, R. J. (1992), ©Macmillan Press, with permission.)

of neurotransmitter release has been associated with Ca^{2+} influx through N-type Ca^{2+} channels *(16,17)*. It is not completely clear at this point which kinds of Ca^{2+} channels are modulated by NPY in DRG neurons. However, experiments in nodose ganglion neurons have clearly shown that NPY effects are specifically directed toward the N-type Ca^{2+} channels *(10)*, and indeed, similar observations have also been made in neurons of the myenteric plexus *(14)* (*see* Section 2.2.). It is of interest to note in this regard that N-type channels are heterogeneous with regard to their sensitivity to ω-conotoxin G-VIA *(18,19)* and that the ω-conotoxin-insensitive, N-like Ca^{2+} current can also be modulated by other receptors *(20)*, although this has not yet been tested for NPY.

A further important point to note about the inhibition of neuronal Ca^{2+} currents by NPY is that, at least in peripheral neurons, it is abolished by treating cells with pertussis toxin *(8)*. This indicates that a pertussis-toxin-sensitive G protein mediates coupling between voltage-sensitive Ca^{2+} channels and the NPY receptors. Inhibition of neuronal Ca^{2+} currents by NPY is also clearly associated with a slowing of the rate of activation of the current (Fig. 2)

(12). This again is a feature that is normally found in peripheral neurons where Ca^{2+} channels are inhibited by neurotransmitter receptors through a G-protein-mediated coupling mechanism *(21)*. Indeed, the slowing of the rate of activation can be mimicked by a nonhydrolyzable analog of GTP, such as GTPγS. There are several pertussis-toxin-sensitive G proteins that are found in the nervous system. However, G proteins of the G_o type are found in the greatest abundance. Several lines of evidence suggest that this type of G protein may be responsible for coupling NPY receptors to voltage-sensitive Ca^{2+} channels in DRG cells. Thus, in pertussis-toxin-treated rat DRG cells, infusion of the α-subunit of G_o is found to recouple NPY receptors to voltage-sensitive Ca^{2+} channels in DRG cells *(22)*. Indeed, this G-protein α-subunit performs the task better than a variety of other pertussis-toxin-sensitive G proteins including a_{i1}, a_{i2}, or a_{i3}. Such observations support those made in other laboratories using different paradigms that also suggest a role for G_o in coupling receptors to neuronal Ca^{2+} channels (reviewed in *17*).

Although such experiments establish the fact that NPY receptors seem to couple to neuronal Ca^{2+} channels through pertussis-toxin-sensitive G proteins, they do not indicate whether this coupling is direct or whether diffusible second-messenger molecules are also involved. For example, Y_2-type NPY receptors have also been shown to activate phospholipase C in rat DRG neurons, causing the synthesis of IP_3 and presumably also of diacylglycerol *(23)*. Furthermore, it has been suggested that stimulation of protein kinase C may be an important link in the inhibition of DRG Ca^{2+} channels by a number of neurotransmitters *(17)*. This hypothesis was further tested in DRG neurons in which the activity of protein kinase C was suppressed through chronic treatment of the neurons with phorbol esters *(24)*. This produced a population of neurons in which protein kinase C activity was entirely absent. In these cells, it was observed that there was a change in the sensitivity of the inhibition of neuronal Ca^{2+} currents by NPY. Nevertheless, robust inhibition of Ca^{2+} influx was still observed. It must therefore be concluded that protein kinase C activation does not play an essential role in the coupling of NPY receptors to Ca^{2+} channels in DRG neurons. However, some modulatory role appears to be possible. Such findings are supported by studies carried out on myenteric plexus (discussed in Section 2.2.). In summary, therefore, activation of Y_2 NPY receptors on sensory neurons appears to inhibit neurotransmitter release in a pertussis-toxin-sensitive manner. This inhibition appears to be owing to the ability of the peptide to inhibit Ca^{2+} influx through voltage-sensitive Ca^{2+} channels. Coupling of the Y_2 NPY receptors to Ca^{2+} channels appears to utilize the G protein G_o. Protein kinase C is not an essential mediator of this coupling, but may play a modulatory role.

NPY has been shown to inhibit transmitter release from numerous peripheral neuroeffector junctions *(25)*. As with many such effects, the ability of the inhibitory transmitter to block neurotransmitter release is dependent on how strongly the presynaptic neuron is stimulated *(26)*. Thus, the inhibitory effects of NPY can be overcome by stimulating the nerve more rapidly or with longer trains of action potentials *(13,27)*. Electrophysiological data obtained on the effects of NPY have been able to explain this phenomenon. For example, in Fig. 1, one can see the effect of firing trains of action potentials of greater and greater lengths on the amplitude of the $[Ca^{2+}]_i$ signal recorded in the soma of a rat DRG neuron. At longer trains of action potentials, $[Ca^{2+}]_i$ reaches an asymptote. This is because of the fact that the neuron buffers larger Ca^{2+} loads more efficiently, thus limiting the ability of Ca^{2+} to rise within the neuron. The increased buffering capacity of DRG neurons seems to be provided primarily by mitochondria *(13)*. The differential effects of NPY on Ca^{2+} influx elicited by short and long trains of action potentials can also be seen in this figure. For short trains of action potentials, NPY inhibits Ca^{2+} influx through the period of the train. However, during a long train of action potentials, the inhibited Ca^{2+} signal gradually attains the same amplitude as the control. Such behavior would be directly predicted from the nonlinearity present between Ca^{2+} influx and the amplitude of the Ca^{2+} signal as indicated in Fig. 3. Thus, the ability of strong stimulation to overcome presynaptic inhibitory effects of NPY has its direct correlation in the ability of NPY to modulate Ca^{2+} influx into the neuron directly.

2.2. Myenteric Plexus Neurons

Complimentary experiments to those discussed for DRG neurons have also been carried out with cultured myenteric plexus neurons *(14)*. NPY is also found in high concentrations within the enteric ganglia and probably carries out important roles in controlling the functions of the gastrointestinal tract *(3)*. Myenteric

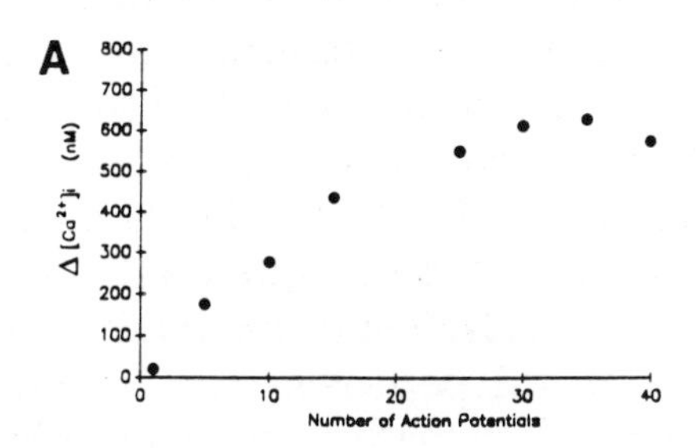

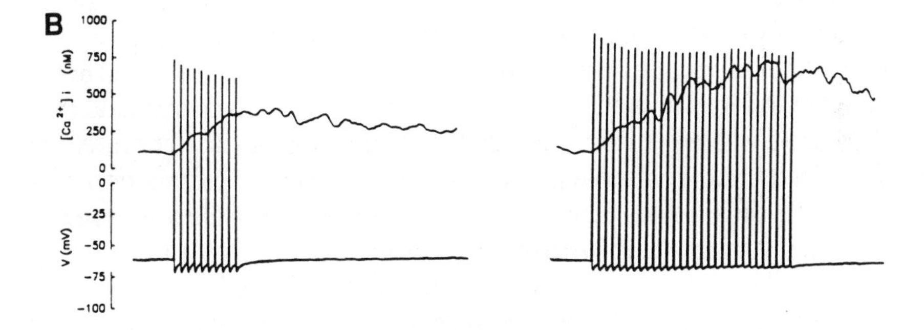

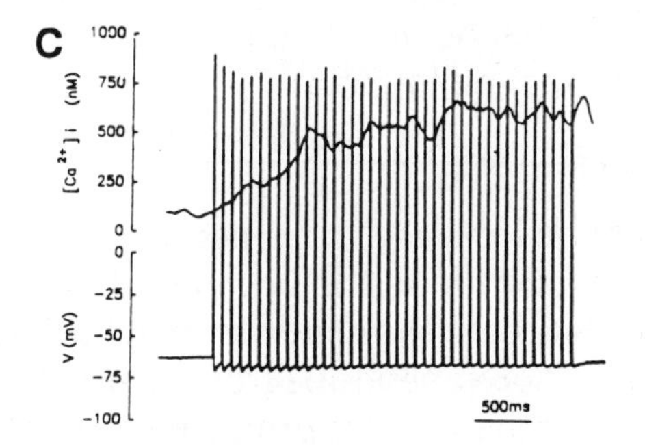

Fig. 3. Simultaneous current clamp and $[Ca^{2+}]_i$ recordings from a single DRG cell. A illustrates the relationship between the number of action potentials fired per train (12.5 Hz) and the amplitude of the $[Ca^{2+}]_i$ increase that resulted. B illustrates individual traces for the data points indicated (i) 10 APs (ii) 30 APs and (iii) 40 APs per train with associated $[Ca^{2+}]_i$ signals. C is mean data points collected from curves generated in 25 DRG cells at a 12.5-Hz firing rate.

plexus neurons appear to contain two types of Ca^{2+} channels corresponding to the N and L types. The former are sensitive to block by ω-conotoxin and the latter to modulation by dihydropyridine drugs. Whole-cell Ca^{2+} currents reported in these neurons are inhibited by NPY, as are those in DRG cells discussed above. Complimentary studies using fura-2 to measure Ca^{2+} influx confirm this conclusion. As in DRG cells, NPY-induced inhibition of Ca^{2+} influx is sensitive to treatment with pertussis toxin. It seems likely that it is the N-type Ca^{2+} channels in these cells that are inhibited by NPY. This conclusion comes from such experiments as those shown in Fig. 4. Here it can be seen that a depolarization-evoked Ca^{2+} influx is partially blocked by a high concentration of the dihydropyridine L-channel blocker nitrendipine. The remaining Ca^{2+} influx is abolished by NPY. The additive effects of these two agents indicate that they are directed toward different populations of Ca^{2+} channels in these cells. Experiments performed on myenteric plexus neurons further indicate that diffusible second-messenger molecules are not necessary in the NPY-induced inhibition of N channels. This is indicated in the experiments shown in Fig. 5. Here the activity of a single N-type Ca^{2+} channel is recorded on an on-cell patch, and NPY is added outside the patch pipet. No inhibition of the channel activity is observed. Thus, it appears that the NPY receptor, G protein, and N-type Ca^{2+} channel must be in close proximity to one another in order for coupling to occur.

2.3. Sympathetic Neurons

NPY is found to be localized in numerous sympathetic neurons, and in fact, it is often found to be colocalized with norepinephrine. Numerous studies have demonstrated the ability of NPY to inhibit the release of norepinephrine from sympathetic neurons *(2,5)*. It seems likely that this effect is produced through inhibition of Ca^{2+} influx in a manner similar to that described above for sen-

Fig. 4. *(opposite page)* Dihydropyridines and NPY block different components of $[Ca^{2+}]_i$ transient. A: 50 m*M* K^+-evoked $[Ca^{2+}]_i$ transients (horizontal bars), as assayed by the fluorescent Ca^{2+} indicator dye fura-2. Left: $[Ca^{2+}]_i$ transient obtained under drug-free conditions, $[Ca^{2+}]_i$ transient 2 was acquired with the DHP antagonist nitrendipine (3 μ*M*; a maximal dose) in the bath. $[Ca^{2+}]_i$ transient 3, both nitrendipine and NPY (100 n*M*) were in the bath and gave the largest inhibition. Prior to $[Ca^{2+}]_i$-

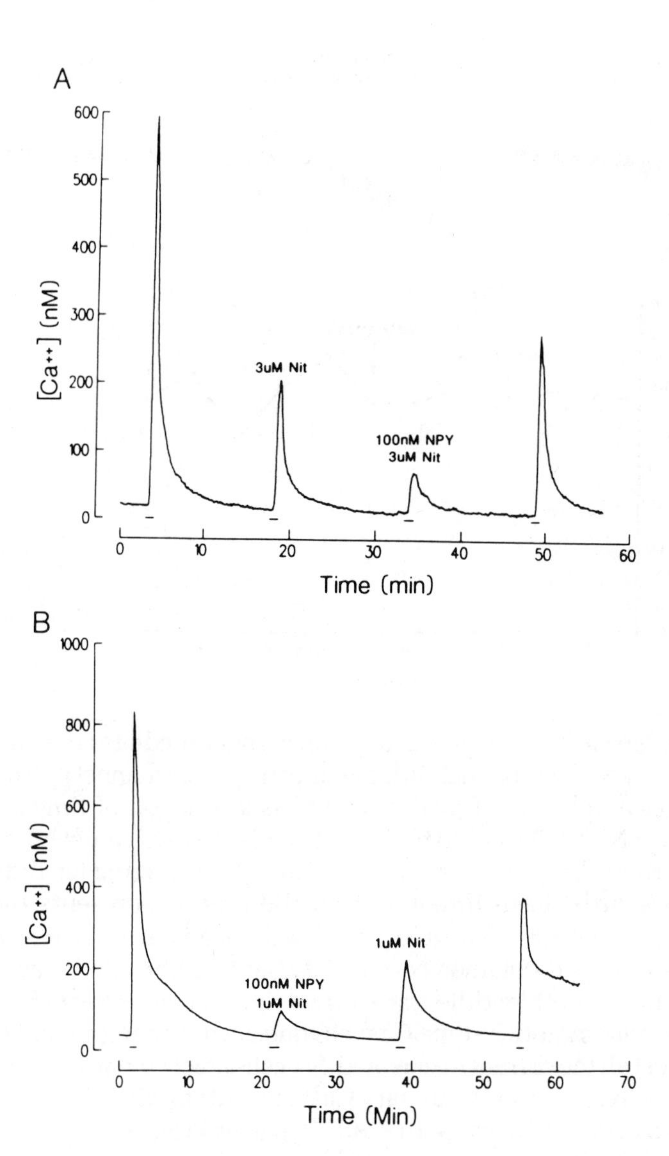

transient 4, drugs were washed out and some recovery occurs. B shows that the order of a similar experiment was not important. The second $[Ca^{2+}]_i$-transient shows that the combination of NPY and nitrendipine produced maximal block. Both nitrendipine and NPY were introduced into the bath 60 s before the high-K^+ depolarization. Combinations of NPY and DHPs were tried in four other cells with similar results. All other conditions are as indicated in the figure. (From Hirning, L. D., Fox, A. P., and Miller, R. J. (1990), ©Elsevier Press, with permission.)

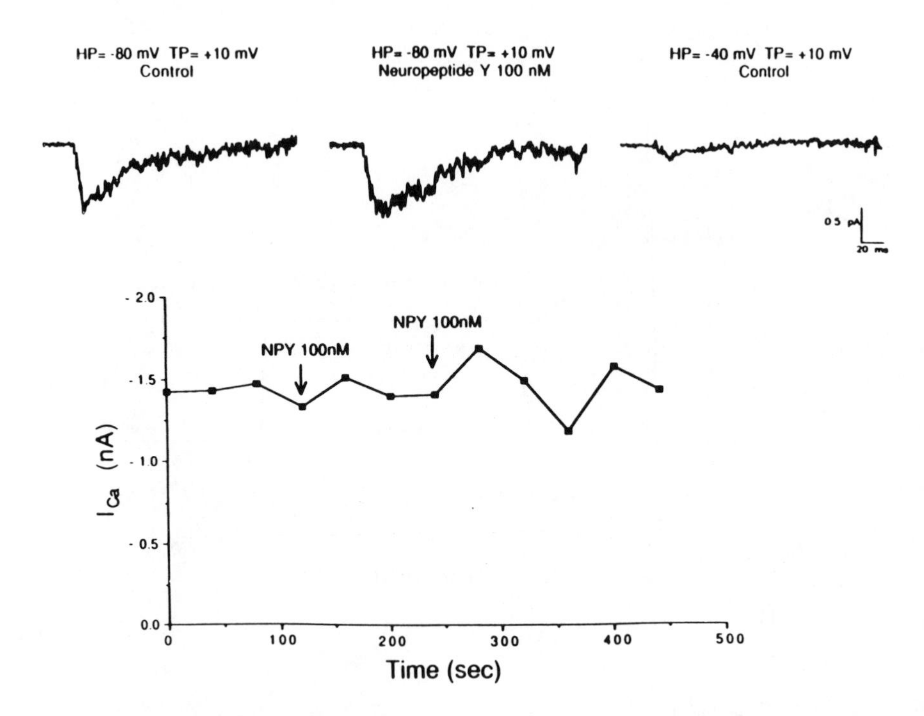

Fig. 5. No easily diffusible second messenger mediates the NPY inhibitor of N-type Ca^{2+} channels in a cell-attached multichannel patch. Graph illustrates amplitude of peak currents as a function of time after the addition of NPY (100 n*M*) to the bath, outside the patch pipet NPY added at each arrow. A mean current was obtained by averaging ten individual current records. Each time-point on the graph represents the peak inward Ca^{2+} current measured from these mean currents. Representative mean currents obtained before and after addition of NPY are shown in the top row (left, middle traces). The right mean current shows that the patch had mainly N-type Ca^{2+} channels; after changing to HP = -40 mV, almost all the current inactivated. Six cell-attached patch experiments showed no response of the unitary Ca^{2+} currents to NPY. (From Hirning, L. D., Fox, A. P., and Miller, R. J. (1990), ©Elsevier Press, with permission.)

sory neurons and neurons of the myenteric plexus. Indeed, recordings from acutely isolated superior cervical ganglion neurons of the rat (Foucart et al., unpublished observations) or frog have shown that NPY can inhibit Ca^{2+} currents in these cells *(15)*. Again, these effects of NPY are sensitive to treatment by pertussis toxin. An interesting observation has been made by Schofield and Ikeda, who showed that the inhibitory effects of NPY in the bullfrog were specifically directed toward the small C-type sympathetic neurons and that no effects were observed on the larger B-type cells. This correlated with the presence of NPY immunohistochemically in C-type rather than B-type sympathetic neurons. In a similar series of studies, Foucart et al. (unpublished observations) have now demonstrated that the inhibitory effects of NPY on Ca^{2+} currents in rat sympathetic neurons seem to be more prevalent in cells with a smaller diameter rather than larger sympathetic neurons. This contrasts with the effects of norepinephrine, for example, which seem to be equally distributed over sympathetic neurons of all sizes.

3. The Central Nervous System

Although NPY and its analogs have been injected, dialyzed, and otherwise introduced into the CNS, resulting in many behavioral, endocrine, and neurochemical changes, comparatively little research has been directed toward examining the actions of NPY on the electrical properties of central neurons. This section will outline what is known about NPY's actions in the areas in which it has been studied and will speculate on the role the peptide could play in the physiology of these areas.

3.1. Hippocampus

The first CNS preparation studied for the actions of NPY was the in vitro hippocampal slice. Hippocampus has a reasonably high concentration of NPY, but has a very high number of receptors, especially concentrated in the strata radiatum and oriens of areas CA1 and CA3 *(28)*, layers of the hippocampus in which most of the synaptic interactions occur. Interestingly, there is again a mismatch in receptor and peptide concentrations in this brain region; hippocampus proper (areas CA1 and CA3) has most of the receptors, with only a moderate amount of peptide, whereas dentate gyrus has very high concentrations of the peptide, with only rather low concentrations of receptor *(28)*.

Transverse hippocampal slices have prolonged viability and mechanical stability, and retain well-characterized, excitatory, and inhibitory synaptic connections that can be conveniently activated with a stimulating electrode placed visually in the pathways of interest. A further advantage is that when a stimulus is applied to an input pathway, such as stratum radiatum in CA1, large field

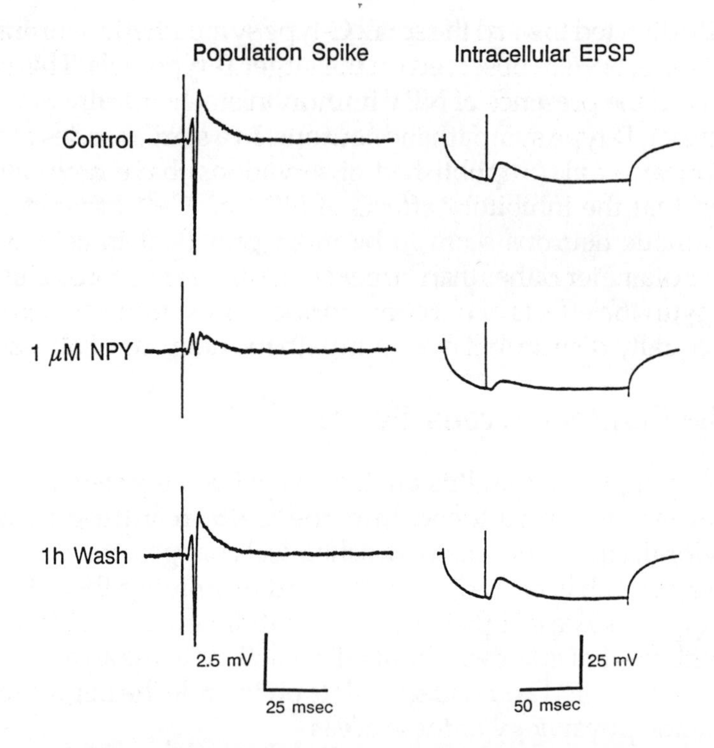

Fig. 6. Inhibition by NPY (1 μ*M*) of orthodromic synaptic responses in area CA1 of the rat hippocampal slice. Left: Population spike response in CA1 evoked by stimulation of stratum radiatum. Right: Excitatory postsynaptic potential recorded intracellularly from a CA1 pyramidal cell near the extracellular electrode measuring the population spike, left. NPY inhibits both (glutamate-mediated) responses reversibly. (From *31*, ©1990 The New York Academy of Sciences, with permission.)

potentials are generated, which can be easily recorded with an extracellular electrode (cf *29*). Also, intracellular recordings can be made from individual neurons, both with conventional sharp microelectrodes and more recently with patch-clamp electrodes, using the tight-seal, whole-cell configuration *(30)*.

An extracellular field potential, evoked from stratum radiatum and recorded in the cell-body layer of CA1 (Fig. 6), consists of two main components: a slower, positive-going wave, upon which is superimposed a sharp negativity. The slow wave (population EPSP; pEPSP) is a field potential generated by the influx of current into the apical dendrites, whereas the fast component (population spike,

PS) is owing to the synchronous discharge of numerous pyramidal cells *(32)*. The response is mediated entirely by synaptic events and is sensitive to antagonists of excitatory amino acid receptors. Bath application of NPY (1 μM) strongly reduces both components of the response; the effect is gradually, but fully, reversible (Fig. 6; *33–35*). When a simultaneous intracellular recording is made from a nearby pyramidal neuron, the stimulus evokes an EPSP, which is reduced by application of NPY, and recovers with the same time-course as the extracellular responses. The response to a hyperpolarizing pulse applied to the neuron via the electrode is unaltered during the inhibition of the EPSP, indicating that the reduction in EPSP amplitude is not the result of an increase in membrane conductance at the soma *(34,35)*. The response to an iontophoretic application of glutamate to distal dendrites of a CA1 pyramidal cell in the presence of reduced extracellular Ca^{2+} and elevated extracellular Mg^{2+} was undiminished in the presence of 1 μM NPY, suggesting that there was no postsynaptic action of NPY and that therefore NPY acted at a presynaptic site *(34)*. Although it was reported *(35)* that under the same conditions, NPY reduced spontaneous firing in pyramidal neurons recorded with an extracellular microelectrode, this was not observed in numerous intracellular recordings *(34)*. An explanation for these observations consistent with the known actions of NPY would be if the cells received a tonic excitatory input sensitive to the actions of NPY, as have been observed in cultures of hippocampal neurons (*see* Section 3.2.).

NPY did not affect passive or active properties of the cell bodies of the presynaptic CA3 pyramidal neurons, nor the excitability of their axons *(36)*. The site of NPY's action, therefore, is at or near the presynaptic terminals themselves. Because these terminals are inaccessible to microelectrode techniques, a pharmacological approach was used to examine the mechanism by which NPY acted at the presynaptic terminal. Two reasonable mechanisms by which NPY might act would be increasing the conductance of a K^+ channel or decreasing the conductance of a Ca^{2+} channel. Application of low concentrations (10 μM) of the K^+ channel blocker, 4-aminopyridine (4-AP), which has a high affinity for presynaptic terminal K^+ channels, blocked the inhibition mediated by 1 μM NPY. However, because blocking K^+ channels at the presynaptic terminals will itself cause an increased influx of Ca^{2+}, the action of 4-AP on the NPY-mediated inhibition was tested with extracellular Ca^{2+} reduced

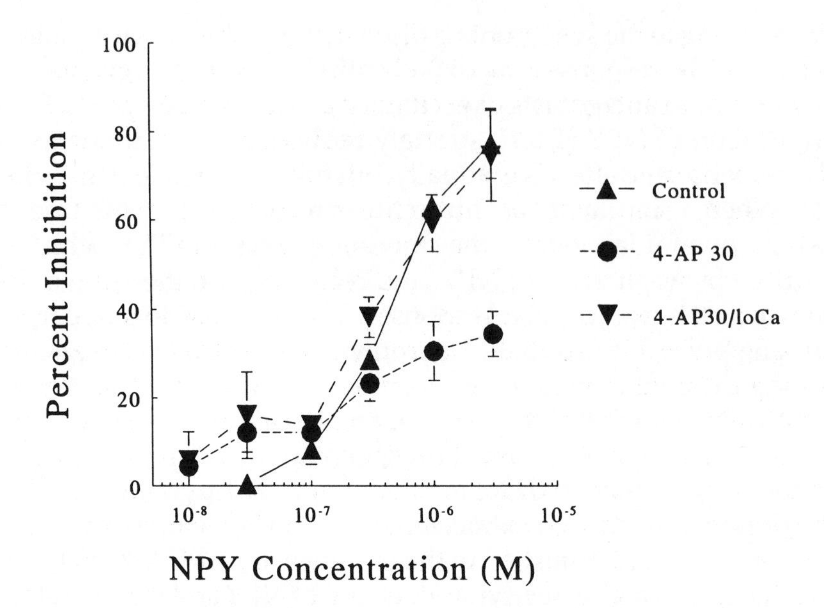

Fig. 7. Effects of 4-aminopyridine and low Ca^{2+} on the NPY-mediated inhibition of the extracellularly recorded population EPSP (pEPSP) in CA1 of rat hippocampal slice. Inhibition was measured as percent decrease in initial slope of control pEPSP under each condition. Ca^{2+} was 1.5 m*M* in control and in 4-AP (30 µ*M*), and 0.75 m*M* in 4-AP/low Ca^{2+}. 4-AP reduced inhibition mediated by NPY, but reducing Ca^{2+} in the presence of 4-AP restored the inhibition mediated by NPY at all effective concentrations. Modified from *37*, Bleakman, D., Harrison, N. I., Colmers, W. F., and Miller, R. J. (1992), ©1992 Macmillan Press, with permission.)

to compensate for the increase caused by 4-AP. Under these conditions, NPY again inhibited synaptic transmission, consistent with an action on presynaptic Ca^{2+} channels. More detailed experiments have demonstrated that the concentration–response curve for NPY in 4-AP, low Ca^{2+} is identical with that of control (Fig. 7; *37*). Further evidence for an action on presynaptic Ca^{2+} influx was obtained with synaptosomes prepared from rat hippocampus, in which $^{45}Ca^{2+}$ influx was inhibited by NPY when the synaptosomes were depolarized with 75 m*M* K^+ (Colmers et al, unpublished).

The pharmacology of the NPY receptor mediating presynaptic inhibition in hippocampus is consistent with a Y_2 receptor subtype. Porcine NPY, PYY, and human NPY are equipotent at the stratum radiatum-CA1 synapse, whereas desamido NPY is inactive. NPY_{2-36} is equipotent with the intact peptide, whereas fragments as short as NPY_{16-36} are active, although there is a sharp drop in activity with fragments shorter than NPY_{13-36} *(38)*. The mechanism whereby the receptors couple to the Ca^{2+} channels in the presynaptic terminal is not clear; however, NPY's inhibition in hippocampus is not affected by membrane-soluble analogs of cyclic AMP *(39)*. Blockers of the arachidonic acid pathway are without effect on NPY's presynaptic actions, whereas phorbol esters do reduce NPY-mediated inhibition, but this may be because of their directly elevating intraterminal Ca^{2+} (Colmers and Klapstein, unpublished observations). In adult rats, it appears that a pertussis-toxin-sensitive G protein is not involved, since slices taken from rats pretreated for 72 h with pertussis toxin intracerebroventricularly were as sensitive as were untreated control preparations to the presynaptic actions of NPY and the $GABA_B$ agonist baclofen, although postsynaptic responses mediated by $5\text{-}HT_{1A}$ and $GABA_B$ receptors were abolished by PTX pretreatment *(40)*. By contrast, similar pretreatment of cultured hippocampal neuron with PTX renders them insensitive to the actions of NPY (*see* Section 3.2.).

NPY acts at some, but not all, of the excitatory synaptic connections within hippocampus. Excitatory, feedforward synaptic connections onto CA1 and CA3 pyramidal cells all respond to NPY, whereas those onto dentate granule cells appear unaffected. In this, the electrophysiological results are consistent with the density of receptors obtained from binding data *(28)*. Interestingly, orthodromically evoked (feedforward) inhibitory synaptic potentials (IPSPs) are inhibited by the same amount as excitatory synaptic potentials, but antidromically evoked (recurrent) IPSPs are not affected, nor are IPSPs evoked by direct focal electrical stimulation of interneurons (when glutamate receptors are blocked; *41*). These results suggest that it is only the feedforward excitatory synapses (which also drive the interneurons) that bear NPY receptors. The distribution of actions suggests that, when NPY is released in hippocampus, it can regulate excitability in a manner that does not alter the properties of normal inhibition. This is supported by the observation that the frequency of spontaneous miniature excitatory postsynaptic currents recorded in CA3 pyramidal cells with whole-cell, patch-clamp electrodes is reversibly reduced by application of NPY (Fig. 8; McQuiston and Colmers, unpublished observations).

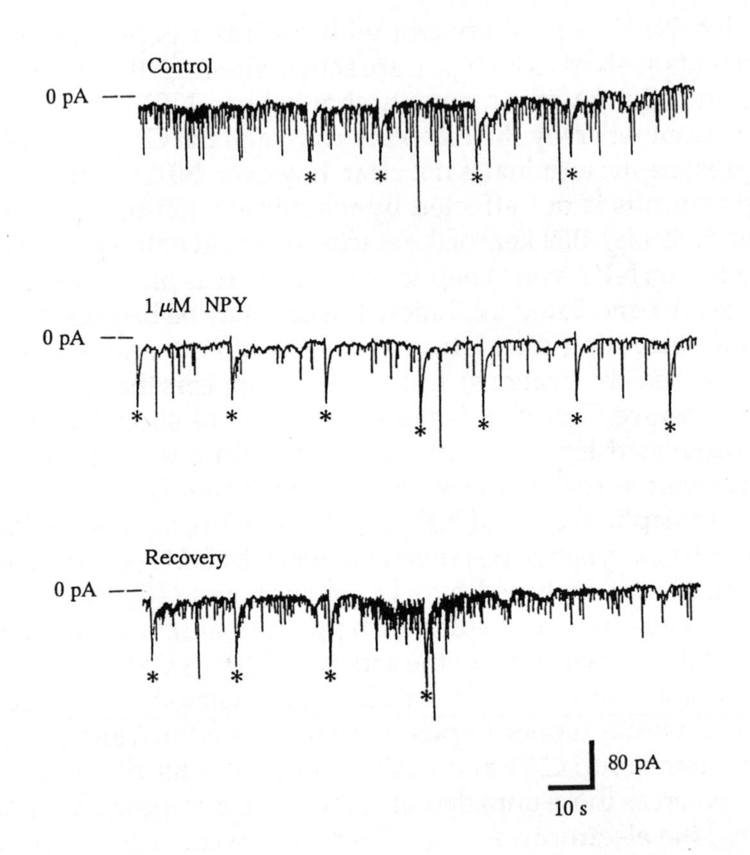

Fig. 8. NPY reversibly reduces the frequency of spontaneous miniature EPSCs in a CA3 pyramidal cell recorded with whole-cell patch clamp in rat hippocampal slice. Extracellular solution contained 100 μM picrotoxin to eliminate miniature synaptic currents mediated by $GABA_A$ receptors. Asterisks (*) represent the response of the neuron to iontophoretic pulses of NMDA applied to the proximal dendrites of the cell (*see* Fig. 9). Recovery was after 60 min of washout. A. R. McQuiston and W. F. Colmers, unpublished observations.

One report of excitatory actions of NPY on dentate granule cells exists *(42)*. However, the method of application of NPY (by pressure ejection into the slice) is subject to artifact because of the possibility of mechanical stimulation of the impaled neuron. There have been no other observations of an excitatory action of NPY in hippocampus.

Recent evidence has indicated that NPY is capable of displacing σ opioid ligands with high affinity *(43)*, and that microiontophoresis of NPY onto CA3 pyramidal neurons in vivo potentiated the excitation elicited by iontophoretic application of NMDA, but not that elicited by iontophoresis of AMPA or kainate, consistent with an action of NPY on a σ site on the NMDA receptor *(44)*. However, when NMDA was applied by microiontophoresis to CA3 pyramidal neurons recorded in whole-cell voltage clamp in in vitro hippocampal slices, 1 μ*M* NPY did not affect the response, at the same time as the EPSC evoked from mossy fibers was reduced by about 50% (Fig. 9; McQuiston and Colmers, 1992). More recently, evidence from other laboratories suggests that the σ-ligand displacement by NPY is not always observed *(46)*. It remains to be convincingly demonstrated that there is an actual interaction between NPY and the NMDA channel itself.

3.2. Hippocampal Neurons in Culture

When embryonic (E16-E17) rat hippocampal pyramidal neurons are placed in culture, they can form excitatory synaptic connections and are spontaneously active *(47)*. Measurements with fura-2 demonstrate that $[Ca^{2+}]_i$ levels fluctuate in these neurons under normal conditions, and can be reduced by the application of tetrodotoxin (TTX), extracellular Ni^{2+}, or supramaximal concentrations of the glutamate receptor antagonists APV and CNQX, indicating that spontaneous glutamatergic synaptic activity causes an elevation in basal $[Ca^{2+}]_i$ (Fig. 10; *47a*). NPY reduces $[Ca^{2+}]_i$ in these cultures; this effect is mimicked by NPY_{13-36}, suggesting the involvement of a Y_2 receptor. When extracellular K^+ is elevated to 50 m*M*, a rapid, transient elevation in $[Ca^{2+}]_i$, is reduced equally by NPY (100 n*M*) or CNQX/APV. However, when excitatory synaptic transmission is blocked with CNQX/APV, NPY no longer affects the K^+-mediated elevations in $[Ca^{2+}]_i$, indicating that postsynaptic Ca^{2+} conductances are not affected by NPY. Current clamp recordings of these neurons reveal barrages of spontaneous EPSPs, which 100 n*M* NPY completely abolishes. Tight-seal, whole-cell, voltage-clamp recordings of Ba^{2+} currents in these neurons showed no postsynaptic actions of NPY, whereas baclofen did inhibit these currents, and Cd^{2+} abolished them entirely. The absence of NPY effects on somatic Ca^{2+} currents was confirmed in

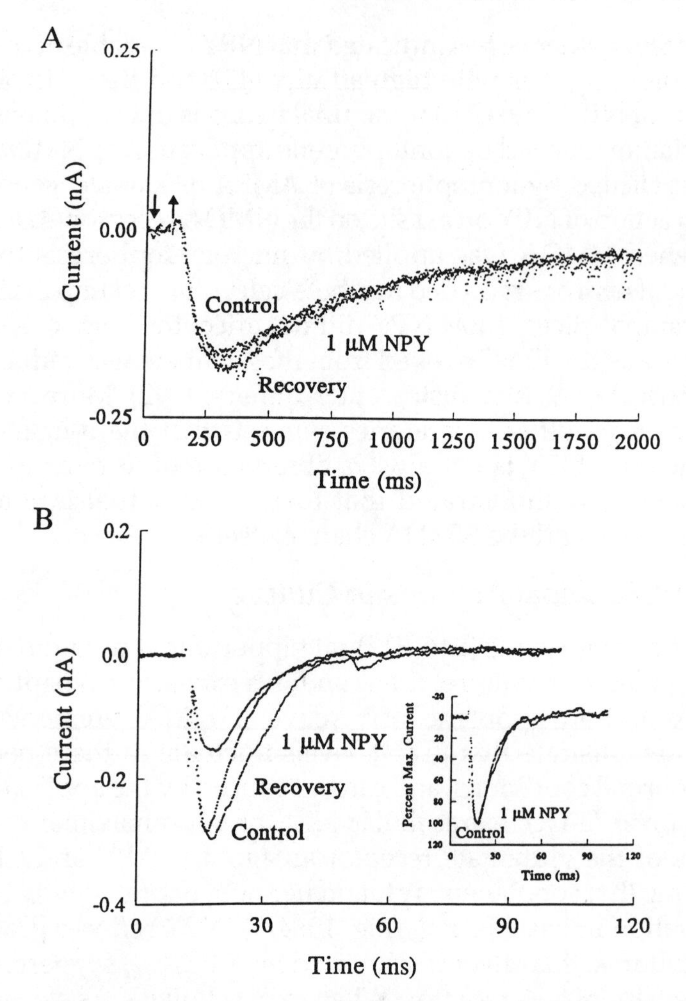

Fig. 9. NPY does not affect NMDA-mediated currents, while inhibiting mossy fiber synaptic responses in a CA3 pyramidal neuron in vitro. Same neuron as in Fig. 8; V_h = –60 mV. Patch pipet contained a K^+ gluconate-based solution. A: Superimposed responses to brief (100 ms, –190 nA) ejection of NMDA (50 m*M*) onto the proximal dendrites of the cell, in control, during application of 1 μ*M* NPY, and after 60 min of washout (recovery). B: Synaptic current evoked in this cell by stimulation of mossy fiber inputs, in control, during application of NPY and after 60 min of washout (recovery). Inset: Synaptic responses in control and NPY in B shown scaled to the same amplitude, indicating NPY does not affect kinetics of synaptic current. (From *45*, ©1992 Elsevier Press, with permission.)

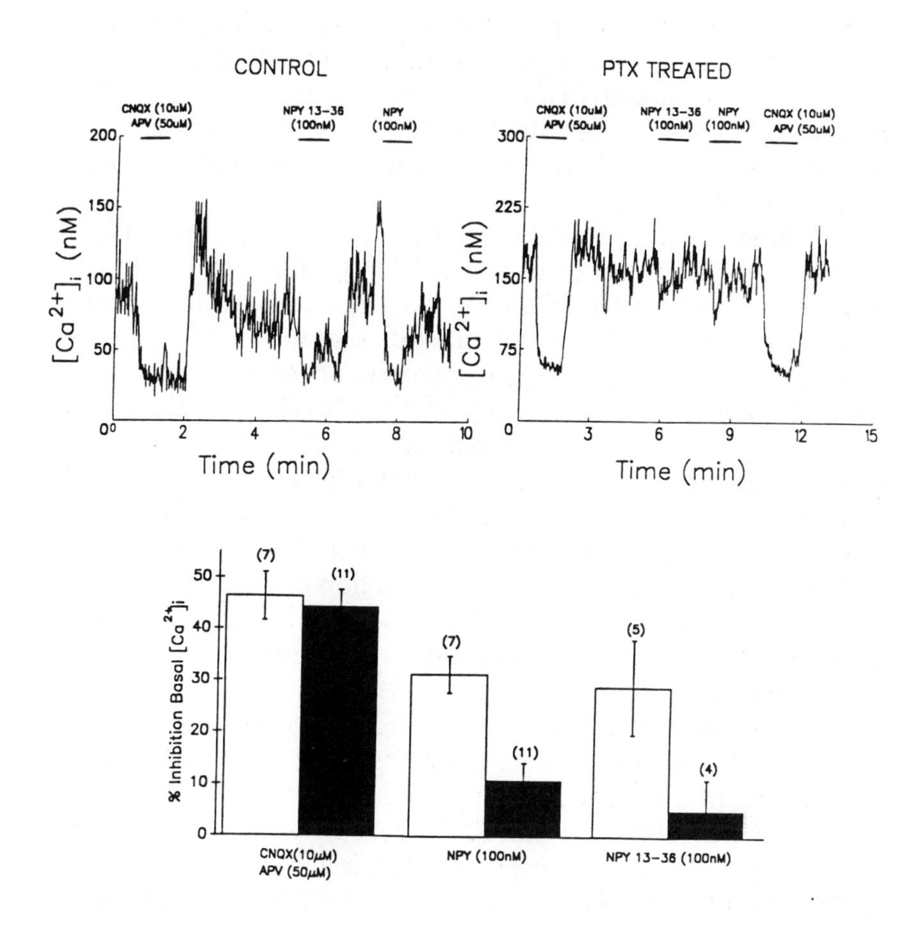

Fig. 10. NPY reduces fluctuations in $[Ca^{2+}]_i$ (recorded in a fura-2-loaded neuron) caused by spontaneous excitatory synaptic potentials in cultured hippocampal pyramidal cells in a pertussis-toxin-sensitive manner. Left: Neuron from a control culture. The glutamate receptor antagonists CNQX and APV greatly reduce spontaneous fluctuations in this neuron, as do NPY_{13-36} and NPY. Right: Neuron from a culture pretreated with pertussis toxin (PTX). NPY and NPY_{13-36} are without effect on the fluctuations in $[Ca^{2+}]_i$ in this cell, although glutamate receptor antagonists still block the responses. Bottom: Percent inhibition of basal $[Ca^{2+}]_i$ by CNQX/APV, NPY (100 n*M*), and NPY_{13-36} (100 n*M*) in cells from control cultures (open bars) and from PTX-pretreated cultures (solid bars). Numbers in parentheses indicate numbers of observations. (From *47a*, ©1992 Macmillan Press.)

experiments in which the pore-forming antibiotic, amphotericin B, was used in the pipet to gain electrical access to the interior of the cell after formation of a gigaseal, preventing washout of "necessary" intracellular constituents *(12)*. Taken together, the above data indicated that NPY acted only to inhibit transmitter release in these neurons.

Action potentials evoked in fura-2-loaded cultured hippocampal cells led to an elevation of $[Ca^{2+}]_i$. In normal saline, the evoked potentials were followed by a series of synaptically mediated depolarizations and action potentials (as the mostly excitatory synaptic circuits in the culture resonated). These secondary excitations also led to an elevation of $[Ca^{2+}]_i$. NPY (100 n*M*) blocked the synaptically mediated events, but did not affect the initial rise in $[Ca^{2+}]_i$ caused by the evoked action potentials (Fig. 11). When the experiment was repeated in the presence of CNQX and APV, NPY had no effect on the elevation of $[Ca^{2+}]_i$ elicited by the action potentials, whereas baclofen did reduce the action potential-evoked rise in $[Ca^{2+}]_i$. Therefore, in cultured hippocampal neurons as well as in the in vitro hippocampal slice, NPY acts presynaptically to inhibit the release of glutamate.

The question of second-messenger involvement was also addressed in these experiments by pretreating the cultures with PTX. Pretreated neurons no longer demonstrated a reduction in basal $[Ca^{2+}]_i$ to NPY, although CNQX/APV was effective in reducing $[Ca^{2+}]_i$ levels in these cells (Fig. 10). It is thus clear that in the cultured embryonic hippocampal neurons, NPY's presynaptic actions are mediated via a PTX-sensitive G protein, whereas it appears that in adult hippocampus this is not the case. A recent report *(48)* has shown that the PTX-insensitive G protein, G_z, is capable of coupling to receptors that also mediate their responses via G_i or G_o. It remains to be demonstrated whether this G protein, or a similarly PTX-insensitive one, is actually present in the presynaptic terminals in rat hippocampus in vivo.

Fig. 11. *(opposite page)* NPY inhibits synaptically mediated elevations in $[Ca^{2+}]_i$ without reducing elevations mediated by action potentials in a cultured hippocampal pyramidal cell. Upper traces represent fura-2 measurements of $[Ca^{2+}]_i$ in the cell whose intracellularly recorded membrane potential is shown in the lower trace. An identical number of action potentials were evoked at 20 Hz by brief (4–8 ms, 400 pA) current pulses

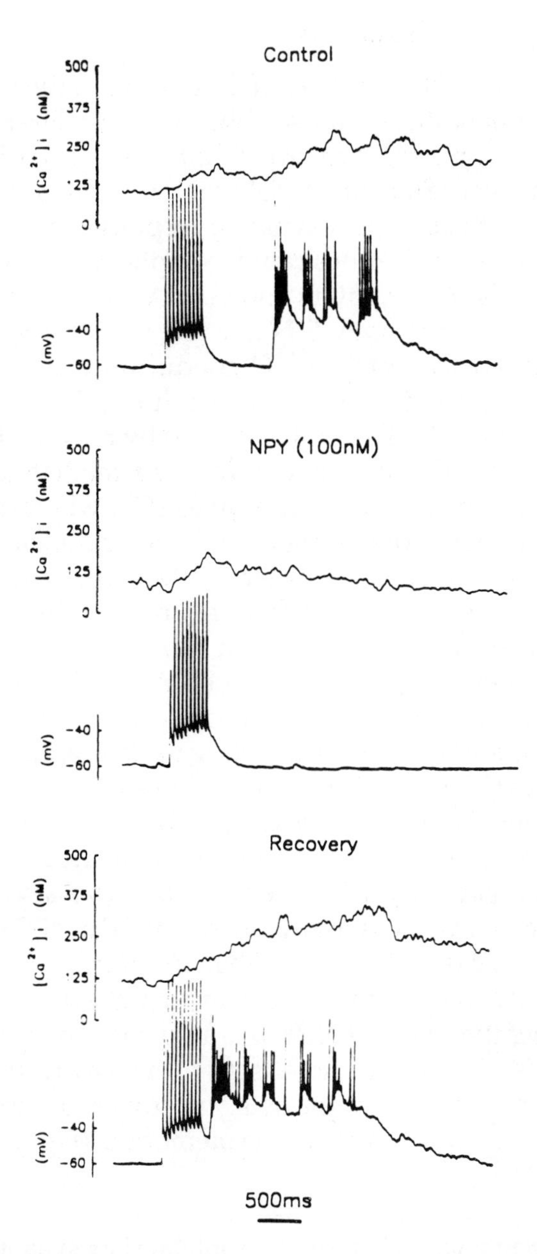

in control (upper panel), in the presence of NPY, and after washout. Note NPY reduces synaptically-mediated responses evoked in the neuronal network formed by the cultured cells, but does not alter the initial elevation in $[Ca^{2+}]_i$ caused by the train of action potentials. (From *47a*, ©1992 Macmillan Press, with permission.)

3.3. Dorsal Raphe Nucleus

The brainstem dorsal raphe (DR) nucleus is the source of the serotonergic innervation of most of the forebrain. The nucleus contains mostly serotonergic neurons that can be identified on the basis of their electrical characteristics in an in vitro slice preparation, as well as the complex synaptic responses evoked by focal electrical stimulation. Besides rapid synaptic responses mediated by several glutamate receptors and $GABA_A$ receptors, there is an IPSP mediated by serotonin (5-HT) acting via $5\text{-}HT_{1A}$ receptors to activate an inwardly rectifying K^+ conductance *(49)*, and a slow EPSP (sEPSP) mediated by noradrenaline (NA) activating α_1 adrenoceptors *(49–51)*. The IPSP is caused by release of 5-HT from DR cells and is similar to other autoreceptor-mediated inhibitions in other nuclei, such as locus coeruleus *(52)*, whereas the NA is released from fibers originating elsewhere. The excitatory amino acid input arises from habenular nuclei *(53)*, whereas GABA responses arise at least in part from neurons within the DR *(54)*. DR also receives an NPY innervation *(55)*, and NPY receptors are concentrated there *(28)*. Application of NPY (1 μM) to DR serotonergic cells in the in vitro slice preparation causes a reduction in the IPSP and in the sEPSP, without affecting the fast, amino acid-mediated responses (*56*; Fig. 12). As in hippocampus, the inhibition was fully reversible, and was mimicked by NPY_{13-36} and the selective Y_2 agonist [Cys^2, Aoc^5-24,D-Cys^{27}]-pNPY *(57)*, indicating the inhibition was mediated by a Y_2 receptor. NPY was also effective at inhibiting the pharmacologically isolated IPSP or sEPSP, although the inhibition of the IPSP was less than in control *(56)*.

Single microelectrode voltage clamp of these neurons was used to test whether the action of NPY was at a pre- or postsynaptic site. By itself, NPY (1 μM) had no effect on the steady-state current-voltage (I-V) relationship of the neurons, suggesting that it did not itself change the conductance of the membrane (Fig. 13). NPY also

Fig. 12. *(opposite page)* NPY and PYY inhibit slow synaptic potentials evoked in dorsal raphe serotonergic neurons. a: Synaptic potential evoked in control. KCl (2*M*) was in electrode. Digital acquisition rate was slowed eight times in center of trace (arrow, scale bar) to capture decay of the slow component. DSP—depolarizing synaptic potential (a mixture of

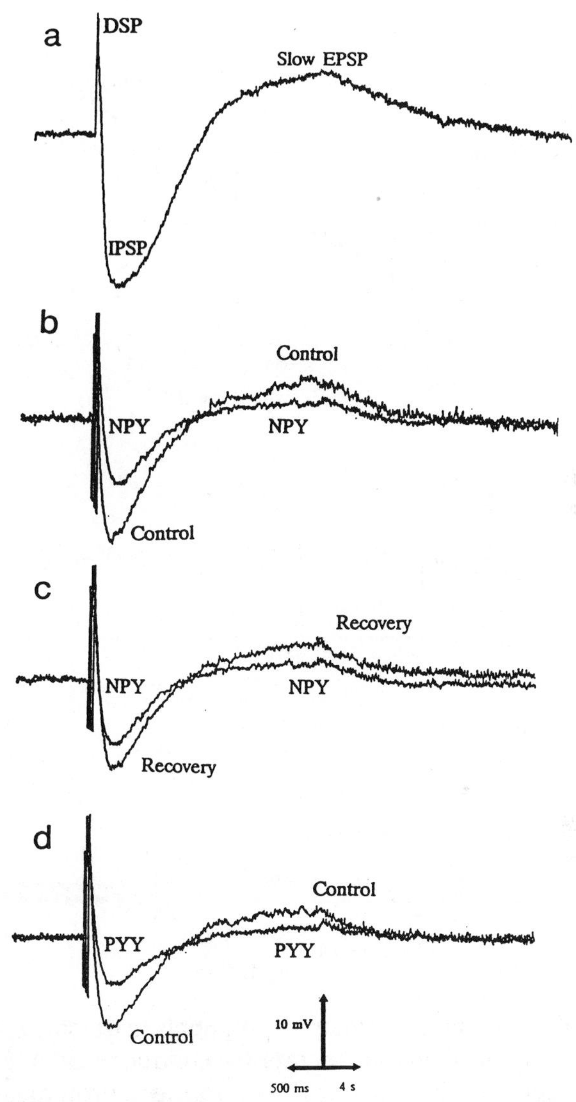

$GABA_A$ and glutamate-receptor mediated responses); IPSP—5-HT_{1A} receptor-mediated inhibitory postsynaptic potential; slow EPSP—α_1 receptor-mediated slow excitatory synaptic potential. b: Response from a different neuron as in a. Shown superimposed are the synaptic response in control and after application of 1 μM NPY. c: NPY response in b, superimposed on recovery (60 min of wash). d: Recovery response in c, superimposed on response in 1 μM PYY. Response recovered from PYY effect on washout (not illustrated). (From *56*, 1992 ©Society for Neuroscience, with permission.)

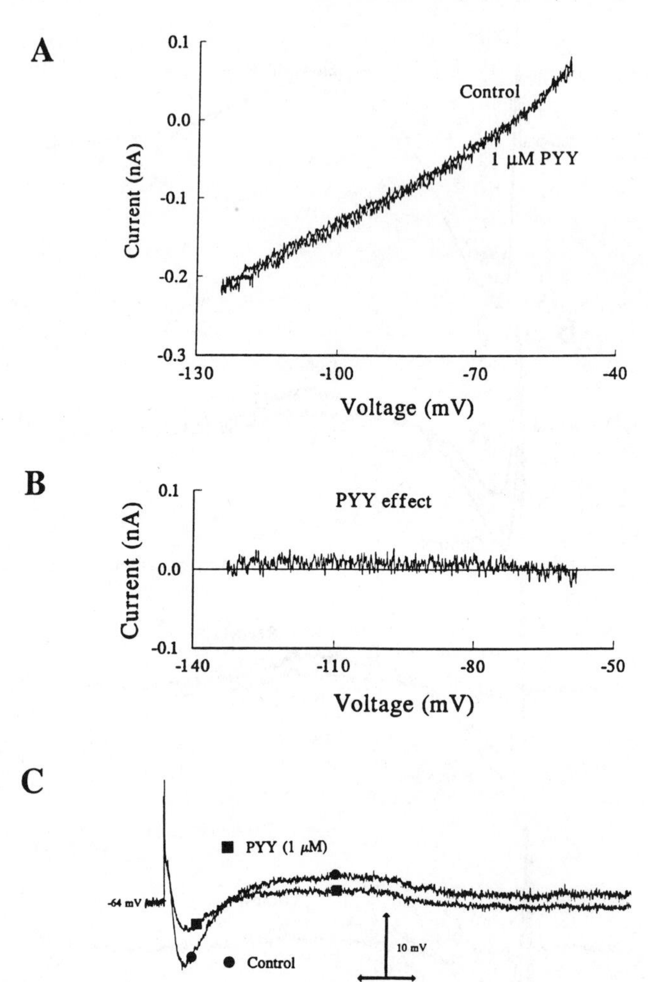

Fig. 13. PYY inhibits slow synaptic potentials in dorsal raphe neurons without affecting postsynaptic membrane conductance. A: Steady-state current–voltage relationship of a dorsal raphe neuron measured with a slow (>20 s) voltage ramp command from –115 to –40 mV in control and after application of 1 μ*M* PYY, shown superimposed. B: Digital subtraction of current–voltage relationships shown in A. PYY caused no change in resting or steady-state voltage-dependent currents. C: Complex synaptic potential evoked (in current clamp) in this cell, just shortly before and after, respectively, the current-voltage ramps were taken in control and PYY. (From *56*, 1992 ©Society for Neuroscience, with permission).

did not affect the steady-state membrane conductances elicited by application of $5\text{-}HT_{1A}$ and α_1 adrenoceptor agonists, even while it inhibited the evoked synaptic responses in the same neurons. As in hippocampus and the peripheral neuroeffector junction, NPY has a presynaptic inhibitory effect. Unlike in hippocampus, it inhibited the slow, rather than the fast, synaptic responses in this nucleus. This is also the first electrophysiological observation of an interaction of NPY with a 5-HT system *(56)*.

In vivo, dorsal raphe neurons are spontaneously active because of a tonic level of α_1 receptor activation by NA *(57a)*. The net effect of NPY receptor activation in dorsal raphe nucleus would be a change in state of the neurons, since spontaneous activity owing to NA would be reduced, but the response to exogenous excitation (from habenula; *54*) would be prolonged because of the reduction in the autoreceptor-mediated IPSP. This change of state is similar to the state of "quiet readiness" induced in hippocampal neurons by NA *(58)*, where the cells are hyperpolarized, but when they reach threshold, will fire for a prolonged period, thereby increasing the "signal-to-noise ratio" for a given response *(56)*.

3.4. Locus Coeruleus

An interesting contrast to the observed effects by NPY on Ca^{2+} currents was observed in the noradrenergic nucleus locus coeruleus (LC). Application of NPY here has been shown to cause a depression in the spontaneous firing rate and (at higher concentrations) a hyperpolarization, which can be blocked by application of α_2 receptor antagonists *(59)*. NPY potentiates the hyperpolarization elicited by selective α_2 agonists that are not subject to uptake, so it does not inhibit the reuptake of NA. Instead, it appears to be potentiating the postsynaptic response to α_2 receptor activation, much as has been observed at the sympathetic neuroeffector junction, where NPY potentiates postsynaptic α_1 adrenoceptor-mediated responses (Grundemar et al., this vol.). Unlike the latter effect, the response in LC to agonists at μ opioid receptors (which activate the same potassium conductance in these cells as α_2 receptors; *60*) is unaffected by NPY, suggesting a specific interaction between NPY and α_2 receptors *(59)*. The subtype of NPY receptor mediating this response was not identified.

4. Conclusions

Neuropeptide Y has a number of potent effects in both the central and the peripheral nervous systems. At this point, it appears that NPY is largely an inhibitory factor in many systems, where its release elicits profound changes in synaptic transmission, either by actions at presynaptic terminals or at the cell soma. In most preparations studied, NPY appears to act by reducing the influx of Ca^{2+} through voltage-dependent channels, although in at least one (amphibian) preparation it can also activate a K^+ conductance *(60a)*. These responses are largely mediated by Y_2 receptors. In many cases, it appears that NPY elicits a physiologically appropriate, highly specific response within a neuronal system; thus, in hippocampus it appears to control excitatory throughput in hippocampus without influencing the properties of the recurrent inhibitory circuits, whereas in DR, it should change the behavioral state of the neurons. These are actions consistent with a role of neuromodulator, where a coordinated response is elicited within a neuronal system.

However, this is unlikely to be the whole story. Clearly, in LC, the effect of NPY is inhibitory, but is entirely different from the observed direct effects on Ca^{2+} or K^+ channels, since it depends on the activation of other receptors for the expression of its effect, reminiscent of its potentiation of vasoconstrictor actions in the vasculature (cf Grundemar et al., this vol.). There is increasing evidence for Y_1 receptors in different areas of the CNS, particularly in frontal neocortex (Grundemar et al., this vol.). Recent evidence from acutely dissociated nodose ganglion neurons shows that they bear both Y_2 and Y_1 receptors, and that activation of Y_1 receptors *potentiates* an N-type Ca^{2+} current, whereas Y_2 receptors inhibit the same current, as in other cells *(61)*. Of particular note is the evidence that Y_1 and Y_2 responses are *both* mediated by G proteins that are pertussis toxin substrates in these cells. In addition, chromaffin cell nicotinic acetylcholine receptor currents can be selectively and equipotently inhibited by NPY and the C-terminal fragment NPY16–36, whereas Ca^{2+} currents in these cells are not affected *(62)*. This is interpreted as being a Y_2 receptor-mediated effect. However, since the NPY16–36 fragment is markedly less potent at activating Y_2 receptors in neurons *(12,38)*, the potency of this particular fragment in this preparation is surprising. It may mean that the NPY receptor in chromaffin cells is a particular subtype of the Y_2 receptor or another receptor type (e.g., *63*), or that the effect is owing to a receptor-independent action of the PP-fold peptides postulated to occur in mast cells *(64)*. Molecular biological approaches to the receptor subtypes present in the chromaffin cells or appropriate pharmacological studies are needed to clarify this question (*see* Grundemar et al., this vol.).

Also, there are many behavioral effects, such as the stimulation of feeding, that are most simply explained by an excitatory or excitomodulatory effect (cf Stanley, this vol.). Indeed, there is evidence from extracellular recordings in vitro that application of NPY to sympathoexcitatory neurons of the rostroventrolateral medulla *(65)* elicits an elevation in spontaneous firing frequency (*see also* McAuley et al., this vol.). Thus, the next important goal is to describe the electrophysiological actions of NPY in the different neuronal systems where it has been found to act, and to examine the mechanisms of action. Ultimately, the goal of these and similar studies is to determine the physiological role for NPY in the neuronal systems on which it impinges. Given the ubiquity and abundance of NPY, the pursuit of this goal should keep many investigators gainfully employed for the foreseeable future.

Acknowledgments

Research in the authors' laboratories has been supported by grants DA02101, MH 40165, and IP30-DK42086 from the National Institutes of Health (R. J. M.) and MT-10520 from the Medical Research Council of Canada (W.F.C.). D. B. was supported by a Fulbright Fellowship. W. F. C. is a Medical Scholar of the Alberta Heritage Foundation for Medical Research. The original version of this chapter was entitled by R. J. M.: "Naughty Peptide Y: An Opera in 4 Acts," which W.F. C. thought was much too good to let go entirely to waste.

References

1. Chronwall, B. M., Di Maggio, D. A., Massari, V. J., Pickel, V. M., Ruggiero, D. A., and O'Donohue, T. L. (1985). The anatomy of neuropeptide Y containing neurons on the rat brain. *J. Neurosci.* **15,** 1159–1181.
2. Lundberg, J. M., Terenius, L., Hokfelt, T., Martling, C. R., Tatemoto, K., Mutt, V., Polak, J., Bloom, S., and Goldstein, M. (1982) Neuropeptide Y (NPY)-like immunoreactivity in peripheral noradrenergic neurons and effects of NPY or sympathetic function. *Acta. Physiol. Scand.* **116,** 479–480.

3. Costa, M. and Furness, J. B. (1989). Structure and neurochemical organization of the enteric nervous system, in *Handbook of Physiology, 6 (The Gastrointestinal System)*, American Physiol. Soc., pp. 97–109.
4. Grundemar, L., Wahlestadt, C., and Reis, D. J. (1991) Long lasting inhibition of cardiovascular responses to glutamate and the baroreceptor reflux elicited by neuropeptide Y injected into the nucleus tract as solitarius of the rat. *Neurosci. Lett.* **122,** 135–139.
5. Lundberg, J. M., Pernow, J., and Lacroix, J. S. (1989) *News in Physiol. Sci.* **4,** 13–17.
6. Wahlestedt, C., Yanaihara, N., and Hakanson, R. (1986) Evidence for different pre and postjunctional receptors for neuropeptide Y and related peptides. *Reg. Pep.* **13,** 307–318.
7. Wahlestedt, C., Grundemar, L., Hakanson, R., Heilig, M., Shen, G. H., Zukowska-Grojek, Z., and Reis, D. J. (1990) Neuropeptide Y subtypes Y1 and Y2. *Ann. NY Acad. Sci.* **611,** 7–26.
8. Walker, M. W., Ewald, D. A., Perney, T. M., and Miller, R. J. (1988). Neuropeptide Y modulates neurotransmitter release and calcium currents in rat sensory neurons. *J. Neurosci.* **8,** 2438–2446.
9. Bongianni, F., Christenson, J., Hokfelt, T., and Grillner, S. (1990) Neuropeptide Y-immunoreactive spinal neurons make close oppositions on axons of primary sensory afferents. *Brain Res.* **523,** 337–341.
10. Wiley, J. W., Gross, R. A., Lu, Y., and MacDonald, R. C. (1990) Neuropeptide Y reduces calcium current and inhibits acetylcholine release in nodose neurons via a pertussis toxin sensitive mechanism, *J. Neurophysiol.* **63,** 1499–1507.
11. Giuliani, S., Maggi, C. A., and Meli, A. (1989) Prejunctional modulatory action of neuropeptide Y on peripheral terminals of capsaicin sensitive sensory nerves. *Brit. J. Pharmacol.* **98,** 407–412.
12. Bleakman, D., Colmers, W. F., Fournier, A., and Miller, R. J. (1991) Neuropeptide Y inhibits Ca^{2+} influx into cultured rat dorsal root ganglion neurons via a Y_2 receptors. *Brit. J. Pharmacol.* **103,** 1781–1789.
13. Thayer, S. A. and Miller, R. J. (1990) Regulation of the intracellular free calcium concentration in single rat dorsal root ganglion neurons in vitro. *J. Physiol.* **425,** 85–115.
14. Hirning, L. D., Fox, A. P., and Miller, R. J. (1990) Inhibition of calcium currents in cultured rat myenteric neurons by neuropeptide Y: Evidence for a direct receptor/channel coupling. *Brain Res.* **532,** 120–130.
15. Schofield, G. G. and Ikeda, S. R. (1988) Neuropeptide Y blocks a calcium current in C cells of bullfrog sympathetic ganglia. *Eur. J. Pharmacol.* **151,** 131–134.
16. Miller, R. J. (1987) Multiple calcium channels and neuronal function. *Science* **235,** 46–52.
17. Miller, R. J. (1990) The receptor mediated regulation of calcium channels and neurotransmitter release. *FASEB J.* **4,** 3291–3300.
18. Lester, H. A., Snutch, T. P. Leonard, J. P., Nargeot, J., Curtis, B. M., and Davison, N. (1989) Expression of mRNA encoding voltage-dependent Ca^{2+} channels in Xenopus oocytes: Review and progress report. *Ann. NY Acad. Sci.* **560,** 174–182.

19. Mogul, D. J. and Fox, A. P. (1991) Evidence for four different types of Ca^{2+} channels in acutely-isolated hippocampal CA3 neurones of the guinea pig. *J. Physiol.* **433,** 259–281.
20. Pennington, N. J., Kelly, J. S., and Fox, A. P. (1991) A study of the mechanism of Ca^{2+} current inhibition produced by serotonin in rat dorsal raphe neurons. *J. Neurosci.* **11,** 3594–3609.
21. Bean B. P. (1989) Neurotransmitters inhibit neuronal calcium currents by changes in channel voltage dependence. *Nature* **340,** 153–156.
22. Ewald, D. A., Pang, I.-H., Sternweis, P. C., and Miller, R. J. (1989) Differential G-protein mediated coupling of neurotransmitter receptors to Ca^{2+} channels in rat dorsal root ganglion neurons in vitro. *Neuron* **2,** 1185–1193.
23. Perney, T. M. and Miller, R. J. (1989) Two different G-proteins mediate neuropeptide Y and bradykinin stimulated phospholipid breakdown in cultured rat sensory neurons. *J. Biol. Chem.* **264,** 7328–7337.
24. Ewald, D. A., Matthies, H. J. G., Perney, T. M., Walker, M. W., and Miller, R. J. (1988) The effect of down regulation of protein kinase C on the inhibitory modulation of dorsal root ganglion neuron Ca^{2+} currents by neuropeptide Y. *J. Neurosci.* **8,** 2447–2451.
25. Lundberg et al. (1990)
26. Duckles, S. P. and Budai, D. (1990) Stimulation intensity as a critical determinant of presynaptic receptor effectiveness. *Trends in Pharmacol. Sci.* **11,** 440–443.
27. Bleakman, D. and Miller, R. J. (1991) Modulation of the intracellular free calcium concentration in rat dorsal root ganglion neurons in vitro (submitted).
28. Martel, J. C., Fournier, A., St-Pierre, S., and Quirion, R. (1990) Quantitative autoradiographic distribution of $[^{125}I]$ Bolton-Hunter neuropeptide Y receptor binding sites in rat brain. Comparison with $[^{125}I]$-peptide YY receptor sites. *Neuroscience* **36,** 255–283.
29. Dingledine, R. (ed.) (1984) *Brain Slices* Plenum, New York, 442p.
30. Hestrin, S., Nicoll, R. A., Perkel, D. J., and Sah, P. (1990) Analysis of excitatory synaptic action in pyramidal cells using whole-cell recording from rat hippocampal slices. *J. Physiol.* **422,** 203–225.
31. Colmers, W. F. (1990) Modulation of synaptic transmission in hippocampus by Neuropeptide Y: Presynaptic actions. *Ann. NY Acad. Sci.* **611,** 206–218.
32. Andersen. P., Silfvenius, H., Sundberg, S. H., Sveen, O., and Wigström, H. (1978) Functional characteristics of unmyelinated fibers in the hippocampal cortex. *Brain Res.* **144,** 11–18.
32a. Andersen, P., Bliss, T. V. P., and Skrede, K. K. (1971) Unit analysis of hippocampal population spikes. *Exp. Brain Res.* **13,** 208–221.
33. Colmers, W. F., Lukowiak, K. D., and Pittman, Q. J. (1985) Neuropeptide Y reduces orthodromically-evoked population spike in rat hippocampal CA1 by a possibly presynaptic mechanism. *Brain Res.* **346,** 404–408.
34. Colmers, W. F., Lukowiak, K. D. and Pittman, Q. J. (1987) Presynaptic action of neuropeptide Y in area CA1 of the rat hippocampal slice. *J. Physiol.* **383,** 285–299.
35. Haas, H. L., Hermann, A., Greene, R. W., and Chan-Palay, V. (1987) Action and location of neuropeptide tyrosine (Y) on hippocampal neurons of the rat in slice preparations. *J. Comp. Neurol.* **257,** 208–215.

36. Colmers, W. F., Lukowiak, K., and Pittman, Q. J. (1988) Neuropeptide Y action in the rat hippocampal slice: site and mechanism of presynaptic inhibition. *J. Neurosci.* **8,** 3827–3837.
37. Klapstein, G. J. and Colmers, W. F. (1992) 4-Aminopyridine and low Ca^{++} differentiate presynaptic inhibition mediated by Neuropeptide Y, baclofen and 2-chloroadenosine in rat hippocampal CA1 *in vitro*. *Br. J. Pharmacol.* **105,** 470–474.
38. Colmers, W. F., Klapstein, G. J., Fournier, A., St-Pierre, S., and Treherne, K. A. (1991) Presynaptic inhibition by Neuropeptide Y (NPY) in rat hippocampal slice *in vitro* is mediated by a Y_2 receptor. *Br. J. Pharmacol.* **102,** 41–44.
39. Klapstein, G. J., Treherne, K. A., and Colmers, W. F. (1990) NPY: presynaptic effects in hippocampal slice *in vitro* are Y_2 receptor-mediated but not via inhibition of adenylate cyclase. *Ann. NY Acad. Sci.* **611,** 457–458.
40. Colmers, W. F. and Pittman, Q. J. (1989) Presynaptic inhibition by Neuropeptide Y and baclofen in hippocampus: insensitivity to pertussis toxin treatment. *Brain Res.* **489,** 99–104.
41. Davies, C. H., Davies, S. N., and Collingridge, G. L. (1990) Paired-pulse depression of monosynaptic GABA-mediated inhibitory postsynaptic responses in rat hippocampus. *J. Physiol.* **424,** 523–532.
42. Brooks, P. A., Kelly, J. S., Allen, J. M., Smith, D. A. S., and Stone, T. W. (1987) Direct excitatory effects of neuropeptide Y (NPY) on rat hippocampal neurones in vitro. *Brain Res.* **408,** 295–298.
43. Roman, F. J., Pascaud, X., Duffy, O., and Junien, J. L. (1991) N-Methyl-D-aspartate receptor complex modulation by neuropeptide Y and peptide YY in rat hippocampus in vitro. *Neurosci. Lett.* **122,** 202–204.
44. Monnet, F. P., Debonnel, G., and de Montigny, C. (1990) Neuropeptide Y selectively potentiates N-methyl-D-aspartate-induced neuronal activation. *Eur. J. Pharmacol.* **182,** 207,208.
45. McQuiston, A. R. and Colmers, W. F. (1992) Neuropeptide Y does not alter NMDA conductances in CA3 pyramidal neurons: a slice-patch study. *Neurosci. Lett.* **138,** 261–264.
46. Tam, S. W. and Mitchell, K. N. (1991) Neuropeptide Y and peptide YY do not bind to brain σ and phencyclidine binding sites. *Eur. J. Pharmacol.* **193,** 121,122.
47. Forsythe, I. D. and Westbrook, G. L. (1988) Slow excitatory postsynaptic currents mediated by N-methyl-D-aspartate receptors on cultured mouse central neurones. *J. Physiol.* **396,** 515–534.
47a. Bleakman, D., Harrison, N. L., Colmers, W. F., and Miller, R. J. (1992) Investigations into neuropeptide Y-mediated presynaptic inhibition in cultured hippocampal neurones of the rat. *Br. J. Pharmacol.* **107,** 334–340.
48. Wong, Y. H, Conklin, B. R., and Bourne, H. R. (1992) G_z-mediated hormonal inhibition of cyclic AMP accumulation. *Science* **255,** 339–342.
49. Yoshimura M. and Higashi H. (1985) 5-Hydroxytryptamine mediates inhibitory postsynaptic potentials in rat dorsal raphe neurons. *Neurosci. Lett.* **53,** 67–74.
49a. Yoshimura, M., Higashi, H., and Nishi, S. (1985) Noradrenaline mediates slow excitatory synaptic potential in rat dorsal raphe neurons *in vitro*. *Neurosci. Lett.* **61,** 305–310.

50. Williams, J. T., Colmers, W. F., and Pan, Z. Z. (1988) Intrinsic and inwardly rectifying conductances in rat dorsal raphe neurones in vitro. *J. Neurosci.* **8,** 3499–3506.
51. Pan Z. Z. and Williams J. T. (1989) Differential actions of cocaine and amphetamine on dorsal raphe neurons in vitro. *J. Pharmacol. Exp. Ther.* **251,** 56–62.
52. Egan, T. M., Henderson, G., North R. A., and Williams, J. T. (1983) Noradrenaline-mediated synaptic inhibition in rat locus coeruleus neurones. *J. Physiol.* **345,** 477–488.
53. Kalen, P., Karlson, M., and Wiklund, L. (1985) Possible excitatory amino acid afferents to *nucleus raphe dorsalis* of the rat investigated with retrograde wheat germ agglutinin and D-[3H]-aspartate tracing. *Brain Res.* **360,** 285–297.
54. Harandi, M., Aguera, M., Gamrani, H., Didier, M., Maitre, M., Calas, A., and Belin, M. F., (1987) γ-aminobutyric acid and 5-hydroxytryptamine interrelationship in the rat nucleus raphe dorsalis: combination of autoradiographic and immunocytochemical techniques at light and electron microscopic levels. *Neuroscience* **21,** 237–251.
55. de Quidt M. E. and Emson P. C. (1986) Distribution of neuropeptide Y-like immunoreactivity in the rat central nervous system—II. Immunohistochemical analysis. *Neuroscience* **18,** 545–618.
56. Kombian, S. B. and Colmers, W. F. (1992) Neuropeptide Y (NPY) selectively inhibits slow synaptic potentials in rat dorsal raphe nucleus *in vitro* by a presynaptic action. *J. Neurosci.* **12,** 1086–1093.
57. Krstenansky, J. L., Owen, T. J., Buck, S. H., Hagaman, K. A., and McLean, L. R. (1989) Centrally truncated and stabilized porcine NPY analogues: design, synthesis and mouse brain receptor binding. *Proc. Natl. Acad. Sci. USA* **86,** 4377–4381.
57a. Baraban, J. M. and Aghajanian, G. K. (1980) Suppression of firing activity of 5-HT neurons in the dorsal raphe by alpha-adrenoceptor antagonists. *Neuropharmacology* **19,** 355–363.
58. Madison, D. V. and Nicoll, R. A. (182). Noradrenaline blocks accommodation of pyramidal cell discharge in hippocampus. *Nature* **299,** 636–638.
59. Illes, P. and Regenold, J. T. (1990) Interaction between neuropeptide Y and noradrenaline on central catecholamine neurones. *Nature* **344,** 62–63.
60. North, R. A. and Williams, J. T. (1985). On the potassium conductances increased by opioids in rat locus coeruleus neurones. *J. Physiol.* **364,** 265–280.
60a. Zidichouski, J. A., Chen, H., and Smith, P. A. (1990) Neuropeptide Y activates inwardly-rectifying K^+-channels in C-cells of amphibian sympathetic ganglia. *Neurosci. Lett.* **117,** 123–128.
61. Wiley, J. W., Gross, R. A., and MacDonald, R. C. (1991) Neuropeptide Y (NPY) Y_1 and Y_2 receptors mediate opposite actions on nodose neuronal calcium currents via pertussis toxin-sensitive pathways. *Soc. Neurosci. Abstr.* **17,** 902.
62. Nörenberg, W., Illes, P., and Takeda, K. (1991) Neuropeptide Y inhibits nicotinic cholinergic currents but not voltage-dependent calcium currents in bovine chromaffin cells. *Pflügers Arch.* **418,** 346–352.
63. Balasubramaniam, A., Sheriff, S., and Fischer, J. F. (1990) Neuropeptide Y(18–36) is a competitive agonist of neuropeptide Y in rat cardiac ventricular membranes. *J. Biol. Chem.* **265,** 14,724–14,727.

64. Grundemar, L. and Håkanson, R. (1991) Neuropeptide Y, Peptide YY and C-terminal fragments release histamine from rat peritoneal mast cells. *Br. J. Pharmacol.* **104,** 776–778.
65. Sun, M.-K. and Guyenet, P. G. (1989) Effects of vasopressin and other neuropeptides on rostral medullary sympathoexcitatory neurons "in vitro". *Brain Res.* **492,** 261–270.

The Role of NPY and Related Peptides in the Control of Gastrointestinal Function

Helen M. Cox

1. Introduction

The aims of this chapter are to provide an overview of the effects that neuropeptide Y (NPY), peptide YY (PYY), and related pancreatic polypeptides have on gastrointestinal function, focusing on their inhibitory effects on epithelial electrolyte and fluid secretion. The functional aspects of each pancreatic polypeptide (PP) is discussed in the light of its known cellular localization and the identification of receptors specific for the respective peptide. Readers requiring additional information concerning the enteric nervous system are referred to the chapter by Sundler in this volume and to Furness and Costa *(1)*. Comprehensive epithelial reviews concerning electrolyte secretion in mammalian small and large intestine are available by Donowitz and Welsh *(2)* and Binder and Sandle *(3)*, respectively.

2. Localization of Pancreatic Polypeptides Within the Rat Gastrointestinal Tract

2.1. Neuropeptide Y

Early immunohistochemical studies first identified the presence of PP-like immunoreactivity in endocrine cells *(4)* and neurons *(5)* of the pancreas and intestine. Subsequently, Vaillant and Taylor *(6)* demonstrated PP-like material in rat intestine and brain using

From: *The Biology of Neuropeptide Y and Related Peptides;*
W. F. Colmers and C. Wahlestedt, Eds. © 1993 Humana Press Inc., Totowa, NJ

antibodies raised against the C-terminal hexapeptide of mammalian PP that crossreacted with both PYY and NPY. NPY-like immunoreactivity was specifically localized within enteric neurons *(7)* using antibodies raised against NPY and absorbed with bovine PP. The greatest density of intrinsic NPY-immunoreactive nerves was observed in the small intestine *(8)*. Few NPY-positive neurons were observed in the rat colon *(9)*, and fewer still in the stomach wall *(10)*. In the rat jejunum, NPY neurons in the myenteric plexus issue short fibers (2 mm) that are descending, whereas submucous NPY axonal processes are longer and ascending. Submucosal NPY neurons primarily innervate the mucosal region, these cells sending dense networks of fibers into the villus area *(8)*. In both ganglial populations, NPY is colocalized with vasoactive intestinal polypeptide (VIP); in fact, all NPY-positive neurons contain VIP (and all VIP-positive cells contain NPY) in the rat small intestine *(8,11)*. The colocalization between these two peptides is not complete in the rat colon. Although all NPY neurons also contain VIP, these cells represent only 25% of the total population of VIP-immunoreactive neurons in this area *(9)*. Markedly different patterns of peptide colocalization are observed in the enteric nervous system (ENS) of different species *(12,13)*, and the present chapter will deal primarily with the rat ENS (making comparisons with other species wherever necessary). For example, myenteric and submucous NPY neurons in the guinea pig small intestine also contain calcitonin gene-related peptide (CGRP), cholecystokinin (CCK), somatostatin (SOM), and choline acetyltransferase (ChAT) *(12)*. In human appendix *(14)* and esophagus *(15)*, however, VIP/NPY colocalization is abundant in nerve fibers and cell bodies. In the rat small intestine (which has been the tissue of choice in functional studies described below), the projection pattern of VIP/NPY nerves differs significantly from those separate neuronal populations containing CGRP, SOM, substance P (SP), or enkephalin *(8)*. Whether any one of these neuronal subdivisions is cholinergic remains to be determined. A close association between the processes of NPY-positive submucous neurons and the epithelial lining has been observed in guinea-pig small intestine at the electron-microscopic level *(16)*. Even though the overall patterns of colocalization are different between the rat and guinea-pig ENS, the presence of NPY within neurons innervating the mucosal region implicates this peptide as a candidate neurotransmitter with the capacity to alter epithelial function.

Although the majority of NPY-positive neurons in the intestine are intrinsic, a significant extrinsic population of NPY-containing fibers exists around vascular elements. These are adrenergic *(17)*, originate from prevertebral ganglia, and disappear following sympathectomy (*7; see* chapter by Sundler, this vol.). Intrinsic ganglionic NPY cell bodies were unaffected by pretreatment with 6-hydroxydopamine, although a reduction in fiber density has been observed in submucosal regions by Wang et al. *(18)*. It is nevertheless clear that NPY is present in both extrinsic, adrenergic, and intrinsic, nonadrenergic neurons innervating the intestinal wall, and that the latter population predominates.

2.2. Peptide YY

In contrast with the neuronal localization of NPY, peptide YY (PYY) is present in endocrine cells of the mucosal lining *(19,20)*. As with the variations in number of NPY-containing neurons, so too the number of PYY endocrine cells varies, being most abundant in the lower intestine (in rat and human) with relatively few positive cells in the small intestine. PYY-containing endocrine cells in the distal intestine also contain the glucagon-like peptide, glicentin *(20)*, and in colorectal L-cells, these two peptides are observed in the same secretory granules *(21)*, despite the fact that they are products of different precursors. The pattern of PYY-containing endocrine cell number compares well with the extracted levels of PYY found along the length of the intestine *(22)*, this being highest in the cecum, large intestine, and rectum. There is evidence for neuronal localization of PYY in the gut of certain species, such as cat and ferret (*see* chapter by Sundler, this vol.), in the rat central nervous system *(23)*, and in rat superior cervical sympathetic ganglia *(24)*. In cat and ferret intestine, PYY-positive neurons are located in myenteric ganglia only and are particularly prominent in the upper intestine, decreasing in number descending down the GI tract. In the rat, however, PYY is only found in endocrine cells, where it is colocalized with glicentin.

2.3. Pancreatic Polypeptide

Relatively little information exists concerning the cellular localization of pancreatic polypeptide (PP)-like immunoreactivity within the mucosal lining of the gastrointestinal tract. There do not appear to be any PP-containing neurons in the rat ENS, although a few scattered endocrine cells have been observed (Sundler, personal communication). The densest localization of PP-containing endocrine cells occurs in the pancreatic islets in a variety of species, including humans *(25,26)*, and the demonstration of cytoplasmic processes of these PP cells is suggestive of a paracrine function for PP *(27)*.

3. Localization of Receptor Binding Sites for the Pancreatic Polypeptides

3.1. NPY/PYY Binding Sites

The presence of prominent networks of NPY-containing fibers in the mucosal region of the rat small intestine implicates a role for NPY in the control of epithelial function. Indeed the identification and characterization of binding sites for PYY and NPY in epithelial plasma membranes from rat small intestine were first described by Laburthe et al. *(28)*. Radioiodinated PYY labeled a single population of sites with an affinity (K_d) of 0.4 n*M* and maximal binding capacity (B_{max}) of 336 fmol/mg. Using a single ligand concentration (0.15 n*M*), the highest levels of binding were obtained in epithelial preparations from the jejunum and duodenum. Significantly lower levels of binding were observed in ileal epithelia, and virtually no specific binding was found in preparations from the stomach, cecum, or colon. This regional distribution of receptors broadly follows the relative density of NPY innervation, but not the density of PYY-containing endocrine cells. Thus, it is important to note that these receptors are likely to be affected by released neurotransmitter, namely, NPY, as well as by PYY released from endocrine cells in the lower intestine. The pharmacological specificity exhibited by jejunal receptors shows that PYY displaces ^{125}I-PYY binding with four times the potency of NPY (IC_{50} of 0.25 n*M* compared with 1 n*M*). Avian, rat, and human pancreatic polypeptides were at least three orders of magnitude less potent than NPY, and this specificity profile was very close to that observed in functional studies (*see* Section 4.).

The distribution of this common population of receptors (with high affinity for both NPY and PYY) along the crypt-villus axis shows a relative concentration of sites in the crypt region *(29)*. In contrast with VIP receptors, which appear to be constant in number along the villus length, NPY/PYY receptor number is eight times greater in crypt epithelia compared with villus epithelia. All NPY/PYY receptors identified in the different epithelial populations exhibited a similar affinity and specificity. Using jejunal epithelial cell preparations, the same group described the attenuation of VIP (and other secretagogs)–stimulated adenosine 3',5'-monophosphate (cAMP) production by NPY and PYY *(30)*. This negative coupling to adenylate cyclase was subsequently observed predominantly in crypt epithelia *(29)*, the proposed site of chloride secretion. The attenuation of secretagog-stimulated cAMP levels by NPY and PYY was most pronounced in crypt epithelial cells from the upper intestine *(29)*, there being no significant effect of either peptide on cAMP levels in epithelia obtained from the cecum or colon *(30)*. As in binding studies, the order of potency PYY ≥ NPY > PP was observed for inhibition of VIP-stimulated cAMP production in jejunal epithelia.

Recently, Raufman and Singh *(31)* have identified PYY and NPY (but not PP) inhibition of secretagog-elevated levels of intracellular cAMP in dispersed chief cells from guinea-pig stomach. PYY was more effective than NPY (half-maximal effects obtained with 1- and 10-n*M* concentrations, respectively), and this inhibition was similar in profile to the attenuation afforded by both peptides of secretin and VIP-stimulated pepsinogen secretion from the same cellular preparations.

A detailed biochemical study of NPY binding sites in small intestine epithelial cells from the rat has described a preponderance of sites within the basolateral domain *(32)*. These binding sites exhibited an apparent affinity of 15 n*M*, and although this value is lower than that described for epithelial NPY/PYY receptors in crude epithelial preparations *(28)*, it is similar to the EC_{50} values observed for NPY in in vitro studies. Two NPY receptor species were identified using crosslinking techniques, exhibiting molecular sizes of 48–55 and 33–35 kDa, respectively *(32)*. The greater molecular-size protein appears to be similar to a 50-kDa binding subunit of the Y_2 receptor from rabbit kidney proximal tubule basolateral membranes *(33)*. Voisin et al. *(34)* have further characterized the ^{125}I-PYY binding sites from rat jejunal crypts. Following solubilization, a single population of sites was observed with a mol wt *(M_r)* of 48,000 and an agonist order of potency (from displacement studies) of PYY > NPY >> PP. This peptide order of potency is very similar to that observed in in vitro functional studies, where antisecretory responses from rat jejunal preparations were measured (*see* Section 4.1.). This PYY-preferring receptor is a glycoprotein with a cleavable fragment of M_r 28,000 observed after denaturing procedures. Whether another distinct population of lower affinity, NPY-preferring binding sites exist in intestinal epithelia and what its functional significance is remain unclear. The presence of such a population in both small *(35)* and large intestine, with no apparent receptor gradient along the crypt/villus axis, requires further characterization.

3.2. Pancreatic Polypeptide-Specific Binding Sites

Specific, high-affinity binding sites for PP have been located predominantly within basolateral plasma membranes from canine duodenum/jejunum *(36)*. PYY and NPY exhibited low affinities for these PP binding sites (64 and 39 n*M*, respectively) compared with an IC_{50} of 1.9 n*M* for porcine PP. Saturation analysis of ^{125}I-PP binding to basolateral membranes revealed two classes of sites with affinities of 1.5 and 26 n*M*, the lower affinity site probably corresponding to those with affinity for PYY and NPY. The physiological role that these sites might subserve has yet to be identified.

4. Functional Significance of NPY and PYY on Epithelial Ion and Fluid Transport

4.1. Antisecretory Effects of NPY and PYY in Rat Intestinal Tract

The first documented study of antisecretory effects of NPY and PYY described their inhibition of prostaglandin-E_2 (PGE_2)-mediated fluid secretion in rat jejunum in vivo *(37)*. NPY appeared to be more effective than PYY in anesthetized rats, there being no effect of either peptide on basal net fluid and ion absorption. Simultaneous infusion of NPY with PGE_2 markedly reduced the reversed net Na and Cl secretion seen with the secretagog alone. Further in vivo studies described NPY's capacity to cause net absorption of fluid (in basal conditions and VIP-pretreated rats), the most prominent effects being seen in the jejunum *(38)*. This regional selectivity was significant, since dense NPY-like innervation and receptor localization have subsequently been described in this gastrointestinal area.

In vitro studies measuring changes in short-circuit current (s.c.c.) across mucosal sheets from chosen gastrointestinal areas have provided detailed information of the antisecretory potential of NPY, PYY, and their structural analogs. Briefly, the details of this technique involve blunt dissection of the tissue of choice, removing the overlying smooth muscle. Preparations are placed between two halves of perspex Ussing chambers with a window size of 0.6 cm^2. Voltage-recording electrodes, one on either side of the tissue, are filled with Krebs-Henseleit (K-H) solution and lead via KCl-agar bridges to calomel cells. Current-passing electrodes placed as far from the tissue as possible (i.e., at each end of the half chamber) are again filled with K-H and lead via KCl-agar bridges to Ag-AgCl electrodes. Prior to tissue placement, fluid resistance between the voltage-recording electrodes is compensated for, and thus, when the mucosal sheet is present, the voltage difference recorded arises only from that across the preparation. Using this technique, information concerning passive or active ion-transport processes can be obtained, but many groups choose to voltage clamp preparations automatically, and under these conditions, s.c.c. measurements result from active, electrogenic transport processes. Each side of the epithelium is bathed in K-H circulated through heat exchangers using a gas lift (95% O_2/5% CO_2) and maintained at 37°C.

Our own studies have on the whole utilized mucosal preparations of rat jejunum, an area noted for its dense NPY innervation. All preparations are allowed to stabilize, and once a basal current is attained, peptides or other agents can be added to either reservoir and any resultant changes in s.c.c. are continuously recorded. Application of increasing concentrations of either NPY or PYY to the basolateral surface evoked prolonged reductions in basal s.c.c. *(39)*. It should be noted at this juncture that apical additions of NPY or PYY had no significant effect on epithelial ion transport. Rat jejunum mucosa exhibited the greatest sensitivity to NPY and PYY (Fig. 1), 100 n*M* of each peptide affecting reductions in s.c.c. of –12.0 0.7 A/0.6 cm^2 ($n = 6$) and –12.2 1.1 A/0.6 cm^2 ($n = 6$), respectively. In the descending colon, maximal responses of –6.5 0.4 A/0.6 cm^2 ($n = 4$) for NPY (1*M*) and –7.3 0.4 A/0.6 cm^2 ($n = 4$) for PYY (1*M*) were obtained, whereas mucosal preparations of fundic stomach were not significantly affected. This pattern of sensitivity correlates closely with the relative density of NPY innervation, from its most dense in the rat small intestine to least dense in the

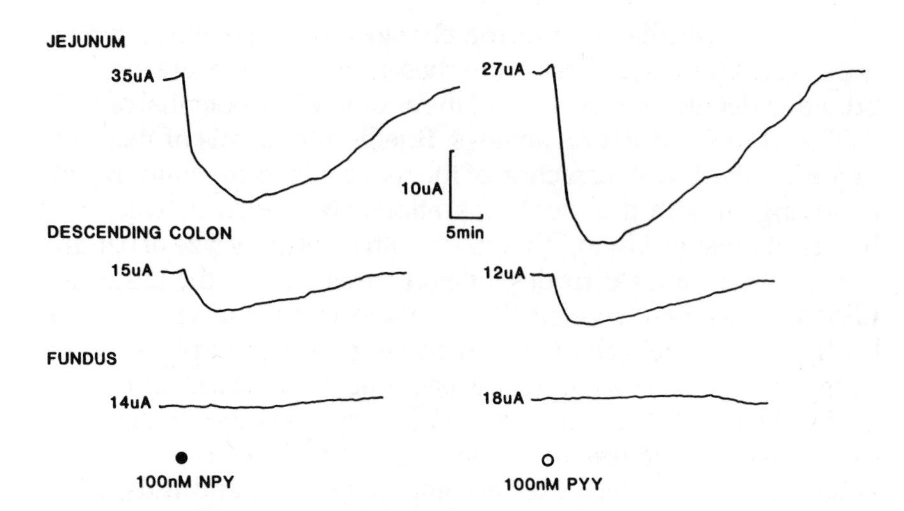

Fig. 1. Reductions in basal short circuit current (s.c.c.) following basolateral application of either NPY (100 n*M*) or PYY (100 n*M*), to mucosal preparations from different areas of the gastrointestinal tract. Baseline currents are given to the left of each recording.

stomach, with the colon intermediary between the two. The size of these antisecretory effects did not correlate with the level of the basal s.c.c., and all responses to NPY and PYY were unaffected by either tetrodotoxin (TTX) or the α_2- antagonist, yohimbine. Thus, despite the similarity between the profiles of antisecretory responses generated by NPY and PYY and the α_2-selective agonist, xylazine (Fig. 2; *40*), all these effects are highly specific and mediated via selective receptors found predominantly within the basolateral epithelial surface. In contrast with the direct effects of NPY in rat jejunum, in porcine distal jejunum, a portion of the NPY-antisecretory response was mediated by noradrenaline *(41)*, whereas in full thickness mouse jejunum, NPY responses were abolished by TTX *(42)*. In the latter case, it was suggested that NPY may be affecting antisecretory responses via σ receptors, although this is by no means clear at present.

NPY's antisecretory capacity also extends to effects on higher centers, and significantly to areas of the CNS involved in the regulation of gastrointestinal function and feeding behavior. Intrahypothalamic injections of picomole amounts of NPY resulted in prolonged inhibition of interdigestive levels of gastric acid output

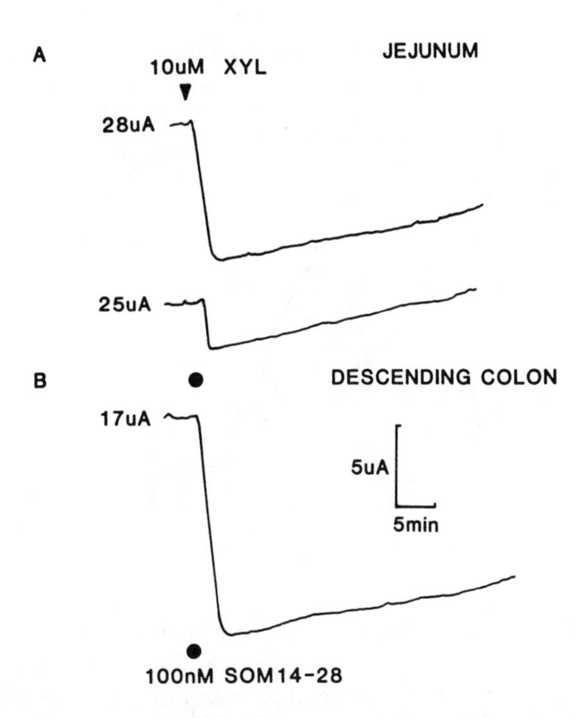

Fig. 2. Reductions in basal s.c.c. following basolateral addition of the α_2-agonist, xylazine (XYL, 10μ*M*, ▼) and somatostatin (SOM_{14-28}, 100 n*M*, ●) to the basolateral compartments bathing jejunum (A) and descending colon (B) preparations.

in the anesthetized rat *(43)*. This attenuation appeared to be independent of changes in blood pressure and was limited to the paraventricular nucleus, an area in which NPY also produces pronounced stimulation of food intake (*44; see* Section 5.3.).

Another major antisecretory neuropeptide is somatostatin (SOM); however, in contrast with NPY and PYY, SOM (in the rat) is significantly more effective in the descending colon compared with the upper intestine (Fig. 2, *45*). In rat jejunum, the EC_{50} for NPY is commonly 10–12 n*M*, whereas PYY is more effective, with an EC_{50} of 1–3 n*M* (see Fig. 3; NPY EC_{50} = 10 n*M*; PYY EC_{50} = 1.5 n*M*). Human PP, however, had very little effect in rat jejunum, reducing basal s.c.c. by only –2.2 0.2 A/0.6 cm^2 (n = 3) at 1 *M* (Fig. 3). These EC_{50} values for NPY and PYY antisecretory effects are compared with those of 15 n*M* for SOM_{14-28} and 13 n*M* for SOM_{1-28} in the descending colon.

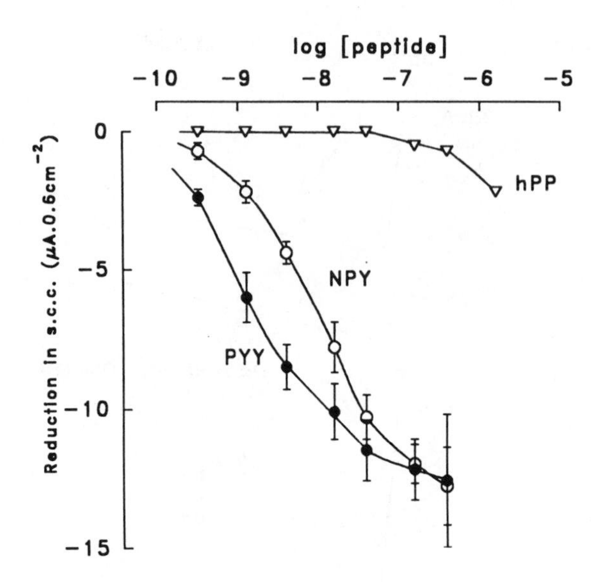

Fig. 3. Concentration–response curves for PYY (●), NPY (○), and human PP (hPP, ∇) in rat jejunum mucosa. Responses were recorded from 0.6 cm^2 areas of voltage-clamped preparations, in a cumulative fashion. Each point represents the mean ± SEM of the following respective *n* values; *n* = 6 for PYY; *n* = 3–8 for NPY, and *n* = 3 for hPP.

4.2. Identification of Ionic Species Responsible for Electrogenic Changes in Rat: A Comparison with Guinea Pig, Rabbit, and Porcine Antisecretory NPY Effects

Use of a range of ion channel blockers and cotransport inhibitors indicated that an inhibition of chloride secretion was a major contributing mechanism to NPY and PYY's electrogenic effects *(39)*. To verify whether these peptide-induced reductions in s.c.c. were due to either an inhibition of Cl secretion or a stimulation of absorption (or that sodium ions were involved), studies of ^{36}Cl and ^{22}Na fluxes were performed (Fig. 4A and B). Unidirectional flux studies identified both a reduction in ^{36}Cl flux from serosa to mucosa (S → M) and a lesser increase in ^{36}Cl absorption (M → S) in the presence of 100 n*M* NPY (Fig. 4A). No significant changes in ^{22}Na movement in either direction (Fig. 4B) were seen following NPY application. In the rat jejunum, NPY reduced the basal s.c.c. by attenuating net Cl secretion, and these responses are therefore described as antisecretory.

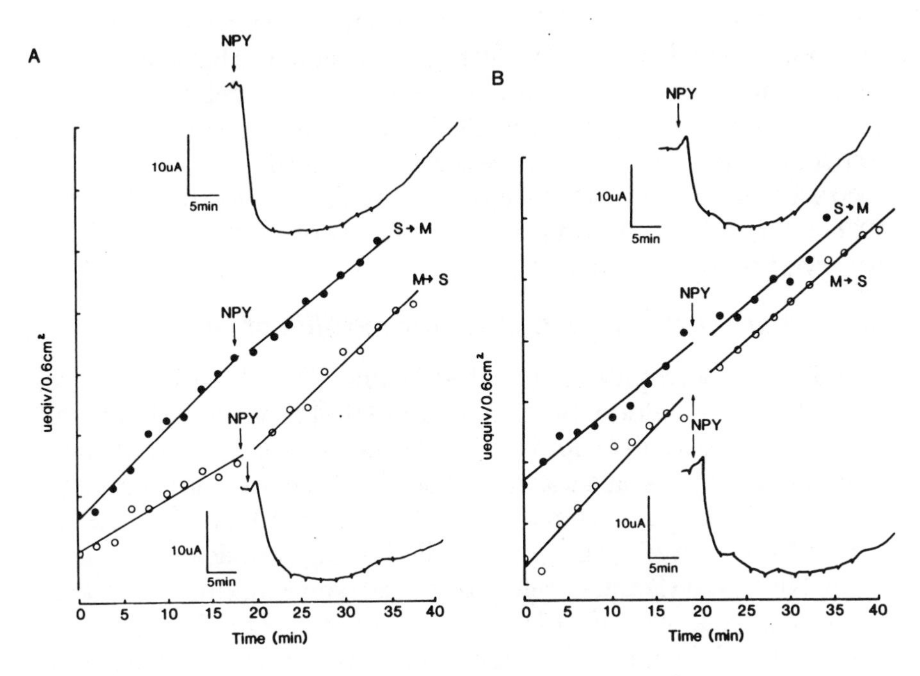

Fig. 4. A. Movement of ^{36}Cl S → M (●) and M → S (○) before and after application of 100 n*M* NPY basolaterally. Trace amounts of ^{36}Cl were added to either the apical (○) or basolateral (●) bathing media, and the preparations allowed to equilibrate for 30 min. Samples of 1 mL were then taken from the cold side at 2-min intervals replacing with Krebs-Henseleit (K-H) buffer each time. The ^{36}Cl appearing in the cold bathing media was calculated as microequivalents and plotted against time. Linear regression analysis was performed on all samples prior to addition of NPY, and those samples obtained subsequent to peptide addition were analyzed separately. The corresponding reductions in s.c.c. are shown for comparison, and these data are representative of four studies performed in each group. B. Movement of ^{22}Na S → M (●) and M → S (○) before and after addition of 100 n*M* NPY basolaterally. Trace amounts of ^{22}Na were added to either apical or basolateral bathing media, and the experimental protocol followed that described for ^{36}Cl experiments.

There are differences in the specific ions that contribute to electrogenic NPY and PYY responses in different species. In rabbit ileum, NPY increased Cl absorption as well as reducing Na absorption *(46)*, although the latter effect was not observed by Friel et al. *(47)*. Here, an enhanced absorption of Na was seen together with a net increase in Cl absorption. In porcine distal jejunum under basal conditions, NPY induced net Cl absorption and this was not affected by the absence of extracellular HCO_3 ions. No significant changes in ^{22}Na flux were observed under control conditions, but following removal of HCO_3, NPY enhanced both unidirectional ^{22}Na fluxes in porcine jejunum. Despite the inconsistencies seen with Na transport, the attenuation of Cl secretion by NPY and PYY is the most common primary ionic event responsible for the observed reductions in s.c.c.

4.3. Broad Spectrum Antisecretory Profile of NPY

The antisecretory effects of NPY and PYY described above were all obtained under basal, nonstimulated conditions. In the rat small intestine in vitro approx 50% of the basal s.c.c. is generated by endogenous prostaglandin formation. Application of the cyclo-oxygenase inhibitor piroxicam reduced basal s.c.c. from 27.8 1.1 A/0.6 cm^2 to 14.8 0.9 A/0.6 cm^2 (n = 91). Following piroxicam pretreatment, NPY (30 n*M*) antisecretory responses were significantly attenuated (from controls of –9.2 0.5 A/0.6 cm^2, n = 47 to –2.3 0.5 A/0.6 cm^2 n = 16). Subsequent investigations, however, found that NPY responses could be restored by intermediate addition of a range of secretagogs, such as VIP, forskolin, and PGE_2 (*see* example traces in Fig. 5; *48*). All of these agents cause secretion by elevating intracellular cAMP either by a receptor-operated mechanism (VIP or PGE_2) or directly, as in the case of forskolin. Once the secretagog-elevated s.c.c. had reached a steady state in piroxicam pretreated tissues, application of NPY produced antisecretory responses that were not significantly different from controls (after PGE_2 and forskolin), although following higher concentrations of VIP, NPY responses were enhanced (*48;* and *see* Fig. 6). The mechanisms underlying this magnification are unknown, but NPY's capacity to reduce secretion is not simply limited to its capacity to attenuate adenylate cyclase activity. A similar restoration of NPY responses was observed following application of the phosphodiesterase inhibitor, isobutyl-1-methyl-xanthine (IBMX) and also after addition of dibutyryl cAMP (dbcAMP), neither secretagog requiring the involvement of adenylate cyclase to increase Cl secretion. Explanations of the restoration of NPY-antisecretory responses under these conditions must rely on the basal turnover of adenylate cyclase, IBMX thereby conserving cAMP and dbcAMP effectively adding to it. Flint et al. *(49)* have observed similar inhibition of VIP and forskolin-stimulated

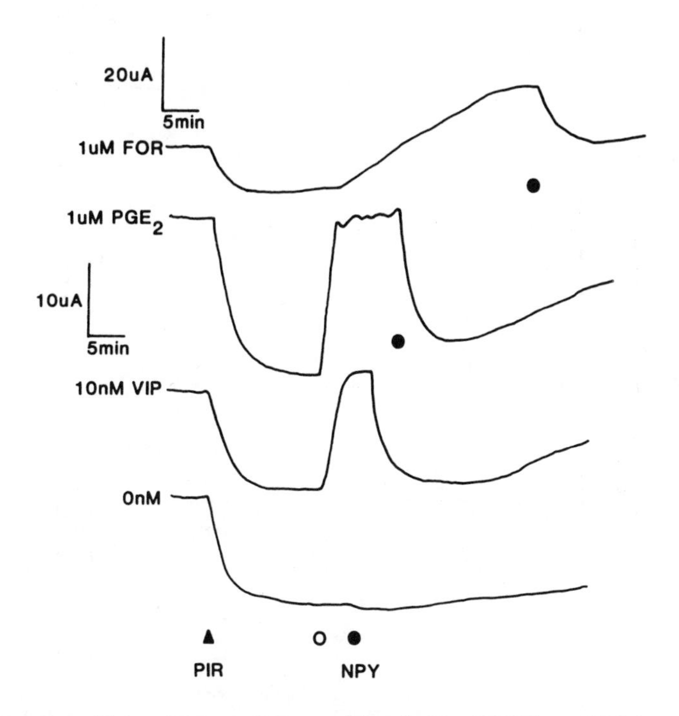

Fig. 5. The effect of VIP, PGE_2, and forskolin (FOR) on NPY responses in rat jejunal preparations pretreated with piroxicam. All tissues were pretreated with 5μ*M* piroxicam (PIR) added apically and basolaterally. Fifteen minutes after this, one of three different secretagogs was added, either 10 n*M* VIP or 1μ*M* PGE_2 added basolaterally, or 1μ*M* FOR added to both sides. The resultant increases in current were allowed to stabilize before 33 n*M* NPY were applied. The recordings were obtained from different animals, and each trace is a single example of a range of concentrations tested with each secretagog. Note the difference in vertical scale for the top trace where the secretagog is 1μ*M* FOR.

secretion by NPY in voltage-clamped preparations of rabbit distal colon. In addition to this attenuation of cAMP-generated secretion, NPY abolished substance P (SP, 100 n*M*) and significantly inhibited carbachol (CCh, 10 μ*M*) secretory responses *(48)*. Both secretagogs affect an increase in intracellular Ca, thereby increasing apical Cl secretion *(50,51)*. Thus, NPY (and presumably PYY) exhibits a broad spectrum antisecretory capacity. The ability to attenuate cAMP-mediated secretory responses is more easily explained than NPY's interactions with SP and CCh secretion, although the latter

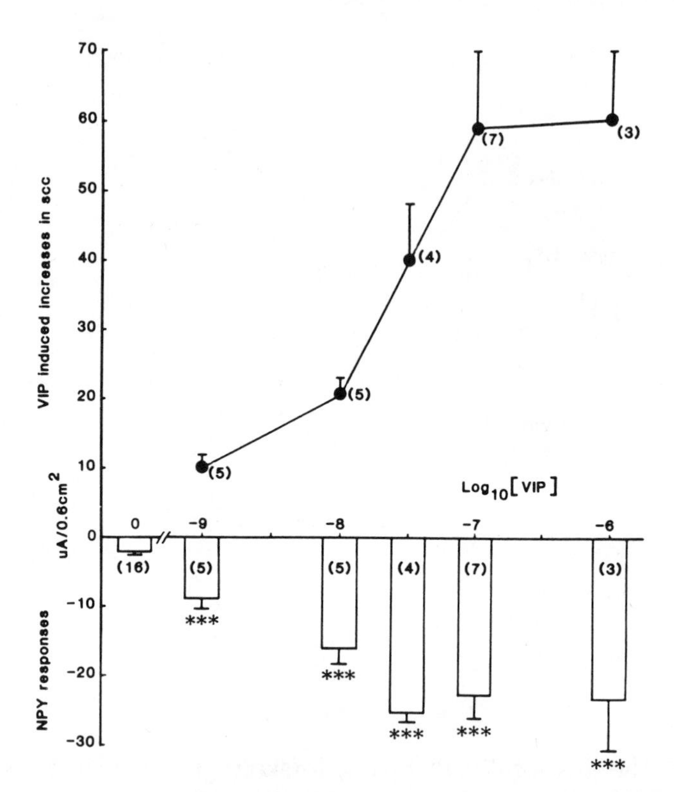

Fig. 6. The restoration of NPY responses after VIP addition to piroxicam-pretreated jejunal epithelia. Piroxicam (5μ*M*) was added to all preparations before one of a range of VIP concentrations was applied basolaterally. Fifteen minutes later, when the elevated s.c.c. had stabilized 33 n*M* NPY were added, and the resultant attenuation of current was recorded and plotted below the respective VIP concentrations. The sizes of all NPY responses were significantly larger than control NPY responses after piroxicam alone (***$p < 0.001$). Values are the mean ± SEM, and n values are shown in parenthesis.

could be mediated by an inhibitory G-protein system coupled to the activation of phospholipase C. Alternatively, crosstalk between the two second-messenger systems could provide an explanation. These broad spectrum effects with NPY are also observed in the descending colon (Cox and Yui, unpublished), and are similar to the extensive antisecretory profiles of SOM *(45,52)* and the α_2-agonists, such as xylazine *(40)*. A similar inhibition of CCh and SP responses by adrenaline has been observed in chicken distal ileum *(53)*, although there was no significant change in agonist-activated Ca entry. Thus, it was suggested that adrenaline was affecting the mechanism through which Ca alters secretion. So, too, if the epithelial cAMP and Ca secretory mechanisms are interconnected, it is possible that the antisecretory agents NPY, SOM, and xylazine may interfere with both systems at a point beyond which commonality between them exists. Clearly detailed measurements of changes in intracellular Ca and cAMP are necessary for a full understanding of these broad-spectrum antisecretory effects. As yet, no information is available concerning the involvement of NPY and PYY effects on the third major messenger system in epithelia, namely guanylate cyclase.

The in vitro studies with NPY and PYY have nevertheless provided explanatory mechanisms for recent clinical observations. Ileostomy patients receiving VIP infusions produced an increased ileal output that was significantly reduced by coadministration of PYY (at plasma levels similar to those measured postprandially, *54*). A PYY infusion alone had no significant effect on nonstimulated ileal output. Clearly, PYY coinfusion is attenuating the elevated intraepithelial cAMP production induced by VIP in a manner similar to that seen with NPY inhibition of both epithelial anion secretion and prestimulated adenylate cyclase activity in vitro. The absence of PYY effects under normal conditions also finds parallels with the virtual loss of NPY responses after lowering basal secretion with piroxicam.

The dependence of NPY antisecretory responses on endogenous eicosanoid production has also been observed in porcine distal jejunum (*41*), and the potency of NPY increased following addition of secretagogs, e.g., forskolin or 8-bromo-cAMP. However, the actions of NPY on both basal and cAMP-induced secretion were abolished by TTX and are thus mediated by another antisecretory agent(s), in contrast with the direct NPY/PYY epithelial effects seen in the rat, rabbit, and guinea-pig intestine.

4.4. Effects of NPY and PYY on Field-Stimulated Electrogenic Ion Secretion

Electrical field stimulation (EFS) of intestinal preparations from rabbit (*55,56*), guinea pig (*57–59*), and human (*60*) results in increases in s.c.c. Admittedly, a variety of different stimulation parameters have been used. Nevertheless, the resultant secretory responses are abolished by TTX, but only partially inhibited by a combination of muscarinic and nicotinic antagonists. The remaining EFS-induced

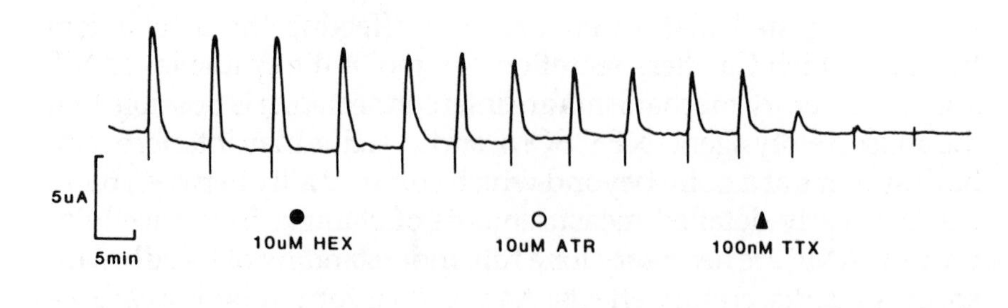

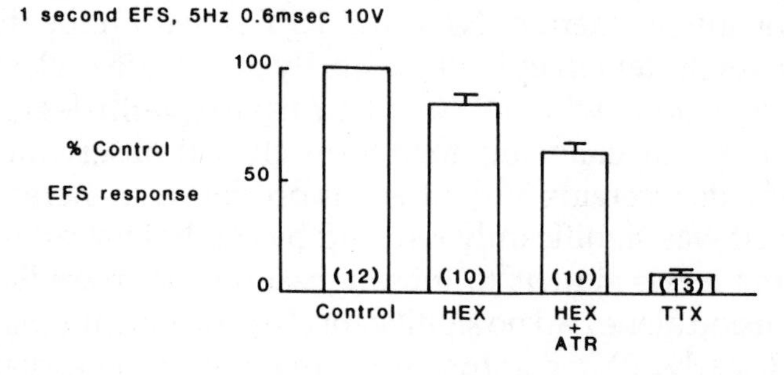

Fig. 7. Secretory responses in rat jejunal preparations generated by a 1-s electrical field stimulation (EFS) of 5 Hz, each pulse lasting 0.6 ms. When these 1-s stimuli were delivered at 7-min intervals, the resultant transient secretory responses were of constant peak height. The effects of basolateral addition of hexamethonium (HEX, 10μ*M*), atropine (ATR, 10μ*M*), and tetrodotoxin (TTX, 100 n*M*) are shown in the upper trace with pooled data below. After pretreatment, the resultant EFS responses were calculated as a percentage of the control responses (in the absence of HEX and ATR) obtained in that tissue. The number of observations is given in parenthesis.

secretion is either totally or partially mediated by neuropeptides, of which SP and VIP are most commonly implicated (with the exception of porcine gut, where neither peptide appears to be involved; *61*). In rat intestine, both jejunum and descending colon (*62*; Cox, unpublished observations), a 1-s period of EFS (bipolar rectangular pulses of 0.6 ms duration and frequency of 5 Hz) resulted in rapid, transient secretory responses that when delivered at 7-min intervals, were unchanged over a period of up to 2 h. These secretory responses were inhibited 50% by simultaneous addition of the nicotinic ganglion blocker, hexamethonium, and the muscarinic antagonist, atropine (Fig. 7, from controls of 7.6 0.6 A/0.6 cm^2 to 3.8 0.3 A/0.6 cm^2 [$n = 79$]). Subsequent application of TTX (100 n*M*) virtually abolished secretory responses (Fig. 7, to 0.6 0.1 A/0.6

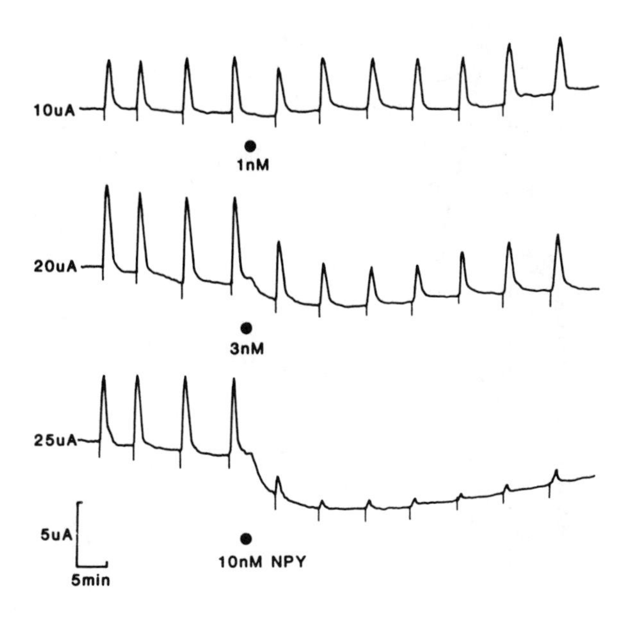

Fig. 8. The effect of NPY on EFS responses in the presence of hexamethonium and atropine (each 10μ*M*). Stimuli (5 Hz, 0.6 ms for 1 s) were delivered at 7-min intervals. Application of hexamethonium and atropine reduced EFS responses 40% (not shown), and the remaining resistant secretory EFS responses were inhibited in a concentration-dependent manner by NPY (applied basolaterally). Baseline currents are given on the left, and these three traces were obtained from adjacent sections of jejunum.

cm^2 [n = 12]). A proportion of the noncholinergic EFS response appears to be mediated by both SP and VIP, as indicated by the 50% inhibition of secretory responses following addition of a cocktail of antagonists for neurokinin (NK) 1, 2, and 3 receptors (GR71251, L659,877, and [DPro^2, DTrp^{6,8}, Nle^{10}]-NKB all at 1 μ*M*) plus a VIP antagonist ([AcTyr^1]hGRF (1-40)-OH, 300 n*M*). These analogs all significantly attenuated respective agonist-induced secretory responses *(63,64)*. The remaining NANC responses (adrenoceptor antagonists have no significant effect on EFS responses) are of unknown origin, but are not mediated by ATP, 5HT, or histamine (Cox, unpublished).

Pretreatment of tissues with increasing concentrations of NPY results in a concentration-dependent inhibition of both the basal s.c.c. and the EFS responses (Fig. 8). By interpolation of the level of

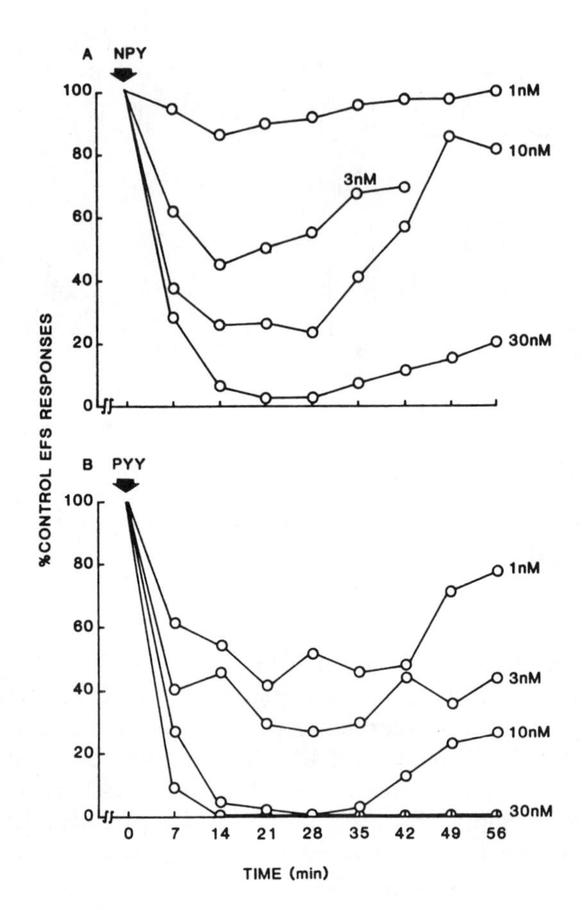

Fig. 9. The effect of increasing concentrations of NPY (A) and PYY (B) on EFS secretory responses in the presence of hexamethonium and atropine. Control noncholinergic responses were obtained in each tissue before addition of either NPY or PYY, and designated 100%. After basolateral peptide application, all subsequent EFS responses were calculated as a percentage of the respective tissue control, and *n* values were (A) 3, 3, 4, and 3 for 1, 3, 10, and 30 n*M* NPY and (B) $n = 3$ for all PYY concentrations. All points represent the mean, and errors are not shown for reasons of clarity, although they were 5–10% of respective means.

inhibition at 14 min, an apparent EC_{50} of 3 n*M* is obtained for NPY as compared with 1 n*M* for PYY (Fig. 9A and B).

The size of EFS secretory responses and their inhibition by NPY (and PYY) were not dependent on the secretory state of the tissue. In the presence of piroxicam, NPY still attenuated EFS responses without any significant effect on the new basal level of s.c.c. *(62)*. Thus, it is possible that the attenuation of EFS responses could be mediated postjunctionally by epithelial mechanisms similar to those seen during the inhibition of CCh and SP responses by NPY. However, certain subtle differences in the order of fragment agonist potencies (Section 4.5.) may indicate the presence of prejunctional NPY/PYY receptors in these jejunal preparations.

NPY also attenuates EFS secretory responses in guinea-pig and porcine intestine *(41,65)*. In the former tissue, NPY had no specific effect on basal s.c.c., but did attenuate secretory responses generated by VIP, bethanechol, and EFS. An EC_{50} of 15 n*M* for NPY inhibition of EFS responses was obtained irrespective of the presence of atropine in the guinea-pig distal colon. An EC_{50} of 5 n*M* was obtained for NPY inhibition of EFS secretory responses in porcine jejunum *(41)*, and this value was 5.6- and 1.5-fold greater than the inhibition observed of basal s.c.c. and forskolin-stimulated current, respectively.

In the rat jejunum, both SOM_{14-28} and SOM_{1-28} attenuated NANC EFS responses to much the same extent; however, both peptides were significantly less effective than NPY (Fig. 10). Following peptide addition, EFS-secretory responses returned to control levels within 40 min. The α_2-agonist, xylazine, which also exhibited a broad spectrum antisecretory profile *(40)*, also attenuated EFS responses. At 10 μ*M*, a reduction in basal s.c.c. accompanied a prolonged inhibition of EFS secretion that was 25% of controls, 70 min after xylazine addition (Fig. 10).

4.5. Use of Peptide Fragments to Classify NPY/PYY Receptors

NPY and PYY exhibit both pre- and postjunctional effects at the sympathetic neuroeffector junction *(66)*, the prejunctional inhibition of noradrenaline (NA) release being observed with a C-terminal fragment, PYY_{13-36}, which had no effect on postjunctional vasoconstriction (*see* Grundemar et al., this vol.). Thus, the presence of two different NPY receptors was postulated and termed Y_1 and Y_2. Y_1 receptors will accept only full-length NPY or PYY, whereas Y_2 receptors will tolerate loss of N-terminal residues, although (in most systems) the resultant C-terminal fragments have a significantly reduced potency compared with the full-length peptides. So, too, in the intestine, C-terminal fragments of NPY do reduce basal s.c.c. in rat jejunum, although with EC_{50} values of 1μ*M*

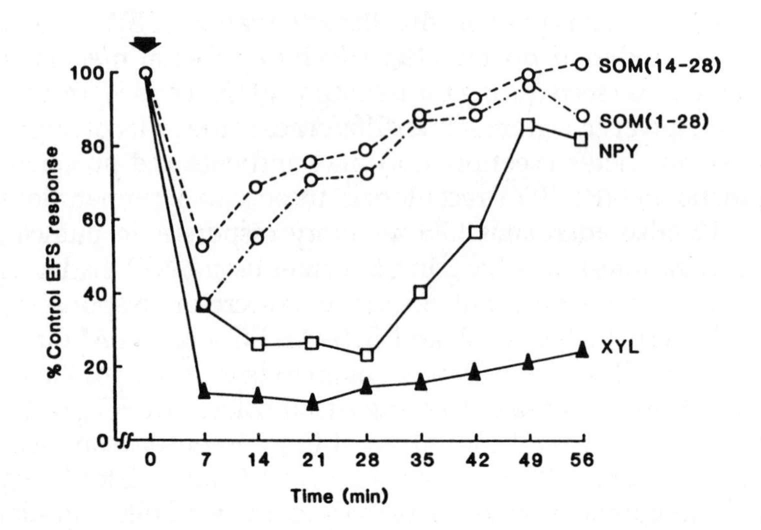

Fig. 10. The effect of four different antisecretory agents on EFS responses in the presence of hexamethonium and atropine (10 µ*M* each). Stimuli (5 Hz, 0.6 ms for 1s) were delivered at 7-min intervals, and tissues received basolateral addition of either xylazine (XYL, 10 µ*M*, *n* = 4), NPY (10 n*M*, *n* = 4), SOM (1–28) (300 n*M*, *n* = 3), or SOM (14–28) (100 n*M*, *n* = 6). Control responses were denoted as 100%, and all subsequent reductions in EFS response were calculated as a percentage of the respective noncholinergic tissue control. Control EFS responses as µA/ 0.6 cm^2 were as follows: 3.2 ± 0.7 for XYL; 4.8 ± 1.2 for NPY; 4.5 ± 0.9 for SOM (1–28); and 5.9 ± 1.4 for SOM (14–28).

or more (Fig. 11; *62*). The order of potency obtained in these functional studies (NPY»NPY [11–36] ≥ [12–36] ≥ [13–36] ≥ [14–36] >> NPY [22–36] ≥ [15–36] ≥[20–36] ≥ [16–36]) was similar to that obtained for displacement of ^{125}I-PYY binding to epithelial membranes *(30)*. The fragment NPY_{26-36}, plus desamido NPY and the C-terminal flanking peptide of NPY (CPON) were all inactive and had no effect on NPY-induced reductions in s.c.c.

Concentration-dependent inhibition of EFS-generated secretion was observed with the same range of fragments (Fig. 12; *62*), whereas des amido NPY, NPY_{26-36}, and CPON were again inactive. NPY was significantly more effective than NPY (11–36) and NPY_{13-36}, and with single concentrations (300 n*M*) of all subsequent shorter fragments, a graded attenuation of EFS responses was obtained with an apparent order: $NPY_{14-36} \geq NPY_{15-36} > NPY_{16-36} \geq$

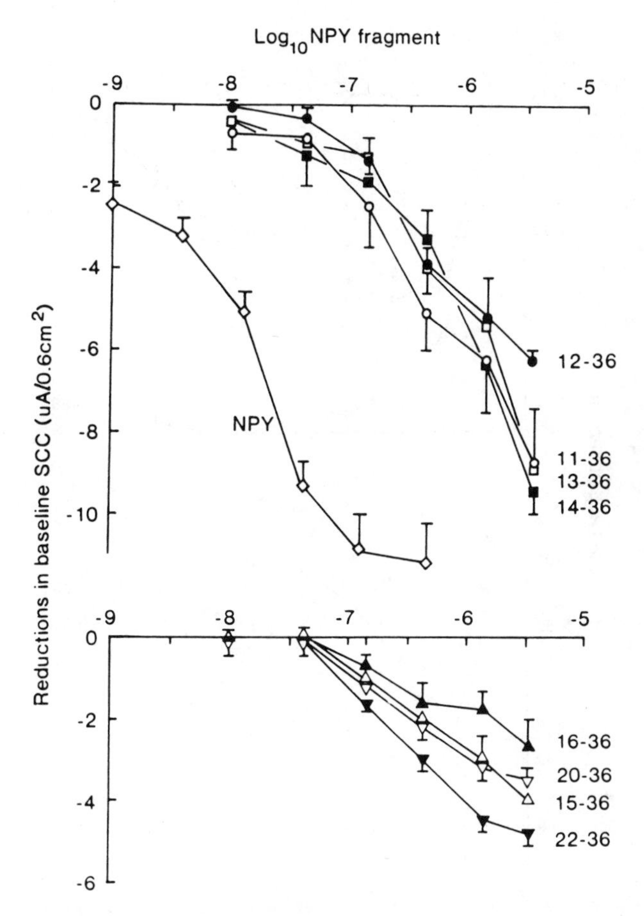

Fig. 11. Antisecretory effects of NPY and fragments on baseline s.c.c. All peptide additions were made to the basolateral surface in a cumulative fashion; reductions in s.c.c. were recorded as $\mu A/0.6\ cm^2$. Each point represents the mean with vertical lines indicating 1 SEM (some errors are omitted for reasons of clarity). The number of observations for each fragment are as follows: NPY, 3–11; NPY_{11-36}, 3–9; NPY_{12-36}, 2–9; NPY_{13-36}, 2–11; NPY_{14-36}, 2–9; NPY_{15-36}, 2–7; NPY_{16-36}, 2–7; NPY_{20-36}, 3–7; NPY_{22-36}, 3–7.

$NPY_{20-36} > NPY_{22-36}$. Since C-terminal fragments attenuate basal and EFS-stimulated secretion, then by definition, the receptors involved are of the Y_2 type. It is important to note that NPY receptors in epithelia from rabbit proximal tubules are also of the Y_2 type *(67)*. Whether, in the gut, these two inhibitory effects are mediated via distinct receptors located pre- and postjunctionally

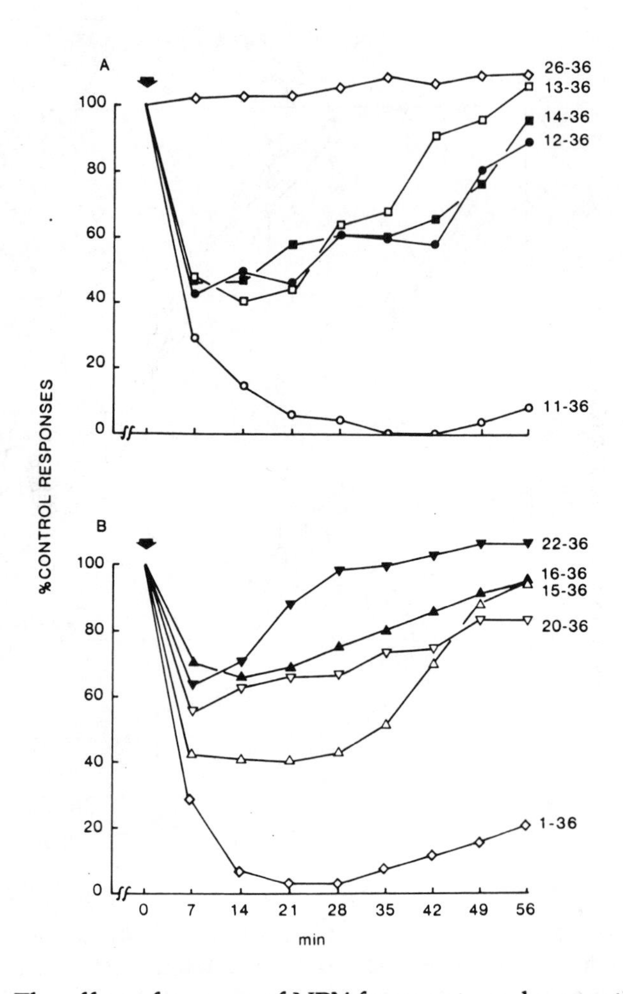

Fig. 12. The effect of a range of NPY fragments on hexamethonium/atropine-resistant EFS responses. Control secretory responses were obtained in each tissue in the presence of hexamethonium and atropine before application of NPY fragments (at 300 n*M* throughout). The figure is split into (A) and (B) for reasons of clarity. Control EFS responses were denoted as 100%, and all subsequent increases in s.c.c. after EFS were calculated as a percentage of each tissue control. Control EFS responses as $\mu A/0.6\ cm^2$ were as follows with *n* values in parenthesis: 2.9 ± 0.4 (4) for NPY_{11-36}; 1.7 0.2 (3) for NPY_{12-36}; 1.8 0.2 (4) for NPY_{13-36}; 3.4 0.8 (4) for NPY_{14-36}; 5.5 0.8 (4) for NPY_{15-36}; 6.5 0.5 (4) for NPY_{16-36}; 4.7 0.3 (4) for NPY_{20-36}; 6.9 1.7 (4) for NPY_{22-36}, and 6.6 0.8 (4) for NPY_{26-36}. All points and values are quoted as means, and errors are not shown for reasons of clarity, but were not >10%.

is not absolutely clear. NPY_{11-36} and NPY_{13-36} at 300 n*M* and 1 μ*M* did not significantly attenuate secretory responses to CCh or SP, whereas at these concentrations, they had severe effects on EFS secretion. This apparently reduced spectrum of fragment activity together with the different order and potency ratios obtained with NPY, NPY_{11-36}, and NPY_{13-36}, may indicate the presence of prejunctional Y_2 receptors; however, further investigations are necessary to confirm this. As yet, no other functional studies have been performed with NPY/PYY fragments in the gastrointestinal tract.

The discovery of a Y_1-selective NPY agonist [Leu^{31}, Pro^{34}] NPY *(68)* has recently allowed definitive characterization of NPY responses. In the rat intestine, [Pro^{34}] NPY (the substitution of isoleucine for leucine at position 31 being of minor significance) was inactive, neither affecting basal s.c.c. nor EFS secretory responses, thus confirming the Y_2 character of NPY and PYY antisecretory effects.

4.6. Structure–Activity Studies with Truncated NPY Analogs and NPY Antagonists

The distinct structural features and sequences common to all three of the pancreatic polypeptides have provided insights into the optimal peptide structure essential for biological activity *(69,70)*. In order to test the importance of the central turn region in bringing the N-terminal polyproline helix and the C-terminal amphiphilic α-helix together, a number of conformationally restricted, truncated analogs were synthesized and examined in different systems *(71,72)*, including the rat intestine *(73)*. These analogs, based on the tertiary structural model of NPY with varying amounts of N- and C-terminal helical regions removed and replaced by a single 8-aminooctanoic acid residue (Aoc), have yielded additional information concerning epithelial Y_2-receptor activation. From the concentration–response relationships shown in Figs. 13 and 14, certain critical structural requirements are highlighted. First, single amino acid substitution to [Pro^{34}] NPY or [Ile^{34}] NPY results in total loss of antisecretory activity and severely attenuated potency, respectively, and neither peptide altered NPY responses. Removal and replacement of the central amino acids $^{5-24}$ with a single Aoc plus the introduction of a cysteine-D-cysteine disulfide bridge via

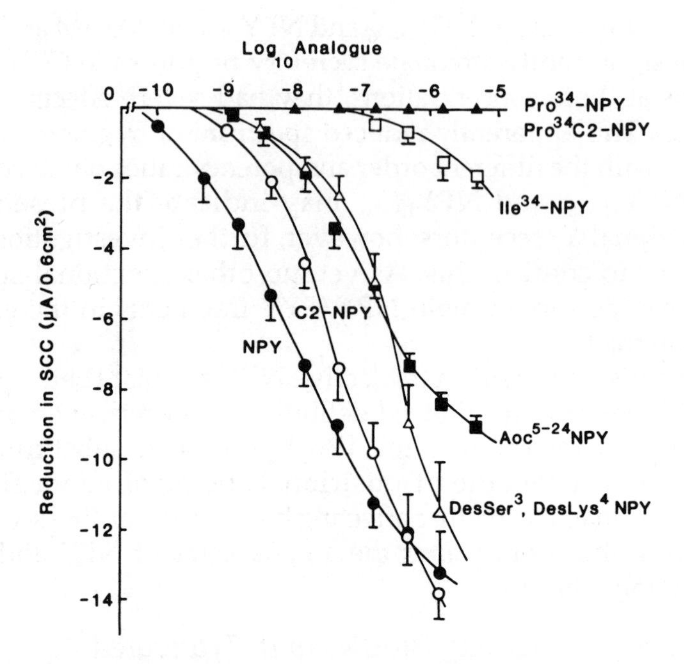

Fig. 13. Concentration–response curves for NPY (●, $n = 3–10$), C2-NPY (○, $n = 4$), [Aoc$^{5-24}$] NPY (■, $n = 6$), [Des Ser^3, Des Lys^4] C2-NPY (Δ, $n = 3–5$), [Ile^{34}] NPY (□ $n = 3$), and [Pro^{34}] C2 NPY and [Pro^{34}] NPY (▲, $n = 3$ each).

positions 2 and 27 (72, and *see* Table 1 *below* for details) to give an analog [Cys^2, Aoc$^{5-24}$, DCys^{27}] NPY (or C2-NPY for short) resulted in a reduction in potency (EC_{50} of 50 n*M*) compared with NPY (EC_{50} 9.3 n*M*). [Des Ser^3, Des Lys^4] C2-NPY, which possesses only 36% of the NPY sequence, exhibited an EC_{50} of 410 n*M*. The third peptide worthy of particular note is the linear version of C2-NPY, i.e. [Aoc$^{5-24}$] NPY (lacking the cysteine-D-cysteine disulfide bridge), which reduced basal s.c.c. with an EC_{50} value of 81.2 n*M* (Table 1). All of these analogs were more potent than C5-NPY ([Cys^5, Aoc$^{7-20}$, DCys^{24}] NPY) and C7-NPY (DCys^7, Aoc$^{8-17}$, Cys^{20}] NPY, both of which retained greater sequence homology with NPY (55 and 67%, respectively). The order of C2-NPY > C5-NPY ≥ C7-NPY obtained in the intestine mucosa is different from that observed in brain and spleen membranes *(71,72)*. Here the order C7-NPY > C5-NPY > C2-NPY shows the loss of potency more closely following the loss of central amino acids from the NPY sequence. The changed order of potency for antisecretory responses in the intestine may be

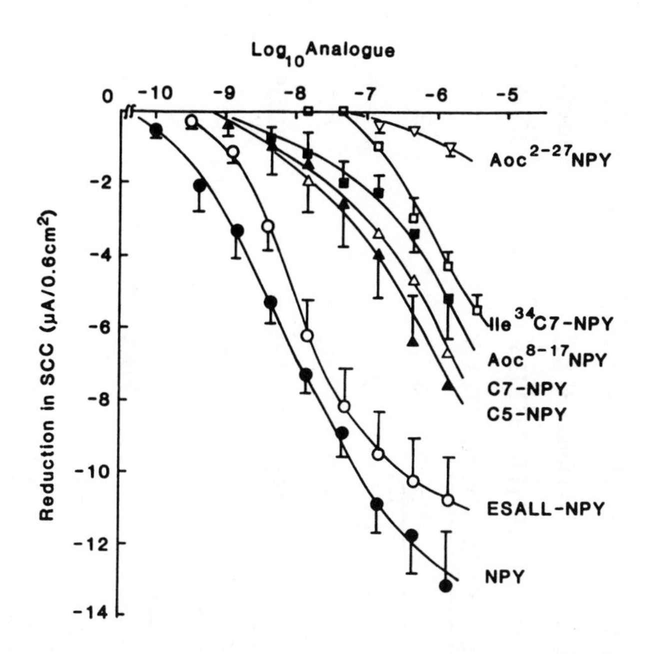

Fig. 14. Concentration–response curves for NPY (●, n = 3–10), ESALL-NPY (○, n = 3–5), C5-NPY (▲, n = 5), C7-NPY (Δ, n = 5), [Aoc^{8-17}] NPY (■, n = 4), [Ile^{34}] C7-NPY (□, n = 3), and [Aoc^{2-27}] NPY (∇, n = 4).

indicative of a different receptor subtype in epithelia, although this is by no means clear at the present time. Significant loss of α-helicity, i.e., in the case of C2-NPY, did not result in the drastic reductions in potency observed in other systems, equally increasing the potential for α-helicity, i.e., ESALL-NPY ([Glu^{16}, Ser^{18}, Ala^{22}, $Leu^{28,31}$] NPY) was without significant effect (EC_{50} of 11.2 n*M*). Taken together, these structure–activity studies highlight the importance of N-terminal residues (1–4) in combination with C-terminal residues (25–36) (which alone would be inactive) the resulting chimera [Aoc^{5-24}] NPY being a relatively potent NPY analog. Beck et al. *(74)* using a slightly different truncated analog, NPY (1–4) Aca (25–36) (containing an ε-aminocaproic acid [Aca] residue, two methylene groups shorter than Aoc), found this analog to be only 3.2-fold less potent than NPY at displacing radiolabeled NPY from renal cortical homogenates. If these renal binding sites are representative of epithelial NPY/PYY receptors, then similar structural

Table 1
Peptide Sequences and Antisecretory Potency[a]

Abbreviated name	Sequence (1 5 10 15 20 25 30 36)	EC_{50} n*M*
NPY	YPSKPDNPGEDAPAEDLARYYSALRHYINLITRQRY	9.3
[Pro[34]] NPY	YPSKPDNPGEDAPAEDLARYYSALRHYINLITRPRY	Inactive
[Ile[34]] NPY	YPSKPDNPGEDAPAEDLARYYSALRHYINLITRIRY	10,000
ESALL-NPY	YPSKPDNPGEDAPAEELSRYYAALRHYLNLLTRQRY	11.2
C2-NPY	YCSK - RHcINLITRQRY	50.0
[Pro[34]] C2-NPY	YCSK - RHcINLITRPRY	Inactive
[Aoc[5–24]] NPY	YPSK - RHYINLITRQRY	81.2
[Des Ser[3], Des Lys[4]] C2-NPY	YC - RHcINLITRQRY	410.0
C5-NPY	YPSKCD - - - - - - - - - - - - - - YSAcRHYINLITRQRY	1000
C7-NPY	YPSKPDc - - - - - - - - - - - ARCYSALRHYINLITRQRY	1000
[Ile[34]] C7-NPY	YPSKPDc - - - - - - - - - - - ARCYSALRHYINLITRIRY	10,000
[Aoc[8–17]] NPY	YPSKPDN - - - - - - - - - - ARYYSALRHYINLITRQRY	1000
[Aoc[2–27]] NPY	Y- INLITRQRY	10,000

[a]Numbers 1, 5, 10, 15, 20, 25, 30, and 36 correspond to the amino acid position in the sequence of NPY. A dashed line represents the residues replaced by a single 8-aminooctanoic acid residue (Aoc). Underlined residues are those that differ from NPY, and the single letter abbreviations are as follows: A, alanine; C, cysteine; c, D-cysteine; D, aspartate; E, glutamate; G, glycine; H, histidine; I, isoleucine; K, lysine; L, leucine; N, asparagine; P, proline; Q, glutamine; R, arginine; S, serine; T, threonine; Y, tyrosine. EC_{50} values are listed as either nanomolar concentrations or as inactive. The latter category indicates that no significant reductions in s.c.c. were obtained within

requirements appear to be necessary for renal and intestinal epithelial Y_2 receptors. In conclusion, therefore, the amino acid residues present in the central core of the peptide, i.e., the α-helical region together with part of the N-terminal polyproline helix, possess a predominantly structural role, restricting peptide conformation and providing the necessary orientation of the terminal regions of NPY for optimal interaction with the Y_2 epithelial receptor.

In recent months, certain peptide *(75)* and nonpeptide *(76)* antagonists have been described. Two C-terminal decapeptide analogs, namely Ac-[3-(2, 6, dichlorobenzyl)-Tyr^{27}, D-Thr^{32}]-NPY_{27-36} (otherwise known as PYX-1) and Ac-[3, 5-di (2, 6, dichlorobenzyl)-Tyr^{27}, D-Thr^{32}]-NPY_{27-36} have been tested in voltage-clamped jejunal preparations for antagonism. Only at 1μ*M* concentrations and higher was significant inhibition of NPY antisecretory responses seen (data not shown). This apparently low affinity is not altogether surprising given that PYX-1 is thought to be a Y_1 receptor blocker *(75)*. The nonpeptide, putative NPY antagonist HE 90481 has also been tested, but had no significant effect on NPY responses, at concentrations up to and including 100μ*M*. New high-affinity antagonists are eagerly awaited in order that the true functional significance of NPY and PYY may be unequivocally established.

5. Effects of Pancreatic Polypeptides on Other Gastrointestinal Functions

5.1. Mesenteric Blood-Flow Effects of NPY and PYY

Both NPY and PYY exhibit the capacity to reduce intestinal blood flow. In the anesthetized cat, Hellström *(77)* found intra-arterial administration of NPY caused prolonged vasoconstriction and consequent reduction of colonic blood flow. In comparison, NA vasoconstrictor responses were short-lived, and both neurotransmitters affected these changes directly on vascular smooth muscle cells. In human mesenteric vein, NPY caused contractions (although not in mesenteric arteries; *78*) and as commonly seen, this effect was resistant to adrenoceptor blockade. A comparison of increased organ vascular resistance following infusion of NPY into the pithed rat showed that the most pronounced vasoconstriction occurred in the mesenteric bed *(79)*, and this was a major contributing factor to the increased total peripheral resistance seen with NPY. Renal and large intestine vascular resistances were significantly increased, although small intestine resistance was not. These together with other changes could possibly result in an increased proportion of cardiac output being delivered to the small intestine. Reduced mesenteric blood flow has also been observed in conscious rabbits (with intact cardiovascular reflexes), although in this study, the greatest vasoconstriction was observed in renal vasculature *(80)*.

Like NPY, PYY is a vasoconstrictor, significantly reducing intestinal blood flow specifically in jejunal and ileal muscularis externa of the dog *(81)*. A redistribution of blood flow away from the muscularis toward the mucosa/submucosa was observed (there being no significant change in whole gut wall blood flow), and this is noteworthy given the antisecretory effects of the peptide in this area. The combination of gastrointestinal vascular and epithelial NPY/PYY effects will therefore facilitate reabsorption of electrolytes and nutrients across the upper intestine.

5.2. Effects of NPY and PYY on Gastrointestinal Motility

In guinea-pig small intestine *(82)* and colon *(83)*, NPY exhibits indirect inhibitory effects. In the distal jejunum/proximal ileum, distension-induced motor activity of the circular smooth muscle was inhibited by NPY (with an approximate EC_{50} of 3 n*M*). Both the cholinergic and noncholinergic components of the ascending enteric reflex (AER) contraction of circular smooth muscle were attenuated by NPY (note the similar inhibitory pattern with anion secretion). In addition, peristaltic movements were abolished, and phasic contractions induced by the nicotinic agonist DMPP (dimethylphenylpiperazinium) were inhibited by NPY. NPY had no direct effect on smooth muscle, and its effects on AER were not altered by α_2-adrenoceptor, opioid, or P_1-purinoceptor blockade. They were, however, attenuated by apamin, implicating Ca-dependent K-channel involvement in the mechanism of NPY action. In colonic longitudinal muscle, however, NPY-induced relaxation involved the stimulation of NA release from sympathetic nerves, which by way of α_2-adrenoceptors present on postganglionic cholinergic neurons, afforded an inhibition of transmission. Thus, it would appear from these two studies that NPY has the capacity to inhibit cholinergic and noncholinergic contraction at the level of submucous and myenteric ganglia in the small and large intestine, respectively. Colonic motility induced by electrical stimulation of feline pelvic nerves was significantly attenuated by close arterial infusion of NPY *(77)*. Only at high (probably pharmacological) doses did NPY have a direct effect on smooth muscle. NPY, again in the cat, produced a biphasic effect (rapid contraction followed by more prolonged relaxation) on lower esophageal sphincter (LOS) tone *(84)*. The initial contraction was cholinergic and contrasts with effects described above in nonsphincter regions of the gut. The longer-lasting relaxation of the LOS was a mixture of direct smooth muscle effects together with both cholinergic and noncholinergic indirect components.

The inhibitory effects of PYY on intestinal (jejunal and colonic) motility were first observed by Lundberg et al. *(19)* in anesthetized cats. In Heidenhain pouch dogs *(85)*, PYY effects were pronounced in the interdigestive state where migrating contractions in the gastric body and antrum of the stomach were inhibited via a mechanism dependent on extrinsic innervation. PYY was suggested to exert its effects at the preganglionic level of the vagus nerve. The peptide had no effect on the time interval between interdigestive migrating contractions. Recent studies have identified (in guinea-pig isolated muscle strips; *86*) a complex mechanism of PYY inhibition of gastric motility. PYY caused a concentration-dependent relaxation of muscle strips (EC_{50} of 6 n*M*) as well as a significant inhibition of electrically stimulated muscle contractions. Atropine abolished PYY-mediated relaxations and the electrically stimulated contractions, and thus it was concluded that the peptide could inhibit antrum gastric motility by attenuating both basal and stimulated cholinergic transmission. In this system, PYY appeared to act presynaptically, inhibiting ACh release by attenuating the cAMP-dependent component of cholinergic transmission that was sensitive to pertussis toxin. These mechanisms could go some way to explaining the delayed gastric emptying observed in human subjects *(87)* after PYY and NPY administration.

In other many and varied smooth muscle preparations, PYY is consistently inhibitory. In the rat small intestine longitudinal smooth muscle, PYY caused relaxation (specifically in the jejunum and ileum) via nonadrenergic and noncholinergic mechanisms, as well as duodenal contractions that were also indirect although cholinergic *(88)*. As part of a biphasic effect in jejunal smooth muscle, small initial contractions were observed that were also noncholinergic. As seen before with NPY, PYY also inhibited electrically evoked contractions of ileal preparations *(88)*. However, this effect could be postjunctionally mediated, since contractions produced by carbachol and histamine were also significantly reduced by PYY. Cholecystokinin-induced contractions of isolated dog ileum were also inhibited by a prejunctional mechanism, PYY inhibiting the release of ACh from cholinergic nerve terminals *(89)* and not altering responses to exogenously added ACh.

5.3. Effects of NPY and PYY on Food Intake and Emesis

NPY is a potent orexigenic and emetic peptide. Hypothalamic and also ventricular injections of NPY result in increased food intake in rats *(44,90)*. The hypothalamus, specifically the paraventricular nucleus (PVN, where there is a dense NPY innervation; *91*) influences autonomic outflow to the cardiovascular and gastrointestinal systems by virtue of direct descending projections linking the PVN to the medulla and spinal cord *(92,93)*. When injected into the PVN, both NPY and PYY elicit dose-dependent eating responses potentiating the rate and duration of eating in satiated rats *(94; see* review by *95*). These effects are slow in onset, but prolonged (lasting for hours at maximal doses), and their profile contrasts with the rapid, transient character of feeding stimulated by noradrenaline or adrenaline *(44)*. These contrasting time-courses are also observed when comparing the inhibition of gastric acid secretion following NA or NPY injection into the PVN *(43)*. The maximal feeding responses following PVN, NA, or NPY are comparable, and, like NA, NPY also stimulates a small, but significant drinking response. NPY levels in the PVN change in a reciprocal manner in response to either deprivation or intake of food *(96)*. The receptors involved in stimulating feeding behavior are probably Y_1, since [Leu^{31}, Pro^{34}] NPY, but not NPY_{13-36} was effective (*see 97* for review; see also Stanley, this volume).

PYY and (to a lesser extent) NPY are potent emetic agents *(96)*. It appears that PYY-induced vomiting is dependent upon intact innervation within the area postrema *(98)*, and there is evidence that PYY increases blood flow in this area *(99)*. By virtue of PYY localization within endocrine cells in the lower intestine, it is therefore possible that PYY acts as a humoral agent stimulating neurons in the area postrema, and this by virtue of projections to brain stem structures results in the vomiting reflex (*see 100,101* for reviews). Whether the redistribution of intestinal blood flow together with changes in intestinal motility observed with emetic doses of PYY *(81)* have a role in the genesis of nausea and vomiting remains to be established.

5.4. Regulation of PYY and PP Release

Many factors afford the release of PYY and PP from their respective locations, i.e., endocrine cells of the large intestine and pancreatic islets, respectively. Significantly increased plasma levels of PYY were observed following ingestion of a mixed meal, and on investigation of the relative effects of different nutrients, fat was found to be the most potent stimulant *(102,103)*. In dogs, intraduodenal administration of oleic acid *(104)* and electrical stimulation of the vagus, or luminal glucose administration (in pigs; *105*) all resulted in increased circulating PYY levels. The time-course of raised PYY concentrations shows significant increases within 10–30 min, reaching a peak at 1–3 h and remaining elevated for several hours after ingestion *(102)*. This contrasts with the more rapid release of PP *(106,107)*, which peaks within 15–30 min after ingestion. The release of PYY from the distal ileum and colon precedes the arrival of luminal contents and secretagogs, and thus, the early phase of hormone secretion is probably mediated by a neuroendocrine mechanism. Other stimulants include vascular administration of gastrin-releasing peptide (GRP; *105*), cholecystokinin (CCK; *104*), and the stimulation of pancreatic secretion *(108)*. The bile acid deoxycholate is also a particularly potent PYY stimulant *(107)*. Both PYY and PP exhibit the capacity to reduce the pancreatic and biliary secretion, as well as prestimulated gastric acid secretion (*102,109,110, also see 107* for review). Thus, it is possible that the postprandial PYY response contributes to the physiological inhibition of acid secretion and the reduced gastric emptying observed after ingestion of fatty foods, and may therefore be a prime candidate involved in "ileal brake" (the phenomenon of reduced intestinal transit and delayed gastric emptying seen in humans following the perfusion of ileum and colon with fat; *111*).

In response to food, the profile of PP release is biphasic, the early peak phase being predominantly mediated by cholinergic (and noncholinergic) vagal and intrinsic pancreatic neurotransmission (*see 107* for review). The second, more prolonged PP response is a combination of numerous stimuli, including neuronal, hormonal, and other factors, such as circulating nutrients. A description of these mechanisms is outside the scope of this chapter, and the reader is referred to a comprehensive review elsewhere *(107)*.

6. Concluding Remarks on the Functional Role of the Pancreatic Polypeptides

Despite the subtle species differences observed and highlighted above, it is clear that NPY, PYY, and PP all exert inhibitory effects on their respective targets within the gastrointestinal tract. The cellular location of each peptide implicates its likely physiological status given the receptor localization and the stimuli that afford peptide release. NPY's presence in intrinsic enteric neurons innervating various targets within the gut wall indicate a neurotransmitter and neuromodulatory role. The predominant localization of PYY within endocrine cells in the large intestine, and its release by luminal food and glucose, as well as a range of peptides that are secreted after ingestion of a meal, all point to the hormonal status of this peptide. The location of PP within pancreatic islets and its release by stimuli associated with food intake also indicate a hormonal or paracrine action. Both NPY and PYY exert potent antisecretory effects either directly via high-affinity receptors located within the basolateral epithelial domain or in combination with indirect mechanisms involving other antisecretory agents (depending on the species used). Circulating plasma PYY levels have been estimated to be in the 0.1–1.0 n*M* range *(112)*, concentrations that in vitro have significant antisecretory effects. Local concentrations of NPY are more difficult to estimate, but could, by extrapolation, be within the low-nanomolar concentration range. The inhibition of gastric emptying and motility in general, together with increased mucosal absorption and pancreatic exocrine secretion observed in response to PYY and NPY, could all contribute to an increased efficiency of digestion and absorption of nutrients, fluid, and electrolytes in healthy individuals. NPY and PYY are also potent stimulants of food intake. The simultaneous effect of food ingestion and delayed gastric emptying would also reduce the likelihood of overwhelming the small intestine with undigested food.

This chapter has focused on the epithelial effects of NPY and PYY in the intestine, and if these two peptides are involved in homeostatic mechanisms, then one might expect similar response profiles in renal epithelia. In fact, high-affinity specific binding sites for NPY have been identified in rabbit proximal convoluted tubules *(113)* with Y_2-receptor binding characteristics *(67,74)*, and antinatriuretic effects in response to NPY have been observed in monkeys *(114)*. However, it is not yet clear whether the sodium retention afforded by NPY is independent of reductions in renal blood flow and glomerular filtration rate.

Both NPY and PYY are potentially important factors in controlling fluid and electrolyte loss, and circulating PYY levels are notably elevated in certain diseases that cause diarrhea *(115)*. Infusion of PYY (reproducing levels observed postprandially) into ileostomy patients hypersecreting in response to VIP infusions resulted in significant reductions in ileal output and transit time *(54)*. Thus, in hypersecretory states, PYY is capable of acting as a natural inhibitor of diarrhea, although interestingly its effects were negligible in a control patient not receiving concomitant VIP infusion. Holzer-Petsche et al. *(116)* have recently shown that NPY infused into healthy volunteers (at concentrations that do not alter blood pressure) causes significant increases in net absorption of water, sodium, potassium, and chloride under basal conditions. Following intraluminal administration of PGE_2, net water secretion was reduced 36% and sodium, potassium, and chloride secretion was attenuated 70–80%. These reductions in both basal and stimulated fluid and electrolyte secretion in humans are readily explained by the mechanisms observed in isolated intestinal preparations (and discussed in Section 4.3.). It is apparent from these clinical studies and in vitro work that the antisecretory effects of NPY and PYY are more pronounced following secretagog pretreatment *(37,48)*. Our own studies also show that extremely high levels of secretion observed following exposure of preparations to the diterpene, forskolin, can result in a loss of NPY/PYY responsiveness. Thus, there may be an upper secretory limit beyond which the peptides are relatively ineffective. It is clear from evidence obtained from a wide range of species that NPY and PYY have a significant physiological role in the control of fluid and electrolyte secretion.

An orally active somatostatin agonist, octreotide, is already available for therapeutic control of postoperative secretory diarrhea and diarrhea associated with carcinoid and VIP-oma syndromes *(117,118)*. Like NPY and PYY, SOM and octreotide exhibit a broad spectrum of antisecretory capacity (*45,119*). Thus, the design of stable, orally active NPY and PYY agonist analogs may also provide important new drugs for the control of hypersecretory states. Future physiological studies will, however, require selective, high-affinity antagonists (preferably with Y_2 selectivity for epithelial preparations) if the functional role of NPY and PYY is to be established unequivocally.

Acknowledgments

The author acknowledges funding from the Medical Research Council. Figure 4A and B is reproduced with the permission of *The Journal of Physiology*. Figure 6 is reproduced from *Pflügers Arch.*, and Figures 8, 11, and 12 from the *British Journal of Pharmacology* with permission from respective publishers, Springer-Verlag GmbH, and Macmillan Press Ltd.

References

1. Furness, J. B. and Costa, M. (1987) *The Enteric Nervous System*. Churchill Livingstone, Edinburgh.
2. Donowitz, M. and Welsh, M. J. (1987) Regulation of mammalian small intestinal electrolyte secretion, in *Physiology of the Gastrointestinal Tract*. L. R. Johnson, ed., Raven, New York, pp. 1351–1388.
3. Binder, H. J. and Sandle, G. I. (1987) Electrolyte absorption and secretion in the mammalian colon, in *Physiology of the Gastrointestinal Tract*. L. R. Johnson, ed., Raven, New York, pp. 1389–1418.
4. Larsson, L. I., Sundler, F., and Håkanson, R. (1976) Pancreatic polypeptide—a postulated new hormone: Identification of its cellular storage site by light and electron microscopic immunocytochemistry. *Diabetologia* **12,** 211–226.
5. Lorén, I., Alumets, J., Håkanson, R., and Sundler, F. (1979) Immunoreactive pancreatic polypeptide (PP) occurs in the central and peripheral nervous system: preliminary immunocytochemical observations. *Cell Tissue Res.* **200,** 179–186.
6. Vaillant, C. and Taylor, I. L. (1981) Demonstration of carboxyterminal PP-like peptides in endocrine cells and nerves. *Peptides* **2,** 31–35.
7. Sundler, F., Moghimzadeh, E., Håkanson, R., Ekelund, M., and Emson, P. (1983) Nerve fibers in the gut and pancreas of the rat displaying neuropeptide Y immunoreactivity. *Cell Tiss. Res.* **230,** 487–493.
8. Ekblad, E., Winther, C., Ekman, R., Håkanson, R., and Sundler, F. (1987) Projections of peptide-containing neurones in rat small intestine. *Neuroscience* **20,** 169–188.
9. Ekblad, E., Ekman, R., Håkanson, R., and Sundler, F. (1988) Projections of peptide containing neurones in the rat colon. *Neuroscience* **27,** 655–674.

10. Ekblad, E., Ekelund, M., Graffner, H., Håkanson, R., and Sundler, F. (1985) Peptide containing nerve fibres in the stomach wall of rat and mouse. *Gastroenterology* **89,** 73–85.
11. Pataky, D. M., Curtis, S. B., and Buchan, A. M. J. (1990) The colocalisation of neuropeptides in the submucosa of the small intestine of normal, Wistar and non-diabetic BB rats. *Neuroscience* **36,** 247–254.
12. Furness, J. B., Costa, M., Gibbins, I. L., Llewellyn-Smith, I. J., and Oliver, J. R. (1985) Neurochemically similar myenteric and submucous neurones directly traced to the mucosa of the small intestine. *Cell Tiss. Res.* **241,** 155–163.
13. Keast, J. R., Furness, J. B., and Costa, M. (1985b) Distribution of certain peptide-containing nerve fibres and endocrine cells in the gastrointestinal mucosa in five mammalian species. *J. Comp. Neurol.* **236,** 403–422.
14. Ekblad, E., Arnbjörnsson, E., Ekman, R., Håkanson, R., and Sundler, F. (1989) Neuropeptides in the human appendix. Distribution and motor effects. *Dig. Dis. Sci.* **34,** 1217–1230.
15. Wattchow, D. A., Furness, J. B., Costa, M., O'Brien, P. E., and Peacock, M. (1987) Distributions of neuropeptides in the human esophagus. *Gastroenterology* **93,** 1363–1371.
16. Fehér, E. and Burnstock, G. (1986) Electron microscopic study of NPY containing nerve elements of the guinea-pig small intestine. *Gastroenterology* **91,** 956–961.
17. Lundberg, J. M., Hokfelt, T., Änggard, A., Kimmel, J., Goldstein, M., and Markey, K. (1980) Coexistence of an avian pancreatic polypeptide (APP) immunoreactive substance and catecholamines in some peripheral and central neurones. *Acta Physiol. Scand.* **110,** 107–109.
18. Wang, Y-N., McDonald, J. K., and Wyatt, R. J. (1987) Immunocytochemical localisation of NPY-like immunoreactivity in adrenergic and nonadrenergic neurons of the rat gastrointestinal tract. *Peptides* **8,** 145–151.
19. Lundberg, J. M., Tatemoto, K., Terenius, L., Hellström, P. M., Mutt, V., Hökfelt, T., and Hamberger, B. (1982) Localisation of PYY in gastrointestinal endocrine cells and effects on intestinal blood flow and motility. *Proc. Natl. Acad. Sci. USA* **79,** 4471–4475.
20. Böttcher, G., Sjölund, K., Ekblad, E., Håkanson, R., Schwartz, T. W., and Sundler, F. (1984) Coexistence of PYY and glicentin immunoreactivity in endocrine cells of the gut. *Regul. Pept.* **8,** 261–266.
21. Böttcher, G., Alumets, J., Håkanson, R., and Sundler, F. (1986) Coexistence of glicentin and peptide YY in colorectal L-cells in cat and man. An electron microscopic study. *Regul. Pept.* **31,** 283–291.
22. Miyachi, Y., Jitsomohi, W., Miyoshi, A., Fujita, S., Mizuchi, A., and Tatemoto, K. (1986) The distribution of polypeptide YY-like immunoreactivity in rat tissues. *Endocrinology* **118,** 2163–2167.
23. Ekman, R., Wahlestedt, C., Böttcher, G., Sundler, F., Håkanson, R., and Panula, P. (1986) Peptide YY-like immunoreactivity in the central nervous system of the rat: occurrence, distribution and partial characterisation. *Regul. Pept.* **16,** 157–168.
24. Häppölä, O., Wahlestedt, C., Ekman, R., Soinila, S., Panula, P., and Håkanson, R. (1990) Peptide YY-like immunoreactivity in sympathetic neurons of the rat. *Neuroscience* **39,** 225–230.

25. Greider, M. H., Gersell, D. I., and Gingerich, R. L. (1978) Ultrastructural localisation of pancreatic polypeptide in the F-cell of the dog pancreas. *J. Histochem. Cytochem.* **26,** 1103–1108.
26. Larsson, L. I., Sundler, F., and Håkanson, R. (1973) Immunohistochemical localisation of human pancreatic-polypeptide (HPP) to a population of islet cells. *Tissue Res.* **156,** 167–171.
27. Rahier, J. and Wallon, J. (1980) Long cytoplasmic processes in pancreatic polypeptide cells. *Cell Tissue Res.* **209,** 365–370.
28. Laburthe, M., Chenut, B., Rouyer-Fessard, C., Tatemoto, K., Couvineau, A., Servin, A., and Amiranoff, B. (1986) Interaction of peptide YY with rat intestinal epithelial plasma membranes: binding of the radioiodinated peptide. *Endocrinology* **118,** 1910–1917.
29. Voisin, T., Rouyer-Fessard, C., and Laburthe, M. (1990a) Distribution of common peptide YY—neuropeptide Y receptor along rat intestinal villus-crypt axis. *Am. J. Physiol.* **259,** G753–G759.
30. Servin, A. L., Rouyer-Fessard, C., Balasubramanian, A., St-Pierre, S., and Laburthe, M. (1989) PYY and NPY inhibit VIP-stimulated adenosine 3', 5'-monophosphate production in rat small intestine: structural requirements of peptides for interacting with PYY-preferring receptors. *Endocrinology* **124,** 692–700.
31. Raufman, J-P. and Singh, L. (1991) Actions of peptide YY and neuropeptide Y on chief cells from guinea-pig stomach. *Am. J. Physiol.* **260,** G820–G826.
32. Nguyen, T. D., Heintz, G. G., Kaiser, L. M., Staley, C. A., and Taylor, I. L. (1990) NPY, differential binding to rat intestinal laterobasal membranes. *J. Biol. Chem.* **256,** 6416–6422.
33. Sheikh, S. P. and Williams, J. A. (1990) Structural characterisation of Y_1 and Y_2 receptors for NPY and PYY by affinity cross-linking. *J. Biol. Chem.* **265,** 8304–8310.
34. Voisin, T., Couvineau, A., Rouyer-Fessard, C., and Laburthe, M. (1991) Solubilization and hydrodynamic properties of active PYY receptor from rat jejunal crypts. *J. Biol. Chem.* **266,** 10,762–10,767.
35. Voisin, T., Rouyer-Fessard, C., and Laburthe, M. (1990b) PYY/NPY receptors in small intestine. *Ann. NY Acad. Sci.* **611,** 343–349.
36. Gilbert, W. R., Frank, B. H., Gavin, J. R., and Gingerich, R. L. (1988) Characterisation of specific pancreatic polypeptide receptors on basolateral membranes of the canine small intestine. *Proc. Natl. Acad. Sci. USA* **85,** 4745–4749.
37. Saria, A. and Beubler, E. (1985) Neuropeptide Y and peptide YY inhibit prostaglandin E_2-induced intestinal fluid and electrolyte secretion in the rat jejunum in vivo. *Eur. J. Pharmacol.* **119,** 47–52.
38. MacFadyen, R. J., Allen, J. M., and Bloom, S. R. (1986) NPY stimulates net absorption across rat intestinal mucosa in vivo. *Neuropeptides* **7,** 219–227.
39. Cox, H. M., Cuthbert, A. W., Håkanson, R., and Wahlestedt, C. (1988) The effect of neuropeptide Y and peptide YY on electrogenic ion transport in rat intestinal epithelia. *J. Physiol.* **398,** 65–80.
40. Cox, H. M. and Cuthbert, A. W. (1989a) Antisecretory activity of the a_2-adrenoceptor agonist, xylazine in rat jejunal epithelium. *Naunyn-Schmiedebergs Arch. Pharmacol.* **339,** 669–674.

41. Brown, D. R., Boster, S. L., Overend, M. F., Parsons, A. M., and Treder, B. G. (1990) Actions of NPY on basal, cAMP-induced and neurally evoked ion transport in porcine distal jejunum. *Regul. Pept.* **29,** 31–47.
42. Rivière, P. J. M., Pascaud, X., Junien, J. L., and Porreca, F. (1990) Neuropeptide Y and JO 1784, a selective σ ligand, alter intestinal ion transport through a common, haloperidol-sensitive site. *Eur. J. Pharmacol.* **187,** 557–559.
43. Humphreys, G. A., Davison, J. S., and Veale, W. L. (1988) Injection of neuropeptide Y into the paraventricular nucleus of the hypothalamus inhibits gastric acid secretion in the rat. *Brain Res.* **456,** 241–248.
44. Stanley, B. G. and Leibowitz, S. F. (1985) Neuropeptide Y injected into the paraventricular hypothalamus: a powerful stimulant of feeding behaviour. *Proc. Natl. Acad. Sci. USA* **82,** 3940–3943.
45. Ferrar, J. A., Cuthbert, A. W., and Cox, H. M. (1990) The antisecretory effects of somatostatin and analogues in rat descending colon mucosa. *Eur. J. Pharmacol.* **184,** 295–303.
46. Hubel, K. A., and Renquist, K. S. (1986) Effect of neuropeptide Y on ion transport by the rabbit ileum. *J. Pharmacol. Exp. Ther.* **238,** 167–169.
47. Friel, D. D., Miller, R. J., and Walker, M. W. (1986) Neuropeptide Y: a powerful modulator of epithelial ion transport. *Brit. J. Pharmacol.* **88,** 425–431.
48. Cox, H. M. and Cuthbert, A. W. (1988) Neuropeptide Y antagonises secretagogue evoked chloride transport in rat jejunal epithelium. *Pflügers Arch.* **413,** 38–42.
49. Flint, J. S., Ballantyne, G. H., Goldenring, J. R., Fielding, L. P., and Modlin, I. M. (1990) NPY inhibition of VIP-stimulated ion transport in the rabbit distal colon. *Arch. Surg.* **125,** 1561–1563.
50. Bolton, J., and Field, M. (1977) Ca-ionophore-stimulated ion secretion in rabbit ileal mucosa. Relation to actions of cAMP and carbamylcholine. *J. Membr. Biol.* **35,** 159–173.
51. Walling, M. W., Brasitus, T. A., and Kimberg, D. V. (1977) Effects of calcitonin and substance P on the transport of Ca, Na and Cl across rat ileum in vitro. *Gastroenterology* **73,** 89–94.
52. Carter, R. F., Bitar, K. N., Zfass, A. M., and Makhlouf, G. M. (1978) Inhibition of VIP-stimulated intestinal secretion and cAMP production by somatostatin in the rat. *Gastroenterology* **74,** 726–730.
53. Chang, E. B., Gill, A. J., Wang, N. S., and Field, M. (1984) α_2-Adrenergic inhibition of Ca^{2+}-dependent intestinal secretion. *Gastroenterology* **86,** 1044.
54. Playford, R. J., Domin, J., Beecham, J., Parmar, K. I., Tatemoto, K., Bloom, S. R., and Calam, J. (1990) Peptide YY: a natural defence against diarrhoea. *Lancet* **335,** 1555–1557.
55. Hubel, K. A. (1978) The effects of electrical field stimulation and tetrodotoxin on ion transport by the isolated rabbit ileum. *J. Clin. Invest.* **62,** 1039–1047.
56. Hubel, K. A. (1984) Electrical stimulus-secretion coupling in rabbit ileal mucosa. *J. Pharmacol. Exp. Ther.* **231,** 577–582.
57. Cooke, H. J. (1984) Influence of enteric cholinergic neurons on mucosal transport in guinea pig ileum. *Am. J. Physiol.* **246,** G263–G267.
58. Keast, J. R., Furness, J. B., and Costa, M. (1985) Investigations of nerve populations influencing ion transport that can be stimulated electrically,

by serotonin and by a nicotinic agonist. *Naunyn-Schmeidebergs Arch. Pharmacol.* **331,** 260–266.
59. Perdue, M. H. and Davison, J. S. (1986) Response of jejunal mucosa to electrical transmural stimulation and two neurotoxins. *Am. J. Physiol.* **251,** G642–G648.
60. Kuwahara, A., Cooke, H. J., Carey, H. V., Mekhjian, H., Ellison, E. C., and McGregor, B. (1989) Effects of enteric neural stimulation on chloride transport in human left colon in vitro. *Dig. Dis. Sci.* **34,** 206–213.
61. Hildebrand, K. R. and Brown, D. R. (1990) Intrinsic neuroregulation of ion transport in porcine distal jejunum. *J. Pharmacol. Exp. Ther.* **255,** 285–292.
62. Cox, H. M. and Cuthbert, A. W. (1990) The effects of neuropeptide Y and its fragments upon basal and electrically stimulated ion secretion in rat jejunum mucosa. *Brit. J. Pharmacol.* **101,** 247–252.
63. Cox, H. M., Tough, I. R., Grayson, K., and Yarrow, S. (submitted) Pharmacological characterisation of neurokinin receptors mediating anion secretion in rat descending colon mucosa. *Naunyn-Schmeidebergs Arch. Pharmacol.*
64. Cox, H. M. and Cuthbert, A. W. (1989b) Secretory actions of VIP, PHI and helodermin in rat small intestine: the effects of putative VIP antagonists upon VIP induced ion secretion. *Regul. Pept.* **26,** 127–135.
65. McCulloch, C. R., Kuwahara, A., Condon, C. D., and Cooke, H. J. (1987) Neuropeptide modification of chloride secretion in guinea-pig distal colon. *Regul. Pept.* **19,** 35–43.
66. Wahlestedt, C., Yanaihara, H., and Håkanson, R. (1986) Evidence for different pre- and post-junctional receptors for NPY and related peptides. *Regul. Pept.* **13,** 307–318.
67. Sheikh, S. P., Sheikh, M. I., and Schwartz, T. W. (1989b) Y_2-like receptors for peptide YY on renal proximal tubular cells in the rabbit. *Am. J. Physiol.* **257,** F978–984.
68. Fuhlendorff, J., Gether, U., Aakerlund, L., Langeland-Johansen, N., Thørgersen, H., Melberg, S. G., Bang Oslen, U., Thastrup, O., and Schwartz, T. W. (1990) [Leu^{31}, Pro^{34}]. Neuropeptide Y: A specific Y_1 receptor agonist. *Proc. Natl. Acad. Sci. USA* **87,** 182–186.
69. Glover, I. D., Barlow, D. J., Pits, J. E., Wood, S. P., Tickle, I. J., Blundell, T. I., Tatemoto, K., Kimmel, J. R., Wollmer, A., Strassburger, W., and Zhang, Y. S. (1984) Conformational studies on the pancreatic polypeptide hormone family. *Eur. J. Biochem.* **142,** 379–385.
70. Krstenansky, J. L. and Buck, S. H. (1987) The synthesis, physical characterisation and receptor binding affinity of neuropeptide Y (NPY). *Neuropeptides* **10,** 77–85.
71. Krstenansky, J. L., Owen, T. J., Buck, S. H., Hagaman, K. A., and McLean, L. R. (1989) Centrally truncated and stabilized porcine NPY analogues: Design, synthesis and mouse brain receptor binding. *Proc. Natl. Acad. Sci. USA* **86,** 4377–4381.
72. McLean, L. R., Buck, S. H., and Krstenansky, J. L. (1990) Examination of the role of amphipathic α-helix in the interaction of NPY and active cyclic analogues with cell membrane receptors and dimyristoylphosphatidylcholine. *Biochemistry* **29,** 2016–2022.
73. Cox, H. M. and Krstenansky, J. L. (1991) The effects of selective amino acid substitution upon neuropeptide Y antisecretory potency in rat jejunum mucosa. *Peptides* **12,** 323–327.

74. Beck, A., Jung, G., Gaida, W., Köppen, H., Lang, R., and Schnorrenberg, G. (1989) Highly potent and small neuropeptide Y agonist obtained by linking NPY 1–4 via spacer to α-helical NPY 25–36. *FEBS Lett.* **244,** 119–122.
75. Tatemoto, K. (1990) Neuropeptide Y and its receptor antagonists. *Ann. NY Acad. Sci.* **611,** 1–6.
76. Michel, M. C. and Motulsky, H. J. (1990) HE90481: A competitive nonpeptidergic antagonist at neuropeptide Y receptors. *Ann. NY Acad.Sci.* **611,** 392–394.
77. Hellström, P. M. (1987) Mechanisms involved in colonic vasoconstriction and inhibition of motility induced by neuropeptide Y. *Acta. Physiol. Scand.* **129,** 549–556.
78. Pernow, J., Svenberg, T., and Lundberg, J. M. (1987) Actions of calcium antagonists on pre- and post-junctional effects of NPY on human peripheral blood vessels in vitro. *Eur. J. Pharmacol.* **136,** 207–218.
79. MacLean, M. R and Hiley, C. R. (1990) Effect of neuropeptide Y on cardiac output, its distribution, regional blood flow and organ vascular resistances in the pithed rat. *Brit. J. Pharmacol.* **99,** 340–342.
80. Minson, R., McRitchie, R., and Chalmers, J. (1989) Effects of neuropeptide Y on the renal, mesenteric and hindlimb vascular beds of the conscious rabbit. *J. Autonom. Nerv. System* **27,** 139–146.
81. Buell, M. G. and Harding, R. K. (1989) Effects of peptide YY on intestinal blood flow distribution and motility in the dog. *Regul. Pept.* **24,** 195–208.
82. Holzer, P., Lippe, I. T., Barthó, L., and Saria, A. (1987) Neuropeptide Y inhibits excitatory enteric neurons supplying the circular muscle of the guinea pig small intestine. *Gastroenterology* **92,** 1944–1950.
83. Wiley, J. and Owyang, C. (1987) Neuropeptide Y inhibits cholinergic transmission in the isolated guinea-pig colon: Mediation through α-adrenergic receptors. *Proc. Natl. Acad. Sci. USA* **84,** 2047–2051.
84. Parkman, H. P., Reynolds, J. C., Ogorek, C. P., and Kicsak, K. M. (1989) Neuropeptide Y augments adrenergic contractions at feline lower esophageal sphincter. *Am. J. Physiol.* **256,** G589–597.
85. Suzuki, T., Nakaya, M., Itoh, Z., Tatemoto, K., and Mutt, V. (1983) Inhibition of interdigestive contractile activity in the stomach by peptide YY in Heidenhain pouch dogs. *Gastroenterology* **85,** 114–121.
86. Wiley, J., Lu, Y., and Owyang, C. (1991) Mechanism of action of PYY to inhibit gastric motility. *Gastroenterology* **100,** 865–872.
87. Allen, J. M., Fitzpatrick, M. L., Yeats, J. L., Darcy, K., Adrian, T. E., and Bloom, S. R. (1984) Effects of PYY and neuropeptide Y on gastric emptying in man. *Digestion* **30,** 255–262.
88. Krantis, A., Potvin, W., and Harding, R. K. (1988) PYY stimulates intrinsic enteric motor neurones in the rat small intestine. *Naunyn-Schmeidebergs Arch. Pharmacol.* **338,** 287–292.
89. Baba, H., Fujimura, M., and Toda, N. (1990) Mechanism of inhibitory action of PYY on cholecystokinin-induced contractions of isolated dog ileum. *Regul. Pept.* **27,** 227–235.
90. Levine, A. J. and Morley, J. E. (1984) Neuropeptide Y: A potent inducer of consummatory behavior in rats. *Peptides* **5,** 1025–1029.
91. Chronwall, B. M., DiMaggio, D. A., Massari, V. J., Pickel, V. M., Ruggiero, D. A., and O'Donohue, T. L. (1985) The anatomy of neuropeptide Y-containing neurons in rat brain. *Neuroscience* **15,** 1159–1181.

92. Sawchenko, P. E. and Swanson, L. W. (1982) Immunohistochemical identification of neurons in the paraventricular nucleus of the hypothalamus that project to the medulla or spinal cord in the rat. *J. Comp. Neurol.* **205,** 260–272.
93. Rogers, R. C. and Hermann, G. E. (1985) Gastric-vasal solitary neurons excited by paraventricular nucleus microstimulation. *J. Auton. Nerv. Syst.* **14,** 352–362.
94. Stanley, B. G., Daniel, D. R., Chin, A. S., and Leibowitz, S. F. (1985) Paraventricular nucleus injections of peptide YY and neuropeptide Y preferentially enhance carbohydrate ingestion. *Peptides* **6,** 1205–1211.
95. Leibowitz, S. F. (1989) Hypothalamic neuropeptide Y and galanin: functional studies of coexistence with monoamines, in *Neuropeptide Y*. V. Mutt, T. Hökfelt, K. Fuxe, and J. M. Lundberg, eds., Raven, New York. pp. 267–281.
96. Sahu, A., Kalra, P. S., and Kalra, S. P. (1988) Food deprivation and ingestion induce reciprocal changes in neuropeptide Y concentrations in the paraventricular nucleus. *Peptides* **9,** 83–86.
97. Kalra, S. P., Sahu, A., Kalra, P. S., and Crowley, W. R. (1990) Hypothalamic neuropeptide Y: a circuit in the regulation of gonadotropin secretion and feeding behaviour. *Ann. NY Acad. Sci.* **611,** 273–283.
98. Harding, R. K. and McDonald, T. J. (1989) Identification and characterisation of the emetic effects of peptide YY. *Peptides* **10,** 21–24.
99. Tuor, U. I., Kondysar, M. H., and Harding, R. K. (1988) Effect of angiotensin II and peptide YY on cerebral circumventricular blood flow. *Peptides* **9,** 141–149.
100. Carpenter, D. O. (1990) Neural mechanisms of emesis. *Can. J. Physiol. Pharmacol.* **68,** 230–236.
101. Kucharczyk, J. and Harding, R. K. (1990) Regulatory peptides and the onset of nausea and vomiting. *Can. J. Physiol. Pharmacol.* **68,** 289–293.
102. Adrian, T. E., Savage, A. P., Sagor, G. R., Allen, J. M., Bacarese-Hamilton, A. J., Tatemoto, K., Polak, J. M., and Bloom, S. R. (1985) Effect of PYY on gastric, pancreatic and biliary function in humans. *Gastroenterology* **89,** 494–499.
103. Pappas, T. N., Chang, A. M., Debas, H. T., and Taylor, I. L. (1986) Peptide YY release by fatty acids is sufficient to inhibit gastric emptying in the dog. *Gastroenterology* **91,** 1386–1389.
104. Greeley, G. H., Jeng, Y-J., Gomez, G., Hashimoto, T., Hill, F. L. C., Kern, K., Kurosky, T., Chuo, H-F., and Thompson, J. C. (1989) Evidence for the regulation of peptide YY release by the proximal gut. *Endocrinology* **124,** 1438–1443.
105. Sheikh, S. P., Holst, J. J., Orskov, C., Ekman, R., and Schwartz, T. W. (1989a) Release of PYY from pig intestinal mucosa; luminal and neural regulation. *Regul. Pept.* **26,** 253–266.
106. Schwartz., T. W. (1983) Pancreatic polypeptide, a hormone under vagal control. *Gastroenterology* **85,** 1411–1425.
107. Taylor, I. L. (1989) Pancreatic polypeptide family: pancreatic polypeptide, neuropeptide Y and peptide YY, in *The Gastrointestinal System II, Handbook of Physiology*, Oxford University Press, Oxford, UK, pp. 475–543.
108. Maouyo, G. D., Taylor, I. L., Gettys, T. W., Greeley, G. H., and Morisset, J. (1991) Peptide YY, a new partner in the negative feedback control of pancreatic secretion. *Endocrinology* **128,** 911–916.

109. Tatemoto, K. (1982) Isolation and characterisation of PYY a candidate gut hormone that inhibits pancreatic exocrine secretion. *Proc. Natl. Acad. Sci. USA* **79,** 2514–2518.
110. Guo, Y-S., Singh, P., Gomez, G., Greely, G. H., and Thompson, J. C. (1987) Effect of PYY on cephalic, gastric and intestinal phases of gastric acid secretion and on the release of gastrointestinal hormones. *Gastroenterology* **92,** 1202–1208.
111. MacFarlane, A., Kinsman, R., Read, N. W., and Bloom, S. R. (1983) The ileal brake: ileal fat slows small bowel transit and gastric emptying in man. *Gut* **24,** 471–472.
112. Unden, A., Tatemoto, K., Mutt, V., and Bartfai, T. (1984) Neuropeptide Y receptor in the rat brain. *Eur. J. Biochem.* **145,** 525–530.
113. Leys, K., Schachter, M., and Sever, P. (1987) Autoradiographic localisation of NPY receptors in rabbit kidney: comparison with rat, guinea-pig and human. *Eur. J. Pharmacol.* **134,** 233–237.
114. Echtenkamp, S. F. and Dandridge. P. F. (1989) Renal actions of neuropeptide Y in the primate. *Am. J. Physiol.* **256,** F524–F531.
115. Adrian, T. E., Savage, A. P., Bacarese-Hamilton, A. J., Wolfe, K., Besterman, H., andBloom, S. R. (1986) PYY abnormalities in gastrointestinal diseases. *Gastroenterology* **90,** 379–384.
116. Holzer-Petsche, U., Petritsch, W., Hinterleitner, T., Eherer, A., Sperk, G., and Krejs, G. J. (1991) Effect of NPY on jejunal water and ion transport in humans. *Gastroenterology* **101,** 325–330.
117. O'Donnell, L. J. and Farthing, M. J. G. (1989) Therapeutic potential of a long acting somatostatin analogue in gastrointestinal diseases. *Gut* **30,** 1165–1172.
118. Sharkey, M. F., Kadden, M. L., and Stabile, B. E. (1990) Severe posthemicolectomy diarrhoea: Evaluation and treatment with SMS 201-995. *Gastroenterology* **99,** 1144–1148.
119. Fassler, J. E., TO'Dorisio, T. M., Mekhjian, H. S., and Gaginella, T. S. (1990) Octreotide inhibits increases in short-circuit current induced in rat colon by VIP, substance P, serotonin and aminophylline. *Regul. Pept.* **29,** 189–197.

Origin and Actions of Neuropeptide Y in the Cardiovascular System

Zofia Zukowska-Grojec and Claes Wahlestedt

1. Introduction

The cardiovascular system is richly innervated by sympathetic nerves containing norepinephrine (NE) and neuropeptide Y (NPY). Although NE is considered a major sympathetic neurotransmitter and a primary mediator of cardiovascular functions, the role of NPY is not yet well defined. Over the last several years, evidence has accumulated to indicate that NPY is a sympathetic cotransmitter mediating vasoconstriction independently of catecholamines, as well as being a modulator of autonomic cardiovascular responses (reviews: *1–4*). NPY is also abundant in epinephrine-containing chromaffin cells of the adrenal medulla *(5)* and, under some conditions, may be secreted into the circulation as an adreno-medullary hormone *(6)*. Finally, our *(7)* and other *(8)* recent data indicate the extraneuronal presence of NPY, e.g., in platelets, where it may subserve autocrine and paracrine functions in platelet–vascular interactions. Thus, there are at least three potential sources for circulating NPY: the sympathetic nerves, the adrenal medulla, and platelets. The first purpose of this chapter is to discuss the release of NPY from different sources into the cardiovascular system in humans and in other mammalian species, in physiological situations, such as stress, and in disease states, such as hypertension.

The second part of this chapter will deal with the cardiovascular effects of NPY, which are many: from directly mediated vasoconstriction and amplification of other vasoconstrictor actions, especially that of norepinephrine, to sympatho-inhibitory and

From: *The Biology of Neuropeptide Y and Related Peptides;*
W. F. Colmers and C. Wahlestedt, Eds.

vasodilator actions resulting from interactions with other mediators, e.g., histamine and norepinephrine. Considering the abundance and potent actions of NPY in the cardiovascular system, and a high degree of conservation of the peptide structure in several mammalian species *(9)*, NPY appears to be an important cardiovascular regulatory factor.

2. Origin of NPY in the Cardiovascular System

2.1. Sympathetic Nerves

2.1.1. Corelease of NPY and NE During Sympathetic Nerve Stimulation

2.1.1.1. Subcellular Localization of NPY and NE

The postganglionic sympathetic nerves contain at least two types of neuronal vesicles, the so-called small and large dense-core vesicles *(10)*, where NE is present and released from; NE is preferentially released from the small, more numerous vesicles whose membranes are rapidly recycled. All sympathetic postganglionic nerves innervating the heart and blood vessels also contain NPY. NPY is apparently present in the large dense-core vesicles, similar to or the same as those containing NE *(11)*, and is coreleased with NE during sympathetic nerve stimulation *(12–15)*. The different subcellular localization of the two transmitters may explain differential regulation of their turnover. It has been suggested that the absence of NPY in the small vesicles indicates that the peptide, unlike NE, is not synthesized in the nerve terminals and not taken up into the storage vesicles *(16)*. As a consequence, NPY release may differ from that of NE in some respect. However, there are also many common features of the release of the two cotransmitters (*see* review: *17*).

2.1.1.2. NPY and Patterns of Sympathetic Nerve Activity

Both NPY and NE release are dependent on influx of extracellular calcium via voltage-dependent N-type calcium channels; thus, release is abolished by calcium-free media and such channels blockers as ω-conotoxin *(18)*. Activation of protein kinase C is required for the release of NE and NPY from the sympathetic nerves *(18)*. NE and NPY releases are guanethidine-sensitive *(13,19)*, indicating their neuronal and exocytotic nature. The release of both is dependent on the frequency of nerve activation; however, although

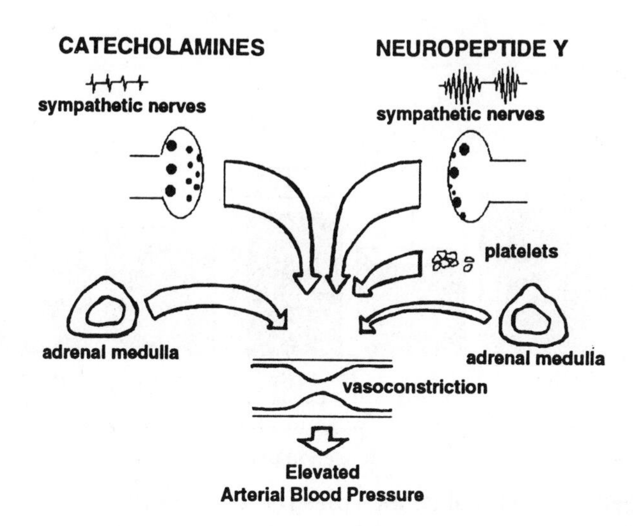

Fig. 1. Sources of catecholamines and neuropeptide Y in the circulation.

NE is released at low nerve activities, NPY release appears to occur at higher frequencies (Fig. 1). Such a pattern of NE and NPY corelease was shown in stimulation of the sympathetic nerves in various preparations, both in vitro, e.g., in isolated perfused canine spleen *(20)* and guinea-pig heart *(14)*, *in situ* in perfused pig spleen *(12,21)*, and in vivo, in conscious calves *(22)*, pithed guinea pigs *(23)*, and rats *(24)*. The ratio of NPY to NE released upon stimulation of the stellate ganglion in the guinea-pig heart was estimated to be about 1:400–500 *(14,15)*.

In some tissues, such as the pig spleen *(12,21,25)*, NPY release was shown to occur preferentially when the sympathetic nerves were stimulated with high-frequency bursts rather than with continuous stimulation. In these studies, the overflow of NPY-like immunoreactivity (-ir) into the splenic vein corresponded to the peak vasoconstrictor effect of stimulation, suggesting the involvement of NPY. Interestingly, it has also been shown that constriction of mesenteric arteries in vitro is augmented by electrical field stimulation at irregular bursts compared to regular nerve activity *(26)*. When sympathetic nerves in the pig spleen were stimulated using an irregular bursting pattern of activity recorded from human muscle sympathetic nerve, the overflow of NPY-ir, but not that of NE, increased more than that elicited by regular nerve activity *(25)*. However, in the same study *(25)*, sympathetic

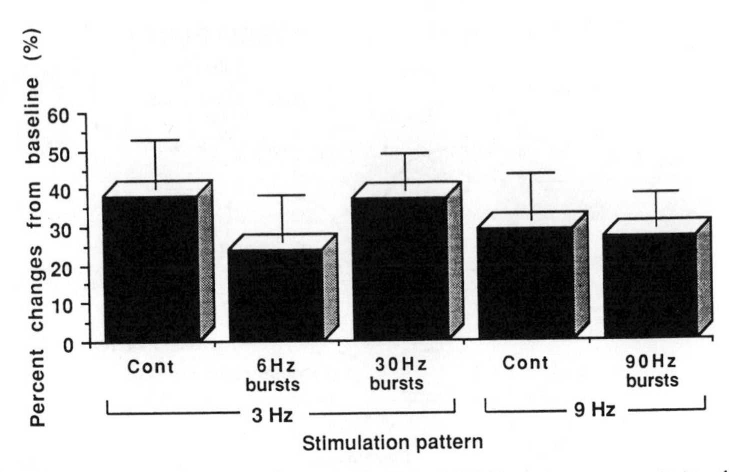

Fig. 2. Changes in renal venous plasma NPY immunoreactivity (-ir) in response to renal nerve stimulation at different patterns in chloralose-anesthetized rats. Plasma NPY-ir levels were increased similarly by continuous and bursting patterns of sympathetic nerve stimulation *(29)*.

nerve stimulation at irregular bursts in the dog gracilis muscle enhanced NPY-ir release only at low, but not at high frequencies. Others were unable to confirm that bursting pattern of nerve stimulation causes greater release of NPY, e.g., in canine heart *(27)* and pithed rats *(24)*. Warner and Levy *(27)* showed that NPY release from cardiac sympathetic nerves (measured indirectly by the degree of attenuation of vagal effect mediated by NPY; *28*) depended on the frequency and duration, but not the pattern of nerve activation.

In our studies, we evoked neuronal, chlorisondamine-sensitive release of NPY by direct stimulation of the total sympathetic outflow in the pithed rats *(24)* and by renal sympathetic nerve stimulation in the anesthetized rats *(29)*. In both cases, plasma NPY-ir spillover increased with higher frequency and long duration of nerve stimulation, and to a similar degree, with bursting and continuous patterns of nerve activation (Figs. 2 and 3). Renal nerve stimulation (1 min, 10 V, Fig. 2) evoked approx 30% increase in NPY-ir in venous renal plasma, and similarly at both 1 and 3 Hz frequencies. Neuronal release of NPY, however, may be obscured in rats by NPY release from platelets that contain large amounts of the peptide releasable during aggregation *(7)* both in vivo and

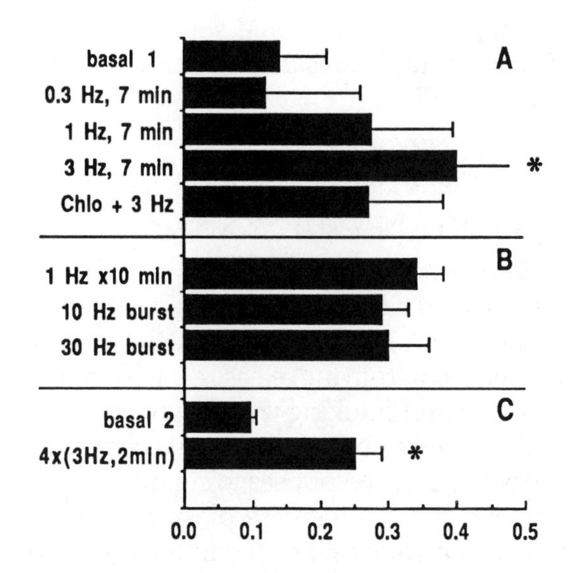

Fig. 3. Plasma NPY-ir responses to stimulation of the total sympathetic outflow in vagotomized pithed rats. Panel A—effect of nerve frequency; panel B—effect of stimulation patterns (panels A and B—blood collected without platelet stabilization); panel C—effect of platelet stabilization on plasma NPY-ir levels, and the effect of repeated stimulation *(24)*.

ex vivo. We have recently developed a platelet stabilization procedure to minimize platelet-derived release of NPY during blood collection *(24)* and reevaluated plasma NPY responses to sympathetic nerve stimulation. Using this method, arterial plasma NPY-ir levels increased 2.5-fold following repeated stimulation (3 Hz, 40 V, 1-ms pulse duration, 4 times, with 1-min intervals in between) of pithed vagotomized rats (Fig. 3).

Thus, if plasma levels truly reflect the release of NPY, it would appear that NE is a primary mediator of sympathetic transmission in states of low sympathetic nerve activity (at rest and during mild activation), and NPY release begins at higher levels of sympathetic nerve activity (Fig. 1). However, the threshold nerve activity for NPY release may differ among tissues and species, e.g., the sympathetic nerves of the pig spleen seem to possess a low threshold (0.59–2 Hz; *25*), whereas the sympathetic nerves in the rat mesenteric bed (unpublished observations from our laboratory) appear to require stimulation at much higher frequencies to release NPY. Furthermore, plasma NPY levels may not reflect neuronally released

NPY. The existence of extraneuronal sources of NPY in species, such as rats, further complicates the assessment of the amounts of NPY released neuronally.

2.1.2. Regulation of NPY Release from Sympathetic Nerves

2.1.2.1. Dissociation of NPY and NE Release

Calcium-dependent exocytotic release of NPY and NE from the sympathetic nerves can be elicited pharmacologically by veratridine and nicotine *(14,17)*, although the ratios of the two transmitters are different from the release elicited by electrical stimulation. In humans, cigaret smoking was shown to elevate circulating levels of NPY-ir, as well as NE *(30)*. In our recent study *(31)* of the sympathetic nerve activity responses to cigaret smoking in human volunteers, nicotine did not increase plasma NPY-ir levels; however, a lower dose of nicotine was used in the cigarets, and plasma NE levels were also unaffected. In the experimental studies *(14,17)*, changes in the overflow of NPY evoked by nerve stimulation and by nicotine or veratridine paralleled those of NE, confirming that the two transmitters are similarly regulated. However, the ratio of NE to NPY may change in the presence of drugs that interfere with the clearance processes for NE, but not for NPY, such as uptake-1 blockers, e.g., desipramine, which shifts the NE:NPY ratio to higher values (approx 1500:1) *(14,17)*. Cocaine, another uptake-1 inhibitor, was actually shown to lower NPY-ir overflow in response to sympathetic nerve stimulation in the isolated perfused guinea-pig heart *(32)*. The inhibitory effect of uptake-1 blockers on NPY release appears to be mediated indirectly by presynaptic α-2 adrenoceptors, stimulated by increased concentrations of NE in the synaptic cleft *(33)*.

The dissociation between NE and NPY release can occur in situations that selectively stimulate the nonexocytotic release of NE, e.g., tyramine and ischemia *(14)*, or following the selective depletion of NE stores by reserpine *(12,13,21,34)*. In the latter case, NPY release is markedly enhanced, and NPY is capable of maintaining physiological pressor responses. In the former case, in the isolated guinea-pig heart model, myocardial ischemia (anoxia) and tyramine both evoked nonexocytotic release of NE in the process of reversed transmembrane amine transport (using the neuronal uptake-1 carrier) without NPY release. The authors *(14)* suggested that since NPY is only released by exocytosis, it may be a better marker of exocytotic release than NE, which can also be released nonexoctotically. This differential NPY and NE release may have clinical implications for the regulation of cardiac function during cardiac ischemia.

2.1.2.2. Presynaptic Inhibition of NPY Release by α-2 Adrenoceptors

By far, the major mechanisms modulating the release of both transmitters are the presynaptic α-2 adrenoceptors. Blockade of α-2 adrenoceptors was shown to augment *(13,15,19,21,35)* and activate with α-2 agonists to inhibit *(15,23)* sympatho-neuronal release of NPY as well as NE. The α-2 adrenergic regulation of NPY release has clearly been demonstrated in isolated organ preparations perfused at constant flow or pressure. In vivo, in humans, phentolamine (a nonselective α-adrenoceptor antagonist) had no effect on basal circulating levels of NPY-ir, but it enhanced plasma NPY-ir responses to intensive exercise *(36)*. However, it has not been determined yet to what degree the enhancement of NPY release in vivo is due to α-2 adrenoceptor blockade and to what degree it is due to a hemodynamic effect of α-1 blockade of lowering vascular resistance or to a reflexive increase of the sympathetic nerve activity. The last mechanism is responsible for an increase in circulating plasma catecholamines in conscious rats (Zukowska-Grojec et al., unpublished observation from our laboratory) and for increased plasma NE responses to sympathetic activation in humans *(37)* treated with a selective α-1 adrenoceptor blocker, prazosin. Similarly, nifedipine, a calcium-channel blocker and a peripheral vasodilator, enhances plasma NPY-ir response to exercise *(38)*. Even in the absence of reflexive changes, an α-1 adrenoceptor blockade may augment NPY-ir overflow from the stimulated tissues by a hemodynamic effect of increasing blood flow. In our studies on pithed adrenal demedullated rats, the α-1 antagonist prazosin was more effective than the α-2 antagonist yohimbine in raising plasma NPY-ir levels during sympathetic nerve stimulation (Shen et al., unpublished observations from our laboratory).

2.1.2.3. Hemodynamic Factors in NPY Overflow into the Circulation

In vivo, the hemodynamic factor may play a major role in determining the overflow of NPY from stimulated sympathetically innervated tissues. Following nerve stimulation, overflow of NPY-ir into the blood or into the perfusate of the isolated tissues is delayed 2–5 min compared to that of NE *(19,34)*. NPY is a larger molecule than NE, and its diffusion through the tissue fluids would be expected to be more restricted. Since the clearance mechanisms for NPY have not been elucidated yet, one can only presume that vasoconstriction reduces NPY spillover into the circulation by exposing the peptide to greater metabolism by tissue proteolytic enzymes and, possibly, to uptake processes. Interestingly, we have observed that plasma NPY-ir levels sometimes decrease during the first few minutes of immobilization stress in rats or during low-frequency sympathetic nerve stimulation in the pithed rats (unpublished observation from our laboratory).

The clearance and metabolism of circulating NPY are not very well known. The half-life of circulating NPY, estimated following administration of exogenous NPY in humans *(38)* and in dogs *(28)*, is approx 20 min. There is no urinary excretion of systemically administered NPY *(38)*, and no net tissue extraction across the human forearm *(38,39)* and the rat kidney *(29)*. However, following ^{125}I-NPY administration in conscious rats (Deka-Starosta et al., unpublished observation), tissue-bound radioactivity was found in several organs (Fig. 4). The highest concentrations of labeled NPY (per gram of tissue) was detected in the spleen, kidneys, and the superior mesenteric artery (Fig. 4). Lower ^{125}I-NPY binding was present in the liver, aorta, and the adrenal medulla. Interestingly, the in vivo labeling of the aorta and adrenal cortex increased after immobilization stress (Fig. 4). It appears that NPY is taken up by (and/or bound to) the vascular smooth muscle cells of rat aorta following in vivo administration, and that NPY binding and/or metabolism is actively regulated by sympathetic nerve activity. We have recently confirmed, both in vivo and in vitro, the presence of specific NPY-binding sites, corresponding primarily to Y1 receptors, on cultured vascular smooth muscle cells from rat aorta and vena cava *(40)*.

2.1.2.4. Presynaptic Inhibition of NPY Release by NPY Autoreceptors

NPY and NE release is crossregulated by their corresponding presynaptic receptors (Fig. 5) *(12,13,15,21,41)*. NPY was shown to have a strong presynaptic inhibitory action on NE release that is independent of α-2 adrenoceptors *(23,42)*. The suppression of nerve stimulation-evoked NE release is thought to be mediated by a spe-

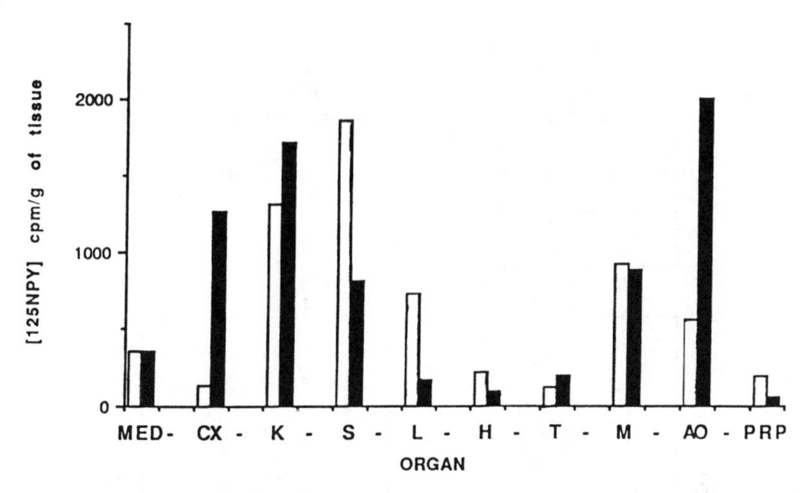

Fig. 4. Distribution of ^{125}I-NPY in peripheral tissues of control (nonstressed) and stressed (2-h immobilization) sham-operated (sham operation for adrenal demedullation) rats; MED—adrenal medulla; CX—adrenal cortex; K—kidneys; S—spleen; L—liver; H—heart; T—thymus; M—superior mesenteric artery; AO—thoracic aorta; PRP—platelet-rich plasma. Stress induced a marked increase in in vivo ^{125}I-NPY binding to the aorta and the adrenal cortex. □ sham control, ■ sham stress.

cific NPY receptor subtype, the Y2 receptor (*43,44; see* Sections 3.2.1.–3.2.4. and Fig. 6) which appears to regulate the release of NPY. Activation of Y2 receptors by peptide YY (PYY), a homologous gut peptide and a natural NPY receptor agonist, reduced the overflow of NPY and NE, elicited upon stimulation of the renal nerves in the pig kidney *(45,46)*. The C-terminal fragments of NPY (and PYY) were characterized as Y2 receptor agonists based on their ability to suppress evoked NE release. Whether they also inhibit NPY release has not been definitively shown, but is presumed to be the case. As discussed later in the chapter, C-terminal fragments of NPY, such as $NPY_{18\text{-}36}$ and $NPY_{22\text{-}36}$, cause hypotension *(47,48)*, which, in some part, may be mediated by the inhibition of NE and NPY release from the sympathetic nerves.

2.1.2.5. Facilitatory Presynaptic Mechanisms

Among the factors that may augment the release of NPY are angiotensin II, known to potentiate the release of NE from the sympathetic nerves and the adrenal medulla. Recently, angiotensin II was shown to enhance sympathetic nerve stimulation-evoked NPY release from the guinea-pig heart *(49)* and the pig kidney *(45,46)*.

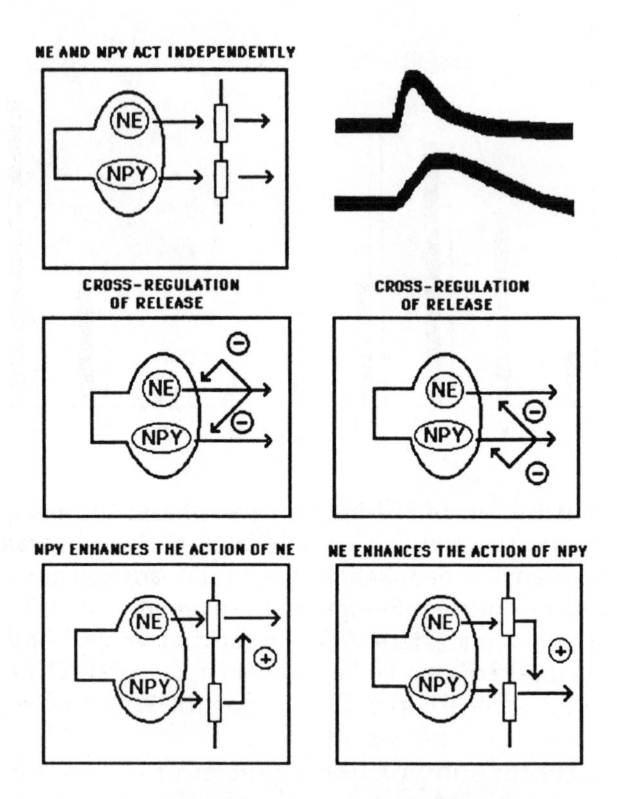

Fig. 5. Actions and interactions of NE and NPY at the sympathetic neuro-effector junctions in the cardiovascular system.

In the case of NE, it has been suggested that epinephrine may augment the release via presynaptic β-2 adrenoceptors *(50)*. Dahlof *(49)* has recently shown that β-2 adrenoceptor agonists also increase plasma NPY responses to stimulation of the sympathetic nerves in the pithed guinea pigs. In contrast to the presynaptic α-2 adrenoceptor inhibition, which appears to affect NPY release immediately, the β-2 adrenoceptor-mediated facilitation develops slowly with time *(49)*.

2.1.3. Circulating Plasma NPY Levels During Stress

2.1.3.1. Intensity of Stressors and NPY Overflow into the Circulation

The issue of NPY as a sympathetic mediator of high-intensity sympathetic nerve stimulation becomes quite evident from the in vivo studies in humans *(50–52)*, pigs *(53)*, and rats *(54–57)*. Plasma

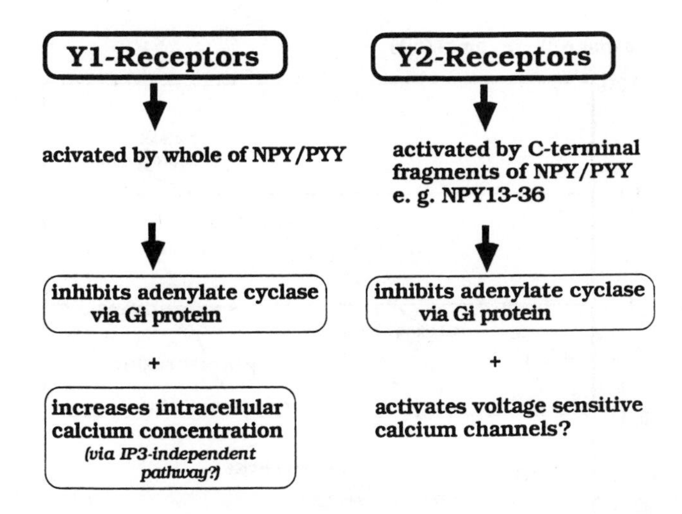

Fig. 6. NPY-receptor subtypes in the cardiovascular system.

levels of NPY-ir do not increase during mild or brief sympathetic nerve activation and stress (*54,55,57,58*; Fig. 7). During exercise in humans, venous plasma NPY-ir does not rise until the work load increases to 75% of the maximal oxygen utilization (or else, duration of exercise increases) *(13,59)*. The release of NPY-ir from the human heart is not enhanced by hypoxia alone (induced by breathing of 12% oxygen), but hypoxia markedly enhances the effects of exercise *(52)*. This may suggest that hypercapnia, which is associated with exercise, is a more potent stimulus for NPY than hypoxia; however, both conditions augment sympathetic nerve activity and have a synergistic effect on NPY release (particularly from the heart). Other stress tests, used to elicit plasma NE responses, such as isometric handgrip and orthostatic tests *(45)*, and head up tilt *(50,51)*, fail to raise plasma NPY-ir levels in humans. Those tests, however, do not lend themselves to maintaining the stressful conditions for longer than a few minutes.

In addition to strenuous exercise, other intense stressors, such as insulin-induced hypoglycemia *(60)* and cold pressor test, were shown to increase plasma NPY-ir both in humans *(50,51)* and in rats *(58)*. In our studies, the cold water test (1 cm of ice-cold water poured on the floor of the rat cage) was one of the most effective "physiological" stimuli increasing plasma NPY-ir levels in conscious intact rats (Fig. 8). However, in rats, the effectiveness of cold expo-

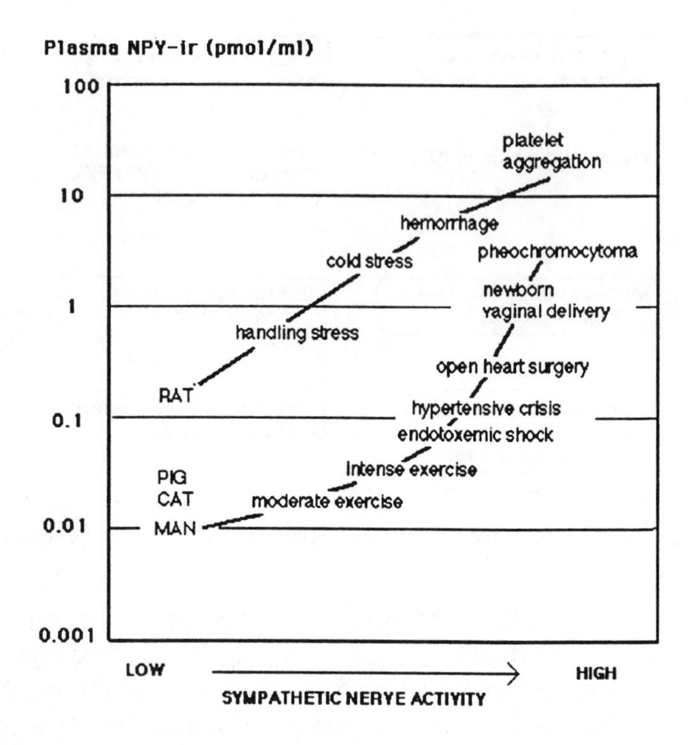

Fig. 7. Circulating plasma NPY-ir levels in various species in conditions with varying sympathetic nerve activity (or adrenomedullary or platelet activation).

sure in raising plasma NPY-ir levels is not only dependent on sympathetic nerves, but is largely related to the activation of platelets (and release of platelet-derived NPY) *(7,24)*.

Recently, Ahlborg and Lundberg *(61)* have reported that prolonged and graded (up to 78% maximal oxygen utilization) arm or leg bicycle exercise similarly elevated arterial plasma NPY-ir levels. In an attempt to evaluate the contributions of different organs to systemic NPY concentrations, the authors have also measured regional blood flow (splanchnic, arm, and leg) and calculated organ "NPY exchanges." Interestingly, the exercising arm or leg did not apparently contribute to systemic NPY, whereas a nonexercising area, splanchnic vascular bed, released NPY in significant levels *(61)*. However, the release of NPY from sympathetic nerves in the exercising skeletal muscle into the blood stream might have been obscured by a hemodynamic factor due to a severalfold increase in arm and leg blood flow *(61)*.

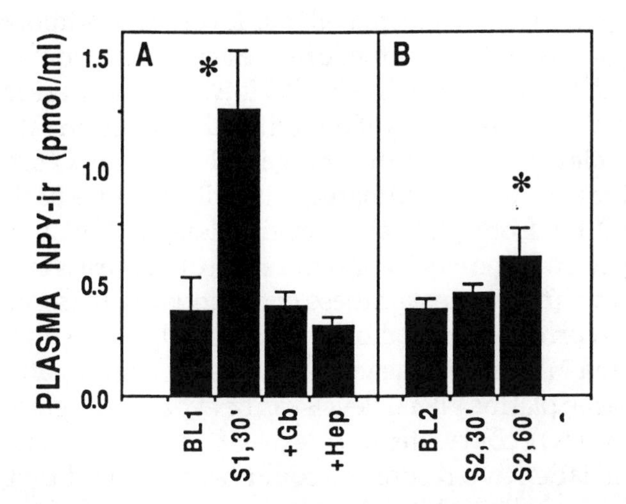

Fig. 8. Plasma NPY-ir levels during cold water stress (S) in rats. Panel A—threefold increase in plasma NPY-ir levels at 30 min of the cold stress (S1, 30 min), when blood was collected without platelet stabilization; these responses were similarly blocked by ganglionic blocker (Gb) or heparin (Hep, an antiaggregatory dose). Panel B—attenuated increase in plasma NPY-ir levels in response to stress (S2, 60 min) when blood was collected with platelet stabilization *(24)*.

2.1.3.2. Relation to Sympathetic Nerve Activity

Generally, stress-induced elevations of plasma NPY-ir levels positively correlate with plasma NE, but not epinephrine *(54,62)*, indicating that the peptide originates from the sympathetic nerves. In humans, the highest levels of circulating plasma NPY-ir were detected in patients with pheochromocytoma *(63)* (Fig. 7), which results from the large production and secretion of the peptide by the chromaffin cells of the tumor. Aside from this specific condition, plasma NPY-ir levels in humans were shown to increase in relation to the sympathetic nerve activation. Using direct sympathetic nerve recording in humans, Eckberg et al. *(63)* have found that plasma NPY levels, like those of NE, increase proportionally to the reflexively increased bursts in muscle sympathetic nerve activity. Some of the highest levels of circulating NPY-ir were found in situations of severe cardiovascular stress, such as in patients with congestive heart failure *(65)*, endotoxic shock *(66)*, hypertensive crisis *(67)*, open heart surgery *(68)*, and in newborns after vaginal delivery *(69)*, as depicted in Fig. 7.

In rats, changes in plasma NPY-ir levels in conditions characterized by increased sympathetic nerve activity are somewhat less pronounced than in humans (Fig. 7). However, there are major differences in the resting levels of NPY in both species. In rats, circulating NPY-ir levels at rest are in a range of 0.1–0.5 pmol/mL, approx 5–10-fold higher than in humans, primarily because of the contribution of NPY from platelets. Nevertheless, a similar pattern of dependence on the intensity and duration of the stress is present in rats and humans. In most of these conditions, increases in plasma NPY levels correlated with those of NE *(54,70)*. We found that apart from plasma NPY-ir levels evoked by platelet aggregation in rats *(7)*, increases in plasma NPY-ir levels can be evoked by hemorrhage (5 mL/kg body wt), cold water test *(24,54–58)*, and electrical footshock *(54,56)*; the latter two responses could be prevented by the treatment with a ganglionic blocker, chlorisondamine, confirming their sympatho-neuronal origin *(56)*. However, under conditions of chronic stress (footshock 15 min a day for 30 d), chlorisondamine fails to lower resting levels and footshock stress-induced increases in circulating NPY-ir levels, indicating that other nonneuronal sources (platelets?) of NPY could have been activated (Fig. 9). Thus, stress-induced increases in circulating NPY are related to the sympathetic nerve activity in rats, but other factors are interacting with it (*see* Sections 2.3. and 2.4.).

Recently, we have found that cold stress-induced plasma NPY-ir responses were greater in sexually mature male than female rats (Fig. 10); this was associated with greater and more protracted hypertensive response to stress in males, suggesting that NPY may be involved in it *(57)*. Since most of the studies on NPY have used male rats, these data cannot be corroborated as yet, but they raise an interesting possibility that gender may determine the neurochemical components (NPY:NE ratio) of sympathetic responses, and by doing so, the hemodynamic responsiveness to stress.

2.1.3.3. NPY as a Primary Sympathetic Neurotransmitter?

If the release of NPY indeed increases only at high frequencies of nerve stimulation, is NPY then a primary sympathetic neurotransmitter? What components of sympathetically-mediated vasoconstriction are mediated by NPY? Normally, most of the vasopressor and vasoconstrictor responses evoked by the sympathetic nerve stimulation at frequencies of up to 3 Hz are blocked by

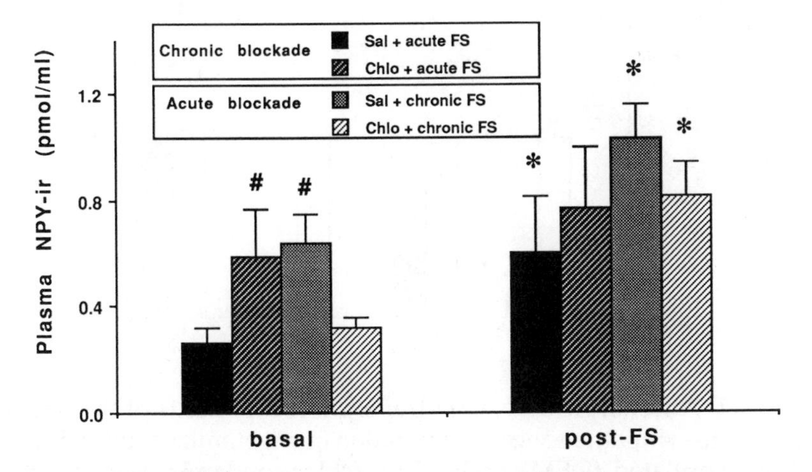

Fig. 9. Plasma NPY-ir responses to acute and chronic stress of footshock (FS: 1.5 mA, 0.5 ms duration, 5 min daily, 30 d). Rats were treated with saline (Sal) or chlorisondamine (Chlo), either acutely (1 h prior to stress) or chronically (30 d). # $p < 0.05$ compared to basal levels in chronic Sal + acute FS; *$p < 0.05$ compared to respective basal levels. Chronic chlorisondamine and chronic stress elevated basal NPY-ir levels. Chlorisondamine blocked stress-induced rise in plasma NPY-ir levels in rats treated chronically with Sal, but not in chronically Chlo-treated rats *(56)*.

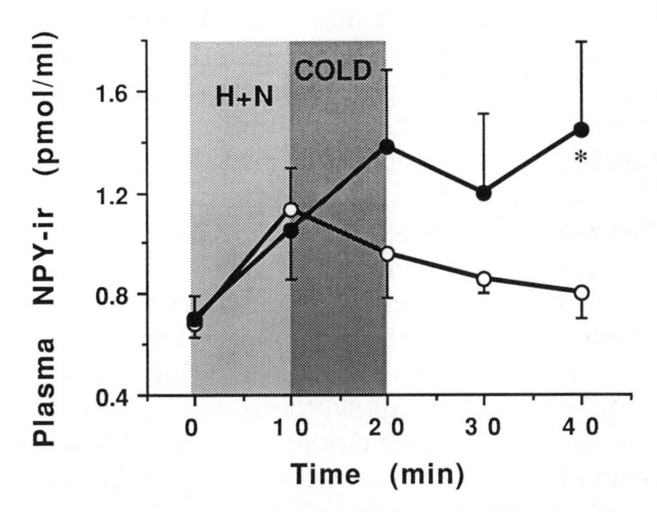

Fig. 10. Sex differences in plasma NPY-ir reponses to stress. COLD—cold (4°C) water stress, H + N—handling and novelty stress. *$p < 0.05$ compared to females *(57)*. –○– females; –●– males.

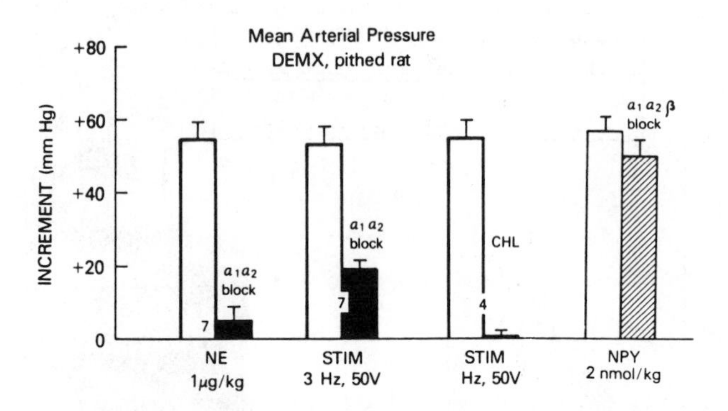

Fig. 11. Adrenergic and nonadrenergic components of the pressor response to sympathetic nerve stimulation in vagotomized, pithed, adrenal demedullated (DEMX) rats; CHL (chlorisondamine)—a ganglion blocker. α1 + 2-adrenergic blockade reveals a nonadrenergic component of the stimulation-induced pressor response, which is mimicked by NPY.

α-adrenoceptor blockade (Fig. 11). This would indicate that NPY may be at best an auxiliary sympathetic cotransmitter under physiological conditions. However, NPY participation in sympathetic neurotransmission increases at higher frequencies of sympathetic nerve activation and in states of NE deficiency. The indirect evidence that the release of NPY may depend on the composition of the neuronal vesicles was provided by studies using reserpinized animals *(12,13,19,21)*. Reserpine causes depletion of NE stores in the sympathetic nerves without directly affecting NPY stores *(13,19)*. In such conditions, sympathetic stimulation in the pig spleen releases only NPY, but not NE *(13)*, whereas pressor responses to nerve stimulation remain unaltered. This suggests that in NE deficiency, NPY may act as a primary sympathetic transmitter capable of maintaining physiological responses. Chronically, reserpinization causes a fall in blood pressure that results in a reflexive increase in the sympathetic nerve activity *(12)*. What follows is an enhanced release of NPY in excess of its synthesis, leading to a depletion of the peptide in the stimulated tissues *(12)*. The NPY depletion can be prevented by pretreatment with a ganglionic blocker or by preganglionic denervation prior to reserpine *(13)*.

Clinically, impaired, deficient adrenergic transmission occurs in patients with idiopathic orthostatic hypotension *(71)*. However, it is not known yet whether the deficiency in NE release is coupled in these patients to enhanced (or diminished) release of NPY, similar to that of reserpinized rats. The depletion of NE stores may develop physiologically under conditions of prolonged release of NE. Rudehill et al. *(53)* showed that during hemorrhage in the anesthetized pig, the first bleeding increased splenic output of NE and only moderately raised NPY overflow; however, the second bleeding resulted in proportionally greater release of NPY compared to that of NE.

2.2. Adrenal Medulla

2.2.1. Release of NPY from Chromaffin Cells

NPY is abundant in chromaffin cells of the adrenal medulla of several mammalian species, including humans *(5,6)*. NPY-ir has been detected by some *(72,73)* in epinephrine-containing chromaffin cells and by others *(74)* in the norepinephrine-containing cells, and was also found to be colocalized with enkephalins *(73,75)*. Additionally, NPY-ir is contained in nerve fibers derived from a plexus in the adrenal capsule and penetrating through adrenal cortex and medulla, many of which innervate blood vessels *(76,77)*. The adrenomedullary content of NPY varies markedly among species, with the highest levels found in the mouse and cat, and the lowest in pig adrenals (intermediate levels in rat adrenals). Furthermore, aging increases adrenomedullary content of NPY (and its oxidized form, methionine sulfoxide NPY) by severalfold *(78,79)*.

Allen et al. *(22)* were the first to report that the electrical stimulation of splanchnic nerves in conscious calves increased adrenal venous (markedly) and systemic (less) plasma NPY-ir levels. Subsequently, several investigators showed the release of NPY could be evoked pharmacologically from the perfused bovine adrenal gland *(80)*, cultured chromaffin cells *(81)*, and by a direct splanchnic nerve stimulation *(82)*. Insulin, which is a known adrenomedullary stimulus activating splanchnic nerves, was found to elevate mRNA encoding NPY in the rat adrenal medulla *(83)*, to produce acute depletion of NPY in the adrenal medulla in rats *(84)*, and to increase circulating plasma NPY-ir levels in humans *(60)*.

In spite of these lines of evidence, the notion of NPY as a adrenomedullary hormone in physiological conditions is questionable. In vivo, the only known situation in which the adrenal medulla secretes large quantities of NPY into the bloodstream is pheochromocytoma *(6,62,85)*. In intact animals with normal adrenal

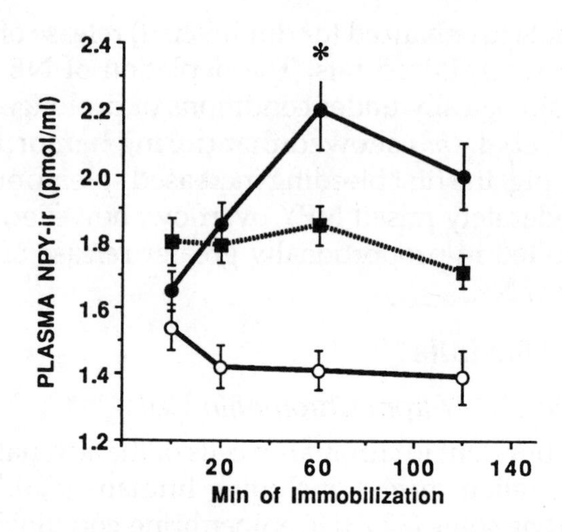

Fig. 12. Plasma NPY-ir responses to immobilization stress in sham-operated -O- (sham) and adrenal demedullated -●- (DEM) rats; -■- Chlo—chlorisondamine pretreatment. Adrenal demedullation increased stress-induced plasma NPY-ir responses compared to those in sham-operated rats (* $p < 0.05$) *(24)*.

medullas, increases in circulating plasma NPY-ir levels evoked by stress *(54,86)* correlate positively with NE, but not with epinephrine, indicating neuronal origin. Bilateral removal of the adrenal medulla does not affect circulating NPY levels at rest *(86)* and enhances, not lowers, plasma NPY-ir responses to immobilization stress in rats (Fig. 12; *24*). In anesthetized dogs, hypotensive hemorrhage, which increased systemic levels of NPY-ir, also raised adrenal venous plasma concentration of the peptide, but adrenal NPY-ir output was significantly decreased because of the concomitant reduction in adrenal blood flow *(82)*. These studies indicated that adrenomedullary secretion of NPY occurred during stress, but these amounts were insufficient to contribute to systemic levels of the peptide. Therefore, under physiological conditions, the adrenomedullary NPY may play a local role, perhaps regulating catecholamine secretion and/or synthesis.

2.2.2. Characterization of Adrenomedullary NPY-Like Immunoreactivity

Not all adrenomedullary NPY-ir appears to be in the form of NPY_{1-36}. The HPLC profile of tumor tissue from patients with pheochromocytoma was shown to contain NPY-ir that had similar

HPLC elution characteristics to those of the authentic peptide *(6)*. However, other NPY-immunoreactive species are present in adrenomedullary extracts *(74,78,80)*. With age, adrenomedullary content of methionine sulfoxide NPY increases in rat medullas *(78)*. Recently, Shimoda et al. (personal communication) have isolated NPY_{1-30} from bovine adrenal medulla and also found it to be present in large quantities in the rat adrenal medullas. The biological activities of these new NPY-like peptides remain to be discovered.

2.3. Extraneuronal NPY: Platelets

2.3.1. Release of NPY During Platelet Aggregation in Rats

The presence and synthesis of NPY extraneuronally was first suggested by findings of NPY mRNA in megakaryocytes from rats and some autoimmune mice *(8)*. We have subsequently found that rat platelets are a rich source of circulating NPY-ir *(7)*. Rat platelet-rich plasma, and platelet pellets prepared from it, contain approximately tenfold higher concentrations of NPY-ir than platelet-poor plasma *(7)*. Collagen, which evokes the secondary, irreversible stage of platelet aggregation and the associated release reaction, causes a dose-dependent release of NPY-ir in parallel, as measured in platelet-poor plasmas rapidly prepared from the aggregating platelet suspension (Fig. 13). Much less of NPY-ir is released during primary, reversible aggregation, such as that induced by adenosine diphosphate in rat platelets prepared from citrated blood (Fig. 13). Also, in vivo, circulating plasma NPY-ir levels increase dose-dependently (up to fourfold) following iv injection of collagen in conscious and pithed normotensive rats (Fig. 14).

2.3.2. Characterization of Platelet-Derived NPY-Immunoreactivity

HPLC characterization of rat platelet-poor plasma reveals a complex pattern consisting of two major NPY-ir peaks *(87,88)*. Typically, platelet aggregation increases NPY immunoreactivity corresponding to rat NPY elevenfold (Fig. 15), accompanied by a smaller increase of the unknown peak of a mol wt larger than 30,000 (bioactivities of the unknown peak are being currently studied). Recently, Ogawa et al. *(80)* have confirmed, using three different HLPC systems combined with radioimmunoassay, that NPY-ir in platelets from Wistar-Kyoto and spontaneously hypertensive (SHR) rats is authentic rat NPY.

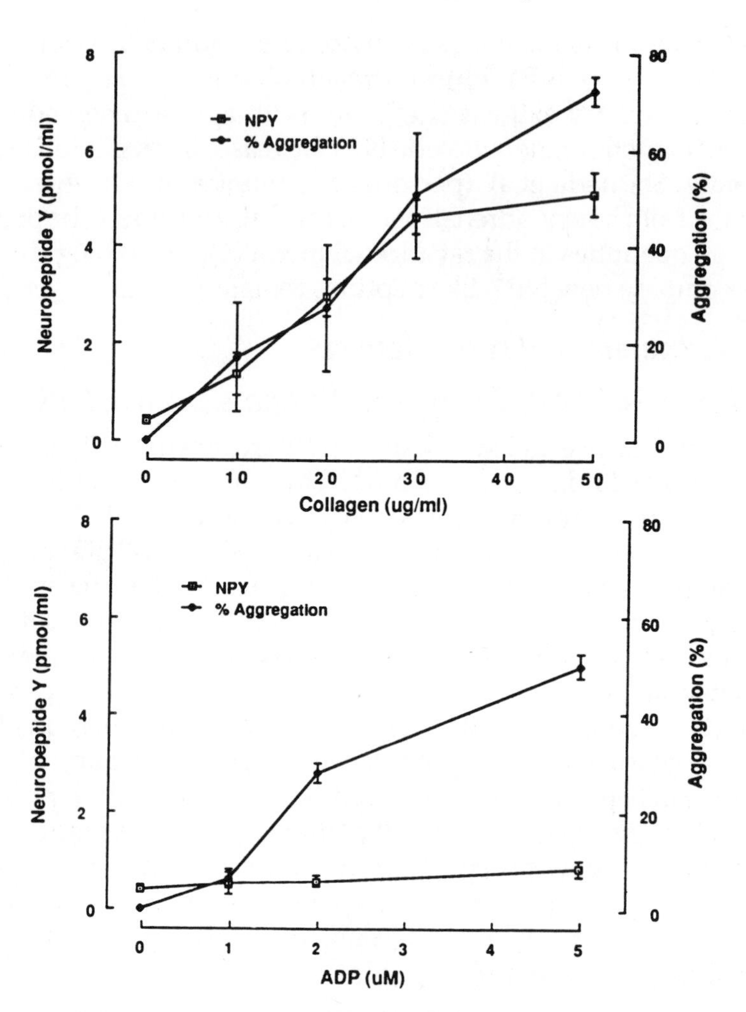

Fig. 13. Release of NPY-ir during aggregation of rat platelets in vitro. Platelet release of NPY-ir paralleled the degree of aggregation induced by collagen, but not adenosine diphospate (ADP) in citrated rat blood indicating that NPY release occurs during secondary phase of aggregation *(7)*.

2.3.3. Species Variability in Platelet-Derived NPY

In rats, high resting circulating NPY-ir levels appear to result from platelet-derived NPY, and this may explain marked differences with other species, such as pigs, guinea pigs, rabbits, cats, dogs, and humans (*see* Fig. 7). The issue of whether the species with low circulating NPY-ir platelet levels do not possess and/or

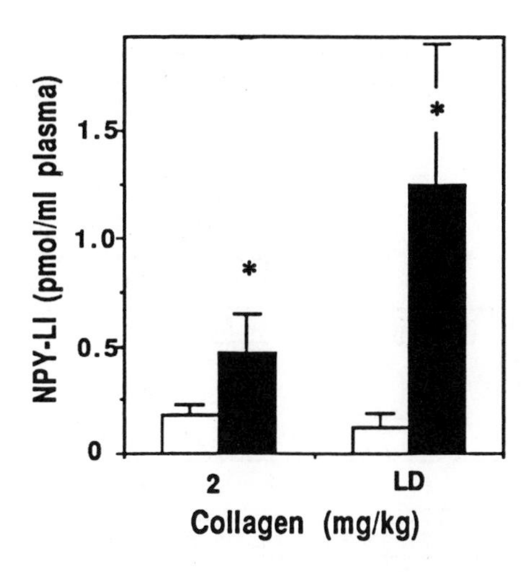

Fig. 14. Collagen-induced release of NPY-ir from rat platelets in vivo. LD—lethal dose of collagen (3–5 mg/kg, iv) *(24)*. □ Basal; ■ stimulated.

release NPY has not been completely resolved. In contrast to rat megakaryocytes, no NPY mRNA was detected in normal human and pig bone marrow *(90)*. It should be noted, however, that low quantities of NPY-ir are detectable in human platelet-rich plasma and platelet pellets *(87)*, but the nature of this immunoreactivity remains unknown. We have also recently found high levels of NPY-ir in rabbit platelet-rich plasma and platelet pellets *(87)*. Conversely, Persson et al. *(91)* were unable to detect NPY mRNA in rabbit spleen, although it is not certain whether splenic expression of NPY gene corresponds to that of megakaryocytes.

Interestingly, it has been shown that although megakaryocytes from normal mice do not contain any NPY mRNA, autoimmune NZB mice express the NPY gene in megakaryocytes *(8)*. Thus, it is possible that the NPY gene is expressed in these cells in all species, including humans, but is normally downregulated by unknown factors (developmental, other?), and might be upregulated in some pathophysiological conditions. For example, in spontaneously hypertensive rats (SHR), platelet content of NPY has been recently shown to be severalfold greater than that of normotensive rats *(89)*; whether this "upregulation" of platelet-derived NPY is genetic and/or secondary to other pathophysiological changes remains to be determined.

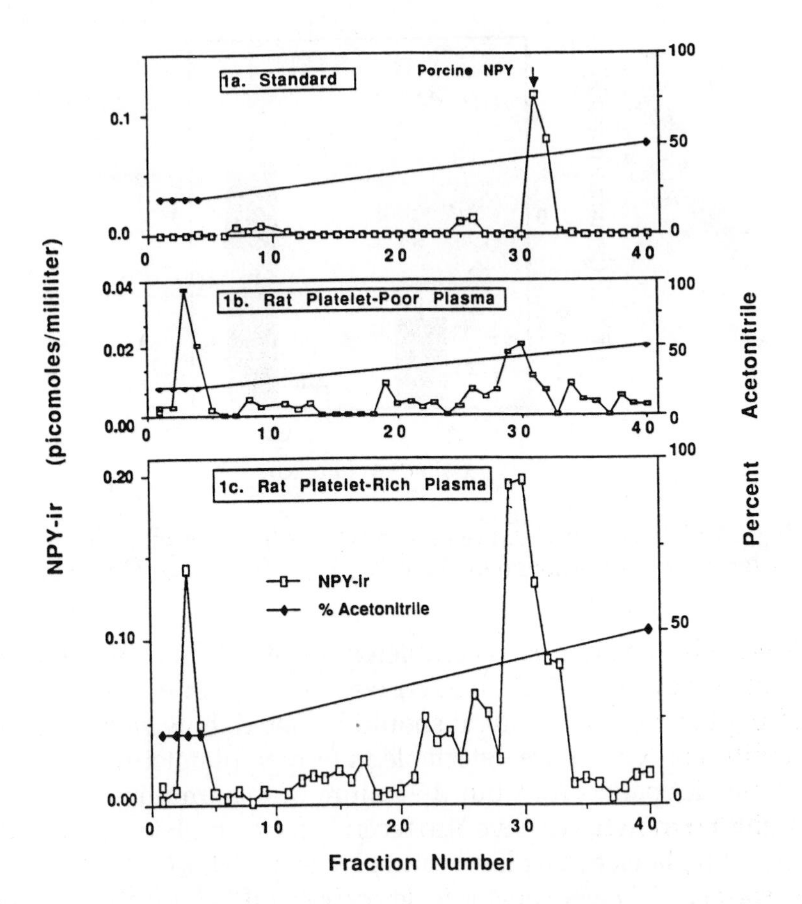

Fig. 15. Reverse-phase HPLC profile of NPY immunoreactivity in rat platelet-poor plasma and platelet-rich plasma aggregated with collagen in vitro.

2.4. Contribution of the Sympathetic Nerves, Adrenal Medulla, and Platelets to Circulating Plasma Levels of NPY-ir

In the species where NPY is a platelet-derived factor, plasma levels of the peptide may be markedly affected by platelet aggregation occurring both in vivo as well as ex vivo, during blood collection and processing. In rats, in which platelet-derived NPY-ir is a major contributor to resting circulating levels, we have adopted a stringent technique of blood collection to avoid artifactual release of NPY from platelets (*see* Figs. 2 and 8). The ability of rat platelets to increase plasma NPY-ir levels to up to tenfold is probably a major factor for a reported marked (15-fold) elevation of the peptide in trunk blood from the decapitated rats *(34)*. High levels of circulating NPY-ir found during hemorrhage *(86)*, myocardial infarction *(34)*, cold stress *(56,58)*, and strenuous exercise *(62)* have been assumed to be purely neurogenic. However, this may not be entirely true since the conclusive studies, e.g., of ganglion blocked, stressed animals, have not always been done. On the other hand, conditions similar to those listed above have been associated with increased platelet aggregation.

One of the most effective stimuli elevating circulating NPY-ir levels in rats is hemorrhage, a cardiovascular stress that activates the sympatho-adrenomedullary system. A loss of about 19% of blood volume causes a 2.5-fold increase in plasma NPY-ir levels, immediately after the bleeding *(58,92)*, although plasma catecholamine levels increase only moderately with this degree of blood loss *(93)*. Moreover, hypotension induced by a vasodilator, sodium nitroprusside, infused into rats is completely ineffective in increasing plasma NPY-ir (*92*, and unpublished data from our laboratory) in spite of a marked adrenergic response. The discrepancy can probably be explained by the release of NPY from platelets, since hemorrhage activates *(94)* and sodium nitroprusside inhibits platelet aggregation *(95)*.

The platelets may also contribute to marked elevation of NPY-ir levels in the cerebrospinal fluid in a model of subarachnoid hemorrhage produced by an injection of clotting blood into cisterna magna of the rabbit *(96)*. The authors of this study suggested that the blood-induced increase in CSF content of NPY-ir is derived from the brain neurons rather than from the sympathetic nerves, since it was not prevented by bilateral cervical ganglionectomy (which removes most cerebral sympathetic innervation). In contrast to rabbits, humans, who may have lesser, if any, platelet release of NPY *(87)*, cerebrospinal NPY-ir levels were found to be only mildly elevated following subarachnoid hemorrhage *(97)*.

In an attempt to dissociate neuronal and platelet-derived NPY, we pretreated rats with a ganglion blocker, chlorisondamine, and heparine. As anticipated, ganglionic blockade completely prevented an increase in plasma NPY-ir levels induced by the cold water stress (Fig. 8). Surprisingly, heparin, at a dose that completely abolished collagen-induced platelet aggregation in vitro (1000 U/mL blood),

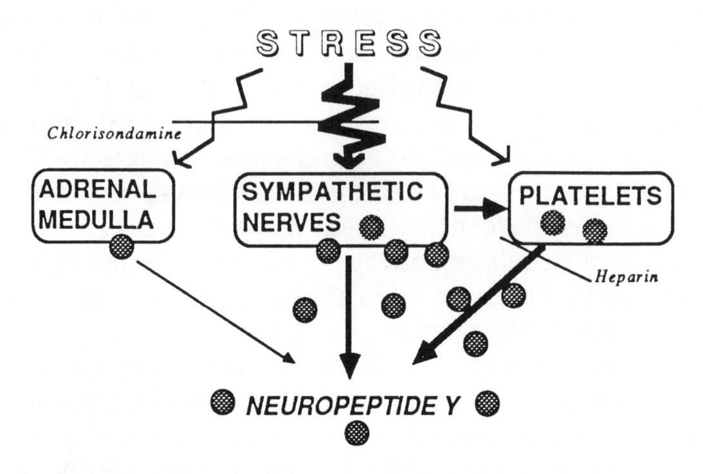

Fig. 16. Sources of NPY release into plasma during stress.

also prevented a cold-induced rise in plasma NPY-ir (Fig. 8). Since chlorisondamine had no direct antiaggregatory action in vitro, we suggested that both the sympathetic nervous system and the platelets play a role. As shown in Fig. 16, the primary event would be the activation of the sympathetic nerves that release catecholamines and NPY, and that would secondarily facilitate platelet aggregation and their release of NPY. In this model, platelets would act as a transducer and an amplifier of neurogenic responses by responding to the sympathetic activation and by releasing NPY directly into the bloodstream. The contribution of platelets to sympathetically induced plasma NPY responses may also be artifactual because of ex vivo platelet aggregation during blood collection, however, we can largely eliminate this component by platelet stabilization during blood sampling (*see* Figs. 3 and 8; *24*).

The three putative sources of NPY, the sympathetic nerves, the adrenal medulla, and platelets, may develop even more complex interactions under chronic conditions. For example, chronic ganglionic blockade (for 30 d, Fig. 9), although preventing the stress-induced increases in plasma NPY-ir, causes doubling of resting circulating levels of the peptide in conscious rats *(56)*. This may suggest that prolonged blockade of the sympathetic nervous system evokes compensatory responses in the remaining sources of NPY (and perhaps catecholamines), most likely, megakaryocytes augmenting their ability to synthesize and release this vasopressor peptide.

The reverse situation seems to occur in autoimmune NZB mice, which have abnormally high expression of NPY mRNA in megakaryocytes, associated with a decreased content of NPY-ir in the sympathetic nerves of the spleen *(91)*.

Finally, the release of NPY has been implicated in altered salt balance states *(4,98)*. Waeber et al. *(98)* have reported that circulating plasma NPY levels were increased markedly by a high-salt diet in conscious uninephrectomized rats. Since plasma catecholamine levels remained unchanged, the authors *(98)* reasoned that NPY release occurred irrespectively of sympathetic nerve activity. Whether this salt-induced increase in plasma NPY was due to the release from the sympathetic nerves, adrenal medulla, platelets, or other unknown sources, or whether salt merely affected NPY clearance, remains to be determined.

3. Actions of NPY in the Cardiovascular System

3.1. Hemodynamic Actions of NPY

3.1.1. Systemic and Regional Hemodynamics

Lundberg and Tatemoto *(99)* were the first to report that NPY (and PYY) increases blood pressure in the cat independently of α-adrenoceptors. Subsequently, pressor effects of NPY have been shown by others in various species, including the rat *(100;* Fig. 11) and humans *(38)*. The pressor response of NPY is characterized by a slower onset and a longer duration than that of norepinephrine (*see* Fig. 5). In conscious rats, systemic administration of NPY produces a long-lasting rise in blood pressure because of a marked increase in total peripheral resistance *(101)*. On a molar basis, NPY was found to be severalfold less potent than angiotensin II and NE in raising total peripheral resistance in anesthetized cats *(102)*, whereas in conscious rats, NPY is sixfold more potent than NE in this respect *(101)*. NPY decreases regional blood flow in the mesenteric and renal beds, and to a lesser extent in the hindquarters (Fig. 17; *100,102*). In pithed rats *(103)*, NPY caused a prolonged rise in total peripheral resistance predominantly because of the increases in vascular resistance in the mesenteric bed, but also in the spleen, the kidneys, skeletal muscles, and large intestines. In anesthetized cats, NPY was 25 times more potent than NE in producing 50% reduction of colonic blood flow *(104)*. In humans, systemic administration of

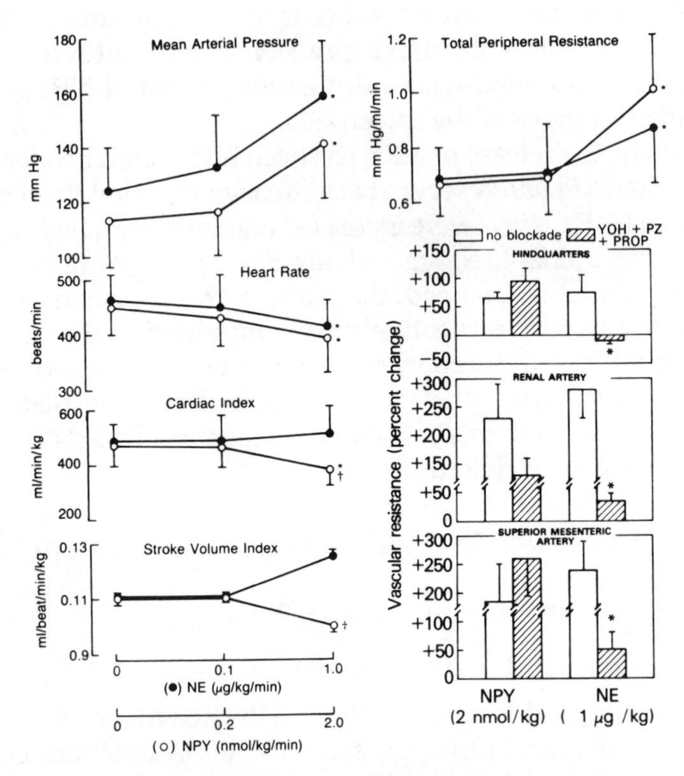

Fig. 17. Hemodynamic effects of NPY and NE in the absence and presence of total adrenergic blockade in conscious rats; YOH—yohimbine; PZ—prazosin; PROP—propranolol.

NPY, which elevated plasma circulating levels of the peptide about 30-fold, did not elicit any changes in blood pressure or heart rate *(38)*. However, infusion of NPY into the brachial artery in normal volunteers *(38)* evoked a dose-dependent, slow-onset, long-lasting increase in venous tone and decreased brachial blood flow. The threshold for these actions of NPY is in the range of low-nanomolar concentrations *(38)*, which in humans might be reached in some pathophysiological states, such as pheochromocytoma (Fig. 7).

At doses that are effective systemically, NPY does not significantly increase pulmonary arterial pressure and vascular resistance *(105,106)*. In vitro, however, NPY constricts both pulmonary and bronchial arteries in the pig *(107)*. In the systemic circulation, two areas are specifically affected, the coronary and cerebral circulation *(see below)*. Intracoronary injections of NPY, at picomolar concentrations that are probably physiological, cause dose-dependent reductions in coronary blood flow without changes in blood pressure and heart rate in anesthetized, open-chest dogs *(108–110)*. These actions are in marked contrast to those of NE, which is much less potent as a coronary vasoconstrictor *(111)* and may indirectly increase blood flow by releasing vasodilator metabolites.

3.1.2. Cardiac Effects

In conscious animals, bradycardia, which accompanies the pressor response to NPY, is of mixed origin, in part reflexly mediated and in part independent of baroreceptors *(101,112)*. The degree of bradycardia evoked by pressor doses of the peptide always appears to be greater than in the case of adrenergic agonists *(101)*. Minson et al. *(112)* have reported that NPY increases the sensitivity of the baroreceptor reflex-mediated changes in heart rate on systemic administration in conscious rabbits, probably because of the reduction in cardiac β-adrenergic tone. Additional, nonbaroreceptor mechanisms are indicated by several lines of evidence. NPY causes an upward shift of the baroreceptor-mediated changes in heart rate (increases maximal responses) in conscious rabbits *(112)* and can evoke bradycardia in the presence of β-adrenergic blockade in conscious rats *(101)*. These effects of NPY could be of central and/or peripheral (direct negative chronotropic interaction with the cholinergic and sympathetic mechanisms) origin. Studies of Potter *(28,113)* have indicated that in anesthetized dogs, NPY, both exogenously administered and released during repeated stimulation of cardiac sympathetic nerves, is able to inhibit vagally induced bradycardia.

Cardiac hemodynamic responses of NPY differ from those of NE (Fig. 17). When both agonists were infused at equipotent pressor doses in conscious rats, stroke volume index was increased by NE and decreased by NPY, whereas cardiac output index was unchanged by NE, but decreased by NPY (Fig. 17). Unlike NE, NPY caused a reduction in left-ventricular dP/dt (myocardial contractility index) and a marked elevation of left-ventricular end-diastolic pressure (LVEDP) in pentobarbital-anesthetized rats, suggesting a negative inotropic effect of the peptide in vivo *(101)*.

The myocardiodepressant effect of NPY was shown in conscious normotensive and hypertensive rabbits, and found to be independent of NPY-induced bradycardia *(114,115)*. However, there are scarce data to indicate a direct negative inotropic action of NPY. Lundberg et al. *(35)* have reported that NPY increases myocardial contraction and heart rate in spontaneously beating guinea-pig atria; however others *(116,117)* were not able to confirm this using the same preparation. NPY has no effect on isolated papillary muscles from cats, guinea pigs, and rats *(116,117)*. Conversely, Balasubramaniam et al. *(118)* have shown that NPY inhibits the contractile force of isolated strips of the rat heart, independently of the reduction in coronary blood flow and heart rate in perfused hearts. In vivo, the negative inotropic effect of NPY may be due to ischemia resulting from coronary vasoconstriction. This was indicated by several studies using Langendorf preparations *(119,120)* and open-chest anesthetized dogs *(116)*, in which intracoronary infusions of NPY caused a decrease in dP/dt preceded by a profound increase in coronary perfusion pressure. Small coronary arteries appear to be particularly sensitive to NPY-induced constriction *(121,122)*. In vivo, NPY markedly reduced coronary blood flow leading to severe myocardial ischemia in patients with angina *(123)* and in dogs *(121)*. Considering the potent vasoconstrictor actions of NPY and its dense innervation of coronary arteries, the peptide may be a mediator of coronary vasospasm, e.g., in Printzmetal's variant angina.

The level of sympathetic activity and interactions of NPY with the release and action of NE additionally influence cardiac effects of the peptide. In pithed rats, which have no sympathetic tone, systemic administration of NPY was found to increase blood pressure, total peripheral resistance, stroke volume, and cardiac output *(103)*. Considering the lack of evidence of direct negative effect of NPY on the heart muscle, the authors *(103)* suggested that NPY raised cardiac output by increasing venous return. This is a likely explanation considering that large veins, e.g., vena cava, possess rich NPY-containing innervation *(76)*, have specific NPY-binding sites *(24)*, and are very sensitive to constrictive effects of the peptide *(124)*. Furthermore, in the areflexic pithed rats, as in isolated cardiac muscles in vitro, NPY may exert direct cardiostimulatory effects. We have recently found that NPY has no cardiodepressant action and actually tends to normalize reduced cardiac output in intact, pentobarbital-anesthesized rats with hypotensive endotoxic shock (Hauser et al., unpublished observation). We speculate, therefore, that in situations when cardiac performance depends to a great extent on cardiac filling, such as in rats with no circulatory reflexes and during endotoxic shock, NPY may have a cardiostimulatory effect.

On the other hand, the negative inotropic effect of NPY in intact rats with active sympathetic tone may be the result of NPY antagonism of cardiac adrenergic responses. To this point, NPY is known to inhibit the release of NE from the heart via a prejunctional effect *(117,119)*, and postjunctionally, it can inhibit β-1 adrenoreceptor-mediated increase in cAMP in cardiac myocytes *(125)*. The sympatho-inhibitory effects of NPY are probably a result of the activation of NPY Y-2 receptors *(see below)*, since they are mimicked by C-terminal fragments of NPY (Y-2 agonists), which also lower cardiac output *(47)*.

3.1.3. Renal Effects

Like NE, NPY causes renal vasoconstriction both in vivo *(100,126)* and in vitro *(127)* independently of α-adrenoceptors (Fig. 18). Both transmitters are abundantly present in the renal sympathetic nerves innervating renal tissues and blood vessels. High-density NPY-specific binding sites were found in the rabbit kidneys in the areas of proximal convoluted tubules and blood vessels *(128)*. Specific NPY-binding sites were either not detectable or were of much lower density in rat and human kidneys *(129)*. Interestingly, however, rats, but not rabbits, appear to be sensitive to the natriuretic effects of NPY *(126)*. Using isolated perfused kidney as a model, Allen et al. *(127)* have found that NPY dose-dependently reduces renal blood flow, without affecting glomerular filtration rate, and elicits a marked natriuretic response (whereas NE causes sodium reabsorption). The natriuretic effect of the renal vasoconstrictor peptide was reproduced in anesthetized uninephrectomized rats *(130)*.

Pfister et al. *(131)* reported that NPY, at nonpressor doses, suppresses renin secretion in rat models with elevated plasma renin activity (adrenalectomy + DOCA therapy or captopril treatment). Subsequently, chronic infusions of nonpressor doses of NPY were shown to prevent the development of renal hypertension in rats, in parallel with the reduction in renin release *(131,132)*. The renin-inhibitory effect appeared to be independent of renal vasoconstriction, since NPY inhibited renin release in vitro, in tissue pieces from hydronephrotic kidneys, by a mechanism sensitive to pertussis toxin and dependent on adenylate cyclase, probably localized to renin-producing cells *(133)*.

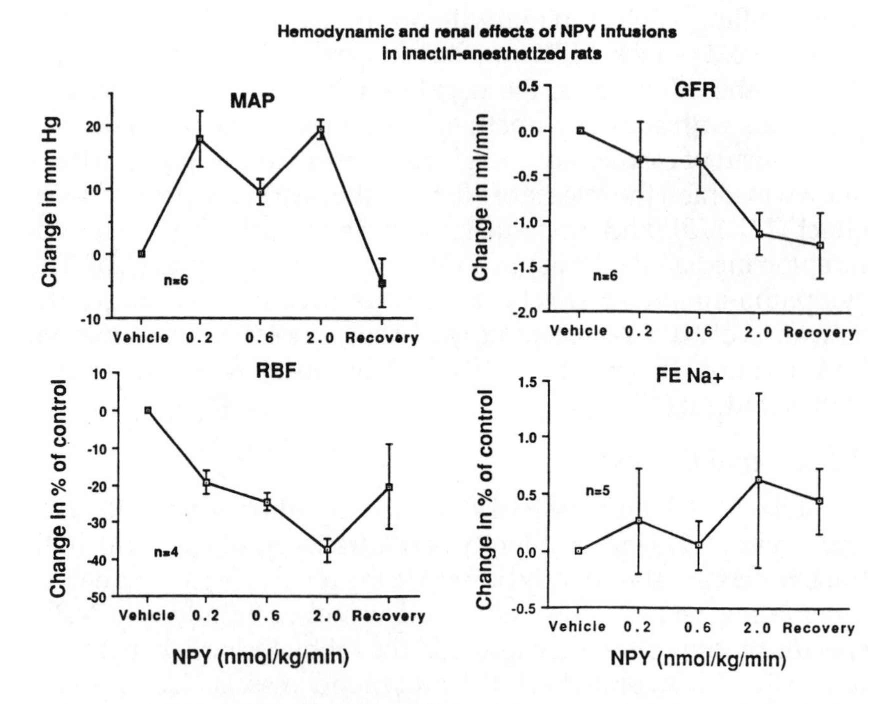

Fig. 18. NPY and the kidneys: Effects of cumulative NPY infusions on renal blood floow (RBF), glomerular filtration rate (GFR), and fractional excretion of sodium (FE Na^+) in anesthetized rats.

Whether the renin-suppressive effect of NPY is associated with natriuresis has not been established yet. Furthermore, recent studies challenge the inhibitory effect of NPY on renin secretion. In animal models where renin release was not stimulated, NPY had variable effects on renin levels dependent on renal hemodynamic responses to the peptide *(126,134)*. With the exception of the anesthetized cat where basal plasma renin activity was shown to decrease during infusions of NPY at nonpressor doses *(135)*, other studies have failed to demonstrate this. In anesthetized primates *(126)*, NPY administered systemically or intrarenally, did not change plasma renin activity or renin secretion rate at any doses tested. In normotensive anesthetized rats, NPY infused at nonpressor doses did not alter glomerular filtration rate, urine volume, or sodium excretion (Cement et al., unpublished observations from our laboratory). Pressor doses of NPY dose-dependently reduced renal blood flow and glomerular filtration rate without significantly changing sodium excretion (Fig. 18); the effects were similar to those of NE, although NPY was a stronger vasoconstrictor on a molar basis. Thus, it appears that inhibition of renin release by NPY is limited to the states where renin secretion is elevated. In all other conditions, the strong hemodynamic action of NPY, renal vasoconstriction, is a predominant factor modulating renin release.

3.2. Vascular Actions of NPY

3.2.1. Direct Vasoconstriction

NPY constricts arteries and veins by a direct, receptor-mediated action on vascular smooth muscle. NPY-induced vasoconstriction is long-lasting, resistant to α-adrenoceptor antagonists *(99,100)*, and is blocked by removal of extracellular calcium *(136–138)* and by calcium-channel blockers *(100,137,139)*. Not all blood vessels are responsive to NPY (*see* Table 1). In many large vessels, e.g., human mesenteric arteries *(157)*, rabbit pulmonary artery *(158)*, and rat aorta *(155)*, NPY is essentially without a constrictive effect. NPY causes potent vasoconstriction in cerebral *(136)*, coronary *(108,121)*, mesenteric and renal *(100)*, and skeletal muscle *(159,160)* blood vessels when applied at nanomolar concentrations. Small, more so than large, human *(123)* and canine *(110,121)* coronary arteries are extremely sensitive to NPY-induced direct vasoconstriction. NPY reduces coronary blood flow in blood vessels from <100 μm to >200 μm in diameter *(110)*, and does so at the concentrations that appear to be physiological (p*M*–n*M*). In the smallest coronary arterioles, the constrictive effect of NPY may actually overcome the autoregulatory vasodilation of these vessels *(110)*. On a molar basis, the potency of NPY as a vasoconstrictor is the greatest in vivo; it is 25-fold greater than NE in reducing cat colonic blood flow *(104)*, threefold greater than NE in rat mesenteric and renal arteries *(100)* and canine splenic arteries *(161)*, and twofold lesser than endothelin in human epicardial coronary vessels *(144)*. Removal of endothelium does not affect the constrictive effect of NPY in blood vessels from humans and other species *(159)*.

Large veins, unlike large arteries, are quite sensitive to the vasoconstrictive effects of NPY. Hence, guinea-pig iliac *(43,162)* vein and vena cava *(124)* are among the most sensitive vascular preparations. The rank order of sensitivity to vasoconstrictor effects of NPY in blood vessels appears to parallel their dependence on extracellular calcium for the contractile response, thus further

Table 1
Contractile Effects of NPY on Arterial and Venous Blood Vessels of Different Species

Arteries			
Cerebral			
Cerebral arteries	Cat	VC + potentiation	*136,140*
	Rabbit	Potentiation	*96*
Basilar artery	Rabbit	No VC	*141*
	Guinea-pig	VC	*142*
Middle cerebral artery	Rat	VC	*143*
Coronary			
Small coronary arteries	Humans	VC	*123*
In vivo	Dog	VC	*109*
			121
In vitro	Dog	VC	*108*
			110
	Guinea pig	VC	*120*
Coronary arteries	Rabbit	+/–VC + potentiation	*111*
Epicardial coronary arteries	Humans	VC	*144*
Pulmonary			
Pulmonary arteries, in vivo	Dog	No VC	*145*
Pulmonary artery, in vitro	Rabbit	No VC; potentiation	*146*
Pulmonary, bronchial, in vitro	Pig	VC	*107*
Skeletal muscle			
Forearm blood flow	Humans	VC	*37*
In vitro	Humans	VC	*39*
	Rabbit	VC	*39*
In vitro	Dog	VC + potentiation	*147*
Femoral artery	Rabbit	No VC; potentiation	*148*
Splanchnic			
Renal artery, in vivo	Rat	VC	*100*
Renal artery, in vitro	Human	VC	*46*
Sup. mesenteric artery, in vivo	Rat	VC	*100*
Sup. mesenteric artery, in vitro	Rat	+/–VC + potentiation	*149*
		VC	*39*
Mesenteric arteries, in vitro	Human	VC, no potentiation	*150*
			39
Mesenteric arterioles	Rat	+/–VC + potentiation	*138*
Gastroepiploic artery	Rabbit	No VC; potentiation	*151*
Other			
Middle ear artery	Rabbit	No VC; potentiation	*152*
Tail artery	Rat	No VC; potentiation	*153*
	Rat	VC + potentiation	*154*
Aorta	Rat	No VC; potentiation	*155*
Omental arteries	Rat	No VC; potentiation	*140*
Uterine (pregnant)	Guinea pig	No VC	*156*
Submandibular artery	Cat	VC	*99*
Skeletal muscle	Human	VC; potentiation	*39*
			43
Veins			
Iliac vein	Guinea pig	VC	*148*
Vena cava	Guinea pig	VC	*124*
	Rat	VC + potentiation	*124*
Pial veins	Cat	VC, no potentiation	*141*
Mesenteric veins	Humans	VC	*157*
			150
Femoral vein	Rat	+/–VC, no potentiation	*146*
Omental veins	Rat	+/–VC, no potentiation	*140*

VC—vasoconstriction.

supporting the notion that NPY-induced vasoconstriction is Ca^{2+} influx dependent *(163)*. Calcium may also be required for binding of NPY to its surface receptors, which may be inhibited by calcium-channel blockers *(163)*.

Direct vasoconstrictor effects of NPY are probably mediated by vascular Y-1 receptors. Specific, high-affinity NPY-binding sites were found in membrane preparations of blood vessels *(164)* and on cultured vascular smooth muscle cells *(40)* (Fig. 19). Wahlestedt et al. *(43)* were the first to suggest that the binding sites for NPY occur in two separate forms, designated Y1 and Y2. The distinction was based on different binding profiles, and effects of NPY/PYY and long C-terminal fragments of NPY/PYY, e.g., NPY_{13-36} *(43)*. This subclassification has been subsequently confirmed by others *(113,165)* and with the use of the selective Y1 agonist, $[Leu^{31}Pro^{34}]$NPY *(166)*. The selective Y1 agonist has proven to be more potent than equimolar doses of NPY in increasing blood pressure *(167)*, indicating that vasoconstrictor receptors are predominantly of the Y1 type. The vascular effects of NPY in vivo appear to be a mixture of the Y1-mediated pressor effect and the Y2-mediated depressor effect.

NPY given iv at a higher dose (10–300 nmol/kg) causes a biphasic pressor–depressor response in conscious and pithed rats (Fig. 20; *48,124*). The hypotensive response resembles hypotension elicited by C-terminal fragments NPY_{18-36} *(168)* and NPY_{22-36} *(48)* (Fig. 20 and *see* Section 3.2.3.). Vasodepressor effects of NPY and NPY-related fragments are indirect, mediated primarily by the release of histamine from mast cells (*48,124*; Fig. 20) and, probably in part, by presynaptic inhibition of NE release via Y2 receptors. Although in vivo activation of Y2 receptors is associated with hypotension, the Y2 receptor subtype is not, in a true sense, a "vasodilator" receptor, since none of the C-terminal fragments of NPY that activate these receptors cause vasodilation in vitro (unpublished observation from our laboratory). In contrast, in some vascular preparations, such as the cutaneous circulation of the rat, where Y2 receptors may occur postsynaptically on the vascular smooth muscle, their activation leads to vasoconstriction *(113)*; elsewhere, C-terminal fragments of NPY do not cause vasoconstriction. Generally, Y2 agonists do not potentiate other vasoconstrictor actions.

In our recent study, we found that cultured rat aortic smooth muscle cells express only Y1-type receptor binding, whereas cells derived from rat vena cava have a mixed population of Y1 and Y2

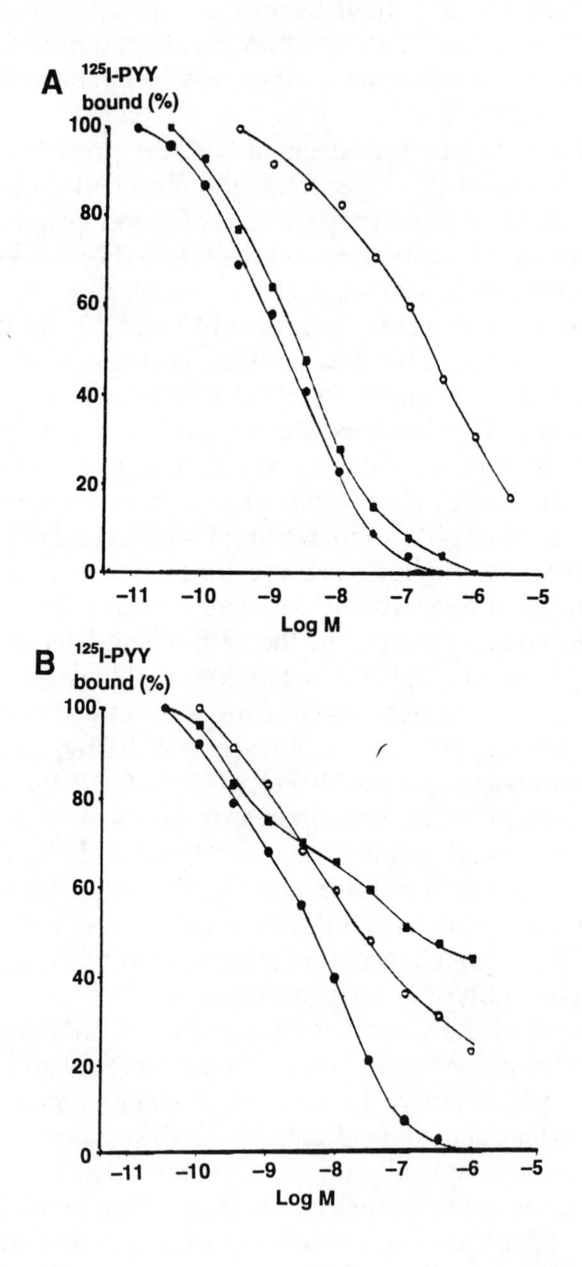

Fig. 19. NPY-binding sites on cultured smooth muscle cells derived from the rat aorta (A) and vena cava (B). ● NPY 1–36; ■ [Pro^{34}]NPY; ○ NPY 13–36.

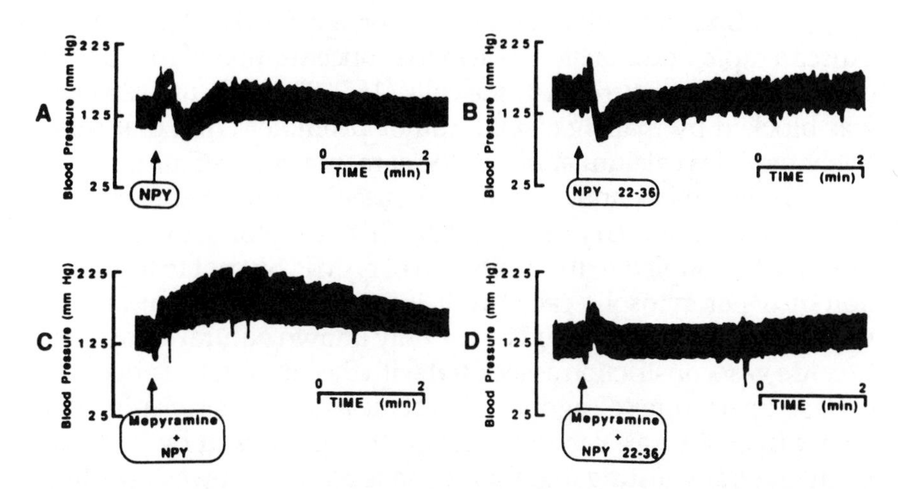

Fig. 20. Blood pressure responses to intravenous administration of high dose (300 nmol/kg) of NPY and $NPY_{22\text{-}36}$ in the presence and absence of mepyramine (histamine H1 receptor blocker) in conscious rats. Hypotensive effects of both peptides was blocked by the H1 antagonist *(48)*.

receptors (*40*; Fig. 19). Interestingly, the vena cava is more sensitive than the aorta to the direct vasoconstrictive effect of NPY, whereas the aorta responds primarily to the potentiating effect of this peptide *(124)*. Certainly, heterogeneity of NPY receptors and NPY interactions with nonmuscle cells (neuronal, mast cells) lead to complex hemodynamic actions of the peptide.

The transmembrane signaling mechanisms that mediate the actions of NPY on vascular smooth muscle are not completely understood (Fig. 6). It was recognized early that NPY receptors are coupled to G-proteins (G_i or G_o), since most NPY actions were blocked by pertussis toxin *(133,138)*. Initially, it was thought that the Y1 receptor may couple to phosphatydylinositol (PI) hydrolysis and mobilization of intracellular Ca^{2+}. However, it is now well accepted that most, if not all, NPY receptors are coupled to the inhibition of adenylate cyclase, and hence, their activation decreases cAMP levels *(167,169)*, similar to that of α-2 adrenoceptors. NPY decreases forskolin-stimulated cAMP levels in neuronal cells bearing only Y1 receptors and in those bearing only Y2 receptor cells *(167)*. However, only Y1 receptor activation appears to induce a rise in cytosolic Ca^{2+} concentration apparently from the IP_3-insensitive pool *(167)*. Similar to NPY, the Y1 receptor agonist, [Leu^{31},Pro^{34}]NPY causes a rapid and transient rise in the concentration of free Ca^{2+} in cultured human neuroblastoma cells *(167)*. This calcium response was blocked by thapsigargin, a tumor promoter *(167)* that selectively depletes calcium stores in the sarcoplasmic reticulum.

On the other hand, in cultured vascular smooth muscle cells from rabbit pulmonary artery *(169)*, NPY receptor activation was found to be coupled to the inhibition of cAMP, but not to the elevation of IP_3 or cytosolic calcium. If this is the mechanism of NPY vascular action, NPY would be the only known natural compound to cause vasoconstriction associated with the inhibition of adenylate cyclase, apart from α-2 agonists. In rat mesenteric arterioles in vitro, NPY causes a weak vasoconstriction that is blocked by pertussis toxin *(138)*, suggesting that this action is mediated by a G-protein-linked receptor.

Some recent studies *(142)* have shown that NPY may cause vasoconstriction (e.g., in guinea-pig basilar artery) by evoking slow depolarization, in a mechanism independent of the endothelium *(170)*. NPY-induced membrane depolarization of cerebral vessels occurred at low-nanomolar concentrations and lasted longer than that evoked by other vasoconstrictors, e.g., NE or histamine *(142)*. Fallgren et al. *(142)* have suggested that NPY constricts cerebral arteries, at least in part, because of a long-lasting depolarization linked to voltage-operated calcium channels, probably of the T-type *(171)*, and in part because of receptor-operated calcium channels. In agreement with this mode of action, NPY-induced vasoconstriction is blocked by calcium-channel blockers and removal of extracellular calcium *(136,137)*. The Y2 receptor subtype may be the type linked to the voltage-sensitive calcium channels *(167)*.

3.2.2. Potentiation of Other Vasoconstrictor Actions

Both in vitro in isolated blood vessels (responsive and unresponsive to NPY) and in vivo in the pithed rat, NPY enhances vasoconstriction induced by sympathetic nerve stimulation *(172)* and α-adrenergic agonists *(160,172)*, indicating that the effect is postjunctional. The amplification is observed at nanomolar NPY concentration, which does not evoke vasoconstriction *per se*. NPY may interact with both α-1 and α-2 adrenergic receptors. In the brain, NPY primarily interacts with α-2 adrenoceptor mechanisms. In the medulla oblongata, the distributions of NPY and α-2 receptors largely overlap and interact both biochemically and functionally, suggesting that these receptors may be expressed in a "coupled" form *(173)*. In blood vessels, NPY potentiates primarily α-1 adrenoceptor-mediated vasoconstriction (Fig. 21A; *174,175*) and pressor responses *(172)*. In addition, this phenomenon occurs also with epinephrine, histamine *(141,151,158)*, angiotensin *(175)*, serotonin, and potassium *(176,177)*; however, in other blood vessels (rabbit basilar artery) NPY failed to enhance vasoconstriction induced by the last two agonists *(141)*. In contrast, vasoconstrictor and pressor responses to vasopressin and an α-2 agonist, B-HT933 *(175,178)* are not augmented by NPY. Therefore, in the peripheral blood vessels, NPY appears to potentiate a mechanism common to receptors coupled to phosphatidylinositol pathway rather than to adenylate cyclase.

NPY has more variable effects on vasoconstriction induced by nerve stimulation. In nonvascular tissues, such as vas deferens *(43,179)*, NPY potently inhibits contractile responses to nerve stimulation as a consequence of the presynaptic inhibition of NE release. In vascular tissues, this action varies among the vessels, and appears to be limited to lower concentrations of NPY ($10^{-9}M$), whereas at higher concentrations of the peptide, the postsynaptic enhancement of vasoconstriction predominates *(160,180)*. The presynaptic inhibitory effect of NPY is resistant to calcium channel blockers *(39,46)*. However, which component(s) of the pressor response to nerve stimulation that NPY potentiates is not clear. We did not detect potentiation by NPY of pressor responses to sympathetic nerve stimulation in the pithed rats *(100)*, whereas Dahlof et al. *(172)*, using a similar experimental model, have found that NPY augments these responses. The differences might have resulted from different modes of stimulation: Dahlof et al. *(172)* used shorter stimulation periods than ours. Similarly, Glover *(181)* reported that NPY potentiates vasoconstriction in response to brief (5-s) nerve stimulation.

According to Vu et al. *(153)*, who studied the rat tail artery, NPY causes greater enhancement of vasoconstriction evoked by short rather than by long trains of stimulation (300 vs 30%). On the other hand, NPY-induced potentiation of vasopressor effects of α-1 agonists is consistently found to be in a range of 25–30% (*146,182*; Fig. 21C). The preferential enhancement of responses to short trains of stimulation may be due to NPY's interactions with other sympathetic nonadrenergic transmitters, such as purines

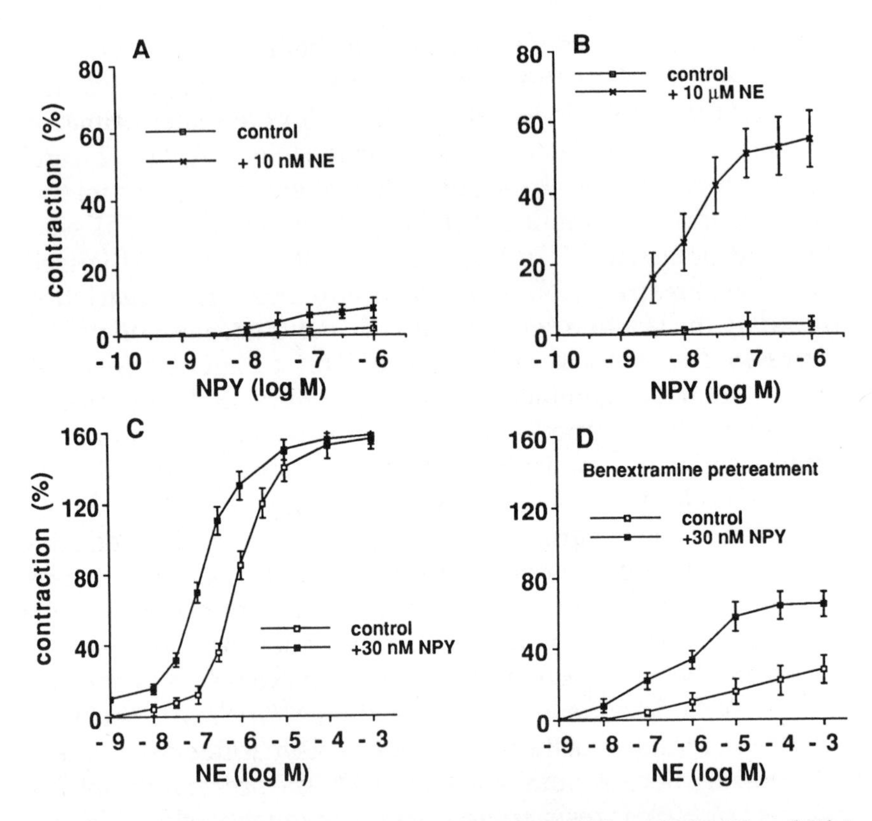

Fig. 21. Contractile responses to NPY (A and B) and NE (C and D) in rabbit isolated pulmonary artery. A: response to NPY in absence and presence of 10 n*M* NE (a threshold NE concentration, elevating tension <5% of maximal KCl, applied 2 min before NPY). B: markedly enhanced NPY responses after exposing blood vessel to 30 μ*M* NE for 30 min, with a addition of 100 μ*M* phentolamine and 1 μ*M* propranolol 10 min before NE (at the time of NPY application, resting tension at baseline level). C: response to NE in the absence and presence of NPY (applied 2 min before NPY) in naive, untreated blood vessel; addition of NPY is manifested as a left shift (potentiation) of NE concentration–response curve. D: responses to NE in blood vessel pretreated with benextramine (an irreversible α-1 adrenoceptor blocker, 100 n*M* for 60 min, followed by extensive washing), in the presence and absence of 30 n*M* NPY (applied 2 min before NE). Thus, when α-adrenoceptor reserve was eliminated, NPY no longer induced a left shift of NE concentration–response curve. Instead, NPY enhanced response to all NE concentrations (compare with A and B). Contraction is expressed in percentage of response to 137.7 m*M* KCl *(146)*.

(183). This explanation, however, is contradicted by the fact that the pressor and vasoconstrictor responses to nerve stimulation, obtained both in the presence and absence of NPY, are completely eliminated by a selective blockade of α-1 adrenoceptors; therefore, they are not apparently purine dependent *(153)*. Finally, it has been suggested that NPY increases the initial, phasic component of vasoconstriction more than it affects the slower, tonic component *(153)*. This action of NPY may be related to its effect on the mobilization of intracellular calcium *(146)*, which is believed to be associated with the phasic component of the contraction *(184)*. In support of this notion, the potentiation of NE-induced vasoconstriction by NPY was shown to be less affected than the direct vasoconstrictor effect of NPY by manipulations on extracellular calcium, but it was abolished by the depletion of intracellular calcium stores *(158)*.

However, other studies *(59,157,172,178)* indicated that NPY preferentially potentiates the component of α-1 adrenergic vasoconstriction that is sensitive to calcium-channel blockers. In some blood vessels, calcium-channel blockers weakly attenuate NPY-induced potentiation of adrenergic vasoconstriction while blocking the direct vasoconstrictor effect of NPY *(152,158)*. Therefore, it cannot be concluded that calcium influx is causally related to NPY-induced potentiation. "Positive" effects of calcium-channel blockers in some vessels may rather be due to their actions on vasoconstriction evoked by NE itself, in vessels highly dependent on Ca^{2+} influx.

The nonspecific, yet selective, nature of NPY-induced potentiation of vasoconstriction (e.g., excluding α-2 adrenergic and vasopressin) suggests that it occurs not at the receptor level, but at some postreceptor sites or that it is mediated by depolarization of vascular smooth muscle cells. In rabbit cerebral arteries, NPY evokes marked depolarization *(185)*. Similarly, histamine and serotonin *(186,187)*, and ouabain and low concentrations of potassium *(185)* cause membrane depolarization while potentiating contraction to NE and other vasoconstrictors. NPY-induced potentiation is prevented by prior depolarization with ouabain *(185)*. Since NPY appears to potentiate the effects of phosphatidylinositol pathway-coupled agonists *(185)*, it is possible that depolarization of vascular smooth muscle cells by NPY (and other agonists) synergistically interacts with the activation of Ca^{2+}-phospholipid-dependent protein kinase C. We *(146)* and others *(169)* have found that NPY has

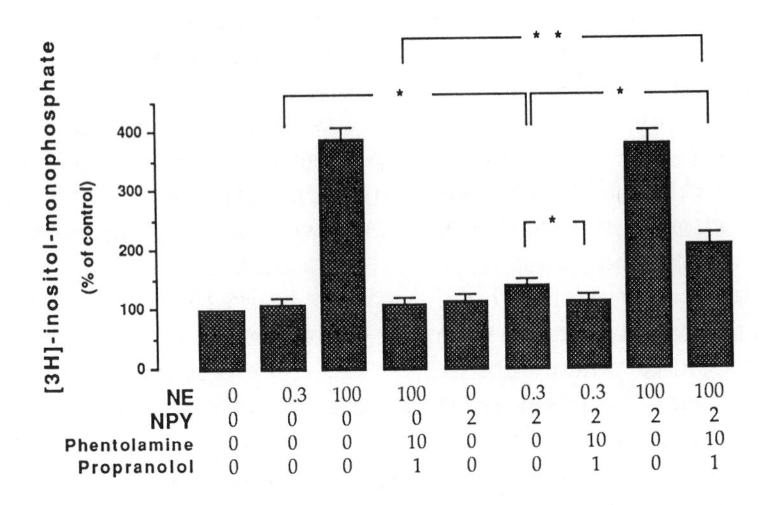

Fig. 22. [^{3}H]inositol monophosphate (IP) accumulation in pieces of rabbit pulmonary artery. Protocol corresponded to experiments on motor activity shown in Fig. 21, A and B. NE stimulated IP accumulation, whereas NPY was ineffective. When NPY was applied after a threshold concentration (0.3 μ*M*) of NE, an increase in IP accumulation was detected. A stronger response to NPY was observed when blood vessel was pretreated with 0.1 m*M* NE (in the presence of phentolamine to completely block NE response; propranolol was routinely applied with phentolamine). Values are means ±SE of 12–18 specimens from three separate experiments *(146)*.

no effect on inositol phosphate accumulation by itself, but potentiates the increase in IP_3 in response to NE (Fig. 22). However, there is no evidence as yet that NPY-induced potentiation of adrenergic vasoconstriction involves enhancement of intracellular calcium release by sarcoplasmic reticulum, as indicated by studies using calcium imaging in vascular smooth muscle cells *(188)*.

In addition to direct interactions of NPY with other vasoconstrictors, the potentiating effect of NPY may involve some indirect actions and other vasoactive substances. In this regard, NPY-induced potentiation was found to be independent of neuronal uptake *(158)*, and some *(189)* have found it independent of the presence of endothelium in the rabbit ear artery. Interestingly, another group of investigators *(190,191)* using the same blood vessel as a model, showed that endothelium removal almost completely abolished the potentiating effect of NPY on vasoconstriction induced by electrical field stimulation and exogenous NE; the reason for this discrepancy is not apparent. On the other hand, NPY appears to interfere with endothelium-dependent relaxation, since the peptide potently inhibits relaxation induced by adenosine and acetylcholine in rabbit cerebral arteries *(185)*, and by vasoactive intestinal peptide in guinea-pig uterine artery *(192)*. Therefore, the inhibition of relaxation by NPY may also contribute to its ability to augment vasoconstriction.

Prostaglandins may be partially involved in NPY-induced sensitization to adrenergic vasoconstriction *(178)*. However, although NPY potentiates both renal output of prostaglandin E_2 and $PGF_{2\alpha}$ and vasoconstriction in response to NE, it potentiates angiotensin action without altering prostaglandin synthesis *(178)*, thus, indicating that other pathways may be involved in the sensitizing effect of NPY. Both the potentiating and contractile effects of NPY were shown to be nonspecifically antagonized by eicosapentanoic acid, independently of prostaglandins *(193)*.

3.2.3. Sympatho-Inhibitory and Hypotensive Effects of NPY

As mentioned above, NPY (and peptide YY) acts on two distinct types of receptors, Y1 and Y2, associated with different actions in the cardiovascular system (Figs. 5 and 6). Initially, the Y2 receptors were presumed to be prejunctional, analogous to α-2 adrenoceptors. However, this is probably an oversimplification, and evidence for postjunctional Y2 receptors, e.g., in coronary blood vessels *(120)*, is forthcoming. The Y2 receptor in rat hippocampal and rabbit kidney membranes was recently affinity labeled and found to be a glycoprotein with a $M_r = 50{,}000$ binding subunit *(194)*.

NPY and C-terminal fragments of NPY (NPY_{18-36}, NPY_{22-36}) given iv at a high doses (>10 nmol/kg) cause vasodepressor responses in conscious and pithed rats *(48)* (Fig. 20). We have recently discovered that these peptides release histamine from mast cells, and their vasodepressor effects are completely prevented by an H1 histamine receptor blocker (Fig. 20) or mast cell degranulation *(48,195)*. Whether or not hypotension induced by high doses of NPY and the Y2 agonists is mediated by Y2 receptors on mast cells is not clear. This action may be due to the high content of positively charged amino acids in the molecules of NPY and its related fragments, since the ability of releasing histamine is shared by other basic peptides *(196)*.

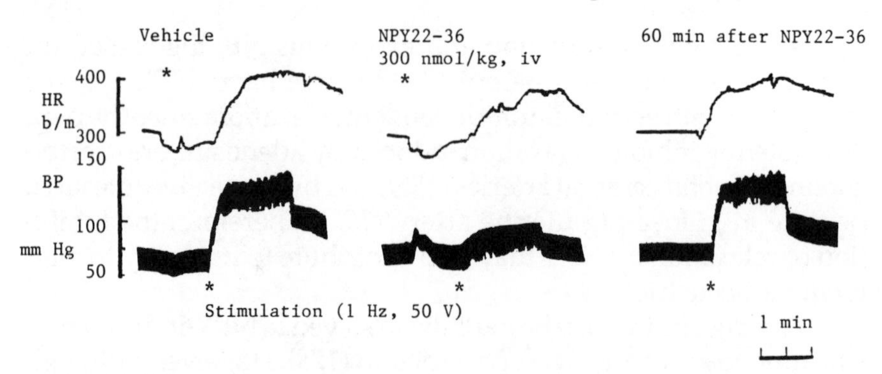

Fig. 23. Blockade by NPY_{22-36} of pressor responses to stimulation of total sympathetic outflow in vagotomized pithed rats.

The hypotension evoked by NPY and the Y2 agonists is mediated by a marked decrease in cardiac output, not by a decrease in systemic vascular resistance *(47)*. As in the case of NPY_{1-36} *(58)*, the cardiodepressant effect of NPY_{18-36} *(47)* does not appear to be due to a direct negative inotropic action. In their recent study in conscious rats, Scott et al. *(47)* have suggested that the reduction of cardiac output by NPY_{18-36} is mediated by coronary vasoconstriction; such an action of C-terminal fragments was shown in guinea-pig hearts *(120)*. Alternatively, hypotension may result from the sympatho-inhibitory action of these peptides via the presynaptic Y2 receptors. In the isolated perfused rat mesenteric bed, C-terminal fragments inhibit both NE release and pressor responses evoked by periarterial nerve stimulation *(149)* similar to the effect of NPY and PYY. To what extent the presynaptic sympatho-inhibitory action of NPY and Y2 agonists contributes in vivo to hypotension is questionable, since plasma levels of catecholamines (and other vasoconstrictors) actually rise because of hemodynamically mediated baroreceptor reflex inhibition *(197)*. In the pithed rats, however, the sympatho-inhibitory effect of NPY_{22-36} is very apparent and results in a dose-dependent reduction of pressor responses to stimulation of the total sympathetic outflow, leading to a complete blockade at the highest dose of 300 nmol/kg (Fig. 23, unpublished observations from our laboratory); whether this is because of presynaptic or ganglionic blockade is a subject of current investigations.

Recently, Brown et al. *(198)* have suggested that NPY_{18-36} may act as an NPY antagonist postsynaptically, since it prevents pressor response to NPY_{1-36} in awake rats pretreated with a H1 histamine blocker. This action of NPY_{18-36} does not appear to be a true NPY receptor antagonism, but rather is due to opposite physiological actions via different receptors. Other C-terminal fragments of NPY, e.g., NPY_{22-36} (also a Y2 agonist), do not seem to share this property.

3.2.4. NPY Interaction with the Adrenergic System

Although most isolated (large) blood vessels are poorly constricted by NPY when compared with NE, NPY is a potent vasoconstrictor in vivo, more potent than NE in many vascular beds. This discrepancy may be the result of the lack of adrenergic tone in isolated vessels. It has been long known that low concentrations of circulating NPY enhance NE pressor effects *per se*. However, we *(58,146)* and others *(199)* have found that this phenomenon is reciprocal; thus, NE affects NPY mechanisms in a similar fashion. Our findings on these interactions are schematically illustrated in Fig. 24 and described below.

The first indication that NPY efficacy as a vasoconstrictor depends on the state of adrenergic tone came from the studies of conscious rats undergoing stress testing. We have found that NPY-induced pressor responses are markedly enhanced in stressed rats with elevated levels of circulating catecholamines *(58)*. Similarly, prolonged infusion of NE to conscious nonstressed rats resulted in marked augmentation of NPY pressor responsiveness (sensitization by 150%; Fig. 24A; *146*). At the same time, NE infusion caused a (homologous) desensitization to pressor effects of NE (a 50% reduction of the pressor response to NE at ED_{30} injected as a bolus during the infusion, Fig. 24A). Addition of NPY, at a low pressor concentration, to the NE infusion resulted in a reversal of desensitization of NE pressor effects (resensitization, Fig. 24A). Interestingly, prolonged infusion of NPY also evoked desensitization of pressor responses to NPY (by 40%), but this was not altered by addition of NE to the infusate (Fig. 24B). Thus, there is a postjunctional crossmodulation between the transmitters, and both NPY and NE amplify each other's actions while desensitizing their own effects. However, only NPY appears to be able to reverse adrenergic desensitization, and not the converse.

We have observed a similar phenomenon in vitro in several isolated blood vessels. Naive rabbit pulmonary artery, which hardly responds to NPY at all (Fig. 21A), becomes strongly responsive

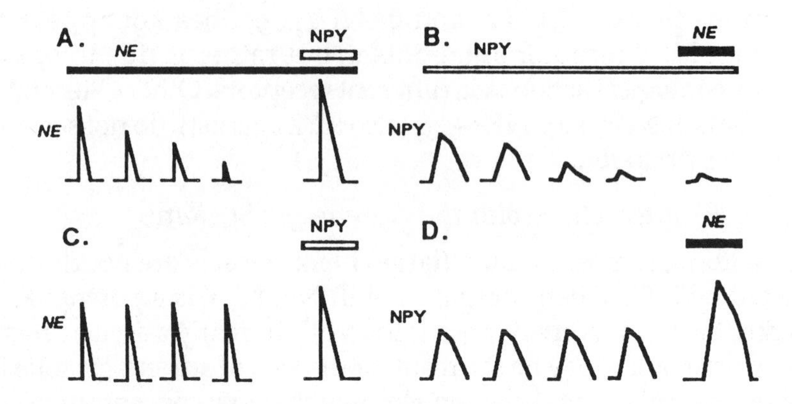

Fig. 24. NE desensitization–NPY sensitization phenomena. Panel A: Adrenergic desensitization—in conscious rats, a prolonged infusion of an ED_{50} pressor dose of NE (filled horizontal bar) developed desensitization to the vasopressor effects of bolus-injected NE (shown as sharp increases in mean arterial pressure). The addition of NPY at a low pressor concentration (open horizontal bar) reversed adrenergic desensitization. Panel B: NPY desensitization—NPY infusion (open horizontal bar) induced desensitization of its own effect, as illustrated by the reduced pressor responses to bolus injections of the peptide (increases in pressure less sharp than those of NE). In contrast to adrenergic desensitization, the addition of NE to the infusate did not resensitize the NPY responsiveness (note: the scheme does not reveal that NPY desensitization was more difficult to achieve than NE desensitization). Panel C: Naive rats—bolus injections of NE did not result in desensitization to CA. The infusion of NPY moderately enhanced the NE response. Panel D: Naive rats—the pressor responses to bolus injections of NPY (ED_{30}) did not desensitize, but were markedly (150%) enhanced by NE infusion *(146)*.

following prolonged exposure of the vessel to relatively high concentrations of NE (Fig. 21B) (or histamine, data not shown), even in the presence of α-1 adrenergic blockade with prazosin *(146)*. Thus, paradoxically, vascular responsiveness to NPY increases in the desensitized ("exhausted") vessels. Whether this phenomenon is related to the upregulation of NPY receptors by the other agonists or to the altered stimulus-response coupling needs to be determined. The NE-induced enhancement of NPY-vasoconstrictor activity consisted of two components. The first component was induced by a threshold concentration of NE (Fig. 21A) and was completely blocked by the α-adrenoceptor antagonist and Ca^{2+}-

entry blocker. The second component required longer preexposure to high (<30 μ*M*, Fig. 21B) NE, and was unaffected by Ca^{2+}-entry blockers, by pharmacological blockers of adrenergic, serotonergic, histaminergic, and dopaminergic receptors, by cyclo- and lipoxygenase inhibitors, or by the removal of endothelium. NE markedly enhanced the efficacy of NPY (Fig. 21B). The magnitude of this enhancement corresponded to the one observed in vivo in conscious rats, and contrasted the moderate effect that NPY had on NE-induced responses (a shift to the left of the NE dose–response curve; Fig. 21C). However, when the α-receptor reserve was removed in rabbit pulmonary artery by the irreversible α-antagonist benextramine (Fig. 21D), NPY also increased the maximum response to NE.

We have proposed that threshold synergism accounts for this cooperativity. Receptor–receptor interactions appear unlikely, since neither NE nor NPY crossregulated adrenergic and NPY receptor numbers and affinities *(146)*. Also, the phenomenon showed a certain nonspecificity (NE could be exchanged for histamine). The common denominator of the interacting agonists seems to be that they all raise intracellular Ca^{2+} levels: NE, by activating phospholipase C with subsequent phosphatidylinositol hydrolysis (Fig. 22) and NPY by a yet unknown pathway involving intracellular Ca^{2+} mobilization. It seems that "crosstalk" between NPY and other vasoconstrictor receptors is between the activation of Ca^{2+}-phospholipid-dependent protein kinase C and the NPY-mediated inhibition of cAMP. Such a synergistic interaction between intracellular signaling systems was previously described in other systems, e.g., rat pinealocytes, for α-1 and β-adrenoceptors *(200)*.

A recent report by Michel et al. *(201)* appears to support our theory of "NE desensitization–NPY sensitization." Using the human neuroblastoma SK-N-MC cell line, these authors found that the preincubation of cells with isoproterenol reduced the number of β-1-adrenoceptors (downregulation) while sensitizing the NPY-stimulated mobilization of intracellular Ca^{2+} *(201)*. In this last model, the desensitization–sensitization phenomena occurred between two distinct receptors of which one, the β-1, is linked to activation and the other, the Y1, to inhibition of adenylate cyclase.

We have concluded from our studies that vascular responsiveness to NPY increases in situations with high sympathetic activity associated with blunted NE responses. Since this phenomenon occurs both in vitro and in vivo, it may represent an important mechanism of amplified vasoconstriction in certain conditions when, following increased release and/or actions of catecholamines, adrenergic desensitization develops. Such conditions may be induced therapeutically by prolonged infusions of adrenergic agonists, e.g., during septic shock *(202)*, or may occur physiologically during prolonged stress or in disease states, such as in spontaneously hypertensive (SHR) rats exposed chronically to high levels of plasma catecholamines *(203)*. Finally, the fact that blood vessels in vitro are largely insensitive to vasoconstrictor action of NPY whereas in vivo they are sensitive (e.g., rat mesenteric arteries) may indicate that NPY, as a weak agonist, requires that the vascular smooth muscle is "primed" by other vasoconstrictors that remove the threshold for the NPY action.

3.2.5. NPY as a Growth Factor for Vascular Smooth Muscle

Growth factors are a diverse group of substances. Sympathetic nerves, catecholamines *(204)*, several vasoconstrictor peptides such as angiotensin and vasopressin *(205)*, and substances released from platelets, such as platelet-derived growth factor *(206)*, are known to stimulate vascular smooth muscle cell growth. To date, there is no indication that NPY belongs to that group, although the peptide is a good candidate for a growth factor: It is derived from sympathetic nerves and platelets, and it is a vasoconstrictor. Recently, we have studied the effects of NPY on primary cultures of vascular smooth muscle cells derived from the rat aorta and vena cava. These vessels were chosen because of their responsiveness to NPY with either vasoconstriction (vein) or potentiation of vasoconstriction (aorta), and the presence of specific NPY receptors (Fig. 19).

In both types of cultured vascular smooth muscle cells, NPY stimulated growth when cells were subcultured in medium containing 0.5% fetal calf serum. Forty-eight-hour treatment with NPY increased the number of cells and [^{3}H]-thymidine incorporation into the cells in a dose-dependent manner, being maximal at the highest dose tested, $10^{-6}M$ (Fig. 25). Similarly, NE stimulated cell growth by increasing [^{3}H]-thymidine uptake at $10^{-5}M$. These findings suggest that the peptide increases DNA replication with cell division, thus leading to hyperplasia. This action may be in line with the type of medial "hypertrophy" that is observed in muscular blood vessels of spontaneously hypertensive rats *(207)*, which can be prevented by neonatal sympathectomy. It has long been

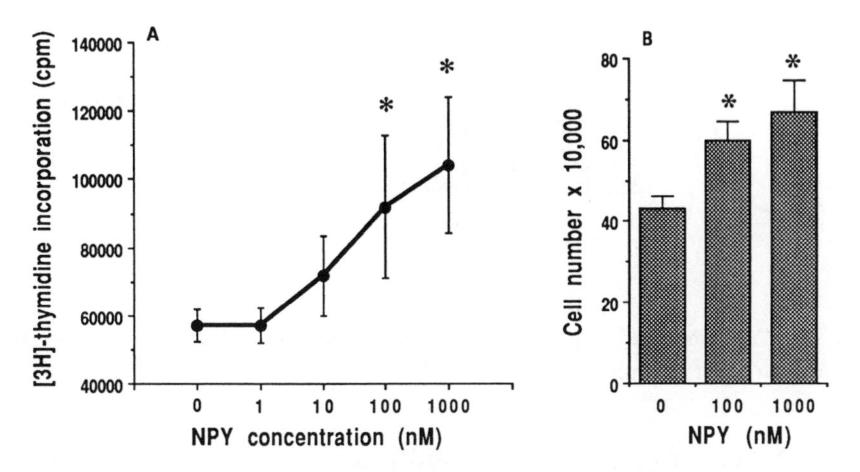

Fig. 25. Mitogenic effect of NPY in the primary culture of rat aortic smooth muscle cells (passage 3–7). NPY increased [^{3}H]thymidine incorporation and cell number (more potently than NE, data not shown) indicating hyperplasia of smooth muscle cells.

postulated that sympathetic nerves have trophic functions on vascular smooth muscle by an action independent of pressure *(208)*, but the nature of this trophic factor has not been elucidated. Dhital et al. *(209)* have suggested that hyperinnervation of cerebral vessels in SHR by sympathetic NPY-containing fibers may play a role in the development of medial hypertrophy. Since NPY is released from the sympathetic nerves and has a mitogenic effect on smooth muscle cells, it may be a mediator of neurogenic vascular hypertrophy in hypertension.

3.3. Role of NPY in Platelet–Vascular Interactions

As discussed earlier (*see* Section 2.3.), platelets of rats *(7)* and some autoimmune mice *(86)* contain large amounts of NPY. In rats, platelet-derived NPY can significantly contribute to circulating levels of the peptide, since platelets release NPY during aggregation directly into the bloodstream. The function(s) of NPY released from platelets is (are) not clear yet. We have been studying the effects of NPY on aggregation of rat and human platelets, and found that NPY does not possess any direct aggregatory activities *(7)*. However, there are conflicting data as to whether or not NPY modulates aggregation either by inhibition or potentiation.

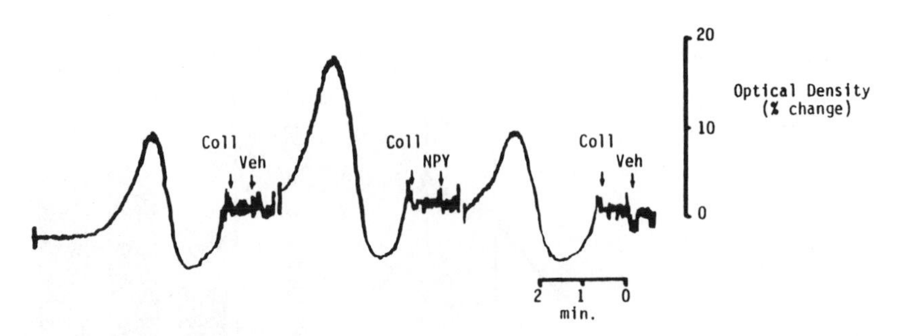

Fig. 26. Potentiation by NPY of aggregation of rat platelets induced by a low dose of collagen (Coll) in vitro.

Myers et al. *(88)* have recently confirmed that NPY has no inhibitory effects on aggregation induced by ED_{80} doses of collagen, adenosine diphosphate, and epinephrine in normal human platelets, in assays using platelet-rich plasma and gel-filtered platelets; specific Y1 (Pro^{34}NPY) and Y2 (NPY_{22-36}) agonists were also ineffective. To test the potentiation of aggregation, human platelet-rich plasma was stimulated with a threshold dose of collagen; preincubation with porcine or human NPY, [Pro^{34}]NPY and NPY_{22-36}, did not augment platelet aggregation. The lack of any apparent activity of NPY was also supported by the failure of NPY to alter basal and illoprost-stimulated cAMP levels in normal human platelets.

Similarly in rats, NPY has no direct aggregatory activity (Fig. 26). However, in the rat platelet-rich plasma, NPY slightly potentiated aggregation induced by a low dose of collagen (Fig. 26) along with a tendency to lower forkolin-stimulated cAMP levels (Myers et al., unpublished observation from our laboratory). These experiments are notoriously difficult to perform, since the sensitivity of platelets to a threshold dose of collagen is highly variable, and endogenous levels of NPY are high in the rats. Further studies using gel-filtered platelets and several aggregatory agonists would be required to determine whether or not NPY potentiates aggregation of rat platelets. The possibility that NPY exerts some actions on platelets is suggested by the presence of binding sites for NPY. Specific binding sites for ^{125}I-NPY were found in the membrane preparations of rat platelets with a K_d of approx 0.8 nM and a B_{max} of approx 470 fmol/mg protein *(87)*.

Although the platelet effects of NPY may be negligible (as in human) or mild (as in rat), platelet-derived NPY may be an important modulator of vascular functions. Other substances released by platelets, such as serotonin *(210)* and platelet-activating factor *(211)*, have also been shown to have more important vasoactive than platelet-targeted functions. In vivo, platelet aggregation in the rat can yield nanomolar concentrations of circulating NPY, which may elicit vasoconstriction and/or potentiation of vasoconstriction induced by other agonists. Based on our studies of stress, we postulated that aggregation of platelets during the sympatho-adrenomedullary activation causes the release of additional amounts of NPY into the bloodstream, and, in that manner, "amplify" the NPY-sympatho-neural response contributing to enhanced vasoconstriction (*see* Fig. 16). In several cardiovascular diseases, such as myocardial infarction and congestive heart failure, both the hyperactivity of the sympatho-adrenomedullary system and platelet aggregation appear to be involved in the pathogenesis. NPY, released from the sympathetic nerves and platelets and acting on blood vessels, may be a missing link between these two systems and vasospasm.

In addition to the regulation of platelet and vascular functions, platelet-derived (and possibly neuronal) NPY may interact with other blood cells and endothelium. Recently, Sung et al. *(211)* have reported that NPY, at nanomolar-micromolar concentrations, stimulates adhesion of neutrophils to human umbilical endothelial cells in culture. The authors *(211)* have suggested that NPY may play an important role in the regulation of hemostasis by microvessels.

3.4. NPY in Cardiovascular Diseases

3.4.1. Pheochromocytoma

Plasma levels of NPY-ir are elevated in several cardiovascular diseases in which vasoconstriction plays an important role in pathogenesis (Fig. 7). The highest circulating NPY levels were found in patients with pheochromocytoma *(6,63,85)*. Manipulations of the tissue, which contains high concentrations of NPY, triggers the release of the peptide into the circulation *(72,212)*. These instances are known to evoke potentially dangerous rises in arterial blood pressure, and the concentrations of plasma NPY yielded in pheochromocytoma are within a range of concentrations that may elicit vasoconstriction in humans *(38)*. However, as yet, there are no studies that would correlate the increases in plasma NPY with changes in blood pressure. Although blood pressure rises respond to α-adrenoceptor blockers in patients with pheochromocytoma, some studies indicate that sodium nitroprusside is more effective *(213)*. It is presumed that NPY contributes to vasopressor effects resistant to α-adrenoceptor blockers.

3.4.2. Ischemic Heart Disease

Several studies have reported increased plasma NPY levels in patients with ischemic heart disease *(65,68)*, especially those with advanced congestive heart failure *(214)*. The severity of heart failure (Killip class), and increased heart and respiratory rates were found to be strongly correlated with plasma NPY levels *(214)*, suggesting that the peptide may be an index of severe cardiovascular disturbances. During cardiopulmonary bypass surgery that entailed aortic occlusion leading to temporary myocardial ischemia, NPY levels in the coronary sinus and in the arterial blood were found to increase threefold, immediately following thoracotomy, indicating that the response may have been related to generalized sympathetic activation *(68)*. In the same study *(68)*, cardiac output of NPY increased tenfold during the reperfusion period together with the elevation of plasma lactate and pyruvate concentrations, suggesting that the release of NPY from the heart is related to ischemia. These authors *(68)* believed that the concentrations of NPY in the coronary sinus blood attained during reperfusion were not likely to exert vasoconstriction; however, no hemodynamic parameters were measured.

We have recently attempted to correlate changes in cardiac and systemic hemodynamics in patients undergoing coronary by-pass surgery with changes in plasma (arterial) levels of NPY (Hauser et al., unpublished observation from our laboratory). Although plasma catecholamine levels increased markedly, there were no significant changes in plasma NPY; arterial plasma NPY levels tended to decrease during surgery and increase 24 h afterward. Furthermore, no correlations were found between plasma NPY (and catecholamine) levels and hemodynamic parameters, such as cardiac output, pulmonary arterial pressure, and systemic arterial pressure. Thus, it appears that arterial NPY levels do not reflect cardiac changes, and the question of whether or not NPY contributes to any cardiovascular alterations induced by myocardiac ischemia in these patients remains unanswered yet.

3.4.3. Essential Hypertension

Only a few studies have evaluated plasma NPY levels in patients with essential hypertension *(85,215,216)*. This is surprising since the hyperactivity of the sympatho-adrenomedullary system is a hallmark of this disease, at least in a certain subgroup of patients (hyperadrenergic). One study of 33 patients *(85)* reported that plasma NPY levels are not different in hypertensives than in normal volunteers; however, it is unclear whether the groups were correctly age matched. Previously, higher plasma catecholamine levels in hypertensives were found only in the younger population of properly age-matched patients and controls *(217)*. More recently, Solt et al. *(216)* have found significantly higher (>200 pg/mL) plasma NPY levels in age-, weight-, and sex-matched patients with essential hypertension after fasting. In this last study, higher plasma NPY levels correlated with the systolic and diastolic blood pressures and ages of patients *(216)*. However, plasma NPY-ir responses to graded bicycle exercise (at 75% work load) were reportedly similar in hypertensive and normotensive subjects *(215)*. Elevated plasma NPY levels were found in severely hypertensive patients (regardless of the origin of hypertension) before as well as 3 wk after the normalization of blood pressure *(67)*. This raises the question as to what is the role of NPY: Is it unrelated to cardiovascular derangements of hypertension, or is it related to some hemodynamic paramenters other than resting arterial blood pressure (e.g., reduced coronary or cerebral blood flow)? Alternatively, circulating plasma NPY levels may only represent a small fraction of NPY released at the sympathetic neuro-effector junctions and, thus, are not a sensitive index of changes in hemodynamics. Measurements of regional plasma NPY spillover may provide a better insight into the problem.

3.4.4. Subarachnoid Hemorrhage

A pathophysiological state in which NPY may play a role as a vasoconstrictor is subarachnoid hemorrhage. Abel et al. *(96)* were the first to report that NPY concentration in the cerebrospinal fluid (CSF) increases sixfold in rabbits with experimental subarachnoid hemorrhage. Moreover, CSF containing high levels of NPY markedly potentiated norepinephrine-induced vasoconstrictrion of cerebral arteries; this effect was eliminated by immunoprecipitation of NPY by specific antibodies, indicating that NPY may contribute to cerebral vasospasm *(96)*. Increased NPY concentrations in the CSF was also found in patients with aneurysmal subarachnoid hemorrhage at 6–11 d posthemorrhage *(97)*. Elevated NPY levels in CSF may be, at least in part, due to enhanced release from sympathetic nerves, as suggested by the findings of NPY depletion in fibers innervating cerebral vessels in rats with subarachnoid hemorrhage *(218)*, and in part due to the release from pial and ependymal fibers *(219)*. In some species, such as rabbits, which may contain platelet-derived NPY *(87)*, aggregation of platelets at the site of blood extravasation may be an additional source of NPY in the CSF. Considering the potent vasoconstrictor activity of NPY and the rich NPY-containing innervation of cerebral blood vessels in many species, including humans, the peptide may be a mediator of a delayed, long-lasting vasospasm characteristic of subarachnoid hemorrhage.

3.4.5. Endotoxic Shock

The role of NPY in endotoxic shock deserves special attention for its potential therapeutic value. Endotoxic hypotensive shock is generally associated with increased sympatho-adrenomedullary activity and elevated plasma catecholamine levels *(220)*. Although adrenergic agonists are frequently used in the treatment of shock *(221)*, their effectiveness is limited by a progressive reduction of vascular responsiveness developed during shock. The mechanisms of the vascular resistance to adrenergic agonists appear to involve desensitization of adrenoceptors because of high circulating catecholamine levels *(221)* and/or impairment of their function resulting from endotoxin *(222)*. Thus, hypotensive endotoxic shock appears to resemble our model of adrenergic pressor desensitization induced by infusion of norepinephrine (*146*; *see* Section 3.2.4.), and, as such, may render blood vessels more responsive to the vasoconstrictor actions of NPY.

Unlike catecholamines, circulating NPY levels are unchanged in several animal models of endotoxic shock *(66,92)*. In humans, circulating NPY levels were found to be somewhat higher than "normal" in patients with sepsis; however, there was no control

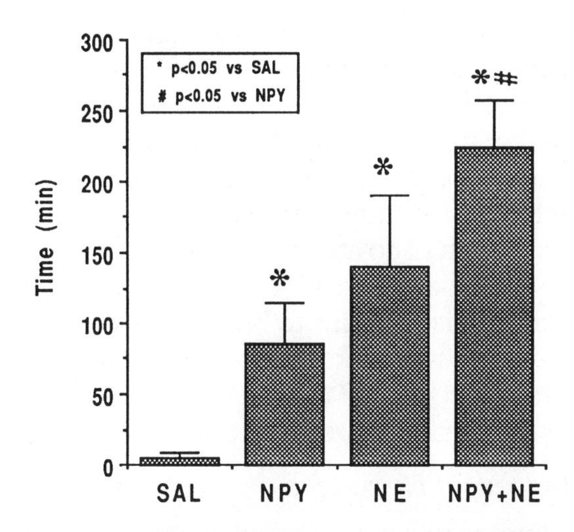

Fig. 27. Survival time after iv administration of a lethal dose of *E. coli* endotoxin (lipopolysacharide, LPS). NPY alone, at a low pressor dose of 0.2 nmol/kg/min, improved survival compared to saline (SAL)-treated rats, and addition of NPY to NE infusion increased the effectiveness of NE (0.1 μg/kg/min).

group in this study *(66)*. Therefore, it appears that in shock there may be a relative deficiency of NPY and an excess of catecholamines, which may favor adrenergic desensitization. Evequoz et al. *(223)* have shown that exogenous administration of NPY is able to reverse reduced pressor responsiveness to NE in rats with mild nonhypotensive endotoxemia. The authors have further confirmed that NPY prevents the fall in blood pressure induced by hypotensive doses of endotoxin in rats without adrenal medullas. This effect of NPY occurred at nonpressor doses of the peptide, and was equipotent or greater than that of epinephrine infusion *(224)*.

We have recently extended these observations to intact rats, and models of more pronounced and prolonged hypotensive endotoxic shock. Rats given 20 mg/kg of lipopolysaccharide (LPS) iv developed severe hypotension and died within 5 min when infused with saline only (Fig. 27). Infusions of NE, NPY, and NE + NPY, started before the LPS and continued for 5 h, all had beneficial effects on blood pressure and survival, but to a different degree. Treatment of rats with NPY + NE reduced the acute hypotensive effect of LPS and markedly prolonged the survival time more than

NE alone (Hauser et al., unpublished observations; Fig. 27). In another experimental model of nonlethal endotoxic shock, LPS administration alone markedly lowered cardiac output and total peripheral resistance (Hauser et al., unpublished observations). Although pressor doses of NPY were used (5 nmol/kg/min), it had no cardiodepressant activity and rather tended to increase cardiac output, whereas NE infusion had no effect (Hauser et al., unpublished observation). Thus, it appears that NPY has beneficial systemic hemodynamic effects and is superior to NE as a vasopressor therapy in hypotensive endotoxic shock. Studies are under way to exclude any deleterious effects of NPY on regional hemodynamics in the areas affected by shock, e.g., renal and splanchnic blood vessels.

3.4.6. Spontaneously Hypertensive Rats (SHR)

SHR are known to have elevated sympatho-adrenomedullary activity, which could be traced to the prenatal period *(225)* and which is maximal during the rapid development of hypertension within the first several weeks of their lives *(203)*. To support this, we found increased circulating NE and epinephrine levels in 4-wk-old SHR compared to age-matched Wistar-Kyoto rats, and a tendency for higher catecholamine levels in the SHR group thereafter (Fig. 28B; *226*). In contrast, plasma NPY levels were similar in age-matched SHR and WKY, although both strains had significantly higher levels at younger than at older age (Fig. 28B). However, SHR have markedly elevated platelet content of NPY compared to that of WKY rats, beginning at 5 wk of age *(89)*. Twenty-four-week-old SHR have approximately twofold higher platelet NPY (normalized for 10^6 platelets/mL plasma) than age-matched WKY rats (Zukowska-Grojec et al., unpublished observations from our laboratory). Considering that platelets can significantly contribute to basal circulating NPY, it is somewhat surprising that plasma levels are not greater in SHR than in WKY rats. However, higher resting NPY levels in arterial plasmas were reported in stroke-prone (SP)-SHR, a substrain that is characterized by a very high blood pressure and a high incidence of stroke *(227)*, compared to normotensive rats. Moreover, these higher NPY levels were resistant to treatment with a ganglionic blocker, pentolinium, indicating that they were not of neuronal origin *(227)*; thus, they might have been derived from platelets. These data suggest that platelet-derived NPY may contribute to higher circulating NPY levels during the advanced stages of hypertension.

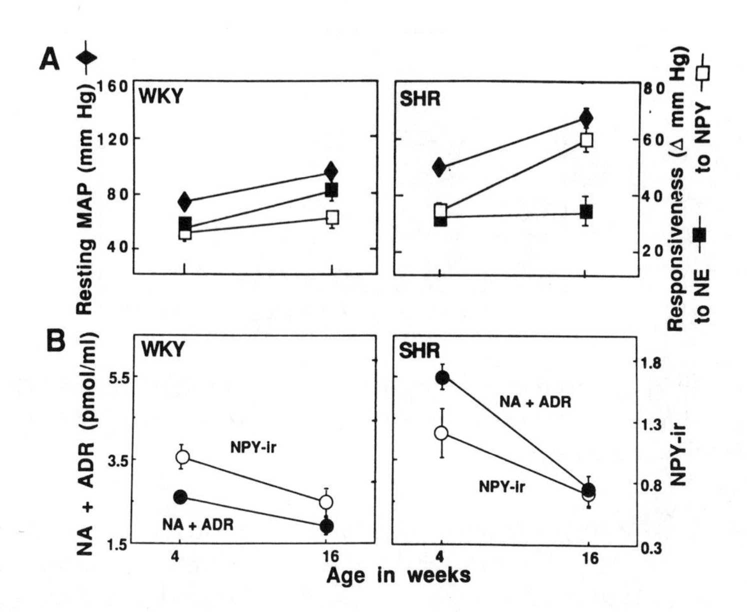

Fig. 28. Responsiveness to and plasma levels of NPY in juvenile and adult spontaneously hypertensive rats (SHR). Panel A: basal mean arterial pressure levels and pressor responses to ED_{50} doses of NPY and NE. Panel B: plasma levels of total catecholamines, norepinephrine (NA), and epinephrine (ADR) and NPY-ir. Pressor responsiveness to NPY increased in parallel with the development of high blood pressure in SHR, whereas responsiveness to NE remained unchanged. Juvenile (4-wk-old) rats of both strains had higher plasma NPY-ir levels than adult rats; there were no strain differences in NPY-ir in spite of higher plasma catecholamine levels in SHR than in normotensive Wistar-Kyoto rats (WKY).

Whether or not SHR also have altered NPY release from the sympathetic nerves has not been determined yet. Several indirect lines of evidence point to the possibility of increased neuronal release of NPY in SHR, at least locally in some vascular beds. Beginning at an early "prehypertensive" age, SHR have denser perivascular innervation with NPY-containing fibers in the mesenteric *(207,228)* and cerebral *(209)* arteries compared to WKY rats. Dhital et al. *(209)* have suggested that hyperinnervation of cerebral vessels in SHR by NPY-containing fibers may play a role in the development of medial hypertrophy, and thus, represent a local neurovascular mechanism protecting the brain against edema and hemorrhage (this possibility is plausible considering recently dis-

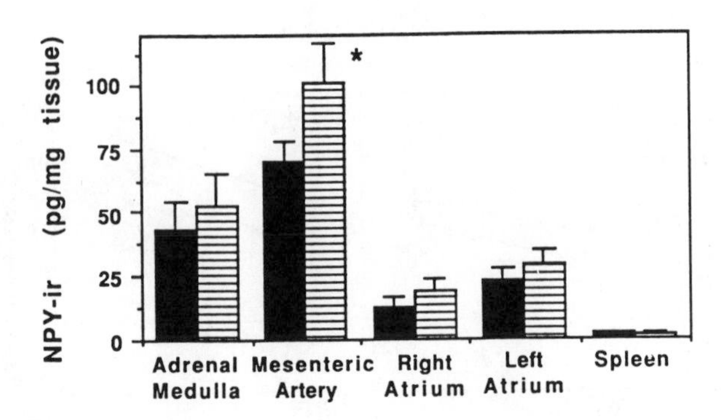

Fig. 29. Tissue content of NPY-ir in 16-wk-old SHR ■ and WKY ▤ rats. SHR had significantly ($p < 0.05$) lower NPY-ir levels in the superior mesenteric artery compared to WKY.

covered growth-promoting effect of NPY on vascular smooth muscle cells; Fig. 25). We (Zukowska-Grojec et al., unpublished observation) have recently found that NPY tissue levels are significantly lower in the mesenteric arteries from 16-wk-old SHR than in those of WKY, and this tendency was also observed in the atria and in the adrenal medulla (Fig. 29). The most likely explanation of lower tissue contents of NPY is that enhanced neuronal release depletes NPY stores, since it occurs with NE in the face of chronic sympathetic overactivity.

Although increased release of NPY in SHR may still be undetermined, some studies indicate that vascular responsiveness to NPY is augmented in SHR compared to WKY rats. Using caudal arteries of adult rats where NPY exerts potentiating effect on adrenergic vasoconstriction, Daly et al. *(229)* have found that NPY antiserum attenuated the contractile response to electrical field stimulation in SHR, but not in WKY, suggesting that the contribution of endogenously released peptide to neurogenic vasoconstriction is greater in hypertensive rats. We have recently studied vascular responsiveness to NPY and the effect of the peptide on NE-induced pressor responses in conscious 4- and 16-wk-old SHR. We reasoned that if NPY becomes a more efficacious vasoconstrictor in blood vessels preexposed to NE (*see* Section 3.2.4.), chronic elevation of circulating catecholamines, present in SHR, may render their vessels hyperresponsive to the peptide. Additionally, if, as some believe *(230)*, adrenergic receptors of SHR have impaired ability to desensitize in the face of chronic catecholamine expo-

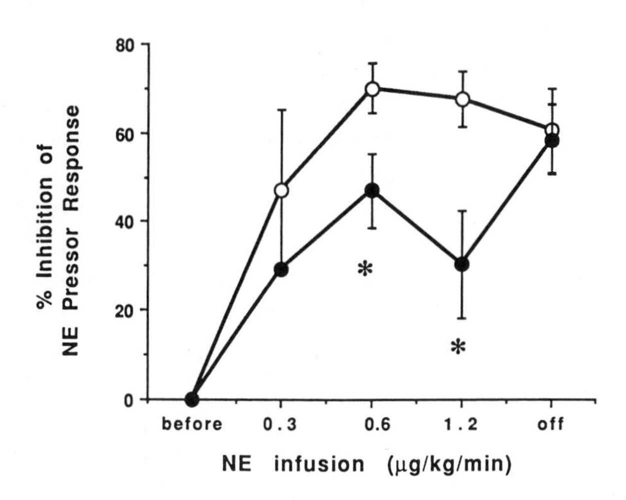

Fig. 30. Impaired ability of adult SHR to develop desensitization to pressor effects of NE. Desensitization expressed as percent inhibition of pressor responses to a test dose of NE (bolus injections of an ED_{30} dose) during ongoing graded desensitizing infusions of NE. Before and following termination of the desensitizing NE infusion, pressor responses to a test dose of NE were similar in SHR and WKY. –O– WKY; –●– SHR.

sure, NPY may be involved in this process (an effect similar to the NPY reversal of adrenergic desensitization observed in the NE infusion studies, *see* Section 3.2.4.).

In spite of higher circulating catecholamine levels, 4- and 16-wk-old SHR exhibit similar or slightly greater pressor responsiveness to NE, compared to that of WKY, and these responses are not altered by ganglionic blockade (Fig. 28A). These "normal" responses of hypertensive rats appear inadequate since chronic exposure to catecholamines would be anticipated to cause compensatory downregulation of receptors. With age and development of hypertension, SHR become hyperresponsive to α-adrenergic agonists. Thus, SHR may have chronically impaired adrenergic desensitization. We tested the ability of SHR to desensitize to pressor effects of NE acutely and found it to be blunted. SHR showed less inhibition of pressor responses to NE (injected as boluses during NE infusion to test the progression of desensitization) and a higher threshold for desensitization than WKY rats (Fig. 30).

Pressor responsiveness to NPY increased in parallel with the development of hypertension leading to twofold greater responses than in WKY at 16 wk of age (Fig. 28A; *226*). Similar findings were recently reported by Miller and Tessel *(231)* who also measured plasma NPY-ir levels and found that higher plasma concentrations were yielded by exogenous administration of the peptide in SHR than in WKY. They hypothesized that SHR are more responsive to the pressor action of NPY, because higher circulating concentrations of the peptide allow for the access to "supersensitive" synaptic NPY receptors *(231)*, although the subtype of these receptors is not known. Hyperresponsiveness to NPY appears to be receptor-specific, since SHR are less responsive to NPY-mediated inhibition of NE release mediated by presynaptic, presumably, Y2 receptors *(232,233)*.

Thus, the development of hypertension in SHR is characterized by the following features: early hyperactivity of the sympathoadrenomedullary system, elevated circulating catecholamines, impaired adrenergic desensitization and hyperresponsiveness to pressor effects of NPY; in addition, SHR have increased content of platelet-derived NPY *(89)* and are hyporesponsive to presynaptic, inhibitory effects of NPY. Chronic pressor NPY influence in SHR may potentiate adrenergic responses and attenuate the development of compensatory adrenergic desensitization, thus promoting the maintenance of high blood pressure. To prove that NPY is responsible, in any part, for the impaired ability of SHR's adrenoceptors to desensitize, specific NPY antagonists would be required. The lack of compensatory adrenergic desensitization in SHR may have implications not only for vasoconstriction, but also for vascular hypertrophy in which sympathetic mechanisms are thought to be involved. Our recent finding of growth-promoting effects of NPY on vascular smooth muscle cells *(40)* suggests that the peptide may play a role in the development of neurogenic vascular hypertrophy in hypertension, providing a missing link between the sympathetic nerves, vasoconstriction, and trophic actions.

References

1. O'Donohue, T. L., Chronwall, B. M., Pruss, R. M., Mezey, E., Kiss, J. Z., Eiden, L. E., Massari, V. J., Tessel, R. E., Pickel, V. M., DiMaggio, D. A., Hotchkiss, A. J., Crowley, W. R., and Zukowska-Grojec, Z. (1985) Neuropeptide Y and peptide YY neuronal and endocrine systems. *Peptides* **6,** 755–768.
2. McDonald, J. K. (1988) NPY and related substances, in *CRC Critical Review in Neurobiology* **4,** 97–135.

3. Potter, E. K. (1988) Neuropeptide Y as an autonomic neurotransmitter. *Pharmacol. Ther.* **37,** 251–273.
4. Waeber, B., Aubert, J.-F., Corder, R., Nussberger, J., Gaillard, R., and Brunner, H. R. (1988) Cardiovascular effects of neuropeptide Y. *Am. J. Hypertension* **1,** 193–199.
5. Allen, J. M., Adrian, T. E., Polak, J. M., and Bloom, S. R. (1983) Neuropeptide Y (NPY) in the adrenal gland. *J. Auton. Nerv. Syst.* **9,** 559–563.
6. Corder, R., Lowry, P. J., Emson, P. C., and Gaillard, R. C. (1985) Chromatographic characterization of the circulating neuropeptide Y immunoreactivity from patients with phaechromocytoma. *Regul. Peptides* **10,** 91–97.
7. Myers, A. K., Farhat, M. Y., Vaz, C. A., Keiser, H. R., and Zukowska-Grojec, Z. (1988) Release of immunoreactive-neuropeptide Y by rat platelets. *Bioch. Biophys. Res. Comm.* **155,** 118–122.
8. Ericsson, A., Schaling, M., McIntyre, K. R., Lundberg, J. M., Larhammar, D., Seroogy, K., Hokfelt, T., and Persson, H. (1987) Detection of neuropeptide Y and its mRNA in megakaryocytes: enhanced levels in certain autoimmune mice. *Proc. Natl. Acad. Sci. USA* **84,** 5585–5589.
9. Larhammar, D., Ericsson, A., and Persson, H. (1987) Structure and expression of the rat neuropeptide Y gene. *Proc. Natl. Acad. Sci. USA* **84,** 2068–2072.
10. Thureson-Klein, A. (1983) Exocytosis from large and small dense-core vesicles on noradrenergic nerve terminals. *Neuroscience* **10,** 245–252.
11. Fried, G., Terenius, L., Hokfelt, T., and Goldstein, M. (1985) Evidence for differential localization of noradrenaline and neuropeptide Y (NPY) in neuronal storage vesicles isolated from rat vas deferens. *J. Neurosci.* **5,** 450–458.
12. Lundberg, J. M., Rudehill, A., Sollevi, A., Thoeodorsson-Norheim, E., and Hamberger, B. (1986) Frequency- and reserpine-dependent chemical coding of sympathetic transmission: differential release of noradrenaline and neuropeptide Y from pig spleen. *Neurosci. Lett.* **63,** 96–100.
13. Lundberg, J. M.., Pernow, J., Franco-Cereceda, A., and Rudehill, A. (1987) Effects of antihypertensive drugs on sympathetic vascular control in relation to neuropeptide Y. *J. Cardiovasc. Pharmacol.* **10 (Suppl. 12),** S51–S68.
14. Haass, M., Hock, M., Richardt, G., and Schomig, A. (1989) Neuropeptide Y differentiates between exocytotic and non-exocytotic noradrenaline release in guinea-pig heart. *Naunyn Schmiedeberg's Arch. Pharmacol.* **340,** 509–515.
15. Haass, M., Cheng, B., Richardt, G., Lang, R. E., and Schomig, A. (1989) Characterization and presynaptic modulation of stimulation-evoked exocytotic co-release of noradrenaline and neuropeptide Y in guinea-pig heart. *Naunyn Schmiedeberg's Arch. Pharmacol.* **339,** 71–78.
16. Hockfelt, T., Johansson, O., Ljungdahl, A., Lundberg, J. M., and Schultznerg, M. (1980) Peptidergic neurons. *Nature* **284,** 515–521.
17. Haass, M. and Schomig, A. (1989) Neuropeptide Y and sympathetic transmission, in *Adrenergic System and Ventricular Arrthytmias in Myocardial Infarction*, Brachmann, J. and Schomig, A., eds., Springer Verlag, Heidelberg, pp. 21–33.
18. Haass, M., Forster, C., Richardt, G., Kranzhofer, R. and Schomig, A. (1990) Role of calcium channels and protein kinase C for release of norepinephrine and neuropeptide Y. *Am. J. Physiol.* **259,** R925–R930.
19. Lundberg, J. M., Anggard, A., Theodeorsson-Norheim, E., and Pernow, J. (1984) Guanethidine-sensitive release of neuropeptide Y-like immuno-

reactivity in the cat spleen by sympathetic nerve stimulation. *Neurosci. Lett.* **52,** 175–180.
20. Schoups, A. A., Saxena, V. K., Tombeur, K., and De Potter, W. P. (1988) Facilitation of the release of norepinephrine and neuropeptide Y by the alpha-2 adrenoceptor blocking agents idazoxan and hydergine in the dog spleen. *Life Sci.* **42,** 517–523.
21. Lundberg, J. M., Rudehill, A., Sollevi, A., and Hamberger, B. (1989) Evidence for co-transmitter role of neuropeptide Y in the pig spleen. *Br. J. Pharmacol.* **96,** 675–687.
22. Allen, J. M., Bircham, M. M., Bloom, S. R., and Edwards, A. V. (1984) Release of neuropeptide Y in response to splanchnic nerve stimulation in the conscious calfs. *J. Physiol. (Lond.)* **357,** 401–408.
23. Dahlof, C., Dahlof, P., and Lundberg, J. M. (1986) Alpha-2 adrenoceptor-mediated inhibition of nerve stimulation-evoked release of neuropeptide Y (NPY)-like immunoreactivity in the pithed guinea-pig. *Eur. J. Pharmacol.* **131,** 279–283.
24. Zukowska-Grojec, Z., Shen, G. H., Deka-Starosta, A., Myers, K. A., Kvetnansky, R., and McCarty, R. (1992) Neuronal, adrenomedullary and platelet-derived neuropeptide Y responses to stress in rats, in *Stress: Neuroendocrine and Molecular Approaches*. Kvetnansky, R., McCarty, R., and Axelrod, J., eds., Gordon and Breach Science Publishers S. A., New York, pp. 197–209.
25. Pernow, J., Schwieler, J., Kahan, T., Hjemdahl, P., Oberle, J., Wallin, B. G., and Lundberg, J. M. (1989) Influence of sympathetic discharge pattern on norepinephrine and neuropeptide Y release. *Am. J. Physiol.* **257,** H866–H872.
26. Nilsson, H., Ljung, B., Sjöblom, N., and Wallin, B. G. (1985) The influence of the sympathetic impulse pattern on contractile responses of rat mesenteric arteries and veins. *Acta. Physiol. Scand.* **123,** 303–309.
27. Warner, M. and Levy, M. N. (1989) Neuropeptide Y as a putative modulator of the vagal effects on heart rate. *Circ. Res.* **64,** 882–889.
28. Potter, E. K. (1987) Cardiac vagal actions and plasma levels of neuropeptide Y following intravenous injection in the dog. *Neurosci. Lett.* **77,** 243–247.
29. Deka-Starosta, A., Garty, M., Chang, P., Goldstein, D. S., Keiser, H. R., Kopin, I. J., and Zukowska-Grojec, Z. (1989) Release of neuropeptide Y and norepinephrine by renal sympathetic nerve stimulation in rats. *Circulation* **80,** II-554.
30. Rudehill, A., Franco-Cereceda, A., Hamsen, A., Stensdotter, M., Pernow, J., and Lundberg, J. M. (1989) Cigarette smoke-induced elevation of plasma neuropeptide Y levels in man. *Clin. Physiol.* **9,** 243–248.
31. Niedermaier, O. N., Smith, M. L., Beightol, L. A., Zukowska-Grojec, Z., Goldstein, D. S., and Eckberg, D. L. (1992) Influence of cigarette smoking on human autonomic function. *Circulation* in press.
32. Archelos, J., Xiang, J. Z., Reinecke, M., and Lang, R. E. (1987) Regulation of release and function of neuropeptides in the heart. *J. Cardiovasc. Pharmacol.* **10 (Suppl. 12),** S45–S50.
33. Zukowska-Grojec, Z., Bayorh, M. A., and Kopin, I. J. (1983) Effect of desipramine on the effects of alpha-adrenoceptor inhibitors on pressor responses and release of norepinephrine into plasma of pithed rats. *J. Cardiovasc. Pharmacol.* **5,** 297–301.

34. Lundberg, J. M.., Pernow, J., Fried, G., and Anggard, A. (1987) Neuropeptide Y and noradrenaline mechanisms in relation to reserpine induced impairment of sympathetic neurotransmission in the cat spleen. *Acta Physiol. Scand.* **131,** 1–10.
35. Lundberg, J. M., Hua, X.-Y., and Franco-Cereceda, A. (1984) Effects of neuropeptide Y (NPY) on mechanical activity and neurotransmission in the heart, vas deferens and urinary bladder of the guinea-pig. *Acta Physiol. Scand.* **121,** 325–332.
36. Pernow, J., Lundberg, J. M., and Kaijser, L. (1988) Alpha-adrenoceptor influence on plasma levels of neuropeptide Y-like immunoreactivity and catecholamines during rest and sympatho-adrenal activation in humans. *J. Cardiovasc. Pharmacol.* **12,** 593–599.
37. Eklund, B., Hjemdahl, P., Seideman, P., and Atterhog, J. H. (1983) Effects of prazosin on hemodynamics and sympatho-adrenal activity in hypertensive patients. *J. Cardiovasc. Pharmacol.* **5,** 384–391.
38. Pernow, J., Lundberg, J. M., and Kaijser, L. (1987) Vasoconstrictor effects in vivo and plasma disappearance rate of neuropeptide Y in man. *Life Sci.* **40,** 47–54.
39. Pernow, J., Ohlen, A., Hokfelt, T., Nilsson, O., and Lundberg, J. M. (1987) Neuropeptide Y: presence in perivascular noradrenergic neurons and vasoconstrictor effects on skeletal muscle blood vessels in experimental animals and man. *Regul. Peptides* **19,** 313–324.
40. Zukowska-Grojec, Z., Colton, C., Yao, J., Abi-Younes, S., Myers, K.A., Koenig, J., and Wahlestedt, C. (1992) Modulation of vascular smooth muscle growth by a sympathetic neurotransmitter, neuropeptide Y. Society for Neuroscience Abstracts, **17,** 285.
41. Donoso, V., Silva, M., St.Pierre, S., and Huidobro-Toro, P. (1988) Neuropeptide Y (NPY), an endogenous presynaptic modulator of adrenergic neurotransmision in the rat vas deferens: structural and functional studies. *Peptides* **9,** 545–553.
42. Westfall, T. C., Martin, J., Chen, X., Ciarleglio, A., Carpentier, S., Henderson, K., Knuepfer, M., Beinfeld, M., and Naes, L. (1988) Cardiovascular effects and modulation of noradrenergic neurotransmission following central and peripheral administration of neuropeptide Y. *Synapse* **2,** 299–307.
43. Wahlestedt, C., Yanaihara, N., and Hakanson, R. (1986) Evidence for different pre- and post-junctional receptors for neuropeptide Y and related peptides. *Regul. Peptides* **13,** 307–318.
44. Wahlestedt, C., Edvinsson, L., Ekblad, E., Hakanson, R. (1987) Effects of neuropeptide Y at sympathetic neuroeffector junctions: Existence of Y_1 and Y_2 receptors, in *Neuronal Messengers in Vascular Function*. Nobin, A. and Owman, C., eds., *Fernstrom Symposium* **10,** pp. 231–242.
45. Pernow, J. and Lundberg, J. M. (1989) Modulation of noradrenaline and neuropeptide Y (NPY) release in the pig kidney *in vivo*: involvement of alpha-2, NPY and angiotensin receptors. *Naunyn Schmiedeberg's Arch. Pharmacol.* **340,** 379–385.
46. Pernow, J., Saria, A., and Lundberg, J. M. (1986) Mechanisms underlying pre- and postjunctional effects of neuropeptide Y in the sympathetic control. *Acta Physiol. Scand.* **126,** 239–249.

47. Scott, N. A., Michel, M. C., Boublik, J. H., Rivier, J. E., Motomura, S., Crum, R. L., Landon, M., and Brown, M. R. (1990) Distinction of NPY receptors *in vitro* and *in vivo*. II. Differential effects of NPY and NPY-(18–36). *Am. J. Physiol.* **259,** H174–H180.
48. Shen, G. H., Grundemar, L., Zukowska-Grojec, Z., Hakanson, R., and Wahlestdt, C. (1991) C-terminal neuropeptide Y fragments are mast cell-dependent vasodepressor agents. *Eur. J. Pharmacol.* **204,** 249–256.
49. Dahlof, P. (1989) Modulatory interactions of neuropeptide Y (NPY) on sympathetic neurotransmission. *Acta Physiol. Scand.* **Suppl. 586,** 1–85.
50a. Dahlof, C. (1981) Studies on beta-adrenoceptor mediated facilitation of sympthetic neurotransmission. *Acta Physiol. Scand.* **Suppl. 500,** 1–147.
50b. Morris, M. J., Russell, A. E., Kapoor, V., Elliott, J. M., West, M. J., Wing, L. M. H., and Chalmers, J. P. (1986) Increases in plasma neuropeptide Y concentrations during sympathetic activation in man. *J. Auton. Nerv. System* **17,** 143–149.
51. Morris, M. J., Elliott, J. M., Cain, M. D., Kapoor, V., West, M. J., and Chalmers, J. P. (1986) Plasma neuropeptide Y levels rise in patients undergoing exercise tests for the investigation of chest pain. *Clin. Exp. Pharmacol. Physiol.* **13,** 437–440.
52. Kaijser, L., Pernow, J., Berglund, B., and Lundberg, J. M. (1990) Neuropeptide Y is released together with noradrenaline from the human heart during exercise and hypoxia. *Clin. Physiol.* **10,** 179–188.
53. Rudehill, A., Olcen, M., Sollevi, A., Hamberger, B., and Lundberg, J. M. (1987) Release of neuropeptide Y upon haemorrhagic hypovolemia in relation to vasoconstrictor effects in the pig. *Acta Physiol. Scand.* **131,** 517–523.
54. Zukowska-Grojec, Z., Konarska, M., and McCarty, R. (1988) Differential plasma catecholamine and neuropeptide Y responses to acute stress in rats. *Life Sci.* **42,** 1615–1624.
56. Zukowska-Grojec, Z. Shen, G. H., Konarska, M., and McCarty, R. (1990) Sources and vasopressor efficacy of circulating neuropeptide Y during acute and chronic stress in rats. *Ann. NY Acad. Sci.* **611,** 412–414.
57. Zukowska-Grojec, Z., Shen, G. H., Capraro, P. A., and Vaz, C. A. (1991) Cardiovascular, neuropeptide Y and adrenergic responses in stress are sexually differentiated. *Physiol. Behav.* **49,** 771–777.
58. Zukowska-Grojec, Z. and Vaz, C. A. (1988) Role of neuropeptide Y (NPY) in the cardiovascular responses to stress. *Synapse* **2,** 293–298.
59. Lundberg, J. M., Martinsson, A., Hemsen, A., Theodorsson-Norheim, E., Svedenhag, J., Ekblom, B., and Hjemdahl, P. (1985) Co-release of neuropeptide Y and catecholamines during physical exercise in man. *Biochem. Biophys. Res. Comm.* **133,** 30–34.
60. Takahashi, K., Mouri, T., Murakami, O., Itoi, K., Sone, M., Ohneda, M., Nozuki, M., and Yoshinaga, K. (1988) Increases of neuropeptide Y-like immunoreactivity in plasma during insulin-induced hypoglycemia in man. *Peptides* **9,** 433–435.
61. Ahlborg, G. , and Lundberg, J. M. (1991) Splanchnic release of neuropeptide Y during prolonged exercise with and without beta-adrenoceptor blockade in healthy man. *Clin. Physiol.* **11,** 343–351.
62. Pernow, J., Lundberg, J. M., Kaijser, L., Hjemdahl, P., Theodorsson-Norheim, E., Martinsson, A., and Pernow, B. (1986) Plasma neuropeptide Y-like

immunoreactivity and catecholamines during various degrees of sympathetic activation in man. *Clin. Physiol.* **6,** 561–578.
63. Grouzman, E., Comoy, E., and Bohuon, C. (1989) Plasma neuropeptide Y concentrations in patients with neuroendocrine tumors. *J. Clin. Endoc. Metab.* **68,** 808–813.
64. Eckberg, D. L., Rea, R. F., Andersson, O. K., Hedner, T., Pernow, J., Lundberg, J. M., and Wallin, B. G. (1988) Baroreflex modulation of sympathetic activity and sympathetic neurotransmitters in humans. *Acta Physiol. Scand.* **133,** 221–231.
65. Maisel, A. S., Scott, N. A., Motulsky, H. J., Michel, M. C., Boublik, L. H., Rivier, J. E., Ziegler, M., Allen, R. S., and Brown, M. R. (1989) Elevation of plasma neuropeptide Y levels in congestive heart failure. *Am. J. Med.* **86,** 43–48.
66. Watson, J. D., Sury, M. R. J., Corder, R., Carson, R., Bouloux, P. M., Lowry, P. J., Besser, G. M., and Hinds, C. J. (1988) Plasma levels of neuropeptide tyrosine Y (NPY) are increased in human sepsis but unchanged during canine endotoxin shock despite raised catecholamine concentration. *J. Endocrinol.* **116,** 421–426.
67. Edvinsson, L., Ekblad, E., and Thulin, T. (1991) Circulating levels of neuropeptide Y and catecholamines in severe hypertension and after treatment to normotension. *Regul. Peptides* **32,** 279–287.
68. Franco-Cereceda, A., Owall, A., Settergren, G., Sollevi, A., and Lundberg, J. M. (1990) Release of neuropeptide Y and noradrenaline from the human heart after aortic occlusion during coronary artery surgery. *Cardiovasc. Res.* **24,** 241–246.
69. Lundberg, J. M., Hemsen, A., Fried, G., Theodorsson-Norheim, E., and Lagerkrantz, C. (1986) Co-release of neuropeptide Y (NPY)-like immunoreactivity and catecholamines in newborn infants. *Acta Physiol. Scand.* **126,** 471–473.
70. Castagne, V., Corder, R., Gaillard, R., and Mormede, P. (1987) Stress-induced changes of circulating neuropeptide Y in the rat: comparison with catecholamines. *Regul. Peptides* **19,** 55–63.
71. Polinsky, R. J., Kopin, I. J., Ebert, M. H., and Weise, V. (1980) The adrenal medullary response to hypoglycemia in patients with orthostatic hypotension. *J. Clin. Endocrinol. Metab.* **51,** 1401–1406.
72. Lundberg, J. M., Hokfelt, T., Hemsen, A., Theodeorsson-Norheim, E., Pernow, J., Hamberger, B., and Goldstein, M. (1986) Neuropeptide Y-like immunoreactivity in adrenaline cells of adrenal medulla and in tumors and plasma of pheochromocytoma patients. *Regul. Peptides* **13,** 169–182.
73. Fried, G., Terenius, L., Brodin, E., Efendic, S., Fahrenkrug, J., Goldstein, M., Hökfelt, T. (1987) Neuropeptide Y, enkephalin and noradrenaline coexist in sympathetic neurons innervating the bovine spleen. *Cell Tissue Res.* **243,** 495–508.
74. Majane, E. A., Alho, H., Kataoka, Y., Lee, C. H., and Yang, H.-Y. T. (1985) Neuropeptide Y in bovine adrenal glands: Distribution and characterization. *Endocrinology* **117,** 1162–1168.
75. Fischer-Colbrie, R., Diez-Guerra, J., Emson, P. C., and Winkler, H. (1986) Bovine chromaffin granules: immunological studies with antisera against neuropeptide Y, [met]enkephalin and bombesin. *Neuroscience* **18,** 167–174.

76. Sundler F., Hakanson R., Ekblad E., Uddman R., and Wahlestedt C. (1986) Neuropeptide Y in the peripheral adrenergic and enteric systems. *Int. Rev. Cytol.* **102,** 243–269.
77. Lundberg, J. M., Terenius, L., Hokfelt, T., and Goldstein, M. (1983) High levels of neuropeptide Y in peripheral noradrenergic neurons in various mammals including man. *Neurosci. Lett.* **42,** 167–172.
78. Higuchi, H. and Yang, H.-Y. T. (1986) Splanchnic nerve transection abolishes the age-dependent increase of neuropeptide Y-like immunoreactivity in rat adrenal gland. *J. Neurochem.* **46,** 1658–1660.
79. Higuchi, H., Yang, H.-Y. T., and Costa, E. (1988) Age-related bidirectional changes in neuropeptide Y peptides in rat adrenal glands, brain, and blood. *J. Neurochem.* **50,** 21,879–21,884.
80. Hexum, T. D., Majane, E. A., Russet, L. R., and Yang, H.-Y. T. (1987) Neuropeptide Y release from adrenal medulla after cholinergic stimulation. *J. Pharmacol. Exp. Ther.* **243,** 927–930.
81. Kataoka, Y., Majane, E. A., and Yang, H.-Y. T. (1985) Release of NPY-like immunoreactive material from primary cultures of chromaffin cells prepared from bovine adrenal medulla. *Neuropharmacology* **24,** 693–695.
82. Briand, R., Yamaguchi, N., and Gagne, J. (1990) Corelease of neuropeptide Y-like immunoreactivity with catecholamines from the adrenal gland during splanchnic nerve stimulation in anesthetized dogs. *Can. J. Physiol. Pharmacol.* **68,** 363–369.
83. Fischer-Colbrie, R., Iacangelo, A., and Eiden, L. E. (1988) Neural and humoral factors separately regulate neuropeptide Y, enkephalin, and chromogranin A and B mRNA levels in rat adrenal medulla. *Proc. Natl. Acad. Sci. USA* **85,** 3240–3244.
84. deQuidt, M. E. and Emson, P. C. (1986) Neuropeptide Y in the adrenal gland: characterisation, distribution and drug effects. *Neuroscience* **19,** 1011–1022.
85. Takahashi, K., Mouri, T., Itoi, K., Sone, M., Ohneda, M., Murakami, O., Nozuki, M., Tachibana, Y., and Yoshinaga, K. (1987) Increased plasma immunoreactive neuropeptide Y concentrations in phaeochromocytoma and chronic renal failure. *J. Hypertens.* **5,** 749–753.
86. Morris, M., Kapoor, V., and Chalmers, J. (1987) Plasma neuropeptide Y concentration is increased after hemorrhage in conscious rats: relative contributions of sympathetic nerves and the adrenal medulla. *J. Cardiovasc. Pharmacol.* **9,** 541–545.
87. Myers, A. K., Wahlestedt, C., Shen, G. H., and Zukowska-Grojec, Z. (1990) Neuropeptide Y (NPY) in mammalian platelets. *FASEB J.* **4,** A882.
88. Myers, A. K., Farhat, M. Y., Shen, G. H., Debinski, W., Wahlestedt, C., and Zukowska-Grojec, Z. (1990) Platelets as a source and site of action for neuropeptide Y. *Ann. NY Acad. Sci.* **611,** 408–410.
89. Ogawa, T., Kitamura, K., Kawamoto, M., Eto, T., and Tanaka, K. (1989) Increased immunoreactive neuropeptide Y in platelets of spontaneously hypertensive rats (SHR). *Biochem. Biophys. Res. Comm.* **165,** 1399–1405.
90. Ericsson, A., Hemsen, A., Lundberg, J. M., and Persson, H. (1991) Detection of neuropeptide Y-like immunoreactivity and messenger RNA in rat platelets: The effects of vinblastine, reserpine, and dexamethasone on NPY expression in blood cells. *Exp. Cell Res.* **192,** 604–611.

91. Persson, H., Ericsson, A., Hemsen, A., Hokfelt, T., Larhammar, D., Lundberg, J. M., McIntyre, K. R., and Schalling, M. (1989) Expression of neuropeptide Y tyrosine (NPY) messenger RNA and peptide in non-neuronal cells, in *Neuropeptide Y.* Mutt, V., Fuxe, K., Hökfelt, T., and Lundberg, J. M., eds., Raven, New York, pp. 43–50.
92. Corder, R., Pralong, F. P., and Gaillard, R. C. (1990) Comparison of hypotension-induced neuropeptide Y release in rats subjected to hemorrhage, endotoxemia, and infusions of vasodepressor agents. *Ann. NY Acad. Sci.* **611,** 474–476.
93. Zerbe, R. L., Feurestein, G., and Kopin, I. J. (1981) Effect of captopril on cardiovascular, sympathetic, and vasopressin response to hemorrhage. *Eur. J. Pharmacol.* **72,** 391–395.
94. Hawinger, J. (1987) Formation and regulation of platelet and fibrin hemostatic plug. *Hum Pathol.* **18,** 11–122.
95. Pasqui, A. L., Capecchi, P. L., Cecatelli, L., Mazza, S., Gistri, A., Laghi Pasisni, F., and DiPerri T. (1991) Nitroprusside *in vitro* aggregation and intracellular calcium translocation. Effect of hemoglobin. *Thromb. Res.* **61,** 113–122.
96. Abel, P. W., Han, C., Noe, B. D., and McDonald, J. K. (1988) Neuropeptide Y: vasoconstrictor effects and possible role in cerebral vasospasm after experimental subarachnoid hemorrhage. *Brain Res.* **463,** 250–258.
97. Suzuki, Y, Sato, S., Suzuki, H., Namba, J., Ahtake, R., Hashigami, Y., Suga, S., Ishihara, N., and Shimoda, S.-I. (1989) Increased neuropeptide Y concentrations in cerebrospinal fluid from patients with aneurysmal suarachnoid hemorrhage. *Hypertension* **20,** 1680–1684.
98. Waeber, B., Corder, R., Aubert, J.-F., Nussberger, J., Gaillard, R., and Brunner, H. R. (1987) Influence of sodium intake on circulating levels of neuropeptide Y. *Life Sci.* 1391–1396.
99. Lundberg, J. M., and Tatemoto, K. (1982) Pancreatic polypeptide family (APP, BPP, NPY and PYY) in relation to sympathetic vasocontriction resistant to alpha-adrenoceptor blockade. *Acta Physiol. Scand.* **116,** 393–402.
100. Zukowska-Grojec, Z., Haass, M., and Bayorh, A. (1986) Neuropeptide Y and peptide YY mediate non-adrenergic vasoconstriction and modulate adrenergic responses in rats. *Regul. Peptides* **15,** 99–110.
101. Zukowska-Grojec, Z., Marks, E. S., and Haass, M. (1987) Neuropeptide Y is a potent vasoconstrictor and a cardiodepressant in rat. *Am.J. Physiol.* **253,** H1234–H1239.
102. Corder, R., Lowry, P. J., Wilkinson, S. J., and Ramage, A. G. (1986) Comparison of the hemodynamic actions of neuropeptide Y, angiotensin II and noradrenaline in anesthetised cats. *Eur. J. Pharmacol.* **121,** 25–30.
103. MacLean, M. R., and Hiley, C. R. (1990) Effect of neuropeptide Y on cardiac output, its distribution, regional blood flow and organ vascular resistances in the pithed rat. *Br. J. Pharmacol.* **99,** 340–342.
104. Hellstrom, P., Olerup, O., And Tatemoto, K. (1985) Neuropeptide Y may mediate effects of sympathetic nerve stimulations on colonic motility and blood flow in the cat. *Acta Physiol. Scand.* **124,** 613–624.
105. Minkes, R. K., Bellan, J. A., Saroyan, R. M., Kerstein, M. D., Coy, D. H., Murphy, W. A., Nossaman, B. D., McNamara, D. B., and Kadowitz, P. J. (1990) Analysis of cardiovascular and pulmonary responses to endothe-

lium-1 and endothelium-3 in anesthetized cat. *J. Pharmacol. Exp. Ther.* **253,** 1118–1125.

106. Lang, S. A., Maron, M. B., Maender, K. C., and Pilati, C. F. (1992) Circulating neuropeptide Y does not produce pulmonary hypertension during massive sympathetic activation. *J. Appl. Physiol.* **73,** 117–122.
107. Martling, C. R., Matran, R., Alving, K., Hokfelt, T. and Lundberg, J. M. (1990) Innervation of lower airways and neuropeptide effects on bronchial and vascular tone in the pig. *Cell Tissue Res.* **260,** 223–233.
108. Aizawa, Y., Murata, M., Hayashi, M., Funazaki, T., Seiki, I., and Shibata, A. (1985) Vasoconstrictor effect of neuropetide Y (NPY) on canine coronary artery. *Jap. Circ. J.* **49,** 584–588.
109. Martin, S. E. and Patterson, R. E. (1989) Coronary constriction due to neuropeptide Y: alleviation with cyclooxygenase blockers. *Am. J. Physiol.* **257,** H927–H934.
110. Komaru, T., Ashikawa, K., Kanatsuka, H., Sekiguchi, N., Suzuki, T., and Takishima, T. (1990) Neuropeptide Y modulates vasoconstriction in the coronary microvessels in the beating canine heart. *Circ. Res.* **67,** 1142–1151.
111. Han, C. and Abel, P. W. (1987) Neuropetide Y potentiates contraction and inhibits relaxation of rabbit coronary arteries. *J. Cardiovasc. Pharmacol.* **9,** 675–681.
112. Minson, R. B., McRitchie, R. J., and Chalmers, J. P. (1990) Effects of neuropeptide Y on baroreflex control of heart rate and myocardial contractility in conscious rabbits. *Clin. Exp. Pharmacol. Physiol.* **17,** 39–49.
113. Potter, E. F., Mitchell, L., McCloskey, M. J. D., Tseng, A., Goodman, A. E., Shine, J., and McCloskey, D. I. (1989) Pre- and postjunctional actions of neuropeptide Y and related peptides. *Regul. Peptides* **25,** 167–177.
114. Minson, R. B., McRitchie, R. J., and Chalmers, J. P. (1987) Effects of neuropeptide Y on left ventricular function in the conscious rabbit. *Clin. Exp. Pharmac. Physiol.* **14,** 263–266.
115. Minson, R. B., McRitchie, R. J., Morris, M. J., and Chalmers, J. P. (1990) Effects of neuropeptide Y on cardiac performance and renal blood flow in conscious normotensive and renal hypertensive rabbits. *Clin. Exp. Hypertens.* **12,** 267–284.
116. Allen, J. M., Gjorstrup, P., Bjorkman, J.-A., Abrahamsson, E. K. T., and Bloom, S. R. (1986) Studies on cardiac distribution and function of neuropeptide Y. *Acta Physiol. Scand.* **126,** 405–411.
117. Wahlestedt, C., Wohlfart, B., and Hakanson, R. (1987) Effects of neuropeptide Y (NPY) on isolated guinea-pig heart. *Acta Physiol. Scand.* **129,** 159–463.
118. Balasubramaniam, A., Grupp, I., Matlib, M. A., Benza, R., Jackson, R. L., Fisher, J. E., and Grupp, G. (1988) Comparison of the effects of neuropeptide Y (NPY) and 4-norleucine-NPY on isolated perfused rat hearts: efefetcs of NPY on atrial and ventricular strips of rat heart and on rabbit heart mitochondria. *Regul. Peptides* **21,** 289–299.
119. Franco-Cereceda, A., Lundberg, J. M., and Dahlof, C. (1985) Neuropeptide Y and sympathetic control of heart contractility and coronary vascular tone. *Acta Physiol. Scand.* **124,** 361–369.
120. Rioux, F., Bachelard, H., Martel, J. C., and St.-Pierre, S. (1986) The vasoconstrictor effect of neuropeptide Y and related peptides in the guinea pig isolated heart. *Peptides* **7,** 27–31.

121. Maturi, A. F., Greene, R., Speir, E., Burrus, C., Dorsey, L. M. A., Markle, D. R., Maxwell, M., Schmidt, W., Goldstein, S. R., and Patterson, R. E. (1989) Neuropeptide Y: A peptide found in human coronary arteries constricts primarily small coronary arteries to produce myocardial ischemia in dogs. *J. Clin. Invest.* **83,** 1217–1224.
122. Svendsen, J. H., Sheikh, S. P., Jorgensen, J., Mikkelsen, J. D., Paaske, W. P., Sejrsen, P., and Haunso, S. (1990) Effects of neuropeptide Y on regulation of blood flow rate in canine myocardium. *Am. J. Physiol.* **259,** H1709–H1717.
123. Clarke, J. G., Kerwin, R., Larkin, S., Lee, Y., Yacoub, M., Davies, G. J., Hackett, D., Dawbarn, D., Bloom, S. R., and Maseri, A. (1987) Coronary artery infusion of neuropeptide Y in patients with angina pectoris. *Lancet* **i,** 1057–1059.
124. Grundemar, L. (1991) Actions of neuropeptide Y on peripheral and central targets. *Acta Physiol. Scand.* Ph.D. Thesis.
125. Kassis, S., Olasmaa, M. Terenius, L., and Fishman, P. H. (1987) Neuropeptide Y inhibits cardiac adenylate cyclase through a pertussin toxin-sensitive G protein. *J. Biol. Chem.* **262,** 3429–3431.
126. Echtenkamp, S. F. and Dandrifge, P. F. (1989) Renal actions of neuropeptide Y in the primate. *Am. J. Physiol.* **256,** F524–F531.
127. Allen, J. M., Raine, A. E. G., Ledingham, J. G. G., and Bloom, S. R. (1985) Neuropeptide Y: a novel renal peptide with vasoconstrictor and natiuretic activity. *Clin. Sci.* **68,** 373–377.
128. Leys, K., Schachter, M., and Sever, P. (1987) Autoradiographic localisation of NPY receptors in rabbit kidney: comparison with rat, guinea-pig and human. *Eur. J. Pharmacol.* **134,** 133–137.
129. Schachter, M., Miles, C. M. M., Leys, K., and Sever, P. S. (1987) Characterization of neuropeptideY receptors in rabbit kidney: preliminary comparisons with rat and human kidney. *J. Cardiovasc. Pharmacol.* **10 (Suppl. 12),** S157–S162.
130. Smyth, D. D., Blandford, D. E., and Thom, S. L. (1988) Disparate effects of neuropeptide Y and clonidine on the excretion of sodium and water in the rat. *Eur. J. Pharmacol.* **152,** 157–162.
131. Pfister, A., Waeber, B., Nussberger, J., and Brunner, H. R. (1986) Neuropeptide Y normalizes renin secretion in adrenalectomized rats without changing blood pressure. *Life Sci.* **39,** 2161–2167.
132. Waeber, B., Evequoz, D., Aubert, J.-F., Fluckiger, J.-P., Juilerat, L., Nussberger, J., and Brunner, H. R. (1990) Prevention of renal hypertension in the rat by neuropeptide Y. *J. Hypertension* **8,** 21–25.
133. Hackenthal, E., Aktories, K., Jakobs, K. H., and Lang, R. E. (1987) Neuropeptide Y inhibits renin release by a pertussis toxin-sensitive mechanism. *Am. J. Physiol.* **252,** F543–F550.
134. Allen, J. M., Hanson, C., and Lee, Y. (1986) Renal effects of the homologous neuropeptides, pancreatic polypeptide (PP) and neuropeptide Y (NPY) in conscious rabbits. *J. Physiol. (Lond.)* **376,** 24P.
135. Corder, R., Vallotton, M. B., Lowry, P. J., and Ramage, A. G. (1989) Neuropeptide Y lowers plasma renin activity in the anesthetized cat. *Neuropeptides* **14,** 111–114.
136. Edvinsson, L., Emson, P., McCulloch, J., Tatemoto, K., and Uddman, R. (1983) Neuropeptide Y: cerebrovascular innervation and vasomotor responses in the cat. *Neurosci. Lett.* **43,** 79–84.

137. Edvinsson, L. (1985) Characterization of the contractile effect of neuropeptide Y in feline cerebral arteries. *Acta Physiol. Scand.* **125,** 33–41.
138. Andriantsitohaina, R., Andre, P., and Stocklet, J.-C. (1990) Pertussis toxin abolishes the effect of neuropeptide Y on rat resistance arteriole contraction. *Am. J. Physiol.* **259,** H1427–H1432.
139. Lundberg, J. M., Hemsen, A., Larsson, O., Rudehill, A., Saria, A., and Fredholm, B. (1985) Neuropeptide Y receptor in pig spleen: binding characteristics reduction of cyclic AMP formation and calcium antagonist inhibition of vasoconstriction. *Eur. J. Pharmacol.* **145,** 21–29.
140. Edvinsson, L., Hakanson, R., Steen, S., Sundler, R., Uddman, R. and Wahlestedt, C. (1985) Innervation of human omental arteries and veins and vasomotor responses to noradrenaline, neuropeptide Y, substance P and vasoactive intestinal peptide. *Regul. Peptides* **12,** 67–79.
141. Edvinsson, L., Ekblad, E. Hakanson, R., and Wahlestedt, C. (1984) Neuropeptide Y potentiates the effect of various vasoconstrictor agents on rabbit blood vessels. *Br. J. Pharmacol.* **83,** 519–525.
142. Fallgren, B., Arlock, P., Jansen, I., and Edvinsson, L. (1990) Neuropeptide Y in the cerebrovascular function: comparison of membrane potential changes and vasomotor responses evoked by NPY and other vasoconstrictors in the guinea pig basilar artery. *Neurosci. Lett.* **114,** 117–122.
143. Tuor, U. I., Kelly, P. A. T., Edvinsson, L., and McCulloch, J. (1990) Neuropeptide Y and the cerebral circulation. *J. Cereb. Blood Flow Metab.* **10,** 591–601.
144. Franco-Cereceda, A. (1989) Endothelin- and neuropetide Y-induced vasoconstriction of human epicardial coronary arteries in vitro. *Br. J. Pharmacol.* **97,** 968–972.
145. Maron, M. B., Lang, S. A., Maender, K. C., and Pilati, C. F. (1991) Effect of pathophysiological plasma concentrations of neuropeptide Y (NPY) on pulmonary hemodynamics. *FASEB J.* **5,** A1428.
146. Wahlestedt, C., Hakanson, R., Vaz, C. A., and Zukowska-Grojec, Z. (1990) Norepinephrine and neuropeptide Y: vasoconstrictor cooperation *in vivo* and *in vitro*. *Am. J. Physiol.* **258,** R736–R742.
147. Revington, M. L. and McCloskey, D. I. (1988) Neuropeptide Y control of vascular resistance in skeletal muscle. *Regul. Peptides* **23,** 331–342.
148. Ekblad, E., Edvinsson, L., Wahlestedt, C., Uddman, R., Hakanson, R., and Sundler, F. (1984) Neuropeptide Y co-exists and cooperates with noradrenaline in perivascular nerve fibers. *Regul. Peptides* **8,** 225–235.
149. Westfall, T. C., Chen, W., Ciarlegio, A., Henderson, K., Del Valle, K., Curfman-Falvey, M. and Naes, L. (1990) *In vitro* effects of neuropeptide Y at the vascular neuroeffector junction. *Ann. NY Acad. Sci.* **611,** 145–155.
150. Hanko, J. H., Tornebtandt, K., Hardebo, J. E., Kahrstrom, J., Mobim, A., and Owman, Ch. (1986) Neuropeptide Y induces and modulates vasoconstriction in intracranial and peripheral vessels of animals and man. *J. Auton. Pharmacol.* **6,** 117–124.
151. Ekblad, E., Hakanson, R., and Sundler, F. (1984) VIP and PHI coexist with an NPY-like peptide in intramural neurones of the small intestines. *Regul. Peptides* **10,** 47–55.
152. Duesler, J. G., Daly, R. N., and Hieble, J. P. (1990) Studies on the mechanism of neuropeptide Y induced potentiation of neurogenic vasoconstriction in the isolated rabbit ear artery. *Am. J. Hypertens.* **3,** 796–799.

153. Vu, H. Q., Budai, D., and Duckles, S. P. (1989) Neuropeptide Y preferentially potentiates responses to adrenergic nerve stimulation by increasing rate of contraction. *J. Pharmacol. Exp. Ther.* **251,** 852–857.
154. Neild, T. O. (1987) Actions of neuropeptide Y on innervated and denervated rat tail arteries. *J. Physiol. (Lond.)* **386,** 19–30.
155. Mabe, Y., Perez, R., Tatemoto, K., and Huidobro-Toro, J. P. (1987) Chemical sympathectomy reveals pre- and postsynaptic effects of neuropeptide Y (NPY) in the cardiovascular system. *Experientia* **43,** 1018–1020.
156. Fried, G. and Thoresen, M. (1990) Effects of neuropeptide Y and noradrenaline on uterine artery blood presure and blood flow velocity in the pregnant guinea-pig. *Regul. Peptides* **28,** 1–9.
157. Lundberg, J. M., Torsell, L., Sollevi, A., Theodorsson-Norheim, E., Pernow, J., Anggard, A., and Hamberger, B. (1985) Neuropeptide Y and sympathetic vascular control in man. *Regul. Peptides* **13,** 41–5.
158. Wahlestedt, C., Edvinsson, L., Ekblad, E., and Hakanson, R. (1985) Neuropeptide Y potentiates noradrenaline-evoked vasoconstriction: mode of action. *J. Pharmacol. Exp. Ther.* **234,** 735–741.
159. Pernow, J. (1989) Actions of the constrictor (NPY and endothelin) and dilator (substance P, CGRP, and VIP) peptides on pig splenic and human skeletal muscle arteries: involvement of the endothelium. *Brit. J. Pharmacol.* **97,** 983–989.
160. Pernow, J., Kahan, T., Hjemdalh, P., and Lundberg, J. M. (1988) Possible involvement of neuropeptide Y in sympathetic vascular control of canine skeletal muscle. *Acta Physiol. Scand.* **132,** 43–50.
161. Corder, R., Lowry, P. J., and Withrington, P. G. (1987) The actions of the peptides, neuropeptide Y and peptide YY, on the vascular and capsular smooth muscle of the isolated, blood perfused spleen of the dog. *Br. J. Pharmacol.* **90,** 785–790.
162. Wahlestedt, C., Grundemar, L., Hakanson, R., Heilig, M., Shen, G. H., Zukowska-Grojec, Z., and Reis, D. J. (1990) Neuropeptide Y receptor subtypes, Y1 and Y2. *Ann. NY Acad. Sci.* **611,** 7–26.
163. Wahlestedt, C. (1987) Neuropeptide Y (NPY): Actions and interactions in neurotransmission. *Acta Physiol Scand.* Ph.D.Thesis.
164. Chang, R., and Lotti, V. J. (1988) Specific [^{3}H]proprionyl-neuropeptide Y (NPY) binding in rabbit aortic membranes: comparison with binding in rat brain and biological responses in rat vas deferens. *Biochem. Biophys. Res. Comm.* **151,** 1213–1219.
165. Michel, M. C., Schlicker, E., Fink, K., Boublik, J., Gothert, M., Willette, R. N., Daly, R. N., Hieble, P., Rivier, J., and Motulsky, H. J. (1990) Distinction of NPY receptors *in vitro* and *in vivo*. I. NPY-(18-36) discriminates NPY receptor subtypes *in vitro*. *Am. J. Physiol.* **259,** E131–E1
166. Fuhlendorff, J., Gether, U., Aakerlund, L., Langeland-Johansen, N., Thogersen, H., Melberg, S. G., Bang-Olsen, U., Tharstrup, O., and Schwartz, T. W. [Leu^{31}, Pro^{34}]Neuropeptide Y—a specific Y1 receptor agonist. *Proc. Natl. Acad. Sci. USA* **265,** 11,706–11,712.
167. Aakerlund, L., Gether, U., Fuhlendorff, J., Schwartz, T. W., and Thastrup, O. (1990) Y1 receptors for neuropeptide Y are coupled to mobilization of intracellular calcium and inhibition of adenylate cyclase. *FEBS Lett.* **260,** 73–78.

168. Boublik, J., Scott, N., Taulane, J., Goodman, M., Brown, M. R., and Rivier, J. (1989) Neuropeptide Y and neuropeptide Y(18-36). *Int. J. Peptide Protein Res.* **33,** 11–15.
169. Reynolds, E. E., and Yakata, S. (1988) Neuropeptide Y receptor-effector coupling mechanisms in cultured vascular smooth muscle cells. *Biochem. Biophys. Res. Comm.* **151,** 919–925.
170. Mejia, J. A., Pernow, J., von Holst, H., Rudehill, A., and Lundberg, J. M. (1988) Effects of neuropeptide Y, calcitonin gene-related peptide, substance P, and capsaicin on cerebral arteries in man and animals. *J. Neurosurg.* **69,** 913–918.
171. Spedding, M. (1987) Three types of Ca^{2+} channel explain discrepancies. *Trends Pharmacol. Sci.* **8,** 115–117.
172. Dahlof, C., Dahlof, P., and Lundberg, J. M. (1985) Neuropeptide Y (NPY): enhancement of blood pressure increase upon alpha-adrenoceptor activation and direct pressor effect in pithed rats. *Eur. J. Pharmacol.* **109,** 289–292.
173. Fuxe, K., Agnati., L. F., Harfstrand, A., Janson, A. M., Neumeyer, A., Andersson, K., Ruggeri, M., Zoli, M., and Goldstein, M. (1986) Morphofunctional studies on the neuropeptide Y/adrenaline costoring terminal systems in the dorsal cardiovascular region of the medulla oblongata. Focus on receptor-receptor interactions in cotransmission, in *Progress in Brain Research,* vol. 68. Hokfelt, T., Fuxe., K., and Pernow, B., eds., Elsevier, **68,** pp. 303–320.
174. Lopez, L. F., Perez, A., St-Pierre, S., and Huidobro-Toro, J. P. (1989) Neuropeptide tyrosine (NPY)-induced potentiation of the pressor activity of catecholamines in conscious rats. *Peptides* **10,** 551–558.
175. Aubert, J. F., Waeber, B., Rossier, K., Geering, J., Nussberger, J., and Brunner, H. R. (1988) Effects of neuropeptide Y on blood pressure response to various vasoconstrictor agents. *J. Pharmacol. Exp. Ther.* **246,** 1088–1092.
176. Andriantsitohaina, R. and Stocklet, J.-C. (1988) Potentiation by neuropeptide Y of vasoconstriction in rat resistance arteries. *Br. J. Pharmacol.* **95,** 419–428.
177. Oshita, M., Kigoshi, S., and Muramatsu, I. (1989) Selective potentiation of extracellular Ca^{2+}-dependent contraction by neuropeptide Y in rabbit mesenteric arteries. *Gen. Pharmacol.* **20,** 363–367.
178. Mohy El-Din, M. M. and Malik, K. U. (1988) Neuropeptide Y stimulates renal prostaglandin synthesis in the isolated rat kidney: contribution of Ca^{++} and calmodulin. *J. Pharmacol. Exp. Ther.* **246,** 479–484.
179. Lundberg, J. M., and Stjarne, L. (1984) Neuropeptide Y (NPY) depresses the secretion of [^{3}H]-noradrenaline and the contractile response evoked be field stimulation in the rat vas deferens. *Acta Physiol. Scand.* **120,** 477–479.
180. Westfall, T. C., Carpentier, S., Chen, X., Beinfeld, M. N. C., Naes, L., and Meldrum, M. J. (1987) Prejunctional and postjunctional effects of neuropeptide Y at the noradrenergic neuroeffector junction of the perfused mesenteric arterial bed of the rat. *J. Cardiovasc. Pharmacol.* **10,** 716–722.
181. Glover, W. E. (1985) Increased sensitivity of rabbit ear artery to noradrenaline following perivascular nerve stimulation may be a response to neuropeptide Y released as a cotransmitter. *Clin. Exp. Pharmacol. Physiol.* **12,** 227–230.
182. Revington, M. L., McCloskey, D. I., and Potter, E. K. (1987) Effects of neuropeptide Y on the pressor responses to phenylephrine and to activation of

the sympathetic nervous system in anesthetized rats. *Clin. Exp. Pharmacol. Physiol.* **14,** 703–710.

183. Burnstock, G. and Kennedy, C. (1986) A dual function for adenosine 5'-triphospate in the regulation of vascular tone. *Circ. Res.* **58,** 319–330.
184. Steinsland, O. S., Furchgott, R. F., and Kirpekar, S. M. (1973) Biphasic vasoconstriction of the rabbit ear artery. *Circ. Res.* **32,** 49–58.
185. Abel, P. W., and Han, C. (1989) Effects of neuropeptide Y on contraction, relaxation, and membrane potential of rabbit cerebral arteries. *J. Cardiovasc. Pharmacol.* **13,** 52–63.
186. Hermsmeyer, K., Abel., and Harder, D. R. (1982) Membrane electrical responses to histamine in muscle cell of cat cerebral artery, in *Cerebral Blood Flow: Effects of Nerves and Neurotransmitters.* Heistad, D. D., and Marcus, M. L., eds. Elsevier, New York, pp. 3–12.
187. Stupecky, G. L., Murray, D. L., and Purdy, R. E. (1986) Vasoconstrictor threshold synergism and potentiation in the rabbit isolated thoracic aorta. *J. Pharmacol. Exp. Ther.* **238,** 802–808.
188. Erne, P. and Hersmeyer, K. (1988) Intracellular Ca^{2+} release in vascular muscle cells by caffeine, ryanodine, norepinephrine, and neuropeptide Y. *J. Cardiovasc. Pharmacol.* **12 (Suppl. 5),** S85–S91.
189. Budai, D., Vu, H. Q., and Duckles, S. P. (1989) Endothelium removal does not affect potentiation by neuropeptide Y in rabbit ear artery. *Eur. J. Pharmacol.* **168,** 97–100.
190. Daly, R. N. and Hieble J. P. (1987) Neuropeptide Y modulates adrenergic neurotransmission by an endothelium dependent mechanism. *Eur. J. Pharmacol.* **138,** 445–446.
191. Hieble, J. P., Duesler, J. G., and Daly, R. N. (1989) Effects of neuropeptide Y on responses of isolated blood vessels to norepinephrine and sympathetic nerve stimulation. *J. Pharmacol. Exp. Ther.* **250,** 523–528.
192. Morris, M. J. (1990) Neuropeptide Y inhibits relaxations of the guinea-pig uterine artery produced by VIP. *Peptides* **11,** 381–386.
193. Juan, H., Sametz, W., Saria, A., and Poch, G. (1988) Eicosapentaeoic acid inhibits vasoconstrictor- and noradrenaline-potentiating effects of neuropeptide Y in the isolated rabbit ear. *J. Auton. Nerv. Syst.* **22,** 237–242.
194. Sheikh, S. P. and Williams, J. A. (1990) Structural characterization of Y1 and Y2 receptors for neuropeptide Y and peptide YY by affinity cross-linking. *J. Biol. Chem.* **265,** 8304–8310.
195. Grundemar, L., Wahlestedt, C., Shen, G. H., Zukowska-Grojec, Z., and Hakanson, R. (1990) Biphasic blood pressure response to neuropeptide Y in anesthetized rats. *Eur. J. Pharmacol.* **179,** 83–87.
196. Mousli, M., Bueb, J.-L., Bronner, C., Rouot, B., and Landry, Y. (1990) G protein activation: a receptor-independent mode of action for cationic amphiphilic neuropeptides and venom peptides. *Trends Pharmacol. Sci.* **11,** 358–362.
197. Brown, M. R., Scott, N. A., Boublik, J., Allen, J. M., Ehlers, R., Landon, M., Crum, R., Ward, D., Bronsther, O., Maisel, A., and Rivier, J. (1989) Neuropeptide Y: biological and clinical studies, in *Neuropeptide Y,* Mutt, V., Fuxe, K., Hökfelt, T., and Lundberg, J. M., eds., Raven, New York, pp. 321–329.
198. Brown, M. R., Feinstein, R. D., Malewicz, K., Broide, D., and Rivier, J. (1991) Neuropeptide Y^{18-36} (NPY^{18-36}): mechanism of cardiovascular actions and antagonism of NPY^{1-36}. *FASEB J.* A1580.

199. Itoi, K., Mouri, T., Takahashi, K., Sasaki, S., Imai, Y., and Yoshinaga, K. (1986) Synergistic pressor action of neuropeptide Y and norepinephrine in conscious rats. *J. Hypertens.* **4(Suppl. 6),** S247–S250.
200. Sugden , D., Vanacek, J., Klein, D. C., Thomas, T. P. and Anderson, W. B. (1985) Activation of protein kinase C potentiates isoprenaline-induced cyclic AMP accumulation in rat pinealocytes. *Nature* **314,** 359–361.
201. Michel, M. C., Feth, F., Racher, W., and Brodde, O.-E. (1991) A human cell line with homogenous populations of beta-1 adrenergic and Y1-like NPY receptors. *FASEB J.* A1067.
202. Schreuder, W. O., Schneider, A. J., Groeneveld, A. B. J. and Thijis, L. G. (1989) Effect of dopamine vs norepinephrine on hemodynamics in septic shock. *Chest* **95,** 1282–1288.
203. Szemeredi, K., Bagdy, G., Stull, R., Keiser, H. R,, Kopin, I. J., Goldstein, D. S. (1988) Sympathoadrenomedullary hyper-responsiveness to yohimbine in juvenile spontaneously hypertensive rats. *Life Sci.* **43,** 1063–1068.
204. Blaes, N. and Boisel, J.-P. (1983) Growth-stimulating effect of catecholamines on rat aortic smooth muscle cells in culture. *J. Cell. Physiol.* **116,** 167–172.
205. Campbell-Boswell, M. and Robertson, L., Jr. (1981) Effects of angiotensin II and vasopressin on human smooth muscle cells in vitro. *Exp. Mol. Pathol.* **35,** 265–276.
206. Blank, R. S. and Owens, G. K. (1990) Platelet-derived growth factor regulates actin isoform expression and growth state in cultured rat aortic smooth muscle cells. *J. Cell Physiol.* **142,** 635–642.
207. Lee, R. M. K., Nagahama, M., McKenzie, R., and Daniel, E. E. (1988) Peptide-containing nerves around blood vessels of stroke-prone spontaneously hypertensive rats. *Hypertension* **11 (Suppl. I),** I-117–I-120.
208. Bevan, R. (1984) Trophic effects of peripheral adrenergic nerves on vascular structure. *Hypertension* **(Suppl. III),** III-19–III-26.
209. Dhital, K. K., Gerli, R., Lincoln, J., Milner, P., Tanganelli, P., Weber, G., Fruschelli, C., and Burnstock, G. (1988) Increased density of perivascular nerves to the major cerebral vessels of the spontaneously hypertensive rat: differential changes in noradrenaline and neuropeptide Y during development. *Brain Res.* **444,** 33–45.
210. Cohen, R. A. (1985) Platelet-induced neurogenic coronary arterial contractions due to accumulation of the false neurotransmitter, 5-hydroxytryptamine. *J. Clin. Invest.* **75,** 286–292.
211a. Sung, C.-P., Arleth, A. J., and Feuerstein, G. (1991) Neuropeptide Y upregulates the adhesiveness of human endothelial cells for leukocytes. *Circ. Res.* **68,** 314–318.
211b. Feuerstein, G., Boyd, L. M., Ezra, D., and Goldstein, R. E. (1984) Effect of platelet-activating factor on coronary circulation of the domestic pig. *Am. J. Physiol.* **246,** H466–H471.
212. Connell, J. M. C., Corder, R., Asbury, J., Macpherson, S., Inglis, G. C., Lowry, P., Burt, A. D., and Semple, P. F. (1987) Neuropeptide Y in multiple endocrine neoplasia: release during surgery for pheochromocytoma. *Clin. Endocrin.* **26,** 75–84.
213. Dagget, P., Verner, I., and Carruthers, M. (1978) Intraoperative management of pheochromocytoma with sodium nitroprusside. *Br. Med. J.* **ii,** 311–313.

214. Hulting, J., Sollevi, A., Ullman, B., Franco-Cereceda, A., and Lundberg, J. M. (1990) Plasma neuropeptide Y on admission to a coronary care unit: raised levels in patients with left heart failure. *Cardiovasc. Res.* **24,** 102–108.
215. Chalmers, J., Morris, M., Kapoor, V., Cain, M., Elliot, J., Russell, A., Pilovsky, P., Minson, J., West, M., and Wing, L. (1989) Neuropeptide Y in the sympathetic control of blood pressure in hypertensive patients. *J. Clin. Exp. Hypertens.* **A11, (Suppl.1),** 59–66.
216. Solt, V. B., Brown, M. R., Kennedy, B., Kolterman, O. G., and Ziegler, M. G. (1990) Elevated insulin, norepinephrine, and neuropeptide Y in hypertension. *Am. J. Hypertens.* **3,** 823–828.
217. Goldstein, D. S. (1981) Plasma norepinephrine in essential hypertension; a study of the studies. *Hypertension* **3,** 48–52.
218. Jackowski, A., Crockard, A., Burnstock, G., and Lincoln, J. (1989) Alterations in serotonin and neuropeptide Y content of cerebrovascular sympathetic nerves following experimental subarachnoid hemorrhage. *J. Cardiovasc. Pharmacol.* **9,** 271–279.
219. McDonald, J. K., Han, C., Noe, B. D., and Abel, P. W. (1988) High levels of NPY in rabbit cerebrospinal fluid and immunohistochemical analysis of possible sources. *Brain Res.* **463,** 259–267.
220. Groves, A. C., Griffith, J., Leung, F., and Meek, R. N. (1972) Plasma catecholamines in patients with serious postoperative infections. *Ann. Surg.* **178,** 102–107.
221. Parker, M. M. and Parillo, J. E. (1983) Septic shock: hemodynamics and pathogenesis. *JAMA* **250,** 3324–3327.
222. McKenna, T. M. (1990) Prolonged exposure of rat aorta to low levels of endotoxin *in vitro* results in impaired contractility: association with vascular cytokine release. *J. Clin. Invest.* **86,** 160–168.
223. Evequoz, D., Waeber, B., Corder, R., Nussberger, J., Gaillard, R., and Brunner, H. R. (1987) Markedly reduced blood pressure responsiveness in endotoxemic rats: reversal by neuropeptide Y. *Life Sci.* **41,** 2573–2580.
224. Evequoz, D., Waeber, B., Aibert, J. F., Fluckinger, J. P., Nussberger, J., and Brunner, H. R. (1988) Neuropeptide Y prevents the blood pressure fall induced by endotoxin in conscious rats with adrenal medullectomy. *Circ. Res.* **62,** 25–30.
225. Teiltelman, G., Ross, R. A., Joh, T. H, and Reis, D. J. (1981) Differences *in utero* in activities of catecholamine biosynthetic enzymes in adrenals of spontaneously hypertensive rats. *Clin. Sci.* **61,** 227s–230s.
226. Zukowska-Grojec, Z., Shen, G., Miskovsky, C., Haass, M., and Wahlestedt, C. (1989) Increased vasporessor action of neuropeptide Y in hyperadrenergic hypertensive state. *Clin Res.* **37,** 307A.
227. Howe, P. R., Rogers, P. F., Morris, M. J., Chalmers, J. P., and Smith, R. M. (1986) Plasma catecholamines and neuropeptide Y as indices of sympathetic nerve activity in normotensive and stroke-prone spontaneously hypertensive rats. *J. Cardiovasc. Pharmacol.* **8,** 1113–1121.
228. Kawamura, K., Ando, K., and Takebayashi, S. (1989) Perivascular innervation of the mesenteric artery in spontaneously hypertensive rats. *Hypertension* **14,** 660–665.

229. Daly, R. N., Roberts, M. I., Ruffolo, R. R., Jr., and Hieble, J. P. (1988) The role of neuropeptide Y in vascular sympathetic neurotransmission may be enhanced in hypertension. *J. Hypertens.* **6 (Suppl. 4),** S535–S538.
230. Schiffrin, E. L. (1984) Alpha-1 adrenergic receptors in the mesenteric vascular bed of renal and spontaneously hypertensive rats. *J. Hypertension* 1984, **2, Suppl 3,** 431–432.
231. Miller, D. W. and Tessel, R. (1991) Age-dependent hyperresponsiveness of spontaneously hypertensive rats (SHR) to the pressor effects of intravenous neuropeptide Y (NPY): The role of mode of peptide administration and plasma NPY-like immunoreactivity. *J. Cardiovasc. Pharmacol.* **18,** 647–656.
232. Westfall, T. C., Han, S. P.,Kneupfer, M., Martin, J., Chen, X., Del Valle, Ciarleglio, A., and Naes, L. (1990) Neuropeptides in hypertension: role of neuropeptide Y and calcitonin gene related peptide. *Brit. J. Clin. Pharmacol.* **30,** 755–825.
233. Tsuda, K., Tsuda, S., Goldstein, M., and Masuyama, Y. (1990) Effects of neuropeptide Y on norepinephrine release in hypothalamic slices of spontaneously hypertensive rats. *Eur. J. Pharmacol.* **182,** 175–179.

Central Cardiovascular Actions of Neuropeptide Y

Moira A. McAuley, Xiaoli Chen, and Thomas C. Westfall

1. Introduction

Elucidation of the putative central hemodynamic actions of neuropeptide Y has been the subject of considerable research, some of which has been stimulated by its actions at the vascular neuroeffector junction. At that site, NPY produces three distinct effects: inhibition of the evoked release of norepinephrine from sympathetic nerves, direct contraction of vascular smooth muscle, and finally, potentiation of the contractile effects of a wide variety of vasoactive agents *(1–3)*. Moreover, the presence and abundance of NPY in central cardiovascular regulatory areas, as well as its colocalization with catecholamines in some of these regions, provide a morphological basis for and implicate this peptide in the modulation of catecholamines and hemodynamic regulation (*see* chapters by Sundler and by Hendry, this vol.).

The importance of the brain stem and higher areas in the regulation of central autonomic function has been the focus of considerable investigation over the last decade with attempts to define the operational microcircuitry involved in these control mechanisms. That the autonomic reflexes involving these regions might be mediated or modulated, at least in part, by peptidergic mechanisms involving NPY as the putative neurotransmitter is a particularly interesting concept—one that is also under investigation by numerous groups and will be considered in this chapter.

From: *The Biology of Neuropeptide Y and Related Peptides;*
W. F. Colmers and C. Wahlestedt, Eds. © 1993 Humana Press Inc., Totowa, NJ

2. NPY: Physiological Significance

2.1. Administration of NPY into the Cerebrospinal Fluid

Early studies addressing the issue of a possible regulatory function of NPY in central cardiovascular control mechanisms involved injection of this peptide into the ventricles of rats. This method permits the peptide to circulate in the cerebrospinal fluid (CSF) and eventually gain access to numerous loci throughout the neuraxes.

Administration of NPY (1250 pmol) into the cisterna magna of conscious rats or into the lateral ventricle of anesthetized rats results in hypotension and bradycardia (*4–6*; Fig. 1). Similarly, intrathecal injection of NPY (0.1–1 nmol) at a level of spinal segment T4 or T10 in the anesthetized or T10 in unanesthetized rats results in a decrease in arterial pressure. This depressor response was associated with a reduction in renal sympathetic nerve activity, as well as prolonged mesenteric and hindquarter vasodilation, resulting in a decrease in total peripheral resistance (Fig. 2; *2,7*). It is of note that this response did not appear to be owing to spinal vasoconstriction and ischemia, since there was no significant reduction in spinal microvascular resistance. In contrast, intracerebroventricular injection of lower doses of NPY (94–470 pmol) increases blood pressure and heart rate in the urethane-anaesthetized rat *(8,9)*. Another report that did not demonstrate any hemodynamic effect of intracisternally administered NPY in concentrations up to 2000 pmol in the conscious rat led to further confusion regarding the central cardiovascular role of NPY *(10)*. It is possible that variation in the anaesthetic state, species, and volume of injectate used in these studies may determine the direction of the response to NPY. Alternatively, these differences in the measured hemodynamic parameters may be the result of the difference in the initial site of action and subsequent circulation of the peptide resulting from injections into the cerebral spinal fluid at the different levels (third ventricle, fourth ventricle, and intrathecal) in these studies. Although these are suitable initial experiments, they demonstrate that further refinement of the injection technique is required to define the nuclei mediating the central cardiovascular actions of NPY and, thus, gain insight into the possible physiological significance of this peptide.

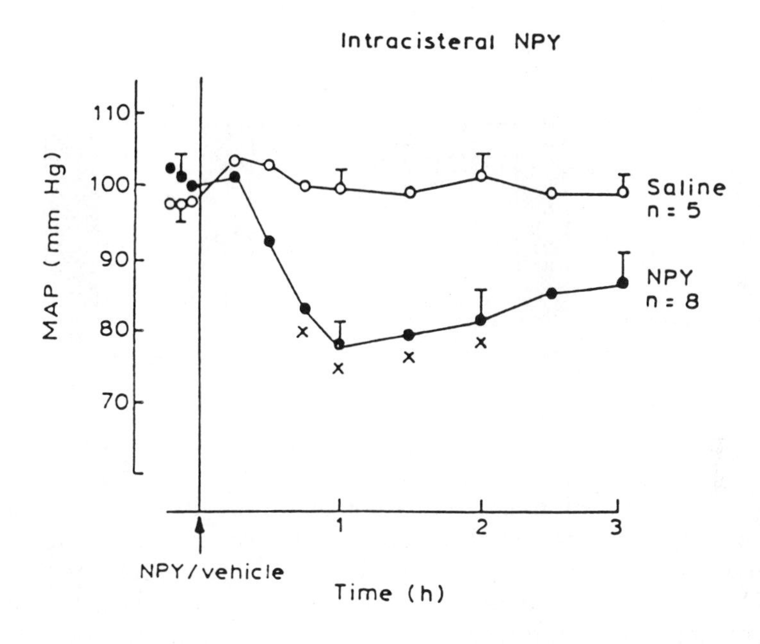

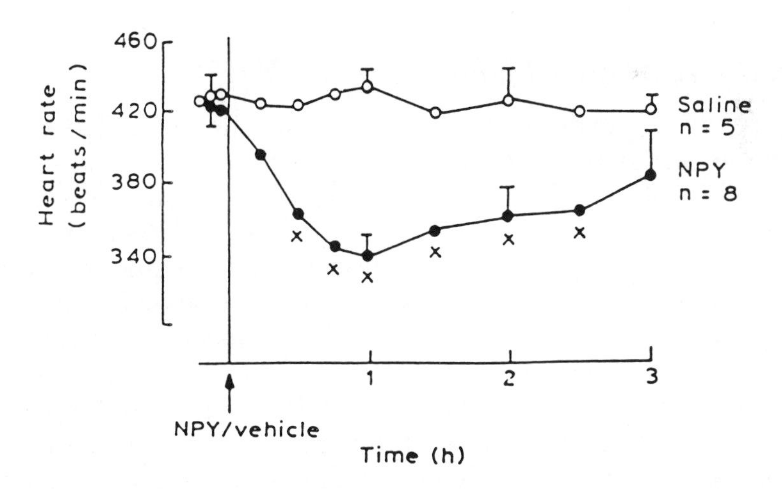

Fig. 1. Effect of intracisternal injection of NPY (1.25 nmol in 5 L saline) or vehicle on mean arterial pressure (MAP) and heart rate in the urethane-chloralose anesthetized rat. Preinjection readings on left of vertical bar. Points represent means ± SEM. *Significant difference from saline control ($p < 0.05$). (Reproduced from 5, with permission).

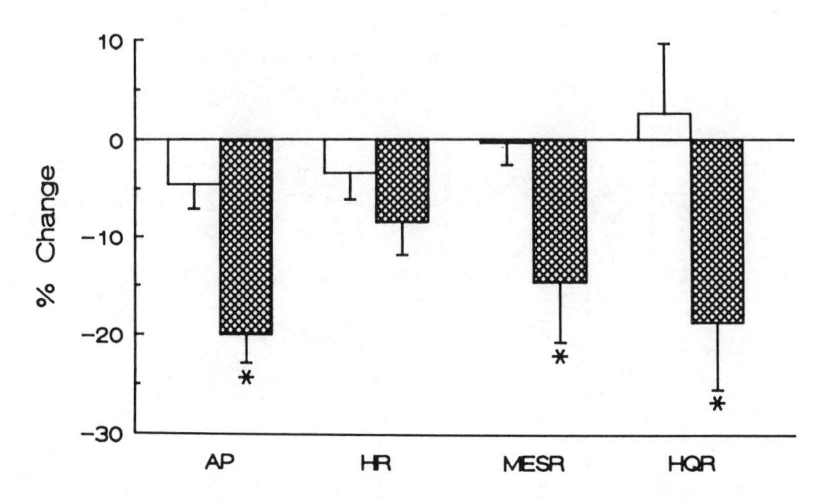

Fig. 2. Effects of intrathecal injection of saline (□) or NPY 4 nmol/kg; (▨) on mean arterial blood pressure (AP), heart rate (HR), mesenteric vascular resistances (MGSR), and hindquarters vascular resistances (HQR) of rats instrumented with Doppler flow probes. Data are plotted as percent changes at peak responses ($n = 8$; *significantly different from saline control value $p < 0.05$). (Reproduced from *7*, with permission).

2.2. Discrete Parenchymal Injections of NPY

Based on the presence of NPY-binding sites as well as NPY-immunoreactive (NPY-ir) cell bodies, fiber networks, and terminals (*see* chapters by Hendry and by Grundemar et al., this vol. for review), investigations have focused on several specific areas throughout the neuraxis to examine the hemodynamic actions of this peptide. In the mammalian brain, a heterogeneous distribution of NPY-ir has been described with highest concentrations in the hypothalamus, limbic system, brain stem nuclei, and cortex *(11–13)*.

The nucleus of the solitary tract (NTS) situated in the dorsomedial medulla is considered to play a major role in the mediation of the baroreceptor reflex. The transmitter of the primary afferent input is uncertain; however, a number of putative transmitters have been postulated, including glutamate and substance P *(14,15)*. NPY-ir perikarya, nerve terminals and NPY-binding sites have also been demonstrated in the dorsal subnuclei of the NTS, with a distribution that corresponds to the location of adrenergic neurons in this region as well as to the site of termination of aortic and carotid

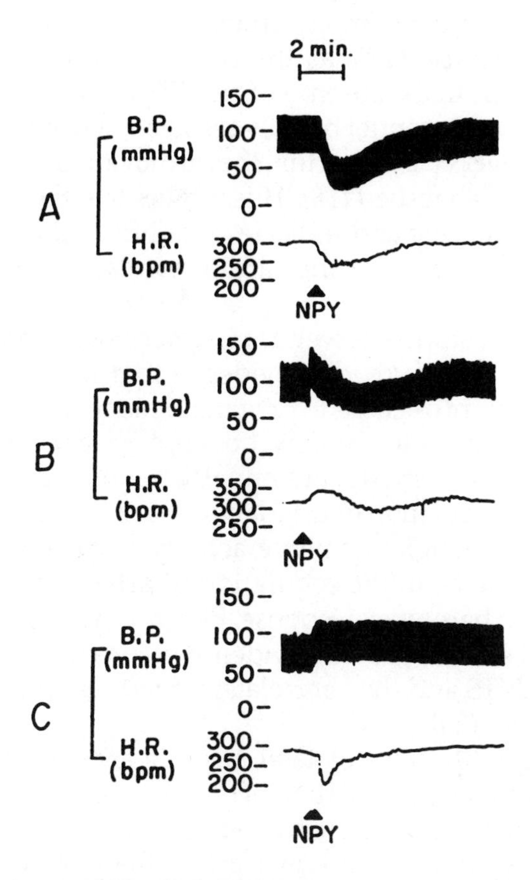

Fig. 3. Cardiovascular effects of unilateral administration of NPY in the (A) nucleus tractus solitarii (NTS), the area postrema (B), and the C1 area (C). NPY was administered at the arrow in a dose of 4.65 pmol. The representative tracings are from three different animals. (Reproduced from *18*, by permission of the American Heart Association).

sinus nerve afferent fibers *(16)*. An early publication by Carter et al. *(17)*, described diametric effects of this peptide at low and high doses following injection into the NTS of the urethane-anesthetized rat. A pressor response was elicited at 470 fmol NPY, whereas a higher dose of NPY, 4.7 pmol, produced a fall in blood pressure. However, these responses were not corroborated by Tseng et al. *(18)* (Fig. 3A), who convincingly demonstrated a reduction in blood pressure and heart rate following microinjection of NPY (46.5 fmol–4.65 pmol) into the NTS of urethane anesthetized rats. These data indicate that the hemodynamic reflexes, which involve the NTS, might include NPY as the putative neurotransmitter or neuromodulator. In the absence of an NPY antagonist, the precise nature of this action cannot be fully resolved; however, in the presence of a selective NPY antiserum, the cardiovascular effects of NPY in the NTS were abolished *(18)*. This implies that the action of NPY in this area is specific and indicates that NPY may participate in the functional circuitry of the cardiovascular reflexes mediated through the NTS.

Anatomical studies as well as pharmacological manipulations of this region have provided evidence that this area, although lacking extensive projections to preganglionic sympathetic neurons, which control vasomotor outflow, does project to several brainstem and forebrain sites with direct sympathetic and vagal connections *(19)*. Indeed, modulation of arterial pressure via brainstem nuclei that control preganglionic nerve activity is considered to be of critical importance in the regulation of arterial pressure in the NTS-mediated baroreflex response. Electrophysiological and neuroanatomical studies provide evidence of a reciprocal connection between the NTS and the ventrolateral medulla, supporting such an involvement *(20)*.

Neurons in the ventrolateral medulla are considered pivotal in the tonic and reflex regulation of blood pressure. On the basis of neuronal content and differential effects on blood pressure, this area can be separated into two regions, the rostral ventrolateral medulla (RVLM) and the caudal ventrolateral medulla (CVLM) *(21–24)*. The former area is dominated by the tonically active C1 vasopressor neuronal cell group, which as well as having reciprocal connections to the NTS also projects to the spinal cord to regulate the activity of the preganglionic sympathetic neurons *(20–22)*. Further characterization of the RVLM through immunohistochemical studies revealed that, in addition to epinephrine, most of the cell bodies and terminals of this region also contain NPY-ir. NPY-binding sites have also been demonstrated in this region. Pharmacological evidence for a putative role of NPY in this area was described by Tseng and colleagues (*18*; Fig. 3C), who observed that NPY microinjected into this region (up to 4.65 pmol) produced a dose-related and sustained increase in blood pressure and heart rate with a rapid but transient decrease in heart rate. In another report, however, concentrations of NPY, 25 and 50 pmol, injected into the same area elicited a small, but not significant increase in blood pressure at the higher dose *(25)*. It may be postulated that NPY in this region acts on the C1 vasopressor neurons to elicit this increase in blood pressure. Since the origin of the majority the NPY-ir terminals in this region is most likely to be a result of projections from the NTS, we could then postulate from the data of Tseng et al. *(18)* that stimulation of the NTS would result in the release of NPY into the C1 region and an increase in blood pressure. However, a characteristic hypotension results from stimulation of the NTS. This paradox should not, however, be allowed to detract from the possible importance of NPY in this circuitry. It is possible that a subpopulation of neurons that contain NPY may provide an excitatory input from the NTS to the C1 area. Correlating the neuroanatomical studies with further studies measuring the release of NPY and other transmitter candidates in the C1 area following stimulation of the NTS, may advance our understanding of the pathways in these regions.

The physiological significance of the presence of NPY within the brain stem is supported by another study that described a pressor response and a twofold elevation of NPY immunoreactivity in spinal subarachnoid space perfusates following chemical stimulation of the C1 neurons *(26,27)*. Whether the increase in blood pressure following chemical stimulation can be attributed to and only to the release of NPY is contentious, since the peak increase in blood pressure associated with stimulation of these C1 neurons preceded the elevation in NPY concentrations in the spinal cord. Moreover, as mentioned earlier, studies in our own laboratory have also shown that following intrathecal injection of NPY, a depressor effect is observed and not an increase in blood pressure, as would be expected from the latter results *(2)*. This effect is thought to be mediated through the Y_2 receptor, since intrathecal injections of C-terminal fragments with specificity for Y_2 receptors produce a similar depressor effect as NPY itself, whereas the putative Y_1-selective analog [Leu^{31} Pro^{34}]NPY has no effect (*28*; Figs. 4 and 5). Although these experiments appear to be contradictory, it has been speculated that exogenously administered NPY acts on a different population of receptors than that innervated by the neurons projecting from the rostral ventrolateral medulla. Alternatively, a concept that was conceived in the peripheral nervous system and may also be applicable in central neurotransmission is that NPY may play a significant role only under certain conditions, such as

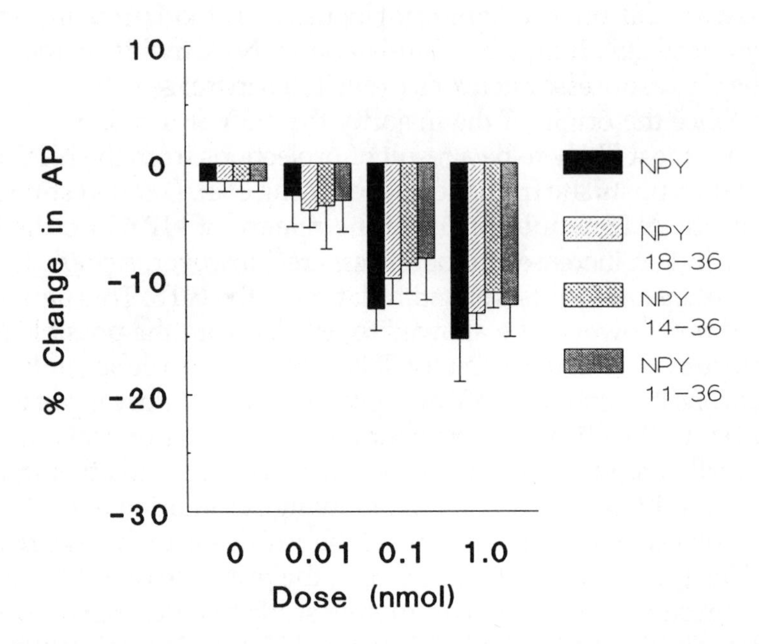

Fig. 4. Dose–response curves for the effects of intrathecal NPY and NPY C-terminal fragments on arterial pressure in anesthetized rats. Data are expressed as percent change in mean arterial pressure ($n = 5$ for each dose). NPY and all NPY fragments tested produced a dose-dependent significant depressor effect. (Reproduced from *28*, with permission.)

intense stimulation. Under these circumstances, the peptide would be released to assist or modulate the action of the cotransmitter. The longer duration of action of NPY compared to epinephrine in the central nervous system, as well as the delay in the release of NPY into the spinal cord subsequent to stimulation of the C1 area, supports this theory.

Measuring transmitter release around discrete nuclei following chemical manipulation at a site distant from the area has proven useful in discerning the possible circuitry in other regions of the brainstem. The majority of pharmacological as well as neuroanatomical evidence indicates that the more caudally situated A1 (noradrenergic) neurons do not project directly to the preganglionic sympathetic neurons in the spinal cord *(21,29,30)*. Instead, it is proposed that these neurons mediate alterations in hemodynamic parameters via an inhibitory pathway to the C1 neurons of the

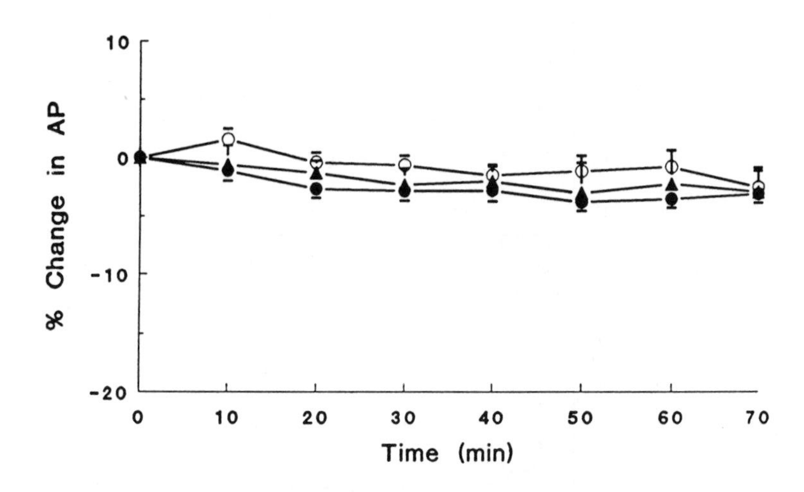

Fig. 5. Effect of the intrathecal administration of the Y_1-specific ligand [Leu[31]Pro[34]] and NPY on arterial pressure. Data are plotted as percent change in mean arterial pressure ± SEM vs time ($n = 5$). No significant differences were observed compared with the saline group. (Taken from *28*, with permission.) -○- saline, -●- Y_1 ligand, 0.1 nmol.

RVLM, which do project to the spinal cord *(31,32)*. An increase in NPY-ir in the spinal cord was observed after inhibition of these vasodepressor neurons, extending these original observations to propose NPY as a possible transmitter in the pathway that mediates the hemodynamic actions of this region *(27)*. Furthermore, discrete microinjections of NPY (25 and 50 pmol) into the A1 region results in a dose-related hypotension most likely owing to activation of A1 neurons, resulting in reduced sympathetic outflow via these C1 neurons (*25*; Fig. 6). A small component of the fall in blood pressure may be associated with the profound and prolonged reduction in heart rate, also observed following NPY administration in this region. Activation of cardiac vagal neurons originating in the nucleus ambiguus has been proposed to be responsible for the bradycardiac effect, with perhaps a contribution at least initially from a decrease in cardiac sympathetic outflow *(35)*.

The area postrema, also located in the caudal medulla, is another region where NPY-ir is present and where this peptide has cardiovascular actions *(18)*. It is a circumventricular organ that receives

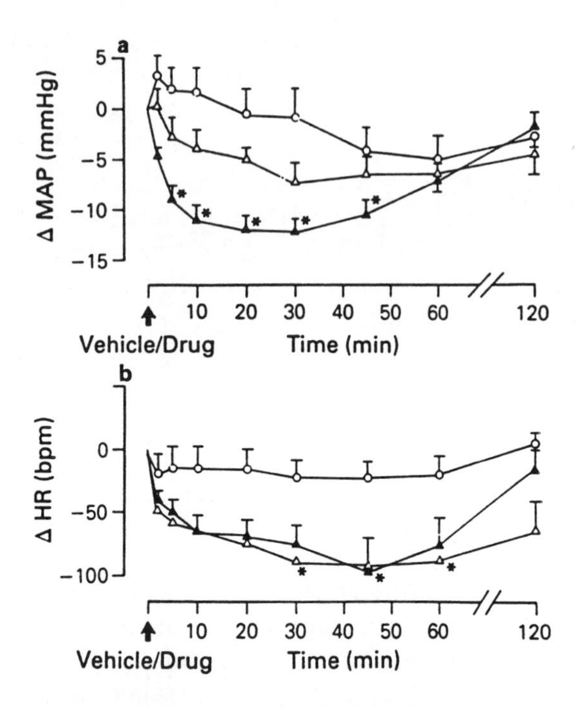

Fig. 6. The effect of 25 pmol (Δ) and 50 pmol (▲) neuropeptide Y (NPY) microinjections into the caudal ventrolateral medulla (CVLM) on (a) mean arterial blood pressure (MAP) and (b) heart rate. Points represent mean changes compared to saline control (□). Vertical bars show SEM. -▲- Y_1 ligand 1.0 nmol. Preinjection values ± SEM are as follows: saline (*n* = 6), MAP = 96 ± 2.6; HR = 369 14; 25 pmol NPY (*n* = 9) MAP = 101 ± 2; HR = 388 ±11; 50 pmol NPY (*n* = 9) MAP = 102 ± 3, HR = 368 ± 9. *Significant differences from control value ($p < 0.05$) (Reproduced from 25, with permission).

afferent input from and projects to brain regions associated with cardiovascular control, including the nucleus of the solitary tract (NTS) and the parabrachial nucleus *(35)*. A biphasic hemodynamic response to NPY (up to 4.65 pmol) was demonstrated in the area postrema, characterized by an initial transient pressor response and tachycardia followed by a gradual hypotension and bradycardia (Fig. 3B; *18*). In addition to its central connections, because this area is devoid of a blood–brain barrier, it may also have an essential function in the humoral modulation of the cardiovascular system. Moreover, it may be sensitive to NPY circulating in the plasma and is thus worthy of further investigation.

NPY is most densely concentrated in the hypothalamus, where NPY-ir fibers and terminals are localized in the arcuate, dorsomedial, paraventricular, anterior paraventricular, and suprachiasmatic nuclei, as well as the medial preoptic area *(12,34)*. Discrete injection of NPY into the paraventricular nucleus of conscious rats with no access to food has been shown to elicit a significant bradycardia and a modest fall in blood pressure *(35)*. It is most likely that the hypotension and bradycardia observed after NPY administration are the result of an inhibitory action of this peptide on paraventricular nuclei, since electrical stimulation of this region evokes an increase in blood pressure *(36)*. This hemodynamic response is most probably mediated via a descending pathway to the motoneurons in the brainstem *(36)*. Microinjection of NPY into this nucleus was also reported to increase food and water intake in the rat *(37)*. This evoked behavioral response may have cardiovascular consequences, since the hypotensive response subsequent to intraventricular NPY administration was reported to be counteracted by an NPY-induced feeding action *(6)*.

An increased plasma level of vasopressin has been observed in response to NPY administered into another hypothalamic region of the conscious rat, the supraoptic nucleus *(38)*. However, the hemodynamic importance of this elevated plasma vasopressin is in contention, since the pressor response to icv administration of NPY in the anesthetized rat was unaffected by iv injection of a specific vasopressin antagonist. Contrary to this, the finding that this pressor response was attenuated in vasopressin-deficient rats (Brattleboro) suggests that central vasopressin pathways may be involved in the response to icv injection of NPY *(8,9)*. Another interesting observation was that microinjections of high doses of NPY (235–2350 pmol) into the posterior hypothalamus elicit a dose-related increase in mean arterial pressure and resistance in the renal and hindquarter vascular bed, as well as a decrease in stroke volume (*39,40*; Fig. 7). These effects were mediated by an increase in centrally mediated sympathetic outflow, since the effects are attenuated by the peripheral administration of α-adrenoceptor antagonists and ganglionic blockers. Moreover, an increase in renal sympathetic nerve activity was also observed following administration of NPY via this route. Vasopressin does not appear to be involved in this response to NPY; however, a role for central histaminergic and cholinergic neurons in these hypothalamic cardiovascular

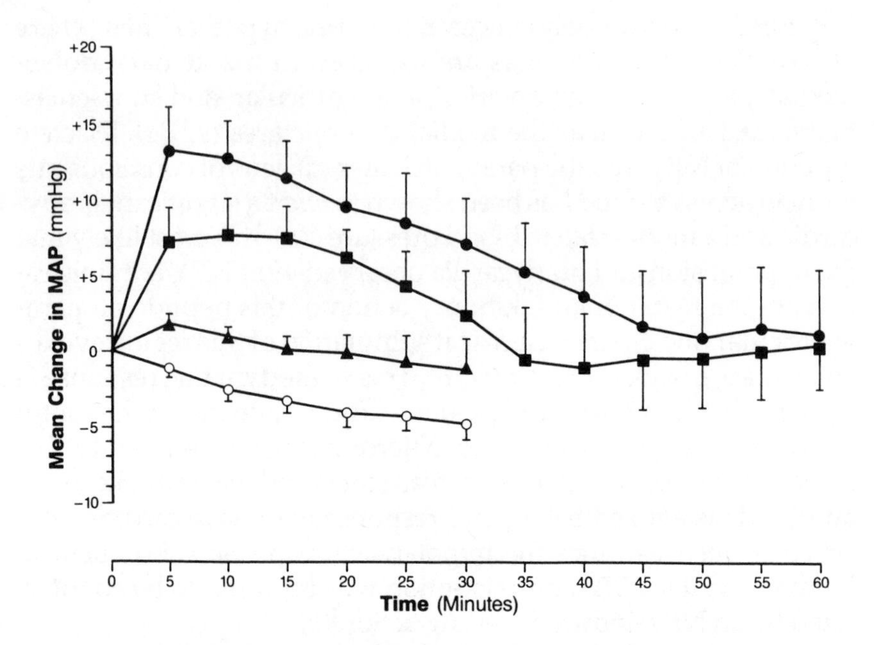

Fig. 7. Time-course of the increase in blood pressure following increasing concentrations of NPY microinjected into the posterior hypothalamic nucleus. Both 0.776 and 2.35 nmol of NPY produced significant increases. (Reproduced from *39*, with permission.) ● 2.35 nmol NPY, ■ 0.776 nmol NPY, ▲ 0.235 nmol NPY, ○ 1 µL vehicle.

actions of NPY has been implicated *(39,40)*. Due to the relatively modest hemodynamic actions of NPY in these hypothalamic regions, it is still unclear if the peptide is associated with important cardiovascular regulation in these areas. A myriad of other effects that may be governed by NPY in the hypothalamus include hormonal secretion, circadian rhythm, and food and water ingestion (*see* chapters by Stanley and by McDonald and Koenig, this vol. for review).

2.3. Existence of Different NPY-Receptor Types in Central Cardiovascular Areas: Functional Relevance?

To date, radioligand binding studies have indicated that at least two subtypes of binding sites for NPY, which differ in affinity and specificity for the monoiodinated radioligands of both the intact neuropeptide Y (1–36) and a long-terminal C-fragment (13–36), are present in the nervous system. These binding sites, as discussed in previous chapters, have been designated Y_1 and Y_2, respectively. Attempts have been made to characterize uniformly these putative receptors in the peripheral nervous system in terms of junctional location and coupling to discrete second-messenger systems, but this has proven to be untenable *(41)*. Based on displacement curves of monoiodinated NPY_{1-36} with NPY_{13-36}, several groups have described binding sites in the central nervous system that have an affinity and specificity characteristic of the Y_2 receptor. Indeed, it has been demonstrated that in porcine hippocampal membranes and rat cortical membranes, the Y2 receptor has been shown to be the principal subtype *(42,43)*. Early studies focused on the participation of the NPY receptor(s) in synaptic transmission, transmitter release, and its coupling to second-messenger systems. In various regions of the central nervous system, NPY can modify cAMP accumulation and inositol phosphate turnover, as well as regulate calcium mobilization in dorsal root ganglion cells, suggesting an involvement of the Y_1-receptor subtype *(46–50)*. Colmers and others (*see* chapters by Bleakman et al. and Grundemar et al., this vol.) have shown that at least in the hippocampus, the Y_2-receptor type may be involved in synaptic transmission. In addition, Perney and Miller *(48)* reported that NPY_{13-36} was an effective agonist in stimulating inositol phosphate production in dorsal root ganglion neurons, whereas this fragment did not alter this system in cortical tissue *(49)*. However, since most of these studies to date investigated the actions of NPY and its analogs on second-messenger systems in regions whose primary role is not associated with cardiovascular function, the physiological relevance of these subtypes in relation to cardiovascular control mechanisms has not been adequately defined. Fuxe and colleagues *(50)*, however, reported that in contrast to the intact peptide NPY_{1-36}, which decreased blood pressure and heart rate when administered into the ventricles of rats, NPY_{13-36} (2–6 nmol) had no hemodynamic effect. This suggests that perhaps only the Y_1 receptor type was responsible for the central cardiovascular effects of NPY. However, recently, the same group published conflicting data that showed that ivt administered NPY_{13-36} (25–3000 pmol) in the conscious unrestrained rat produced an increase, although modest, in blood pressure without an effect on heart rate *(51)*. The reason for this discrepancy between publications from this group was not addressed in the subsequent report by the authors, and thus, the data need to be confirmed. If the latter results are corroborated, it would suggest that these putative receptor subtypes have opposing actions. Thus, the location of these receptors could determine the direction of the response to NPY. As mentioned earlier in this chapter, the NPY receptor subtype that is thought to mediate the depressor effects following intrathecal administration of NPY is the Y_2 receptor, since C-terminal fragments mimic the effect of NPY, whereas the Y_1-selective agent [$Leu^{31}Pro^{34}$]NPY does not (Figs. 4 and 5). Further resolution of the actions of NPY will be advanced when the distribution and function of these putative subtypes of NPY receptors, especially in areas of cardiovascular significance, have been fully characterized.

2.4. NPY: Interactions with Brain Catecholamines

As described elsewhere in this volume, in the peripheral nervous system, NPY is predominantly colocalized with norepinephrine in sympathetic neurons, and interacts with the function of norepinephrine at the pre- and postsynaptic levels. A pervasive colocalization of NPY with other transmitters has also been demonstrated in the CNS, and this has prompted investigation of possible interactions between NPY and these colocalized transmitters. It is of particular interest that many of the central neuronal systems containing catecholamines, considered to be of fundamental importance in the central regulation of blood pressure, also contain NPY *(12,13,55)*. Many studies have concentrated on the colocalization of NPY with catecholamines in the brainstem, where the possible interaction between NPY receptors and α_2 adrenoceptors has been examined. The involvement of α_2 adrenoceptors in the central neural regulation of the cardiovascular system has been well documented, and it is proposed that norepinephrine and epinephrine are the endogenous ligands for these brainstem adrenoceptors *(53,54)*. Norepinephrine, epinephrine, and the α_2 adrenoceptor agonist clonidine reduce blood pressure subsequent to central administration *(25,55,56)*. Injection of NPY into similar areas also produces a hypotensive effect, as discussed previously. The similarity of the actions of catecholamines and NPY suggests that the coexisting peptide may cooperate with the endogenous catecholamines in blood pressure regulation. Evidence of a central interaction between adrenoceptors and the putative NPY receptors derives

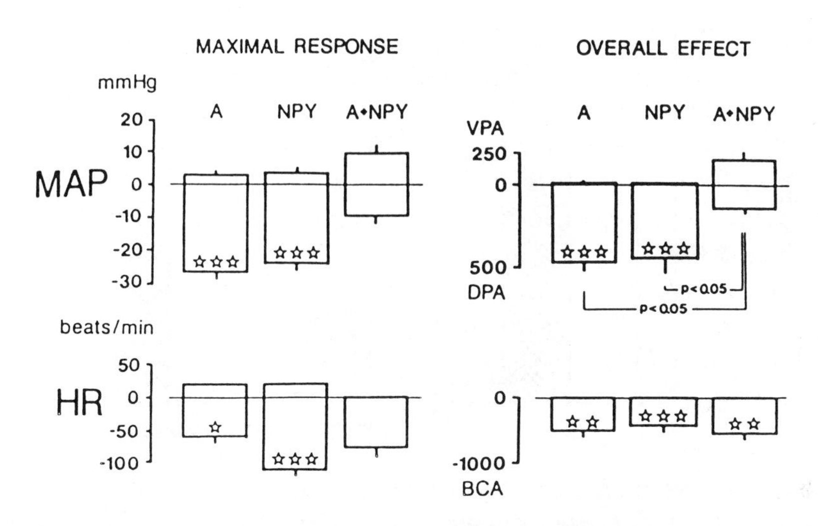

Fig. 8. Interaction between centrally administered NPY and epinephrine on mean arterial blood pressure and heart rate in the conscious freely moving male rat. The left panels show the maximal response (peak), and the right panels show overall effects on blood pressure and HR responses during the 1-h time period after drug administration. Doses used were epinephrine (A) 0.45 mmol ivt; NPY 0.25 nmol ivt, A + NPY, 0.45 + 0.25 nmol ivt. (Reproduced from *60*, with permission.)

from a series of quantitative autoradiographic studies by Harfstrand et al. *(57)*. These authors reported that NPY, in vitro or after ivt administration, reduced the density of α_2 adrenergic binding sites in the medulla oblongata. Comparative studies revealed a decrease in the binding of iodinated NPY in the medulla oblongata subsequent to administration of clonidine into the ventricles or following incubation with clonidine in vitro *(57,58)*. These findings indicate the existence of a receptor–receptor interaction and generally support earlier observations, using radioligand-binding techniques, that NPY can alter the binding characteristics of α_2 adrenoceptors *(59)*. Thus, evidence of a reciprocal modulation of α_2 adrenoceptors and NPY-binding sites exists in the medulla, and it is proposed that this interaction may operate physiologically. In agreement with this view, Fuxe et al. (*60*; Fig. 8) demonstrated that ivt coadministration of equipotent doses of epinephrine and NPY counteracted the hypotensive actions of one another. It is likely that this interaction occurs at postreceptor processes and not at the

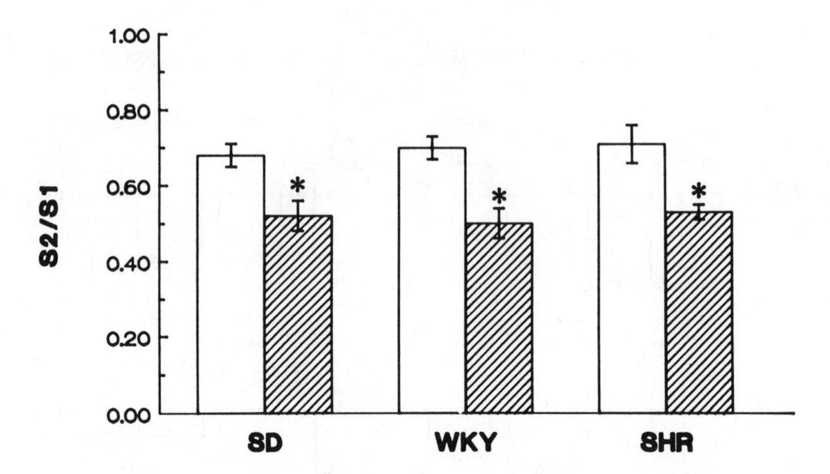

Fig. 9. Effect of NPY on the potassium-evoked release of ^{3}H-NE from slices of the anterior hypothalamus of Sprague Dawley (SD), Wistar Kyoto (WKY), and spontaneously hypertensive (SHR) rats. Slices were exposed twice (S_1 and S_2) to a high K^+ buffer (30 m*M*) for 5 min. NPY was added 5 min before and during S_2. Results are expressed as the mean S_2/S_1 rates (fractional release) ± SEM of 6–10 animals/group. NPY produced a significant decrease in the S_2/S_1 ratios in all cases. No differences were seen between SD, WKY, and SHR (Reproduced from *62*, with permission). □ Control, ▨ NPY 10–6*M*.

receptor level, since blockade of the α_2 adrenoceptor with intracisternally administered RX 781094 attenuated the fall in blood pressure associated with clonidine, but did not alter the hypotensive action of NPY *(61)*. Moreover, that these agents act through a common mechanism of action is further supported by the observation that simultaneous intracisternal injection of NPY and clonidine in anesthetized rats does not lead to any additional lowering of blood pressure compared to that seen with either drug alone *(61)*.

The possibility of central functional interactions has been further supported by reports of an NPY-mediated alteration in catecholamine release or turnover in several brain regions, including the hypothalamus, the cortex, the hippocampus, and the brainstem *(62–67)*. NPY_{13-36} can also elicit alterations in tissue catecholamine levels in certain brain regions *(67)*. NPY was shown to attenuate the potassium-evoked release of [^{3}H]-norepinephrine from slices of the rat anterior and posterior hypothalamus, as well as from the A_2-NTS region of Sprague Dawley rats (Fig. 9). Similar attenuation

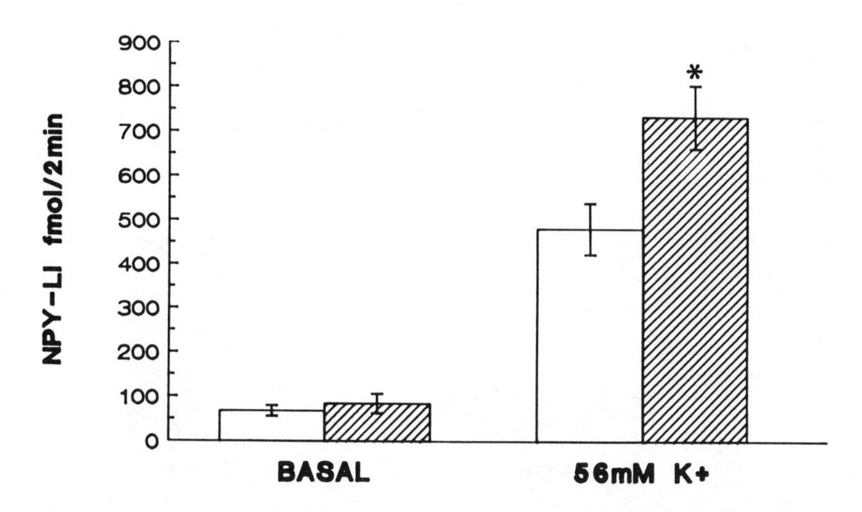

Fig. 10. Effect of yohimbine ($10^{-6}M$) on the potassium-evoked release of NPY from slices of the rat hypothalamus. Slices were preincubated in Krebs-Ringer bicarbonate buffer for 30 min preceding a 2-min basal and 2-min stimulation with 56 m*M* potassium buffer in the presence or absence of yohimbine. Results represent the mean ± SEM of 15 animals in the control group and six animals in the yohimbine group. Basal release for the control group was 67.3 ± 11.9 fmol and for the yohimbine group 84 ± 2.8 fmol. Stimulated release for the control was 480 ± 58 fmol and yohimbine 733 ± 71 $p < 0.05$. (Reproduced from *62*, with permission.) □ Control, ▨ YOHIM $10^{-6}M$.

of the evoked release of norepinephrine was seen by C-terminal NPY fragments, suggesting this effect may be mediated by Y_2 receptors *(62,63)*. An interaction between catecholamines and NPY in the hypothalamus is further supported by the observation that α- and β-adrenoceptor antagonists decreased and increased, respectively, the evoked release of NPY as measured by RIA (Figs. 10–12). These results provide further support for a reciprocal relationship between the regulation of the release of norepinephrine and NPY, with NPY decreasing the release of norepinephrine, and norepinephrine acting through α- and β-adrenoceptors to decrease and increase NPY release, respectively. The inhibitory effect of NPY on NE release was not altered in the presence of α- or β-adrenoceptor antagonists.

In contrast to the effects observed in the hypothalamus, Chen et al. *(7)* observed that intrathecal pretreatment with the α_2 adrenoceptor antagonist, yohimbine attenuated the depressor effect of

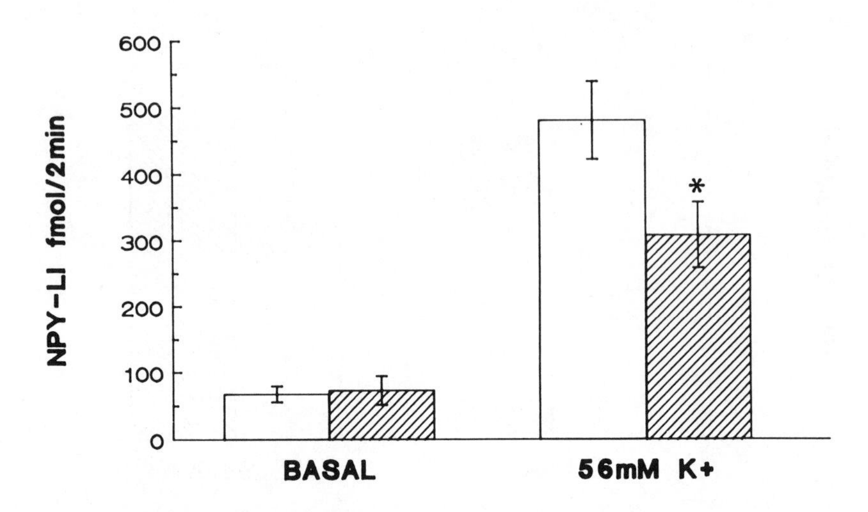

Fig. 11. Effect of clonidine on the potassium-evoked release of NPY from slices of the rat hypothalamus. Clonidine was administered in a manner similar to that in Fig. 9. Results represent the mean ± SEM of 15 animals in the control group and eight animals in the clonidine group. Basal release was 67.3 ± 11.9 fmol for control and 72.8 ± 21.8 fmol for the clonidine group. Control stimulated release was 480.1 ± 58.1 fmol and clonidine 307.7 ± 49.6 fmol. $^*p < 0.05$. (Reproduced from *62*, with permission.) □ Control, ▨ CLON $10^{-6}M$.

intrathecal NPY. These results suggest that spinal α_2 adrenoceptors are involved in mediating the depressor effect of NPY. Furthermore, pretreatment of rats with reserpine (1 mg/kg, ip) for 2 d or with intrathecal 6-hydroxydopamine (20 µg/10 µL) for 7 d before experimentation led to significant depletion of spinal norepinephrine and epinephrine (Table 1) and prevented the depressor effect of intrathecal NPY (Fig. 13). These data suggest that at least part of the depressor effect of intrathecal NPY is mediated via alterations in catecholamine release from catecholaminergic nerve terminals and is not owing to a direct inhibitory action on sympathetic preganglionic neurons in the IML columns. It is well known that norepinephrine is an important neurotransmitter in the sympatho-excitatory descending pathways, which maintain tonic sympathetic nerve activity, and that this effect is mediated by activation of post-sympathetic α_1 adrenoceptors on sympathetic preganglionic neurons *(68)*. Thus, it is possible that activation of presynaptic NPY receptors as well as α_2 adrenoceptors following intrathecal NPY

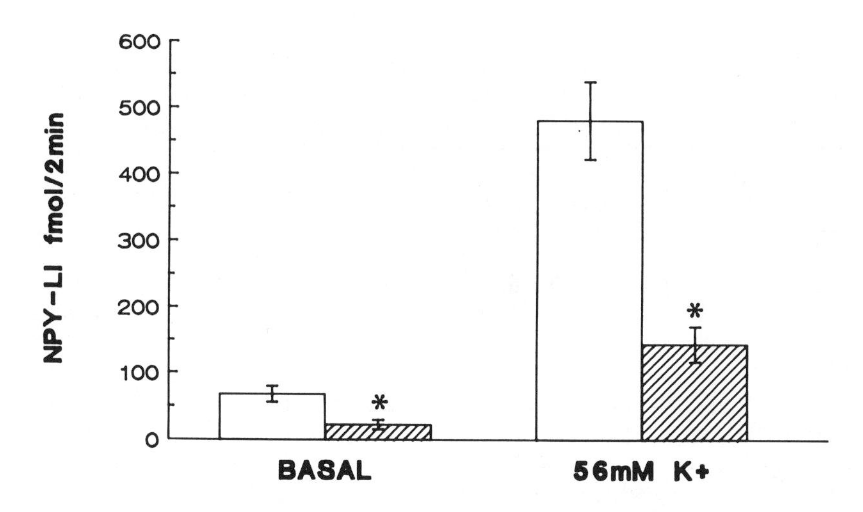

Fig. 12. Effect of propranolol $10^{-6}M$ on the potassium-evoked release of NPY from slices of the rat hypothalamus. Propranolol was administered in a manner similar to that in Figs. 9 and 10. Results are the mean ± SEM for 15 control animals and six propranolol treated animals. Basal release of NPY for the control group was 67 12 and for propranolol 22 ± 7 fmol. Control-stimulated release of NPY was 480 ± 58 and release + propranolol 143 ± 26.9 fmol. Propranolol produced a significant decrease in both basal and stimulated release $p < 0.05$ (Reproduced from *62*, with permission). □ Control, ▨ PROP $10^{-6}M$.

produces suppression of norepinephrine release, which subsequently inhibits sympathetic nerve activity and arterial pressure. In support of this postulate, NPY was observed to attenuate the potassium-evoked release of norepinephrine from the perfusates of the rat spinal cord (*69;* Fig. 14).

In contrast to these effects in the spinal cord, independent mechanisms for NPY have been proposed in other areas. Vallejo and Lightman *(8)* found that intracerebroventricular administration of NPY still evoked a pressor response after depletion of central catecholamines by 6-hydroxydopamine. Interactions between adrenoceptor agonists and NPY on blood pressure responses have, however, been reported in discrete regions of the brain, such as the NTS and locus coeruleus, but not in the C1 or A1 areas of the rostral and caudal medulla, respectively *(9,17,25)*. In the former study, the hypotensive response to norepinephrine was significantly modified by prior or simultaneous injection of a

Table 1
Effects of Pretreatment of SD Rats
with Intrathecal 6-Hydroxydopamine 6-OHDA
on Monoamine Levels in the Spinal Cord, Brain, and Heart (n = 9)

	Monoamine, ng/g tissue		
	Spinal cord	Brain	Heart
NE			
Control	553.78 ± 66.15	184.70 ± 24.33	361.11 ± 0.06
6-OHDA	105.22 ± 11.82	189.71 ± 17.9	356.61 ± 0.02
% Change	–81.0%	2.71%	–1.25%
EPI			
Control	5.94 ± 1.20	2.06 ± 0.16	21.64 ± 2.54
6-OHDA	3.17 ± 0.74	2.14 ± 0.16	22.96 ± 6.42
% Change	–46.51%	3.68%	6.07%
DA			
Control	18.48 ± 1.34	149.99 ± 9.33	3.48 ± 0.54
6-OHDA	18.59 ± 1.89	138.66 ± 2.65	3.28 ± 0.34
% Change	0.61%	–7.00%	–5.90%
5-HT			
Control	598.53 ± 71.17	357.32 ± 30.12	313.17 ± 20.15
6-OHDA	571.02 ± 46.54	375.34 ± 46.6	331.77 ± 93.95
% Change	–4.6%	5.05%	5.94%

[a]6-OHDA was injected in a dose of 20 μg and measurements made 7 d later by HPLC-EC detection.

dose of NPY, which alone had no effect. Thus, although interactions between costored transmitters or their respective receptors may be present in some central areas, such an interaction cannot be considered ubiquitous, and in some regions, colocalization may reflect a provision for separate functions required or evoked under different conditions. In the sympathetic neurons in the periphery where NPY and norepinephrine are costored, differential release of these agents has been observed, lending support to this postulate. Low-frequency stimulation preferentially releases norepinephrine, whereas high-frequency stimulation is needed to release both norepinephrine and NPY *(72)*. It has been proposed that NPY is released only under conditions of high frequency to support the colocalized transmitter and improve transmitter economy. The participation of NPY in neurotransmission may become increasingly important in pathological states such as hypertension.

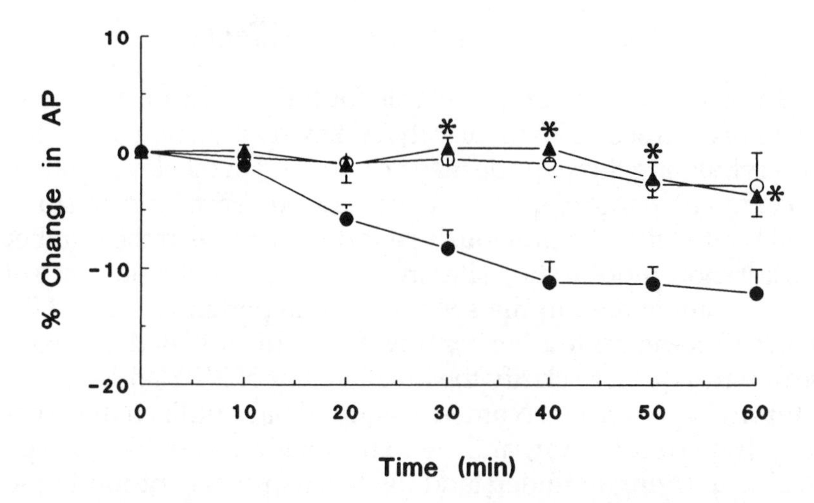

Fig. 13. Effect of intrathecal NPY (0.1 nmol) on arterial pressure in rats with and without 6-hydroxydopamine pretreatment (20 µg). Experiments were carried out 7 d later. Data are expressed as percent changes in arterial pressure ± SEM. The depressor effect of NPY was blocked in rats pretreated with 6-OHDA. $p < 0.05$ (n - 7) compared to saline controls. –O– Saline, –●– NPY (Saline), –▲– NPY (6-OHDA).

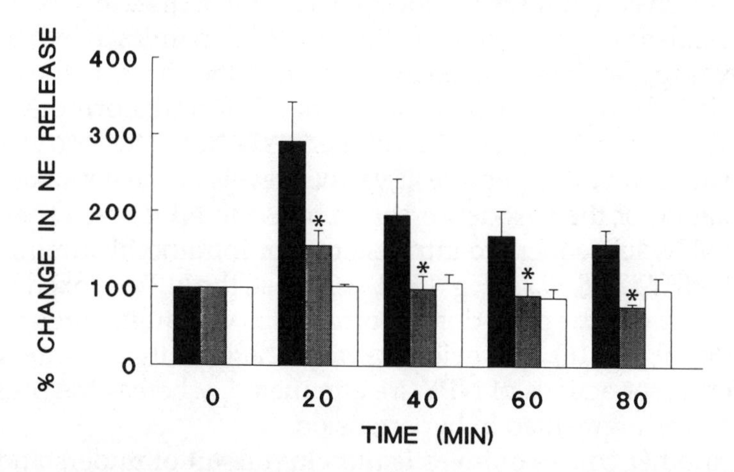

Fig. 14. Effect of NPY ($10^{-7}M$) on basal and potassium (30 mM)-evoked norepinephrine (NE) release in the spinal cord. Data are expressed as percent change in NE release. Each bar represents the mean of 5 samples obtained from an equal number of animals. NPY produced a significant attenuation of potassium-evoked release $p < .05$. (Reproduced from *69*, with permission). ■ K^+30 mM; ▒ NPY $10^{-7}M$, + K^+30 mM; □ NPY $10^{-7}M$.

3. NPY and Hypertension: Cause or Effect?

In addition to the proposed role for NPY in the normal functioning of central cardiovascular dynamics, dysfunction of mechanisms regulating this peptide may be involved in the development and/or maintenance of essential hypertension. A preliminary study in 1983 stated that the previously reported increase in the number of α_2 adrenoceptor-binding sites in the rat brain in the presence of NPY was attenuated in the spontaneous hypertensive rat (SHR) compared to normotensive controls *(59)*. This initiated vigorous research in a quest to clarify the involvement of NPY and adrenoceptors in hypertension. Numerous studies have utilized the genetically hypertensive rat, and by extrapolation, this has proven valuable in trying to understand the mechanisms contributing to or resulting from human essential hypertension. Radioimmunoassay studies described lower concentrations of this peptide in the brainstem and spinal cord of the SHR compared to the age-matched normotensive control *(71)*. Although the differences in the concentrations of this peptide have been observed in various areas of the SHR brain, systematic investigation has not been undertaken to determine whether these changes are associated with the etiology of hypertension or merely a consequence of the disease. However, the postulated involvement of altered NPY dynamics in hypertension was further investigated by comparing the effects of the exogenously administered peptide in the SHR and normotensive animals. These results revealed differences between the two groups in the actions of this peptide at various levels of the neuraxis. An attenuation of the vasodepressor response to NPY was observed in the SHR subsequent to intracisternal or intrathecal administration of NPY (*2,28,60,72*; Fig. 15). In contrast, the NPY-evoked pressor response in the posterior hypothalamus was augmented in the SHR (*68,69*; Fig. 16). This evidence would suggest that at some sites the depressor actions of NPY are attenuated, whereas the pressor actions are augmented in hypertension.

Another course of investigation in pursuit of understanding the involvement of NPY in the mechanisms of hypertension has been to study this peptide's interactions with catecholamines. Alterations in catecholamine pathways in hypertension have been recognized for many years. In the medulla oblongata (a region containing a number of important cardiovascular regulatory neurons,

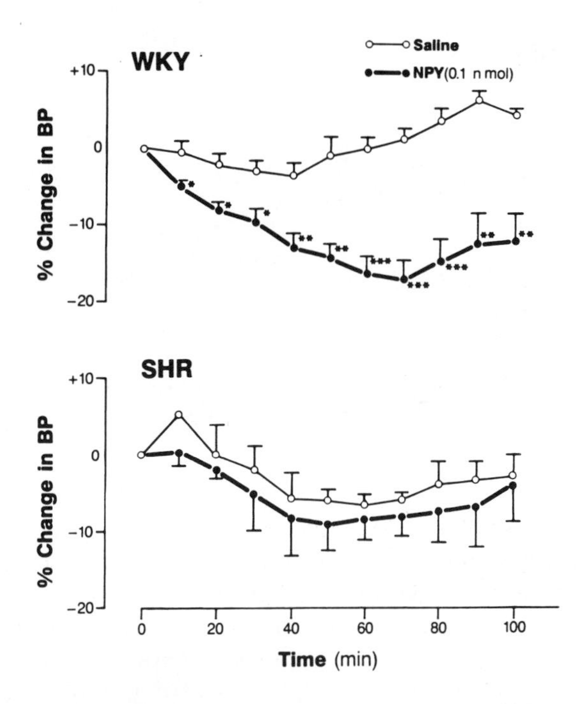

Fig. 15. The effect of saline or NPY on mean arterial blood pressure of WKY (top panel) or SHR (bottom panel). The mean blood pressure in mmHg for the four groups immediately prior to NPY injections was WKY saline 87 0.5; WKY-NPY 86 6; SHR saline 109 8; SHR - NPY 113 4. Each point is the mean ± SEM of at least five observations. Data are plotted as % change in blood pressure vs time in minutes. $^{*}p < 0.05$ compared to saline; + $p < .05$ compared to NPY alone (Reproduced from *72*, with permission).

such as the NTS, the A1, and the C1) of the SHR, a decreased number of α_2 adrenoceptors has been reported compared to normotensive controls, as well as a reduction in the autoregulatory function of this receptor *(65,73,74)*. The finding that NPY is colocalized with catecholamines in a subpopulation of these neurons and interacts functionally with these transmitters would suggest that the association of this peptide with catecholamines may indeed be altered in hypertension *(13,17)*. Several studies have supported this postulate, including Agnati et al. *(59)*, who demonstrated that the NPY-mediated increase in the number of α_2 adrenoceptors in the medulla oblongata was reputedly diminished in the SHR *(59)*.

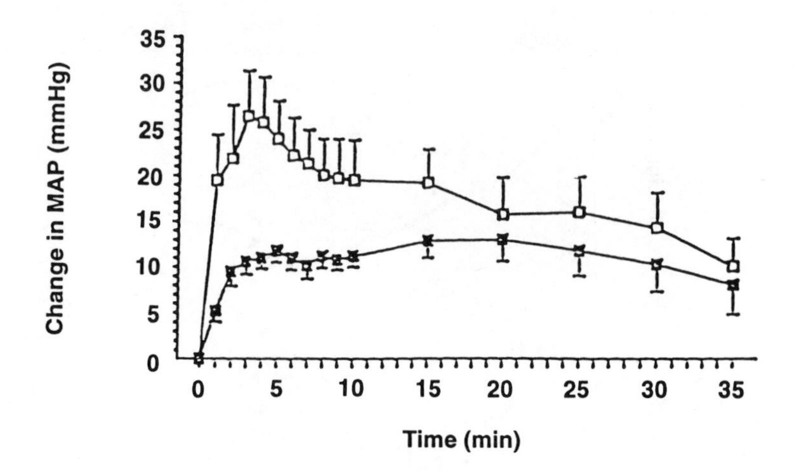

Fig. 16. The effect of unilateral microinjections of NPY (2.35 nmol) into the posterior hypothalamic nucleus on mean arterial blood pressure of anesthetized Sprague Dawley (■ $n = 16$) or SHR (□ $n = 8$). Data are plotted as change in mean arterial pressure in mmHG vs time. Each point is the mean ± SEM. NPY produced a significant greater pressor response in SHR compared to saline controls. (Reproduced from *63*, with permission).

Most others have compared the ability of NPY to alter catecholamine release in the normotensive rat and the hypertensive rat. Although Tsuda et al. *(65)* reported that the inhibitory effect of NPY on norepinephrine release was diminished in a slice preparation of the medulla oblongata of the SHR compared with the normotensive strain, such a difference between these stains was not observed in the anterior or posterior hypothalamus (Fig. 9; *2,3,62,63*). Although technical differences may contribute to these disparate results, it is quite possible that these differences reflect the importance of the medulla oblongata compared to, for example, the hypothalamus, in central hemodynamics and hypertension. Moreover, such findings may open up other avenues in the understanding and treatment of hypertension.

4. Summary

NPY has been shown to produce pressor or depressor effects as well as other hemodynamic effects depending on the site of its release or administration in the neural axis of the central nervous system. The precise physiological role of central NPY on cardiovascular regulation or function is still unclear. However, there is considerable evidence that the peptide does participate in cardiovascular regulation following its release in brain stem areas, the spinal cord, and possibly the hypothalamus. Alterations in the regulation of NPY synthesis, release, and interaction with receptors may exist in various models of hypertension, and thus, contribute to the production or maintenance of elevated blood pressure.

Acknowledgments

Research in the author's laboratory has been supported by grants from the US Public Health Service HL 26319 (TCW), HL 35202 (TCW), and NIDA 02668 (TCW).

References

1. Westfall, T. C., Carpentier, S., Chen, X., Beinfeld, M. C., Naes, L., and Meldrum, M. J. (1987) Prejunctional and postjunctional effects of neuropeptide Y at the noradrenergic neuroeffector junction of the perfused mesenteric arterial bed of the rat. *J. Cardiovas. Pharmacol.* **10,** 716–722.
2. Westfall, T. C., Martin, J., Chen, X., Ciarleglio, A., Carpentier, S., Henderson, K., Knuepfer, M., Beinfeld, M., and Naes, L. (1988) Cardiovascular effects and modulation of noradrenergic neurotransmission following central and peripheral administration of Neuropeptide Y. *Synapse* **2,** 299–307.
3. Westfall, T. C., Chen, X., Ciarleglio, A., Henderson, K., Del Valle, K., Curfman-Falvey, M., and Naes, L. (1990) *In vitro* effects of neuropeptide Y at the vascular neuroeffector junction. *Ann. NY Acad. Sci.* **611,** 145–155.
4. Fuxe, K., Agnati, L. J., Harfstrand, A., Zini, I., Tatemoto, K., Pich, E. M., Hokfelt, T., Mutt, V., and Terenius, L. (1983) Central administration of neuropeptide Y induces hypotension, bradypnea and EEG synchronization in the rat. *Acta Physiol. Scand.* **118,** 189–192.
5. Macrae, I. M. and Reid, J. L. (1988) Cardiovascular significance of neuropeptide Y in the caudal ventrolateral medulla of the rat. *Brain Res.* **456,** 1–8.
6. Härfstrand, A. (1986) Intraventricular administration of neuropeptide Y (NPY) induces hypotension, bradycardia and bradypnea in the awake unrestrained male rat. Counteraction by NPY-induced feeding behaviour. *Acta Physiol. Scand.* **128,** 121–123.
7. Chen, X., Knuepfer, M. M., and Westfall, T. C. (1990) Hemodynamic and sympathetic effects of spinal administration of neuropeptide Y in rats. *Am. J. Physiol.* **259,** H1674–H1680.
8. Vallejo, M. and Lightman, S. L. (1986) Pressor effect of centrally administered neuropeptide Y in rats: Role of sympathetic nervous system and vasopressin. *Life Sci.* **38,** 1859–1866.

9. Vallejo, M. and Lightman, S. L. (1986) Neuropeptide Y alters the hemodynamic responses to noradrenaline in the locus coeruleus of rats. *Neurosci. Lett.* **26 (Suppl.)**, S237.
10. Petty, M. A. and Reid (1981) Opiate analogs, substance P and baroreceptor reflexes in the rabbit. *Hypertension* **3 (Suppl. I)**, I142–I147.
11. Adrian, T. E., Allen, J. M., Bloom, S. R., Ghatei, M. A., Rossor, M. N., Roberts, G. W., Crow, T. J., Tatemoto, K., and Polak, J. M. (1983) Neuropeptide Y distribution in human brain. *Nature* **306**, 584–586.
12. Chronwall, B. M., DiMaggio, D. A., Massari, V. J., Pickel, D. A., Ruggiero, D. A., and O'Donohue, T. L. (1985) The anatomy of neuropeptide Y containing neurons in rat brain. *Neuroscience* **15**, 1159–1181.
13. Everitt, B. J., Hokfelt, T., Terenius, K., Tatemoto, K., Mutt, V., and Goldstein, M. (1984) Differential coexistence of neuropeptide Y (NPY)-like immunoreactivity with catecholamines in the central nervous system of the rat. *Neuroscience* **11(2)**, 443–462.
14. Reis, D. J., Perrone, M. H., and Talman, W. T. (1981) Evidence that glutamic acid is the neurotransmitter of baroreceptor afferents terminating in the nucleus tractus solitarius (NTS). *J. Auton. Nerv. Syst.* **3**, 321–334.
15. Gillis, R. A., Helke, C. J., Hamilton B. L., Norman, W., and Jocobowitz, M. (1980) Evidence that substance P is a neurotransmitter of baro- and chemoreceptor afferents in nucleus tractus solitarius. *Brain Res.* **181**, 476–481.
16. Härfstrand, A., Fuxe, K., Terenius, L., and Kalia, M. (1987) Neuropeptide Y-immunoreactive perikarya and nerve terminals in the rat medulla oblongata: Relationship to cytoarchitecture and catecholamine cell groups. *J. Comp. Neurol.* **260**, 20–35.
17. Carter, D. A., Vallejo, M., and Lightman, S. L. (1985) Cardiovascular effects on neuropeptide Y in the nucleus tractus solitarius of rats: Relationship with noradrenaline and vasopressin. *Peptides* **6**, 421–425.
18. Tseng, C-J., Mosqueda-Garcia, R., Appalsamy, M., and Robertson, D. (1988) Cardiovascular effects of neuropeptide Y in rat brainstem nuclei. *Circul. Res.* **64**, 55–61.
19. Loewy, A. D. and Burton, H. (1978) Nuclei of the solitary tract: efferent projections to the lower brainstem and spinal cord of the cat. *J. Comp. Neurol.* **181**, 421–450.
20. Ciriello, J. and Caverson, M. M. (1986) Bidirectional cardiovascular connections between ventrolateral medulla and nucleus of the solitary tract. *Brain Res.* **367**, 273–281.
21. Ross, C. A., Ruggiero, D. A., Joh, T. H., Park, D. H., and Reis, D. J. (1984a) Rostral ventrolateral medulla: selective projections to the thoracic autonomic cell column from the region containing C_1 adrenaline neurons. *J. Comp. Neurol.* **228**, 168–184.
22. Ross, C. A., Ruggiero, D. A., Park, D. H., Joh, T. H., Sved, A. F., Fernandez-Pardal, J., Saavedra, J. M., and Reis, D. J. (1984b) Tonic vasomotor control by the rostral ventrolateral medulla: Effect of electrical or chemical stimulation of the area containing C_1 adrenaline neurons on arterial pressure, heart rate and plasma catecholamines and vasopressin. *J. Neurosci.* **4**, 474–494.
23. Blessing, W. W. and Reis, D. J. (1982) Inhibitory cardiovascular function of neurons in the caudal ventrolateral medulla of the rabbit: relationship to the area containing A_1 noradrenergic cells. *Brain Res.* **253**, 161–171.

24. Willette, R. N., Punnen-Grandy, S., Krieger, A. J., and Sapru, H. N. (1987) Differential regulation of regional vascular resistance by the rostral and caudal ventrolateral medulla in the cat. *J. Auton. Nerv. Syst.* **18,** 143–151.
25. McAuley, M. A., Macrae, I. M., and Reid, J. L. (1989) The cardiovascular actions of clonidine and neuropeptide-Y in the ventrolateral medulla of the rat. *Br. J. Pharmacol.* **97,** 1067–1074.
26. Chalmers, J., Kapoor, V., Mills, E., Minson, J., Morris, M., Pilowsky, P., and West, M. (1986) Do pressor neurons in the ventrolateral medulla release amines and neuropeptides? *Can J. Physiol. Pharmacol.* **65,** 1598–1604.
27. Pilwosky, P. M., Morris, M. J., Minson, J. B., West, M. J., Chalmers, J. P., Willoughby, J. O., and Blessing, W. W. (1987) Inhibition of vasodepressor neurons in the caudal ventrolateral medulla of the rabbit increases both arterial pressure and the release of neuropeptide Y-like immunoreactivity from the spinal cord. *Brain Res.* **420,** 380–384.
28. Chen, X. and Westfall, T. C. (1991) Characterization of the receptor subtype mediating the depressor effect of intrathecal NPY. *J. Cardiovasc. Pharm.* (in press).
29. Ross, C. A., Armstrong, D. M., Ruggiero, D. A., Pickel, V. M., Joh, T. H., and Reis, D. J. (1981a) Adrenaline neurons in the rostral ventrolateral medulla innervate thoracic spinal cord: a combined immunocytochemical and retrograde transport demonstration. *Neuroscience* **25,** 257–262.
30. Blessing, W. W., Goodchild, A. K., Dampney, R. A. L., and Chalmers, J. P. (1981a) Cell groups in the lower brainstem of the rabbit projecting to the spinal cord, with special reference to catecholamine-containing neurons. *Brain Res.* **221,** 35–55.
31. Willette, R. N., Punnen, S., Krieger, A. J., and Sapru, H. N. (1984) Interdependence of rostral and caudal ventrolateral medullary areas in the control of blood pressure. *Brain Res.* **321,** 169–174.
32. Granata, A. R., Numao, Y., Kumada, M., and Reis, D. J. (1986) A_1 noradrenergic neurons tonically inhibit sympathoexcitatory neurons of C_1 area in rat brainstem. *Brain Res.* **377,** 127–146.
33. Shapiro, R. E. and Miselis, R. R. (1985) The central neural connections of the area postrema of the rat. *J. Comp. Neurol.* **234,** 344–364.
34. Sawchenko, P. E., Swanson, L. W., Grzanna, R., Howe, P. R. C., Bloom, S. R., and Polack, J. M. (1985) Co-localization of neuropeptide Y immunoreactivity in brainstem catecholaminergic neurons that project to the paraventricular nucleus of the hypothalamus. *J. Comp. Neurol.* **241,** 138–153.
35. Harland, D., Bennet, T., and Gardiner, S. (1988) Cardiovascular actions of Neuropeptide Y in the hypothalamic paraventricular nucleus of the conscious Long Evans and Brattleboro rats. *Neurosci. Lett.* **85,** 239–243.
36. Ciriello, J. and Calaresu, F. R. (1980) Role of paraventricular and supraoptic nuclei in central cardiovascular regulation in the rat. *Am. J. Physiol.* **239,** R137–R142.
37. Stanley, B. G. and Leibowitz, S. F. (1984) Neuropeptide Y: Stimulation of feeding and drinking by injection into the paraventricular nucleus. *Life Sci.* **35,** 2635–2642.
38. Willoughby, J. O. and Blessing, W. W. (1987) Neuropeptide Y injected into the supraoptic nucleus causes secretion of vasopressin in the unanesthetized rat. *Neurosci. Lett.* **75,** 17–22.

39. Martin, J. R., Beinfeld, M. C., and Westfall, T. C. (1988) Blood pressure increases after injection of neuropeptide Y into posterior hypothalamic nucleus. *Am. J. Physiol.* **254,** H879–H888.
40. Martin, J. R., Knuepfer, M. M., Beinfled, M. C., and Westfall, T. C. (1989) Mechanism of pressor response to postrior hypothalamic injection of neuropeptide Y. *Am. J. Physiol.* **257,** H918–H926.
41. Wahlestedt, C., Grundeman, L., Hadansow, R., Heilig, M., Shen, G. H., Zukowska-Grogec, L., and Reis, P. (1990) Neuropeptide Y receptor subtypes, Y_1 and Y_2. *Ann. NY Acad. Sci.* **611,** 7–26.
42. Sheikh, S. P., Hakanson, R., and Schartz, T. W. (1989) Y_1 and Y_2 receptors for neuropeptide Y. *FEBS Lett.* **245,** 209–214.
43. Unden, A., Tatemoto, K., Mutt, V., and Bartfai, T. (1984) Neuropeptide Y receptor in the rat brain. *Eur. J. Biochem.* **145,** 525–530.
44. McAuley, M. A., Macrae, I. M., Farmer, R., and Reid, J. L. (1991) Effects of neuropeptide Y on forskolin, alpha$_2$- and beta-adrenoceptor regulated cAMP levels in the rat brain slice. *Peptides* **12,** 407–412.
45. Westlind-Danielsson, A., Unden, A., Abens, J., Andell, S., and Bartfai, T. (1987) Neuropeptide Y receptors and the inhibition of adenylate cyclase in the human frontal and temporal cortex. *Neurosci. Lett.* **744,** 237–242.
46. Westlind-Danielsson, A., Andell, S., Abens, J., and Bartfai, T. (1988) Neuropeptide Y and peptide YY inhibit adenylate cyclase activity in the rat striatum. *Acta Physiol. Scand.* **132,** 425–430.
47. Hinson, J., Rauh, C., and Coupet, J. (1988) Neuropeptide Y stimulates phospholipid hydrolysis in rat brain miniprisms. *Brain Res.* **446 (2),** 379–382.
48. Perney, T. M. and Miller, R. J. (1989) Two different G-proteins mediate neuropeptide Y and bradykinin-stimulated phospholipid breakdown in cultured rat sensory neurons. *J. Biol. Chem.* **264(13),** 7317–7327.
49. Wahlestedt, C., Skagerberg, G., Ekman, R., Heilig, M., Sundler, F., and Hakanson, R. (1987) Neuropeptide Y (NPY) in the area of the hypothalamic paraventricular nucleus activates the pituitary-adrenocortical axis in the rat. *Brain Res.* **417,** 33–38.
50. Fuxe, K., Harfstrand, A., and Agnati, L. (1987) Neuropeptide Y- Adrenaline in central cardiovascular regulation. *Neuroscience* **22,** S391.
51. Fuxe, K., Aguirre, J. A., Agnati, L. F., von Euler, G., Hedlund, P., Covenas, R., Zoli, M., Bjelke, B., and Eneroth, P. (1990a) Neuropeptide Y and central cardiovascular regulation. *Ann. NY Acad. Sci.* **611,** 111–133.
52. Chalmers, J. P. (1975) Brain amines and models of experimental hypertension. *Circ. Res.* **36,** 469–479.
53. Unnerstall, J. R., Kopajtic, T. A., and Kuhar, M. H. (1984) Distribution of a$_2$-agonist binding sites in the rat and human central nervous system: analysis of some functional anatomic correlates of the pharmacologic effects of clonidine and related adrenergic agonists. *Brain Res. Rev.* **319,** 69–101.
54. Haeusler, G. (1982) Central alpha-adrenoceptors involved in cardiovascular regulation. *J. Cardiovasc. Pharmacol.* **4,** S72–S76.
55. Zandberg, P., DeJong, W., and DeWied, D. (1979) Effects of catecholamine-receptor stimulating agents on blood pressure after local application in nucleus tractus solitarius of medulla oblongata. *Eur. J. Pharmacol.* **55,** 43–56.
56. Kubu, T. and Misu, Y. (1981) Pharmacological characterization of the alpha-adrenoceptors responsible for a decrease of blood pressure in the nucleus

tractus solitarii of the rat. *Naunyn-Schmiedeberg's Arch. Pharmacol.* **317,** 120–125.

57. Härfstrand, A., Fuxe, K., Agnati, L., and Fredholm, B. (1989) Reciprocal interactions between a_2-adrenoceptor agonist and neuropeptide Y binding sites in the nucleus tractus solitarius of the rat. *J. Neural Transm.* **75,** 83–99.
58. Fuxe, K., Agnati, L. F., Harfstrand, A., Janson, A. M., Neumeyer, A., Andersson, K., Ruggeri, M., Zoli, M., and Goldstein, M. (1986) Morphofunctional studies on the neuropeptide Y/adrenaline co-storing terminal systems in the dorsal cardiovascular region of the medulla oblongata. Focus on receptor-receptor interactions in co-transmission. *Prog. Brain. Res.* **68,** 303–320.
59. Agnati, L. F., Fuxe, K., Benfenati, F., Battistini, N., Harstrand, A., Tatemoto, K., Hokfelt, T., and Mutt, V. (1983) Neuropeptide Y *in vitro* selectively increases the number of a_2-adrenergic binding sites in membranes of the medulla oblongata of the rat. *Acta Physiol. Scand.* **118,** 293–295.
60. Fuxe, K., Agnati, L. F., Harfstrand, A., Zoli, M., von Euler, G., Grimaldi, R., Merlo Pich, E., Bjelke, B., Eneroth, P., Benfenati, F., Cintra, A., Zini, I., and Martire, M. (1990b) On the role of neuropeptide Y in information handling in the central nervous system in normal and physiopathological states. *Ann. NY Acad. Sci.* **579,** 28–67.
61. Härfstrand, A., Fuxe, K., Agnata, L. F., Ganten, D., Eneroth, P., Tatemoto, K., and Mutt, V. (1984) Studies on neuropeptide Y catecholamine interactions in central cardiovascular regulation in the a-chloralose anaesthetized rat. Evidence for a possible new way of activating the α_2-adrenergic transmission line. *Clin. and Exper. Hyperten.—Theory and Practice* **A6 (10 and 11),** 1947–1950.
62. Westfall, T. C., Han, S. P., Chen, X., Del Valle, K., Curfmann, M., Ciarleglio, A., and Naes, L. (1990a) Presynaptic peptide receptors and hypertension. *Ann. NY Acad. Sci.* **604,** 372–388.
63. Westfall, T. C., Han, S. P., KNuepfer, M., Martin, J., Chen, X., Del Valle, K., Ciarleglio, A., and Naes, L. (1990) Neuropeptides in hypertension: Role of neuropeptide Y and calcitonin gene related peptide. *Brit. J. Clin. Pharmacol.* **30,** 755–825.
64. Tsuda, K., Yokou, H., and Goldstein, M. (1989) Neuropeptide Y in norepinephrine release in hypothalamic slices. *Hypertension* **14,** 81–86.
65. Tsuda, K., Tsuda, S., Masuyama, Y., and Goldstein, M. (1990) Norepinephrine release and neuropeptide Y in medulla oblongata of spontaneously hypertensive rats. *Hypertension* **15,** 784–790.
66. Vallejo, D. A., Carter, S. B., and Lightman, S. L. (1987) Neuropeptide Y alters monoamine turnover in the rat brain. *Neurosci. Lett.* **73,** 155–160.
67. Heilig, M., Vecsei, L., Wahlstedt, C., Alling, C. and Widerlov, E. Effects of centrally administered neuropeptide Y and NPY 13-36 on brain monoaminergic systems of the rat. *J. Neural Transm.* **79,** 193–208.
68. McCall, R. B. (1988) Effects of putative neurotransmitters on sympathetic preganglionic neurons. *Ann. Rev. Physiol.* **50,** 553–564.
69. Chen, X., Knuepfer, M. M., and Westfall, T. C. (1988) Alteration in the evoked release of NE and peripheral vascular resistance following intrathecal NPY. *Soc. Neurosci. Abs.* **14(1),** 150.
70. Lundberg, J. M., Rudehill, A., Sollevi, A., Theodrosson-Norheim, E., Pernow, J., Hamberger, B., and Goldstein, M. (1986) Frequency and reserpine-

dependent chemical coding of sympathetic transmission: Differential release of noradrenaline and neuropeptide Y from pig spleen. *Neurosci. Lett.* **63,** 96–100.

71. Maccarrone, C. and Jarrett, B. (1985) Differences in regional brain concentrations of neuropeptide Y in spontaneously hypertensive (SH) and Wistar-Kyoto (WKY) rats. *Brain Res.* **345,** 165–169.
72. Chen, X., Henderson, K., Beinfeld, M. C., and Westfall, T. C. (1988) Alterations in blood pressure of normotensive and hypertensive rats following intrathecal injections of neuropeptide Y. *J. Cardiovasc. Pharm.* **12,** 473–478.
73. Nomura, Y., Kawata, K., Kitamura, Y., and Watanabe, H. (1987) Effects of pertussis toxin on the a_2-adrenoceptor-inhibitory GTP-binding protein-adenylate cyclase system in rat brain: Pharmacological and neurochemical studies. *Eur. J. Pharmacol.* **134,** 123–129.
74. Qualy, J. M. and Westfall, T. C. (1988) Release of norepinephrine from the paraventricular hypothalamic nucleus of hypertensive rats. *Am. J. Physiol.* **254,** H993–H1003.

Neuropeptide Y Actions on Reproductive and Endocrine Functions

John K. McDonald and James I. Koenig

1. Introduction

In this chapter, we have reviewed the current literature concerning the function of neuropeptide Y (NPY) in endocrine and reproductive systems. The reader is referred to other reviews and recent symposia on NPY for consideration of other areas *(1–4)*. The reader is also referred to other chapters in this volume, particularly those by Hendry and by Stanley.

2. Distribution of NPY

NPY is probably the most abundant peptide in the nervous system, and is distributed in key endocrine and neuroendocrine regulatory centers. Because of its distribution in the hypothalamus, pituitary, and various endocrine glands, NPY is in a unique position to act as a modulator of neurohormone and hormone secretion and synthesis.

2.1. Hypothalamus

NPY is found in very high concentrations in several hypothalamic nuclei that regulate a wide variety of neuroendocrine functions (*5–8; see also* Hendry, this vol.). The paraventricular, dorsomedial, arcuate, and periventricular nuclei, as well as the lateral hypothalamic area and median eminence, are all densely innervated by NPY-containing fibers. These fibers arise from at least two sources. The arcuate nucleus, which has one of the highest concentrations

From: *The Biology of Neuropeptide Y and Related Peptides;*
W. F. Colmers and C. Wahlestedt, Eds.

of NPY-immunoreactive cell bodies in the brain, projects fibers to the paraventricular nucleus and probably to other hypothalamic nuclei and the medial preoptic nucleus *(9,10)*. Previous studies have not revealed coexistence of NPY in dopaminergic neurons in the arcuate nucleus *(11,12)*, however, Ciofi et al. *(13)* recently reported that a few tyrosine hydroxylase-immunoreactive neurons in the arcuate nucleus were also NPY-immunoreactive in lactating rats. This observation raises the possibility that NPY may coexist in tubero-infundibular dopaminergic neurons, although additional evidence that these cells are not adrenergic neurons, which display immunoreactivity to dopamine β-hydroxylase or phenylethanolamine *N*-methyltransferase, would strengthen this result. Many other NPY-immunoreactive fibers in the hypothalamus arise from neurons in brain stem adrenergic and noradrenergic nuclei *(14–19)*. Thus, a large percentage of the NPY-containing fibers in the hypothalamus probably also contain epinephrine or norepinephrine.

NPY-containing fibers have been observed in close association to perikarya that synthesize luteinizing hormone-releasing hormone (LHRH) *(20–23)*, thyrotropin-releasing hormone (TRH) *(24)*, corticotropin-releasing hormone *(25)*, vasopressin *(26)*, and somatostatin. This anatomical proximity of NPY fibers to other neuropeptide-containing neuronal cell bodies implicates a functional role for NPY in the regulation of these neurons.

NPY also coexists in some of the somatostatin- and growth hormone-releasing hormone (GHRH) containing neurons in the arcuate nucleus *(27–30)*. It is quite possible that NPY coexists in other arcuate neurons that contain γ aminobutyric acid (GABA) and perhaps also proopiomelanocortin (POMC) neurons. Subpopulations of NPY-containing neurons in the arcuate nucleus also display receptors for glucocorticoids *(31,32)*, whereas others accumulate radiolabeled estrogen *(33)* and are responsive to changes in these endocrine feedback signals. NPY-immunoreactive neurons in the brain stem display receptors for glucocorticoids *(31)*, thus increasing the potential for multiple endocrine signals to reach hypothalamic regulatory and hypophysiotropic neurons.

2.2. Hypothalamo–Hypophysial Portal System

NPY-containing fibers surround capillaries of the hypothalamo-hypophysial portal system in the median eminence and infundibulum in several species *(30,34,35)*, providing support for the notion of NPY neurosecretion into portal vessels. The specific origin of these fibers is unknown at present, although both the arcuate nucleus and brain stem catecholaminergic nuclei probably contribute. Meister et al. *(12)* observed that neonatal monosodium glutamate treatment eliminated NPY-labeled cell bodies from the ventromedial arcuate nucleus, but had little effect on the distribution or density of NPY-immunoreactive fibers in the rat median eminence. These authors reported some coexistence of NPY and dopamine β-hydroxylase in median eminence fibers and suggested that NPY-containing fibers in the median eminence arise from medullary catecholaminergic neurons *(12)*. Other investigators have measured a 50% reduction in median eminence levels of NPY following neonatal monosodium glutamate treatment *(36)*. Taken together, these studies by Meister et al. *(12)*, Ciofi et al. *(13,30)*, and McDonald et al. *(36)* suggest that most of the NPY fibers in the median eminence probably arise from neurons in brain stem catecholaminergic nuclei, although NPY-containing neurons in the arcuate nucleus also contribute.

In the rhesus monkey, NPY-labeled fibers are very densely distributed around portal vessels in the upper infundibular stem *(35)*. The immunocytochemical evidence strongly supports the hypothesis that this is a major site for NPY release into the portal circulation. Additional fibers travel inferiorly in the infundibular stalk and surround vessels in the lower infundibular stalk, probably releasing NPY into short portal vessels. In the rat, the NPY innervation is robust, although not as dense as in the rhesus monkey *(13,34)*. NPY-labeled fibers traverse the median eminence from the subependymal to external layers and are located in close proximity to portal capillaries in the external zone. Ultrastructural immunohistochemical analysis reveals NPY-positive fibers adjacent to the perivascular space surrounding fenestrated capillaries *(21)*. Recently, Ciofi et al. *(13)* reported that peripheral injection of fluorogold, which was taken up by neurosecretory axons in the median eminence, labeled some NPY- and TH-positive neurons in the arcuate nucleus and, surprisingly, at the periphery of the paraventricular nucleus, suggesting that these cells may release NPY into the portal vasculature. These anatomical results support the concept that NPY is secreted into the hypothalamo–hypophysial portal vessels and bathes the endocrine cells of the anterior pituitary gland. High levels of NPY have been measured in the plasma obtained from these portal vessels in male and female rats and may have direct actions on pituitary hormone release (discussed in Section 3.2.) *(34,37)*.

2.3. Pituitary Gland

2.3.1. Neural Lobe

Other NPY-containing fibers in the infundibular stalk do not contact the portal vasculature, but continue inferiorly, destined for the neural lobe. The rhesus monkey displays marked developmental changes in this system. NPY fibers increased markedly during the first few years of life, but declined in later adult life *(35)*. The function of these fibers in the neural lobe, their origin, and the significance of the developmental changes in fiber density are unknown at present. They may modulate the release of neurohypophysial peptides, such as vasopressin and/or oxytocin, or perhaps they simply release NPY into the general circulation. Osmotic challenge affects NPY levels in the neural lobe of the rat, suggesting that NPY may play some role in the homeostatic regulation of blood volume or electrolyte concentrations (discussed in Section 5.) *(38)*. A potential role for NPY in the regulation of oxytocin release has yet to be examined.

2.3.2. Intermediate Lobe

NPY-immunoreactive fibers and cell bodies have been observed in the intermediate lobe of the rhesus monkey *(35)*. The fibers traverse the border between the neural and intermediate lobes, and may be derived from nerve fibers supplying the neural lobe. In the frog, which displays a dense NPY innervation of the intermediate lobe, NPY acts as an inhibitor of melanocyte-stimulating hormone (MSH) secretion in vitro *(39)*. This effect may be exerted indirectly through folliculo-stellate cells, which display a higher concentration of NPY-binding sites than melanotropes *(40)*. The function of NPY-containing fibers in the intermediate lobe in mammals remains to be elucidated.

2.3.3. Anterior Lobe

In addition to the high concentrations of NPY in the hypophysial-portal plasma, which bathes the cells of the anterior pituitary gland, recent reports indicate that pituitary thyrotropes produce NPY and contain its specific mRNA *(41)*. Since thyroidectomy markedly increased NPY mRNA in thyrotropes, NPY may exert some physiological function in these cells. Pituitary NPY mRNA levels also change following alterations in plasma levels of estrogen *(42)*. Additional research is required to determine if NPY derived from cells of the anterior pituitary may exert a paracrine or autocrine role to regulate thyrotropes, gonadotrophs, or other cell types. Locally produced NPY, NPY secreted from axons in the median eminence into the portal circulation, and NPY-containing fibers that innervate the neural and intermediate lobes probably all contribute to the regulation of the pituitary gland. Thus, there is the potential for multiple roles for NPY as a major regulator of the pituitary gland.

3. Function of NPY in the Reproductive System

The first reported neuroendocrine action of NPY actually involved studies of the endocrine responses to avian (a) and bovine (b) pancreatic polypeptide (PP) following injection into the third cerebral ventricle *(43)*. The observation that injection of aPP or bPP into the third cerebral ventricle of ovariectomized rats dramatically reduced plasma levels of luteinizing hormone (LH) in a dose-dependent manner led to the hypothesis that PP suppressed LHRH secretion from the hypothalamus and stimulated additional investigations *(43–46)*. NPY was isolated and characterized at this time *(47)*, and it was soon demonstrated that NPY accounted for previous reports of PP-like immunoreactivity in the brain *(48)* because of cross-reactivity of PP antisera with NPY. Accordingly, the studies described below will focus on the NPY literature. (For additional reviews, *see 130–132*.)

3.1. Hypothalamus

3.1.1. Modulation of LHRH Secretion

3.1.1.1. In Vivo Investigations and Gonadal Steroid Feedback

When NPY became commercially available, its central nervous system effects were examined by injecting porcine NPY through a chronically implanted third cerebroventricular cannula in conscious, freely moving, ovariectomized rats and sampling blood through an intraatrial cannula *(46,49)*. These experiments demonstrated that NPY exerted a potent and sustained, dose-dependent inhibitory effect on LH secretion from the anterior pituitary gland. Of particular interest was the observation that NPY had no effect on plasma levels of follicle-stimulating hormone (FSH), supporting the idea of a separate releasing factor for FSH. It was hypothesized that these central effects of NPY were mediated by suppression of LHRH release, thereby decreasing LH secretion from the anterior pituitary gland *(46,49)*.

This hypothesis was confirmed in a subsequent investigation of the ability of NPY to suppress pulsatile LH release in ovariectomized rats *(50)*. Using a similar experimental paradigm with more frequent (10 min) blood sampling, central injection of NPY (5.0 μg) suppressed the pulse frequency and amplitude of LH secretion for several hours. A lower dose (0.5 μg) also reduced pulsatile LH release with signs of recovery from suppression after 50 min. Pituitary glands of these animals responded vigorously to a peripheral injection of LHRH (10 ng/100 g body wt) given iv at a time of maximal suppression of LH secretion by NPY, with a 1239% increase in plasma levels of LH. This response to LHRH was 10× greater than seen in control ovariectomized animals receiving saline in the third ventricle instead of NPY. These results demonstrate that NPY inhibition of LH secretion is not caused by the inability of gonadotrophs to release LH. The intracellular stores o' LH may have increased significantly during NPY inhibition (LHRH secretion into the pituitary portal vessels. The hyperresponsiveness of the gonadotrophs may also indicate that following suppression of LHRH release, NPY from the cerebroventricular injection may have reached the anterior pituitary and greatly augmented the stimulatory effect of exogenous LHRH.

Kaynard and colleagues found that third cerebroventricular injection of NPY into ovariectomized rhesus monkeys suppressed mean LH levels and LH pulse frequency, although estradiol-treated ovariectomized monkeys showed little response to NPY *(51)*. The estrogen treatment restored plasma estrogen levels to those measured during the early follicular phase of the menstrual cycle. In these monkeys, NPY did not exert stimulatory effects on LH release and the high dose (15 μg) tested actually decreased LH pulse frequency. These results contrast to some extent with the reported stimulatory effects of NPY on LH release in estrogen-treated rats and rabbits *(44,52,53)*, although the level of estrogen replacement is probably an important factor, since NPY does not stimulate LHRH secretion from the rat median eminence in vitro when estrogen replacement produces low physiological levels in the plasma *(54)*. Perhaps treatment of ovariectomized monkeys with higher levels of estrogen would reveal stimulatory effects of NPY on LH secretion.

Administration of NPY through a push–pull perfusion cannula into the hypothalamus of ovariectomized rabbits decreased pulsatile LHRH release, thereby offering more direct proof of the hypothesis that NPY modulates LHRH secretion *(52)*. In contrast, Woller and Terasawa *(55)* reported that push–pull perfusion of the stalk-median eminence of gonadectomized female and male rhesus monkeys with NPY stimulated LHRH release in a dose-related manner. The authors concluded that NPY, in addition to norepinephrine, may be an important regulator of pulsatile LHRH secretion *(55)*. These data show that LHRH terminals in the stalk-median eminence of the rhesus monkey are responsive to NPY in the absence of gonadal steroids in a similar manner to the response to norepinephrine. It is well established that the pulsatile release of LH is controlled by the episodic release of LHRH from the hypothalamus. The results of these studies strongly support the hypothesis that NPY is a component of the pulse-generating system in the basal hypothalamus.

Administration of estrogen to ovariectomized rats changes the effects of NPY from inhibition of LH secretion to a mild and transient stimulation of LH release *(44)*. Push–pull perfusion of NPY into the hypothalamus of gonadally intact rabbits also increases LHRH release *(52)*. Finally, injection of NPY antiserum into the cerebroventricular system of estrogen–progesterone-treated rats blocked the afternoon LH surge *(56)*. Taken together, these data strongly implicate NPY in the gonadally steroid-dependent modulation of LHRH secretion.

3.1.1.2. In Vitro Investigations of LHRH Release and Gonadal Steroid Feedback

The stimulatory effects of NPY on LHRH secretion from the hypothalamus may be partially exerted on LHRH neurosecretory terminals in the median eminence *(54,57)*. Median eminences were obtained from ovariectomized rats that received sc implants of capsules containing oil vehicle or estradiol benzoate in oil, which produced physiological plasma levels of estradiol. Median eminences were collected between 9:00–10:00 AM when plasma LH levels were low, but median eminence content of LHRH and the basal release of LHRH were significantly increased. Incubation of individual median eminences from estrogen-treated ovariectomized rats with NPY (0.1–10 μ*M*) resulted in a dose-related increase in LHRH release. This effect was directly related to the level of estrogen replacement. It is interesting to note that influx of extracellular calcium was not required for NPY-stimulated LHRH release. Decreasing the calcium concentration in the medium or incubation in calcium-free medium containing EGTA had no effect on the capability of NPY to stimulate LHRH secretion; however, translocation of intracellular calcium was necessary. Median eminences from rats treated with proestrous levels of estrogen showed the greatest stimulatory effect of NPY on LHRH release, whereas median eminences from ovariectomized rats receiving oil vehicle were unresponsive to NPY. NPY also did not modify depolarization-induced LHRH release. These results suggest that the median eminence is probably a site of NPY stimulation of LHRH secretion into the portal vessels in rats with plasma levels of estrogen similar to those preceding the preovulatory surge of LH. Since NPY-induced inhibition of basal LHRH secretion was not observed using median eminences from ovariectomized and vehicle-treated rats, the site of NPY-negative regulation of LHRH secretion in the absence of estrogen may be elsewhere, perhaps at the level of LHRH perikarya in the medial preoptic nucleus and diagonal band of Broca. Alternatively, NPY may regulate inhibitory circuits in the medial basal hypothalamus by acting through neurons containing other transmitters or peptides, or may form an ultrashort loop feedback circuit involving NPY cell bodies in the arcuate nucleus.

Similar ovarian-dependent effects of NPY on LHRH secretion have been reported using medial basal hypothalamic and anterior hypothalamic fragments from rhesus and Japanese macaques *(58)*. NPY stimulated LHRH release from tissues from gonadally intact monkeys and had no effect when fragments were obtained from ovariectomized animals. No effect of NPY on β-endorphin secretion was measured.

Crowley and Kalra *(59)* reported that progesterone enhanced NPY-stimulated LHRH release from medial basal hypothalamic fragments in vitro. Sabatino et al. *(57)* tested the hypothesis that progesterone would increase the efficacy of NPY to stimulate LHRH secretion from median eminence fragments obtained from

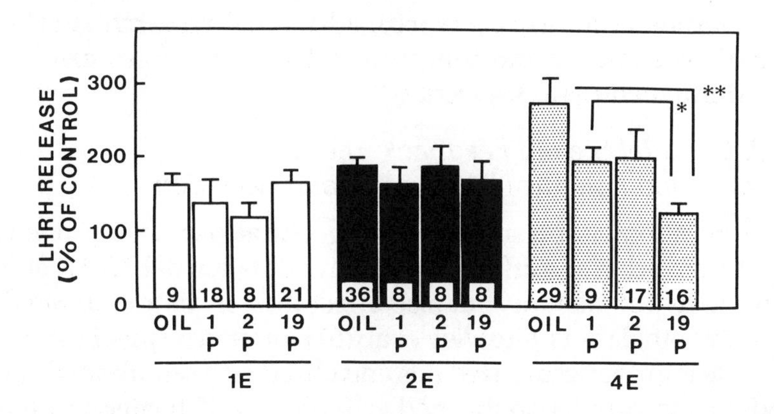

Fig. 1. Effects of NPY ($10^{-5}M$) on LHRH release from median eminence fragments in vitro obtained from ovariectomized rats treated with various doses of estrogen (4E, 2E, or 1E) and progesterone (P–1, 2, or 19 mg) or oil vehicle. Estradiol benzoate was administered in Silastic capsules for 3 d before P injection on 8:00 AM; animals were sacrificed at 11:00 AM. Each bar represents the mean ± SEM of LHRH release, expressed as a percent of control. $^{**}p < 0.01$, OIL vs 19 mg P in the 4E group; $^{*}p < 0.05$, 1 mg P vs 19 mg P in the 4E group. NPY significantly increased LHRH release in all groups when compared to LHRH release during the control period ($p < 0.05$) not shown. From *(57)*. Reprinted by permission from *Neuroendocrinology* and S. Karger AG, Basel.

ovariectomized and estrogen-treated rats. Ovariectomized rats were treated with estrogen capsules for 3 d, as described above, followed by progesterone (1, 2, or 19 mg) or vehicle injection 3 h before sacrifice between 11:00 and 11:30 AM. Although the 2- and 19-mg doses of progesterone increased the median eminence content and basal release of LHRH, the stimulatory effect of NPY on LHRH secretion was not increased. In fact, the efficacy of NPY was slightly lower in rats receiving 2 mg and significantly reduced in animals injected with 19 mg of progesterone (Fig. 1). Although LHRH levels in the median eminence are rising at this time, 3 h after injection and several hours before the afternoon LH surge, NPY is less effective in stimulating LHRH secretion than in the absence of progesterone. Perhaps progesterone serves to attenuate or focus the stimulatory effects of NPY to a particular period *(57)*. In view of this study and another that demonstrates dynamic changes in medial basal hypothalamic levels of NPY and LHRH following progesterone administration *(60)*, additional research is clearly needed to examine the temporal effects of progesterone on NPY-stimulated LHRH secretion.

3.1.2. Gonadal Steroid Feedback Effects on Hypothalamic NPY Content and Secretion

Alterations in plasma levels of gonadal steroids change not only the direction and efficacy of NPY modulation of LHRH release from the hypothalamus, but also affect the concentration of NPY in the hypothalamus. Crowley et al. *(61)* injected 50 μg of estradiol benzoate into ovariectomized rats and observed a significant decline in NPY concentration in the median eminence 48 h later, implying release of NPY into the portal vessels. Progesterone (2.5 mg) injection into similarly treated animals caused a significant increase in median eminence levels of NPY followed by a gradual decline. LHRH concentrations displayed a similar change and were consistent with alterations in peptide levels that preceded the afternoon surge in LH. Estrogen treatment also decreased NPY concentrations in the interstitial nucleus of the stria terminalis and arcuate nucleus, whereas addition of progesterone further changed NPY levels in the interstitial nucleus of the stria terminalis, ventromedial nucleus, and to a lesser extent, in the preoptic and arcuate nuclei *(61)*.

Sabatino et al. *(54)* reported that implantation of Silastic capsules containing estradiol benzoate in oil for 3 d, which produced physiological plasma levels of estradiol, significantly increased the median eminence content of NPY only at the highest dose of estrogen, whereas LHRH levels were more responsive. Injection of 1 or 2 mg of progesterone at 8:00 AM into rats receiving the highest physiological dose of estrogen had no significant effect on NPY levels in the median eminence at 11:00 AM. Only the pharmacological dose of progesterone (19 mg) significantly increased NPY levels. These results suggest that estrogen may be the primary ovarian steroid regulator of NPY levels, although the role of progesterone deserves careful evaluation, since dynamic changes in NPY levels may occur throughout the day, especially before the LH surge *(60–62)*.

Brann et al. *(60)* have recently examined the temporal effects of low levels of estrogen priming in ovariectomized immature rats combined with progesterone, triamcinolone acetonide, or cortisol administration on the concentrations of NPY and LHRH in the medial basal hypothalamus (MBH) and preoptic area (POA). Estrogen was administered (2 μg sc estradiol benzoate) at 5:00 PM on days 27 and 28, and the other compounds (1 mg/kg body wt) at 9:00 AM on day 29. This low level of estrogen replacement was insufficient to cause any significant change in plasma levels of LH or FSH compared to ovariectomized vehicle-treated controls. Estrogen treatment alone had no significant effect on NPY or LHRH levels in the POA or MBH between 9:30 AM–1:00 PM. Progesterone treatment caused an afternoon surge of LH and FSH, and also significantly increased levels of NPY and LHRH in the MBH and POA with a maximum occurring at 12:00 PM for both peptides (Fig. 2). Triamcinolone acetonide, a synthetic corticosteroid that probably acts through a progesterone receptor, caused an afternoon surge in LH and FSH, and had an effect similar to progesterone on levels of NPY and LHRH in the MBH. In contrast, cortisol had no effect on peptide levels in the MBH, but significantly increased both peptides in the POA. Taken together, these results strongly suggest that NPY and LHRH systems are responsive to progesterone treatment of rats given nonstimulatory doses of estrogen. The parallel changes in NPY and LHRH before the afternoon gonadotropin surge support the concept that NPY is an integral component in the neuroendocrine mechanisms that mediate gonadal steroid-induced surges *(60)*.

Sahu et al. *(64)* reported that NPY levels in individual hypothalamic nuclei respond differently to castration, testosterone replacement, and bilateral transection of the brain stem projections from the dorsal tegmentum to the hypothalamus. These projections include noradrenergic and adrenergic fibers, and some of them also contain NPY. Their results indicate that testosterone-sensitive brain stem neurons transmit NPY-containing fibers to the median eminence, whereas testosterone-receptive NPY containing neurons in the arcuate nucleus innervate the arcuate and ventromedial nuclei *(64)*.

Sahu et al. *(63)* proposed that endogenous opioids in the medial basal hypothalamus exert an inhibitory tone on NPY-containing neurons, thereby influencing the net stimulatory effect of NPY on LHRH secretion. Naloxone infusion into estrogen-primed ovariectomized rats significantly increased NPY concentrations in the median eminence before the normal afternoon rise in NPY levels. This increase in NPY also accompanied an early increase in

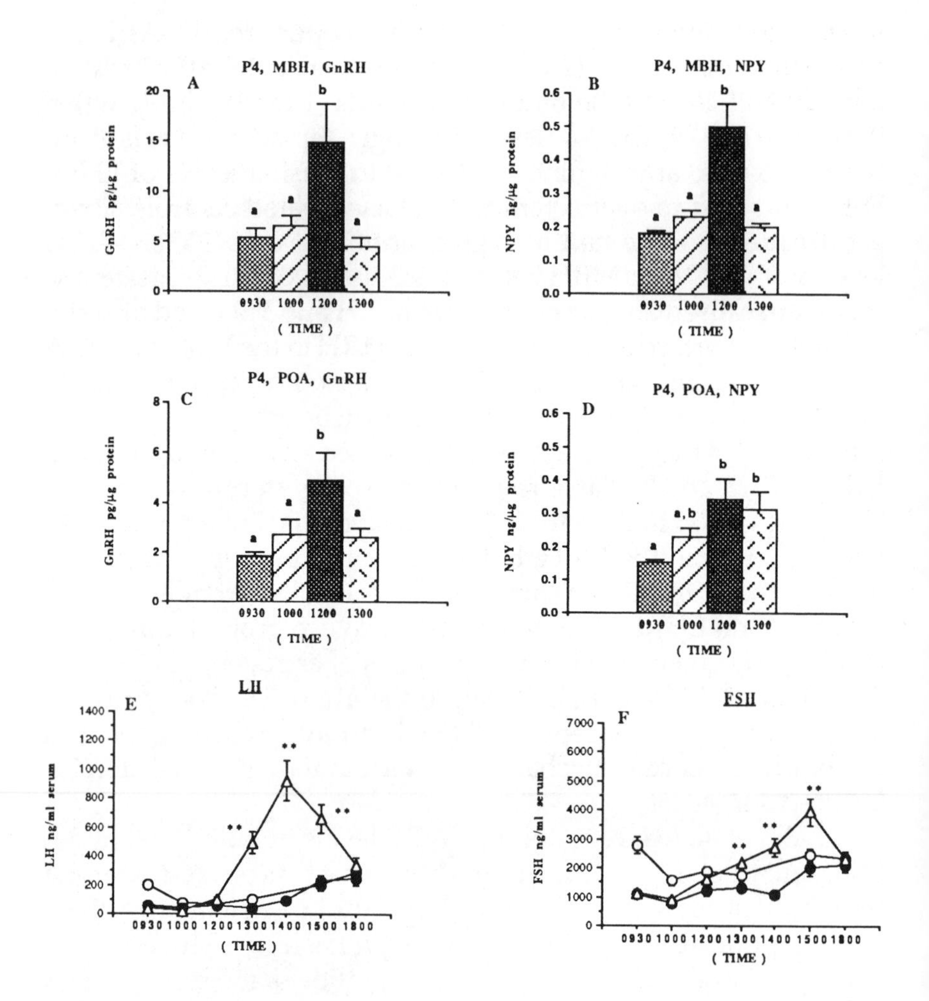

Fig. 2. Effect of progesterone (P_4) on concentrations of NPY and gonadotropin-releasing hormone (GnRH) in the medial basal hypothalamus (MBH) and preoptic-suprachiasmatic area (POA), and associated changes in serum LH and FSH levels. Twenty-six-day-old ovariectomized (OVX) immature rats ($n = 6$) were injected with estradiol (E_2; 2 µg/rat) on days 27 and 28. On day 29, the animals received progesterone (P_4; 1 mg/kg body wt sc) at 9:00 AM, and groups of animals were sacrificed at various time-points for tissue measurements of NPY and GnRH, and serum LH and FSH determinations. *$p < 0.01$; **$p < 0.01$ vs E_2. From *(60)*. Reprinted by permission from *Neuroendocrinology* and S. Karger AG, Basel. OVX —o—; E2 —●—; P_4 —Δ—.

plasma LH. Application of high doses of naloxone (0.5 and 1.0 mg/mL) in vitro stimulated NPY and LHRH secretion from medial basal hypothalamic fragments *(63)*. Opioid neurons may be one of the many neuropeptide/neurotransmitter inputs to NPY neurons in the arcuate nucleus that modulate NPY secretion. Elucidation of the factors that affect NPY secretion and the possible modulatory influence of estrogen and progesterone on the efficacy of these factors are essential steps in understanding how NPY regulates LHRH secretion.

3.1.3. NPY Secretion into the Hypothalamo-Hypophysial Portal Vasculature

High levels of NPY in the portal plasma were first reported in a study using intact, urethane-anesthetized male rats *(34)*. NPY levels in portal plasma were approximately three times greater than in peripheral plasma obtained at the same time. The elution behavior of NPY immunoreactivity in the portal plasma determined by reverse-phase HPLC indicated that rat portal plasma NPY eluted slightly earlier than the porcine NPY standard, since rat NPY differs from porcine NPY only at position 17 and is slightly less hydrophobic. Sutton et al. *(37)* provided the initial report of elevated levels of NPY in portal plasma obtained from female rats. Portal plasma levels of NPY declined significantly on the morning of proestrous when compared to diestrus levels followed by a significant increase on the afternoon of proestrous. Furthermore, these changes in NPY levels were parallel to alterations in LHRH levels, implying possible cosecretion of these neuropeptides in the median eminence. Sutton et al. *(65)* observed a similar relationship at the time of pubertal onset in female rats. Secretion of NPY into the hypophysial portal plasma increased significantly on the afternoon of the day before vaginal opening. LHRH levels also increased at this time, but to a greater extent. These results support the concept of release of both NPY and LHRH from the median eminence in a similar temporal framework. This release pattern may be crucial to maximizing the effectiveness of LHRH on the afternoon of proestrous before the LH surge, since NPY augments the efficacy of LHRH on the gonadotrope *(66)*. In view of the extremely dense innervation of the portal vessels in the rhesus monkey, it is probable that NPY exerts a similar role in nonhuman primates and perhaps also in humans. NPY may have a developmental role in the onset of first ovulation and may be a key neuroendocrine regulator of LH release during subsequent reproductive cycles.

3.2. Anterior Pituitary

In addition to the hypothalamic action of NPY to modulate LHRH secretion in an estrogen-dependent manner, NPY also acts directly on gonadotrophs of the anterior pituitary, thereby exerting a dual control on the regulation of LH secretion.

3.2.1. Modulation of LH Secretion

The first indication that NPY acted on the anterior pituitary gland was reported in 1984. NPY (1–10 μ*M*) was applied to perifused, dispersed rat anterior pituitary cells from ovariectomized animals, suspended in a syringe column with Biogel P-2, and stimulated secretion of LH, FSH, and growth hormone *(46,49)*. Apparently, the method of maintaining the cells has a major effect on the response to NPY, since no effect on hormone release was observed when NPY was applied to overnight cultures of freely floating cells, long-term cultures of dispersed cells, and hemipituitary fragments *(67,68)*. However, other investigators have reported that NPY perifusion of quartered anterior pituitaries from ovariectomized rabbits caused a transient stimulation of LH and FSH secretion, and a sustained stimulation of LH and FSH release when pituitaries were obtained from gonadally intact animals *(53)*. Pau et al. *(58)* observed that NPY stimulated LH release from anterior pituitary fragments in vitro from intact rhesus macaques, but had no effect on pituitary glands from ovariectomized animals. These results suggest that ovarian steroids may modulate the efficacy of NPY on gonadotropin secretion from the anterior pituitary in addition to their modulatory role in NPY-induced LHRH release from the hypothalamus.

NPY also potentiates the effect of LHRH on LH secretion in vitro *(66)*. These investigators have observed a slight, but significant stimulatory effect of NPY on LH release from rat hemipituitary fragments and a marked enhancement of the response to LHRH. Application of NPY to 3 d cultures of anterior pituitary cells had no effect on LH secretion, but potentiated the response to LHRH probably by increasing Ca^{2+} entry into gonadotrophs *(69)*. Since nitrendipine blocked the effect, these investigators speculated that

NPY may act through dihydropyridine-sensitive, L-type voltage-sensitive Ca^{2+} channels in the initial portion of the response to LHRH *(69)*. In contrast, Shangold and Miller *(70)* reported that NPY (1 n*M*–1 μ*M*) had no effect on LHRH-stimulated LH and FSH secretion from 3 d cultures of rat gonadotropes. LHRH also stimulated intracellular Ca^{2+} flux, and NPY inhibited the secondary plateau phase of this response. This effect was reduced by prior treatment with pertussis toxin. The authors concluded that NPY probably activates potassium conductance in gonadotropes. Perhaps the ability of NPY to augment LHRH-induced LH secretion is species specific. Pau et al. *(58)* reported that NPY did not alter the efficacy of LHRH in vitro using anterior pituitaries from Japanese macaques.

Peripheral injection of NPY antiserum into ovariectomized rats treated with estrogen and progesterone blocked the LH surge, suggesting that NPY may augment the efficacy of LHRH in vivo *(37)*. This hypothesis was directly tested by Bauer-Dantoin et al. *(71)* in proestrous rats injected with pentobarbital to eliminate LHRH secretion and block the LH surge. These investigators injected LHRH (15, 150, or 1500 ng/pulse) or saline iv, together with concurrent pulses of NPY (1 or 10 μg/pulse), every 30 min from 2:00–6:00 PM. Injection of NPY alone had no effect on LH levels. In contrast, coadministration of NPY and LHRH greatly potentiated the effectiveness of LHRH in producing LH surges (Fig. 3). This study performed in vivo directly supports the hypothesis that NPY acts as a neurohormonal modulator to amplify the sensitivity of the gonadotrope to LHRH during the generation of the preovulatory LH surge. Taken together, the data support the concept that NPY modulates reproductive hormone secretion through two separate mechanisms, one at the hypothalamus to alter LHRH release and the other at the gonadotrope to increase LHRH-induced LH secretion.

3.2.2. FSH, Prolactin

As mentioned above, perifusion of dispersed rat anterior pituitary cells with NPY increased FSH release as well as LH and growth hormone. No effect on FSH release was observed following third cerebroventricular injection of NPY or PP. Kerkerian et al. *(67)* reported that in castrated and anesthetized male rats, injection of 15 μg of NPY into the lateral ventricle decreased LH levels, but had no effect on FSH. Other investigators reported that sc injection

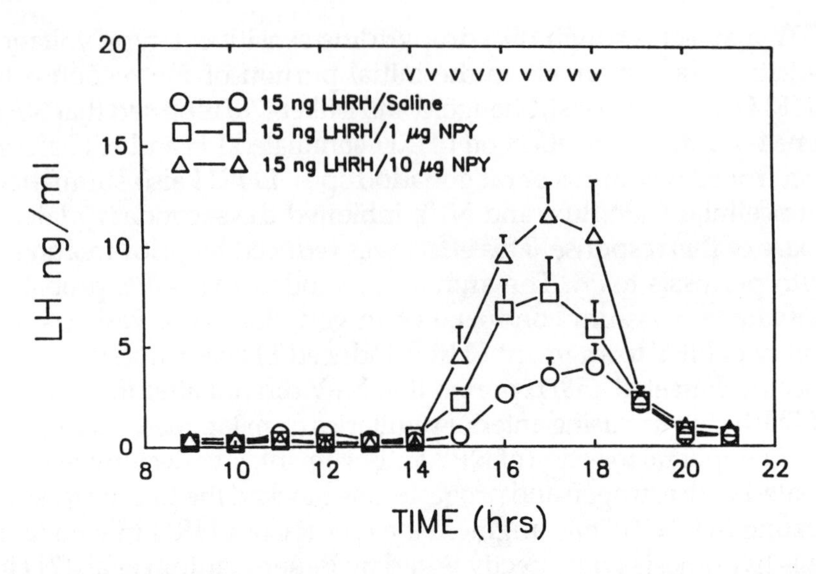

Fig. 3. Potentiation of LH response to 15 ng LHRH/pulse by NPY. Coadministration of 1 μg NPY/pulse with 15 ng LHRH/pulse to pentobarbital-blocked rats resulted in a mean LH plasma level that was significantly higher than the mean produced by the same dose of LHRH alone ($p < 0.001$). Coadministration of 10 μg NPY/pulse with 15 ng LHRH/pulse resulted in a mean LH level that was significantly higher than the mean produced by the same dose of LHRH alone ($p < 0.001$) and the mean LH level produced by coadministration of 1 μg NPY/pulse with 15 ng LHRH/pulse ($p < 0.05$). From *(71)*. Reprinted with permission from *Endocrinology*.

of NPY (10 μg) doubled plasma LH levels, but had no effect on FSH or TSH *(68)*. McCann et al. *(72)* were unable to alter plasma FSH levels in ovariectomized and ovariectomized estrogen–progesterone-treated rats by infusing antiNPY serum into the third cerebral ventricle. Khorram et al. *(53)* observed that perifusion of quartered anterior pituitaries from ovariectomized rabbits with NPY stimulated FSH and LH release, and that estrogen treatment increased the magnitude of the response. These data support the notion of a separate releasing factor for FSH and that the effect of NPY on FSH release may be selective for the anterior pituitary gland.

There is some evidence that NPY affects prolactin secretion, although this issue has not been addressed in detail. Kerkerian et al. *(67)* injected 300 μg of NPY into the jugular vein in intact male rats, and reported a sustained reduction in LH and no effect on

FSH, but a fourfold elevation in plasma levels of prolactin. Such high doses of NPY may have produced a stress effect, accounting for the elevated prolactin levels. Immunoneutralization studies of brain NPY failed to demonstrate significant changes in rat plasma prolactin concentrations 3 h after injection, but later samples showed a lowering of prolactin levels *(72)*, suggesting only minimal involvement of NPY in the regulation of prolactin secretion.

3.3. Gonads

The ovary is richly endowed with NPY-containing fibers located in the interstitium, associated with individual follicles, and surrounding blood vessels *(73,74)*. In the rat, these fibers are carried to the ovary via the plexus nerve and display a high degree, but not an absolute colocalization with norepinephrine *(74)*. Neonatal treatment with guanethidine or antiserum to nerve growth factor (NGF) virtually eliminates NPY-containing fibers from the ovary *(75,76)*. Although a vasomotor role for NPY in the ovary is likely, no endocrine role has yet been discovered *(77)*. Preliminary studies revealed that NPY had no effect on basal or FSH-stimulated estrogen and progesterone production from cultured granulosa cells in vitro *(74)*. However, in view of the interstitial and perifollicular distribution of NPY-immunoreactive fibers, a potential endocrine role for NPY in the ovary must be considered. Jorgensen and colleagues recently provided evidence for the existence of NPY and proNPY in human follicular fluid *(78)*, although it remains to be determined if NPY is actually produced in follicles, simply released from intraovarian nerves, or is transmitted to follicular fluid from the plasma.

The testis is sparsely innervated with NPY-containing fibers that are primarily associated with blood vessels and the tunica albuginea *(79)*. No endocrine role has been reported at this time.

4. Interactions Between NPY and the Hypothalamic–Pituitary–Adrenal Axis

4.1. Hypothalamic Actions of NPY

Initial studies conducted by Harfstrand et al. *(80)* revealed that intracerebroventricular infusions of picomolar amounts of NPY increased plasma concentrations of ACTH, corticosterone, and aldosterone in male rats. Subsequently, Wahlestedt et al. *(81)*, Haas and George *(82)*, Inoue et al. *(83)*, Leibowitz *(84)*, and Albers et al.

(85) confirmed the stimulatory action of NPY on the activity of the hypothalamic–pituitary–adrenal axis. Wahlestedt et al. *(81)* convincingly showed that the site of NPY action was the hypothalamic paraventricular nucleus, which contains the cell bodies of the corticotropin-releasing hormone (CRH) neurons, since direct injections of NPY into this nucleus caused prompt and robust increases in plasma ACTH levels. The hypothesis that the stimulatory effect of NPY on ACTH secretion is mediated by CRH is supported by two lines of evidence. First, immunohistochemical studies by Liposits et al. *(86)* demonstrated the presence of NPY-containing nerve terminals impinging on CRH-containing neurons in the paraventricular nucleus, and second, Inoue et al. *(83)* showed that the CRH antagonist, α-helical CRH, attenuated the stimulatory effect of NPY on ACTH secretion in the dog. Moreover, Haas and George *(82)* noted that infusions of NPY into the rat third cerebral ventricle selectively reduced the concentration of CRH in the median eminence, suggesting that NPY stimulated CRH secretion from nerve terminals in that area. An action of NPY in the median eminence was supported by the report of Tsagarakis et al. *(87)* showing a stimulatory action of NPY on the secretion of CRH from hypothalamic fragments in vitro. Therefore, NPY appears to have direct stimulatory actions on CRH neurons in the paraventricular nucleus and possibly also on CRH-containing nerve terminals in the median eminence, thereby increasing the secretion of CRH into the hypophysial portal circulation. However, although there is an abundance of NPY in the hypothalamus, there is a paucity of receptors for NPY in this region, raising the question of receptor mismatches and the true site of NPY action *(88–90)*. Additional studies are required to resolve this discrepancy fully.

Norepinephrine (NE) and epinephrine (E) have been found to have biphasic effects on the release of hormones of the hypothalamic–pituitary–adrenal axis *(91)*, and immunohistochemical studies have revealed the coexistence of NPY with NE and E in several neuronal populations within the hypothalamus and in extrahypothalamic regions *(16,17,92)*. The effects of NPY on hormone secretion could potentially be mediated by altering the release of NE and/or E from hypothalamic nerve terminals. Harfstrand et al. *(80)* originally reported a biphasic effect of NPY on catecholamine turnover in the hypothalamus, measured by changes in the intensity of catecholamine histofluorescence. In most regions, intraventricular infusion of a low dose of NPY (<250 pmol) reduced the levels of dopamine or NE histofluorescence. However, doses >250 pmol increased catecholamine turnover. Yokoo et al. *(93)* found reduced radiolabeled NE release from synaptosomal preparations in vitro after incubation with NPY consistent with the studies described above. In vivo hypothalamic dialysis studies have noted a regional specificity in NPY-induced catecholamine release. The administration of NPY (100 pmol) into the ventromedial nucleus lowered the release of NE and its metabolite MHPG, whereas a similar infusion into the lateral hypothalamus increased NE release *(94)*. Since the doses of NPY required to alter ACTH and corticosterone secretion are the same as those reported to affect catecholamine turnover, it may be that NPY-induced ACTH secretion is the consequence of an indirect action of NPY on catecholamine-containing nerve terminals in the paraventricular nucleus or median eminence.

Many NPY-containing cells in the hypothalamus and the NPY/NE containing neurons in the brainstem also contain glucocorticoid receptors *(31,32)*. Consequently, the activity of NPY neurons may be influenced by alterations in plasma glucocorticoid concentrations. Corder et al. *(95)* demonstrated that bilateral adrenalectomy significantly changed catecholamine levels in the brain, whereas NPY concentrations were unaffected. In contrast, Rivet et al. *(96)* showed that dexamethasone incubation with hypothalamic tissue in vitro increased the tissue levels of NPY. Consistent with these results is a recent report by Dean and White *(97)* demonstrating a relative decrease in NPY mRNA in the hypothalamus 4 d after removal of the adrenal glands. These findings suggest that glucocorticoids are obligatory for normal expression of the NPY gene in the hypothalamus, and that a dissociation between peptide and mRNA levels may occur. Recently, Sabol and Higuchi *(98)* reported a biphasic action of glucocorticoids on NPY gene expression in a pheochromocytoma (PC12) cell line treated with NGF. PC12 cells represent a unique pluripotent system. Incubation of PC12 cells under normal conditions yields cells with the appearance of adrenal medullary cells. However, on incubation with NGF, a developmental switch occurs, and the cells differentiate into a neuronal phenotype. Early in this phenotypic shift, NPY gene expression is markedly stimulated, and glucocorticoids enhance the expression. Later in the response, after cellular differentiation has occurred, the glucocorticoids reduce NPY expression. The exact mechanism for this reversal is unknown. These observations suggest that a glucocorticoid response element (GRE) may be present in the 5' upstream regulatory region of the NPY gene and may be involved in mediating these effects.

4.1.1. NPY Mediation of Stress-Induced ACTH Secretion

Lesions of the ventral NE bundle, which contain axons of the brain stem catecholamine-containing cells that project to the hypothalamus, disrupt stress-induced ACTH secretion in the rat (as reviewed by *99*). Since NPY coexists in some NE neurons of the brain stem, NPY may play a role in the activation of ACTH secretory mechanisms during stress. This hypothesis was tested by Inui et al. *(100)* using an insulin hypoglycemia stress model. AntiNPY serum or normal rabbit serum was infused into the third cerebral ventricle of dogs. Immediately thereafter, the animals were challenged with an iv bolus of insulin. The resulting hypoglycemia activated ACTH secretion. Although normal rabbit serum failed to alter ACTH or cortisol responses to insulin-induced hypoglycemia, antiNPY serum eliminated these hormonal responses, suggesting that NPY is important in activating stress-induced ACTH secretion. Furthermore, if NPY is involved in the activation of stress-induced ACTH secretion, then neurons containing NPY might be activated by stress, and the concentration of the peptide may change in critical hypothalamic regulatory areas. Hypothalamic NPY-containing neurons have been found to respond to changes in plasma glucose levels after insulin administration *(101)*. The concentrations of NPY in the basal hypothalamus and preoptic regions of the rat brain are reduced 30 and 90 min after the infusion of insulin, concomitant with maximal activation of the hypothalamo–pituitary–adrenal axis. Furthermore, NPY levels in the median eminence are increased at these same times *(101)*. These results suggest that stress may deplete NPY stores in the basal hypothalamus and preoptic area, and enhance transport to the median eminence, where NPY could be released into the portal circulation or affect CRH secretion into the portal blood. Together these studies indicate that stress-induced activation of ACTH secretion owing to increased secretion of CRH and vasopressin into the portal circulation *(102,103)* may be mediated by NPY mechanisms.

The action of NPY in the paraventricular nucleus may also be related to the regulation of energy metabolism (*see also* Stanley, this vol.). Several investigators have reported that NPY infusions into the paraventricular nucleus or adjoining third cerebral ventricle potently increased food intake in conscious male and female rats (for reviews, *see 84,104,* and Stanley, this vol.). The infusion of peptide YY (PYY), a structurally related peptide, was more potent than NPY in stimulating food intake. Activation of NPY receptors by local injection of the peptide selectively increased carbohydrate ingestion rather than fat or protein intake, and increased plasma glucose concentrations. Both streptazotocin-induced hyperglycemia and food deprivation increased NPY levels in the paraventricular and arcuate nuclei, an effect reversed by insulin administration *(104–108,126)*. This finding is consistent with decreased hypothalamic levels of NPY following insulin-induced hypoglycemia *(101; see preceding paragraph)*. NPY levels are elevated in the paraventricular nucleus and arcuate nucleus of spontaneously diabetic rats. The increase in NPY peptide levels in this model of diabetes is associated with increased NPY mRNA levels in the arcuate nucleus *(109)*. The Zucker obese rat also represents a rat model of obesity and hyperphagia. In the obese rats, NPY mRNA levels in the hypothalamus in general, and specifically in the arcuate nucleus, are markedly elevated compared with lean controls *(110)*. These results, in addition to those discussed in the insulin hypoglycemia model, indicate that NPY-containing neurons may be involved in monitoring and maintaining the energy balance of animals, and may be an important link in understanding the etiology of obesity.

4.2. Pituitary Effects of NPY

Immunohistochemical studies revealed moderate to dense innervation of the median eminence by NPY-containing nerve fibers (*see* Sections 2.1. and 2.2.). High concentrations of NPY-immunoreactive material have been identified in the hypophysial portal circulation of the rat *(34)*, suggesting that NPY released from nerve terminals in the median eminence may have direct effects on the pituitary gland. However, Chabot et al. *(111)* observed no effect of NPY on the secretion of β-lipotropin, a fragment of POMC, from pituitary cells in culture. In addition, Danger et al. *(39)* reported an inhibitory effect of NPY on MSH secretion in nonmammalian spe-

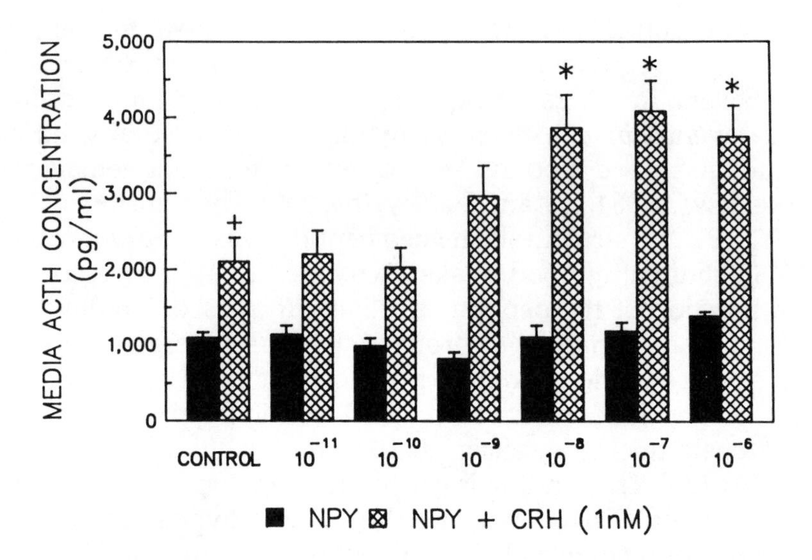

Fig. 4. Effect of various doses of NPY (10 p*M* –1 μ*M*) alone or in combination with CRH (1 n*M*) on ACTH secretion from dispersed anterior pituitary cells in vitro. Anterior pituitary cells were enzymatically dispersed and plated at a density of 250,000 cells/well. Cells were exposed to the peptides for 3 h. Shown are the group means from 6–12 wells/point ± SEM. NPY had no effect by itself, but significantly enhanced the ACTH secretory response to CRH. (*$p < 0.05$ compared to the appropriate control). From *(101)*. Reprinted by permission from *Ann. NY Acad. Sci.*

cies, an action not seen in mammals. We have reexamined the action of NPY on ACTH secretion from dispersed monolayer pituitary cell cultures and, in agreement with Chabot et al. *(111)*, have been unable to induce ACTH secretion with NPY (1 n*M*–1 μ*M*; *101*). However, when ACTH secretion is activated by coincubation with CRH (1 n*M*), NPY produces a marked, dose-related augmentation of the effect of CRH (*101*; Fig. 4). Therefore, NPY may modulate hypothalamo–pituitary axis activity both within the hypothalamus by altering CRH neuronal activity and release, and also at the level of the pituitary corticotrope by enhancing the action of CRH. This latter site of action may be important during stress, when peripheral concentrations of NPY are increased *(112)* and the pituitary is bathed with high levels of centrally derived CRH and perhaps both peripherally and centrally derived NPY.

4.3. Actions of NPY in the Adrenal

High concentrations of NPY are also present in the adrenal medulla. In the adrenal, NPY coexists with E and NE as in some brain areas *(113,114)*. NPY in this organ appears to have weak actions on adrenal cortical hormone secretion *(83)*, but rather alters the secretion of the adrenal medullary substances *(115)*. Although cholinergic stimulation increases the secretion of E and NE from the gland, it also induces the secretion of NPY *(116)*. Of interest is the observation that exogenous NPY reverses the influence of cholinergic stimulation on E and NE secretion *(115)*.

Transformed adrenal medullary carcinoma cells or pheochromocytoma cells (PC12) produce NPY and express the NPY gene. However, when these cells are incubated in the presence of NGF, they begin to express a neuronal phenotype. Associated with this phenotypic change is a dramatic induction of NPY gene expression and peptide production *(99,117,118)*. These studies suggest that mechanisms controlling the NGF-induced phenotypic changes also regulate the expression of the NPY gene.

The regulation of the hypothalamo–pituitary–adrenal axis is a complex, finely orchestrated interplay among numerous neuropeptides and neurotransmitters. Clearly, NPY must be considered as an important component of this regulatory scheme. NPY acts at multiple sites within the brain and also at the pituitary to augment the secretion of CRH and ACTH, respectively. NPY plays a central role in the activation of ACTH release during stress, although the physiological importance of these interactions requires further consideration. NPY appears to have limited actions on adrenal cortical hormone secretion, but its existence in sympathetic nerves and its ability to modulate catecholamine secretion from the adrenal medulla indicate that NPY may be an important integrative peptide regulating both cardiovascular (*see* Zukowska-Grojec et al., this vol.) and endocrine systems.

5. Modulation of Vasopressin Secretion by NPY

NPY-containing neurons are known to project axons from brain stem catecholaminergic nuclei and the arcuate nucleus to the hypothalamic paraventricular and supraoptic nuclei *(9,16,17,92)*. The presence of NPY-containing synaptic boutons impinging on vasopressinergic (VP) neurons in the paraventricular and supraoptic nuclei have been reported *(26,119)*. Microinfusion of NPY has been shown to excite supraoptic neurons *(120)*. Willoughby and Blessing *(121)* reported that infusions of NPY into the supraoptic nucleus increased the plasma concentrations of VP in conscious, unrestrained rats, whereas infusion into the amygdala had no effect. Furthermore, Leibowitz *(84)* reported that infusions of NPY into the paraventricular nucleus augmented plasma VP concentrations. NPY also plays an important role in the developmental organization of the hypothalamo–neurohypophysial system. Vallejo et al. *(122)* administered antiNPY serum or normal rabbit serum to male neonatal rats. After reaching adulthood, the rats were treated with hypertonic saline, a powerful stimulus for VP secretion. In control rats treated with normal rabbit serum, a robust VP response was observed, but in the antiNPY-treated rats, the response was attenuated. These studies revealed a critical role for NPY in the organization and regulation of VP secretion.

NPY may modulate fluid homeostasis via other mechanisms as well. Edwards et al. *(123)* reported that the preoptic region, part of the anterior ventral third ventricle (AV3V) area known to be obligatory for normal fluid homeostasis, is innervated by NPY-containing neurons. These axons appear to be derived from catecholaminergic cell bodies in the caudal ventrolateral medulla and appear to modulate the activity of neurons that sense fluid volume and regulate aldosterone and VP secretion. Harfstrand et al. *(80)* found that NPY is a powerful stimulus for aldosterone secretion. Therefore, NPY may modulate fluid volume by enhancing both VP and aldosterone secretion. In combination with its vasoconstrictive actions, this peptide acts to maintain blood volume and pressure.

It might be anticipated from the foregoing discussion that brain neurons containing NPY would be activated by dehydration, or other stimuli that alter blood pressure or volume. Recent studies by Hooi et al. *(38)* demonstrated that dehydration alters NPY-neuronal activity. Animals maintained on 2% NaCl for 2 or 7 d displayed increased plasma osmolality and VP concentrations. Seven days after initiation of this treatment, the concentrations of NPY in the basal hypothalamus, preoptic region, median eminence, and neural lobe of the pituitary were dramatically increased. Moreover, Brattleboro rats, which lack VP, also exhibited a similar

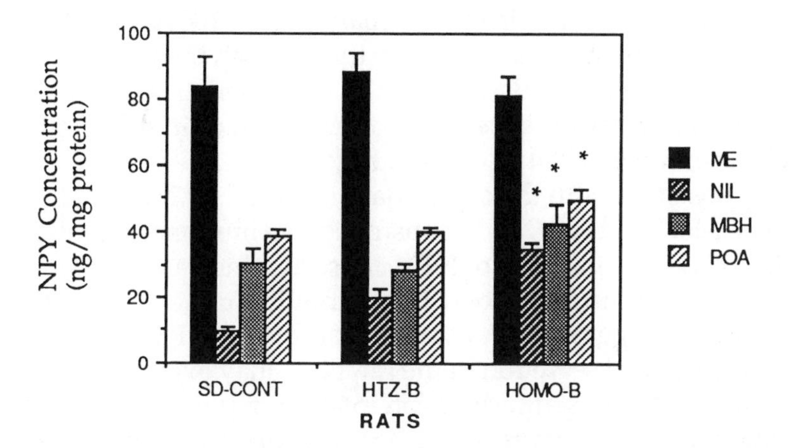

Fig. 5. NPY concentrations in the median eminence (ME), neurointermediate lobe (NIL), medial basal hypothalamus (MBH), and preoptic area (POA) of Sprague-Dawley control rats (SD), heterozygous-Brattleboro (HTZ-B) rats, and homozygous-Brattleboro (HOMO-B) rats. HOMO-B rats displayed significantly increased NPY levels in the NIL, MBH, and POA when compared to NPY levels in similar tissues in HTZ-B and SD controls. ($^*p < 0.05$).

increase in tissue levels of NPY (Fig. 5). These studies suggest that NPY plays a key role in monitoring and regulating fluid homeostasis in the rat, and probably in other mammalian species as well.

6. Action of NPY in the Regulation of Growth-Hormone Secretion

In 1985, McDonald et al. reported that the intracerebroventricular infusion of 2 µg of NPY reduced plasma concentrations of GH in the freely-moving, ovariectomized female rat. Harfstrand et al. *(80)* subsequently confirmed this effect in males after infusion of 250 or 750 pmol of the peptide. However, in contrast to the inhibitory effects exerted by NPY within the brain, McDonald et al. *(46)* demonstrated that perifusion of anterior pituitary cells with NPY in vitro increased GH secretion. Chabot et al. *(111)* also reported a stimulatory action of NPY on GH release in vitro. Peng et al. *(124)* showed that the stimulatory effect of NPY was not restricted to mammalian species. These investigators found a stimulatory effect of NPY on GH release from goldfish anterior pituitary cells in vitro. However, in mammalian species, 100 n*M*–1 μ*M* doses are required for stimulation, but in the goldfish the ED_{50} is 0.5 n*M*. This difference in the dose–response characteristics could be due to species differences in receptor distribution or in the number of NPY receptors per cell in the pituitary.

Consistent with the earlier data reported by McDonald et al. *(45,46)*, McCann et al. *(72)* demonstrated that immunoneutralization of CNS NPY by infusion of NPY antiserum increased plasma GH concentrations in conscious, freely moving intact and castrated female rats, suggesting a physiological role for NPY in the regulation of basal GH secretion in the rat. NPY may exert its actions by regulating somatostatin secretion into the hypophysial portal circulation as originally proposed *(46)*. In support of this hypothesis, McCann et al. *(72)* reported that NPY (1–100 n*M*) stimulated somatostatin release from median eminence fragments in vitro. The β-adrenergic antagonist propranolol attenuated this effect, suggesting the involvement of β-adrenergic receptors in this response. Immunohistochemical studies revealed that NPY coexists with GHRH in neurons of the rat and human arcuate nucleus *(28,29)*; consequently, a modulatory role for NPY on GHRH secretion should also be considered. Although NE receptor activation increases GH secretion *(125)*, NPY appears to reduce NE release within the hypothalamus (*see* Section 4.1.). Therefore, the effects of NPY to reduce GH secretion could be mediated by a withdrawal of stimulatory noradrenergic tone from the GHRH-containing neurons in the arcuate nucleus. In other systems, NPY presynaptically inhibits noradrenergic potentials in the dorsal raphe nucleus (*see* Bleakman et al., this vol.).

7. Regulation of the Hypothalamo–Pituitary–Thyroid System by NPY

In light of the reported ability of NPY to release somatostatin from the median eminence and the inhibitory effects of somatostatin on thyroid-stimulating hormone (TSH) secretion, it might be expected that NPY would indirectly lower plasma TSH concentrations following intraventricular infusion. Harfstrand et al. *(80)* reported a decline in plasma TSH levels after an intracerebroventricular infusion of NPY, but only at high doses (250 or 750 pmol); however, McDonald et al. *(45)* reported no significant effect of intraventricularly injected BPP on plasma TSH levels compared to saline-treated control rats. Consistent with weak effects of NPY on TSH secretion are data showing that passive immunoneutralization of CNS NPY had no effect on plasma TSH levels *(72)*. These results are of particular interest in view of the immunohistochemical studies of Toni et al. *(24)* demonstrating a close association of NPY-containing nerve terminals with TRH-containing cell bodies in the paraventricular nucleus. The inhibitory effect of NPY on TSH secretion could also be related to the ability of NPY to reduce hypothalamic catecholamine turnover. Since activation of NE mechanisms increases TSH secretion in the rat *(125)*, and reports suggest that NPY neurons reduce NE turnover in the hypothalamus *(80)*, then the net stimulatory drive to TRH neurons in the paraventricular nucleus may be reduced. Further studies will be required to verify this theory.

Although no direct effect of NPY on TSH secretion from the pituitary has been reported *(68)*, Jones et al. *(41)* demonstrated the presence of NPY in a subpopulation of rat pituitary thyrotropes. It appears that NPY peptide and gene expression in these cells are dependent on the levels of thyroid hormone. Pituitary glands from thyroidectomized rats had higher levels of NPY mRNA and NPY peptide than pituitary tissue from euthyroid or hyperthyroid animals. The function of NPY of pituitary origin has yet to be established, although a role as a paracrine regulator of pituitary function should be considered.

NPY has also been demonstrated in sympathetic nerves innervating the thyroid gland *(127)*. Peripheral injection of norepinephrine reduced the stimulatory effect of TSH on thyroid hormone secretion in mice, as reflected by radioiodine levels in the plasma *(128)*. Combined administration of NPY and norepinephrine further decreased plasma levels of thyroid hormone. The effect of norepinephrine was attenuated by α-adrenergic blockade; however, infusion of NPY in phentolamine-treated animals potentiated the effect of TSH *(128)*. In contrast to the results in mice, Huffman and Hedge *(129)* were unable to change plasma concentrations of T3 and T4 by peripheral infusion of NPY in rats. In this case, NPY, a potent vasoconstrictor, did not alter blood flow to the gland. These studies suggest that NPY is strategically positioned to modulate the activity of the hypothalamo–pituitary–thyroid axis. At this time, however, only weak effects have been described, and additional physiological studies are required.

8. Summary

NPY is distributed throughout the reproductive neuroendocrine axis, and exerts important regulatory effects at the hypothalamus and anterior pituitary gland to modulate reproductive hormone secretion. NPY alters the secretion of LHRH from the hypothalamus and modifies the efficacy of LHRH to stimulate LH secretion from the anterior pituitary gland. Many of the effects of NPY at the hypothalamus and pituitary are modulated by gonadal steroid feedback. Estrogen and progesterone also affect NPY levels in the hypothalamus, and these changes are often parallel to alterations in LHRH levels. NPY is released into the hypothalamo–hypophysial portal system, and the release pattern is changed by gonadal steroids. Taken together, the data strongly suggest that NPY acts as an important physiological regulator of the reproductive neuroendocrine axis.

Convincing evidence has been presented showing the potential physiological importance of NPY in the central nervous system regulation of ACTH and VP secretion in mammalian species. In the case of ACTH secretion, NPY appears to play a critical role in the transmission of stress-related information to CRH-containing neurons in the paraventricular nucleus and facilitates the release of CRH into the hypophysial circulation. Interestingly, it also appears that NPY may play a role in the developmental organization of hypothalamic–neurohypophysial systems. Whether similar effects occur in hypothalamic–hypophysial systems remains to be established. Determination of the physiological significance of NPY in the regulation of GH and TSH secretion requires further investigation. The intraventricular infusion of NPY was found to reduce GH and TSH secretion in the rat. These responses are similar in direction to the changes observed during stress. Therefore, the physiological role of NPY in the hypothalamic–hypophysial system may be to coordinate and relay sensory information denoting stress to the hypothalamus and activate the appropriate neuroendocrine responses essential for the survival of the organism.

9. Directions for Future Research

In view of the broad distribution of NPY in neuroendocrine systems that control reproduction, growth, metabolism, water balance, and stress, one of the challenges that faces investigators is to determine how physiological and pathophysiological stimuli affect the function of NPY within each system and also between systems. Certain stimuli clearly affect several neuroendocrine systems. Stress has striking effects on reproduction and growth and may represent a key to understanding some of the integrative functions of NPY. The colocalization of NPY in brain stem catecholaminergic neurons and their axonal connections with CRH neurons of the paraventricular hypothalamus, and the existence of glucocorticoid receptors on neurons in the arcuate nucleus point to NPY as an important neuropeptide in the integration of stress signals. The fact that some of these NPY-containing neurons also contact LHRH perikarya and transmit signals to LHRH neurosecretory terminals in the median eminence shows that a stress stimulus may have widespread physiological effects that involve NPY in the modulation of communication among different neuroendocrine axes. The elucidation of the means by which NPY-containing neurons integrate various stimuli and communicate with other systems is an exciting avenue for future research.

Acknowledgments

The support of the National Institutes of Health (DK 40788, HD 19731, HD 26833, RCDA HD 00727, and BRSG SO7-RR-05364) and the March of Dimes Birth Defects Foundation (Basil O'Connor Scholar Research Award 5-525 and Basic Research Grant 1-1118) is gratefully acknowledged. We would also like to thank all of our collaborators who are referenced in the investigations described above.

References

1. McDonald, J. K. (1988) NPY and related substances, in *CRC Critical Reviews in Neurobiology,* vol. 4. Nelson, J., ed., CRC Press, Boca Raton, FL, pp. 97–135.
2. Potter, E. K. (1988) Neuropeptide Y as an autonomic neurotransmitter. *Pharmacol. Ther.* **7,** 251–273.
3. Mutt, V., Fuxe, K., Hokfelt, T., and Lundberg, J. M., eds. (1989) *Neuropeptide Y. Karolinska Institute Nobel Conference Series.* Raven, NY.
4. Allen, J. M. and Koenig, J. I., eds. (1990) Central and peripheral significance of neuropeptide Y and its related peptides. *Ann. NY Acad. Sci.* **611.**
5. Chronwall, B. M., DiMaggio, D. A., Massari, V. J., Pickel, V. M., Ruggiero, D. A., and O'Donohue T. L. (1985) The anatomy of neuropeptide-Y-containing neurons in rat brain. *Neuroscience* **15,** 1159–1181.

6. DeQuidt, M. E. and Emson, P. C. (1986a) Distribution of neuropeptide Y-like immunoreactivity in the rat central nervous system-I. Radioimmunoassay and chromatographic characterisation. *Neuroscience* **18,** 527–543.
7. DeQuidt, M. E. and Emson, P. C. (1986b) Distribution of neuropeptide Y-like immunoreactivity in the rat central nervous system-II. Immunohistochemical analysis. *Neuroscience* **18,** 545–618.
8. Sabatino, F. D., Murnane, J. M., Hoffman, R. A., and McDonald, J. K. (1987) The distribution of neuropeptide Y-like immunoreactivity in the hypothalamus of the adult golden hamster. *J. Comp. Neurol.* **257,** 93–104.
9. Bai, F. L., Yamano, M., Shiotani, Y., Emson, P. C., Smith, A. D., Powell, J. F., and Tohyama, M. (1985) An arcuato-paraventricular and -dorsomedial hypothalamic neuropeptide Y-containing system which lacks noradrenaline in the rat. *Brain Res.* **331,** 172–175.
10. Gustafson, E. L. and Moore, R. Y. (1987) Noradrenaline and neuropeptide Y innervation of the rat hypothalamus are differentially affected by 6-hydroxydopamine. *Neurosci. Lett.* **83,** 53–58.
11. Hokfelt, T., Meister, B., Melander, T., and Everitt, S. (1987) Coexistence of classical transmitters and peptides with special reference to the arcuate nucleus-median eminence complex, in *Hypothalamic Dysfunction in Neuropsychiatric Disorders.* Goodwin, F. K. and Costa, E., eds., Raven, New York, pp. 21–34.
12. Meister, B., Ceccatelli, S., Hokfelt, T., Anden, N.-E., Anden, M., and Theodorsson, E. (1989) Neurotransmitters, neuropeptides and binding sites in the rat mediobasal hypothalamus: effects of monosodium glutamate (MSG) lesions. *Exp. Brain Res.* **76,** 343–368.
13. Ciofi, P., Fallon, J. H., Croix, D., Polak, J. M., and Tramu, G. (1991) Expression of neuropeptide Y precursor-immunoreactivity in the hypothalamic dopaminergic tubero-infundibular system during lactation in rodents. *Endocrinology* **128,** 823–834.
14. Hunt, S. P., Emson, P. C., Gilbert, R., Goldstein, M., and Kimmel, J. R. (1981) Presence of avian pancreatic polypeptide-like immunoreactivity in catecholamine and methionine-enkephalin-containing neurons within the central nervous system. *Neurosci. Lett.* **21,** 125–130.
15. Hokfelt, T., Lundberg, J. M., Lagercrantz, H., Tatemoto, K., Mutt, V., Lindberg, J., Terenius, L., Everitt, B. J., Fuxe, K., Agnati, L., and Goldstein, M. (1983) Occurrence of neuropeptide Y (NPY)-like immunoreactivity in catecholamine neurons in the human medulla oblongata. *Neurosci. Lett.* **36,** 217–222.
16. Everitt, B. J., Hokfelt, T., Terenius, L., Tatemoto, K., Mutt, V., and Goldstein, M. (1984) Differential co-existence of neuropeptide Y (NPY)-like immunoreactivity with catecholamines in the central nervous system of the rat. *Neuroscience* **11,** 443–462.
17. Sawchenko, P. E., Swanson, L. W., Grzanna, R., Howe, P. R. C., Bloom, S. R., and Polak, J. M. (1985) Colocalization of neuropeptide Y immunoreactivity in brainstem catecholaminergic neurons that project to the paraventricular nucleus of the hypothalamus. *J. Comp. Neurol.* **241,** 138–153.
18. Blessing, W. W., Howe, P. R. C., Joh, T. H., Oliver, J. R., and Willoughby, J. O. (1986) Distribution of tyrosine hydroxylase and neuropeptide Y-like

immunoreactive neurons in rabbit medulla oblongata, with attention to colocalization studies, presumptive adrenalin-synthesizing perikarya, and vagal preganglionic cells. *J. Comp. Neurol.* **248,** 285–300.

19. Holets, V. R., Hokfelt, T., Rokaeus, A., Terenius, L., and Goldstein, M. (1988) Locus coeruleus neurons in the rat containing neuropeptide Y, tyrosine hydroxylase or galanin and their efferent projections to the spinal cord, cerebral cortex and hypothalamus. *Neuroscience* **24,** 893–906.
20. Guy, J., Li, S., and Pelletier, G. (1988) Studies on the physiological role and mechanism of action of neuropeptide Y in the regulation of luteinizing hormone secretion in the rat. *Reg. Pept.* **23,** 209–216.
21. Calka, J. and McDonald, J. K. (1989) Neuropeptide-Y and LHRH in the rat hypothalamus: anatomical relationships. *Soc. Neurosci. Abstr.* **15,** 187.
22. Norgren, R. B. and Lehman, M. N. (1989) A double-label pre-embedding immunoperoxidase technique for electron microscopy using diaminobenzidine and tetramethylbenzidine as markers. *J. Histochem. Cytochem.* **37,** 1283–1289.
23. Tsuruo, Y., Kawano, H., Kagotani, Y., Hisano, S., Daikoku, S., Chihara, K., Zhang,T., and Yanaihara, N. (1990) Morphological evidence for neuronal regulation of luteinizing hormone-releasing hormone-containing neurons by neuropeptide Y in the rat septo-preoptic area. *Neurosci. Lett.* **110,** 261–266.
24. Toni, R., Jackson, I. M. D., and Lechan, R. M. (1990) Neuropeptide-Y-immunoreactive innervation of thyrotropin-releasing hormone-synthesizing neurons in the rat hypothalamic paraventricular nucleus. *Endocrinology* **126,** 2444–2453.
25. Liposits, Z., Sievers, L., and Paull, W. K. (1988) Neuropeptide-Y and ACTH-immunoreactive innervation of corticotropin releasing factor (CRF)-synthesizing neurons in the hypothalamus of the rat. An immunocytochemical analysis at the light and electron microscopic levels. *Histochemistry* **88,** 227–234.
26. Iwai, C., Ochiai, H., and Nakai, Y. (1989) Electron-microscopic immunocytochemistry of neuropeptide Y immunoreactive innervation of vasopressin neurons in the paraventricular nucleus of the rat hypothalamus. *Acta Anat.* **136,** 279–284.
27. Chronwall, B. M., Chase, T. N., and O'Donohue, T. L. (1984) Coexistence of neuropeptide Y and somatostatin in rat and human cortical and rat hypothalamic neurons. *Neurosci. Lett.* **52,** 213–217.
28. Ciofi, P., Croix, D., and Tramu, G. (1987) Coexistence of hGHRF and NPY immunoreactivities in neurons of the arcuate nucleus of the rat. *Neuroendocrinology* **45,** 425–428.
29. Ciofi, T., Croix, D., and Tramu, G. (1988) Colocalization of GHRF and NPY immunoreactivities in neurons of the infundibular area of the human brain. *Neuroendocrinology* **47,** 469–472.
30. Ciofi, P., Tramu, G., and Bloch, B. (1990) Comparative immunohistochemical study of the distribution of neuropeptide Y, growth hormone-releasing factor and the carboxyterminus of precursor protein GHRF in the human hypothalamic infundibular area. *Neuroendocrinology* **51,** 429–436.
31. Hisano, S., Kagotani, Y., Tsuruo, Y., Daikoku, S., Chihara, K., and Whitnall, M. H. (1988) Localization of glucocorticoid receptor in arcuate nucleus of the rat hypothalamus. *Neurosci. Lett.* **95,** 13–18.

32. Harfstrand, A., Cintra, A., Fuxe, K., Aronsson, M., Wilstrom, A.-L., Okret, S., Gustafsson, J.-A., and Agnati, L. F. (1989) Regional differences in glucocorticoid receptor immunoreactivity among neuropeptide Y immunoreactive neurons of the rat brain. *Acta Physiol. Scand.* **135,** 3–9.
33. Sar, M., Sahu, A., Crowley, W. R., and Kalra, S. P. (1990) Localization of neuropeptide-Y immunoreactivity in estradiol-concentrating cells in the hypothalamus. *Endocrinology* **127,** 2752–2756.
34. McDonald, J. K., Koenig, J. I., Gibbs, D. M., Collins, P., and Noe, B. D. (1987a) High concentrations of neuropeptide Y in pituitary portal blood of rats. *Neuroendocrinology* **46,** 538–541.
35. McDonald, J. K., Tigges, J., Tigges, M., and Reich, C. (1988) Developmental study of neuropeptide Y-like immunoreactivity in the neurohypophysis and intermediate lobe of the rhesus monkey *(Macaca mulatta). Cell Tiss. Res.* **254,** 499–509.
36. McDonald, J. K., Collins, P., and Reich, C. A. (1986) Neonatal injections of MSG inhibit the development of neuropeptide Y in the rat hypothalamus and posterior pituitary. *Soc. Neurosci. Abstr.* **12,** 1523.
37. Sutton, S. W., Toyama, T. T., Otto, S., and Plotsky, P. M. (1988) Evidence that neuropeptide Y (NPY) released into the hypophysial-portal circulation participates in priming gonadotrophs to the effects of gonadotropin releasing hormone (GnRH). *Endocrinology* **123,** 1208–1210.
38. Hooi, S. C., Richardson, G. S., McDonald, J. K., Allen, J. M., Martin, J. B., and Koenig, J. I. (1989) Neuropeptide Y (NPY) and vasopressin (AVP) in the hypothalamo-neurohypophysial axis of salt-loaded and/or Brattleboro rats. *Brain Res.* **486,** 214–220.
39. Danger, J. M., Tonon, M. C., Cazin, L., Jenks, B. G., Fasolo, A., Pelletier, G., and Vaudry, H. (1990) Regulation of MSH secretion by neuropeptide Y in amphibians. *Ann. NY Acad. Sci.* **611,** 302–316.
40. DeRijk, E. P. C. T., Cruijsen, P. M. J. M., Jenks, B. G., and Roubos, E. W. (1991) [^{125}I]Bolton-Hunter neuropeptide-Y-binding sites on folliculo-stellate cells of the pars intermedia of *Xenopus laevis*: a combined autoradiographic and immunocytochemical study. *Endocrinology* **128,** 735–740.
41. Jones, P. M., Ghatei, M. A., Steel, J., O'Hallorhan, D., Gon, G., Legon, S., Burrin, J. M., Leonhardt, U., Polak, J. M., and Bloom, S. R. (1989) Evidence for neuropeptide Y synthesis in the rat anterior pituitary gland and the influence of thyroid hormone status: comparison with vasoactive intestinal peptide, substance P and neurotensin. *Endocrinology* **125,** 334–341.
42. O'Halloran, D. J., Jones, P. M., Ghatei, M. A., Domin, J., and Bloom, S. R. (1990) The regulation of neuropeptide expression in rat anterior pituitary following chronic manipulation of estrogen status: a comparison between substance P, neuropeptide Y, neurotensin, and vasoactive intestinal peptide. *Endocrinology* **127,** 1463–1469.
43. McDonald, J. K. and Lumpkin, M. D. (1983) Third ventricular injections of pancreatic polypeptides decrease LH and growth hormone secretion in ovariectomized rats. 65th Annual Meeting of The Endocrine Society, 152.
44. Kalra, S. P. and Crowley, W. R. (1984) Norepinephrine-like effects of neuropeptide Y on LH release in the rat. *Life Sci.* **35,** 1173–1176.

45. McDonald, J. K., Lumpkin, M. D., Samson, W. K., and McCann, S. M. (1985a) Pancreatic polypeptides affect luteinizing and growth hormone secretion in rats. *Peptides* **6,** 79–84.
46. McDonald, J. K., Lumpkin, M. D., Samson, W. K., and McCann, S. M. (1985b) Neuropeptide Y affects secretion of luteinizing hormone and growth hormone in ovariectomized rats. *Proc. Natl. Acad. Sci. USA* **82,** 561–564.
47. Tatemoto, K., Carlquist, M., and Mutt, V. (1982) Neuropeptide Y—a novel brain peptide with structural similarities to peptide YY and pancreatic polypeptide. *Nature (Lond.)* **296,** 659–660.
48. DiMaggio, D. A., Chronwall, B. M., Buchanan, K., and O'Donohue, T. L. (1985) Pancreatic polypeptide immunoreactivity in rat brain is actually neuropeptide Y. *Neuroscience* **15,** 1149–1157.
49. McDonald, J. K., Lumpkin, M. D., Samson, W. K., and McCann, S. M. (1984) Third ventricular injections of neuropeptide Y decrease LH and growth hormone secretion in ovariectomized rats. *Soc. Neurosci. Abstr.* **10,** 1214.
50. McDonald, J. K., Lumpkin, M. D., and DePaolo, L. V. (1989) Neuropeptide Y suppresses pulsatile secretion of luteinizing hormone in ovariectomized rats: possible site of action. *Endocrinology* **125,** 186–191.
51. Kaynard, A. H., Pau, K. Y. F., Hess, D. L., and Spies, H. G. (1990) Third-ventricular infusion of neuropeptide Y suppresses luteinizing hormone secretion in ovariectomized rhesus macaques. *Endocrinology* **127,** 2437–2444.
52. Khorram, O., Pau, K. Y. F., and Spies, H. G. (1987) Bimodal effects of neuropeptide Y on hypothalamic release of gonadotropin releasing hormone in conscious rabbits. *Neuroendocrinology* **45,** 290–297.
53. Khorram, O., Pau, K. Y. F., and Spies, H. G. (1988) Release of hypothalamic neuropeptide Y and effects of exogenous NPY on the release of hypothalamic GnRH and pituitary gonadotropins in intact and ovariectomized does in vitro. *Peptides* **9,** 411–417.
54. Sabatino, F. D., Collins, P., and McDonald, J. K. (1989) Neuropeptide Y stimulation of luteinizing hormone-releasing hormone secretion from the median eminence in vitro by estrogen dependent and Ca^{2+} independent mechanisms. *Endocrinology* **124,** 2089–2098.
55. Woller, M. J. and Terasawa, E. (1991) Infusion of neuropeptide Y into the stalk-median eminence stimulates in vivo release of luteinizing hormone-releasing hormone in gonadectomized rhesus monkeys. *Endocrinology* **128,** 1144–1150.
56. Wehrenberg, W. B., Corder, R., and Gaillard, R. C. (1989) A physiological role for neuropeptide Y in regulating the estrogen/progesterone induced luteinizing hormone surge in ovariectomized rats. *Neuroendocrinology* **49,** 680–682.
57. Sabatino, F. D., Collins, P., and McDonald, J. K. (1990) Investigation of the effects of progesterone on neuropeptide Y-stimulated luteinizing hormone-releasing hormone secretion from the median eminence of ovariectomized and estrogen-treated rats. *Neuroendocrinology* **52,** 600–607.
58. Pau, K.-Y. F., Kaynard, A. H., Hess, D. L., and Spies, H. G. (1991) Effects of neuropeptide Y on the *in vitro* release of gonadotropin-releasing hormone, luteinizing hormone, and beta-endorphin and pituitary responsiveness to gonadotropin-releasing hormone in female macaques. *Neuroendocrinology* **53,** 396–403.

59. Crowley, W. R. and Kalra, S. P. (1987) Neuropeptide Y stimulates the release of luteinizing hormone-releasing hormone from medial basal hypothalamus in vitro: modulation by ovarian hormones. *Neuroendocrinology* **46,** 97–103.
60. Brann, D. W., McDonald, J. K., Putnam, C. D., and Mahesh, V. B. (1991) Regulation of hypothalamic gonadotropin-releasing hormone and neuropeptide Y concentrations by progesterone and corticosteroids in immature rats: correlation with luteinizing hormone and follicle stimulating hormone release. *Neuroendocrinology* **54,** 425–432.
61. Crowley, W. R., Tessel, R. E., O'Donohue, T. L., Adler, B. A., and Kalra, S. P. (1985) Effects of ovarian hormones on the concentrations of immunoreactive neuropeptide Y in discrete brain regions of the female rat: correlation with serum luteinizing hormone (LH) and median eminence LH-releasing hormone. *Endocrinology* **117,** 1151–1155.
62. Sahu, A., Jacobson, W., Crowley, W. R., and Kalra, S. P. (1989) Dynamic changes in neuropeptide Y concentrations in the median eminence in association with preovulatory luteinizing hormone release in the rat. *J. Neuroendocrinol.* **1,** 83–87.
63. Sahu, A., Crowley, W. R., and Kalra, S. P. (1990a) An opioid-neuropeptide-Y transmission line to luteinizing hormone (LH)-releasing hormone neurons: a role in the induction of LH surge. *Endocrinology* **126,** 876–883.
64. Sahu, A., Kalra, P. S., Crowley, W. R., and Kalra, S. P. (1990b) Functional heterogeneity in neuropeptide-Y-producing cells in the rat brain as revealed by testosterone action. *Endocrinology* **127,** 2307–2312.
65. Sutton, S. W., Mitsugi, N., Plotsky, P. M., and Sarkar, D. K. (1988b) Neuropeptide Y (NPY): a possible role in the initiation of puberty. *Endocrinology* **123,** 2152–2154.
66. Crowley, W. R., Hassid, A., and Kalra, S. P. (1987) Neuropeptide Y enhances the release of luteinizing hormone (LH) induced by LH-releasing hormone. *Endocrinology* **120,** 941–945.
67. Kerkerian, L., Guy, J., Lefevre, G., and Pelletier, G. (1985) Effects of neuropeptide Y (NPY) on the release of anterior pituitary hormones in the rat. *Peptides* **6,** 1201–1204.
68. Rodriguez-Sierra, J. F., Jacobowitz, D. M., and Blake, C. A. (1987) Effects of neuropeptide Y on LH, FSH and TSH release in male rats. *Peptides* **8,** 539–542.
69. Crowley W. R., Shah, G. V., Carroll, B. L., Kennedy, D., Dockter, M. E., and Kalra, S. P. (1990) Neuropeptide-Y enhances luteinizing hormone (LH)-releasing hormone-induced LH release and elevations in cytosolic Ca^{2+} in rat anterior pituitary cells: evidence for involvement of extracellular Ca^{2+} influx through voltage-sensitive channels. *Endocrinology* **127,** 1487–1494.
70. Shangold, G. A. and Miller, R. J. (1990) Direct neuropeptide Y-induced modulation of gonadotrope intracellular calcium transients and gonadotropin secretion. *Endocrinology* **126,** 2336–2342.
71. Bauer-Dantoin, A. C., McDonald, J. K., and Levine, J. E. (1991) Neuropeptide Y potentiates luteinizing hormone (LH)-releasing hormone-stimulated LH surges in pentobarbital-blocked proestrous rats. *Endocrinology* **129,** 402–408.
72. McCann, S. M., Rettori, V., Milenkovic, L., Riedel, M., Aguila, C., and McDonald, J. K. (1989) The role of neuropeptide Y (NPY) in the control of

anterior pituitary hormone release in the rat, in *Neuropeptide Y, Karolinska Institute Nobel Conference Series.* Mutt, V., Fuxe, K., Hokfelt, A., and Lundberg, J. M., eds., Raven, New York, pp. 215–228.
73. Papka, R. E., Cotton, J. P., and Traurig, H. H. (1985) Comparative distribution of neuropeptide tyrosine-, vasoactive intestinal polypeptide-, substance P-immunoreactive, acetylcholinesterase-positive and noradrenergic nerves in the reproductive tract of the female rat. *Cell Tiss. Res.* **242,** 475–490.
74. McDonald, J. K., Dees, W. L., Ahmed, C. E., Noe, B. D., and Ojeda, S. R. (1987b) Biochemical and immunocytochemical characterization of neuropeptide Y in the immature rat ovary. *Endocrinology* **120,** 1703–1710.
75. Lara, H. E., McDonald, J. K., and Ojeda, S. R. (1990) Involvement of nerve growth factor in the regulation of female sexual development. *Endocrinology* **126,** 364–375.
76. Lara, H. E., McDonald, J. K., Ahmed, C. E., and Ojeda, S. R. (1990) Guanethidine-mediated destruction of ovarian sympathetic nerves disrupts ovarian development and function in rats. *Endocrinology* **127,** 2199–2209.
77. Jorgensen, J. C., Sheikh, S. D., Forman, A., Norgard, M., Schwartz, T. W., and Ottesen, B. (1989) Neuropeptide Y in the human female genital tract: localization and biological action. *Am. J. Physiol.* **257,** E220–E227.
78. Jorgensen, J. C., O'Hare, M. M. T., and Andersen, C. Y. (1990) Demonstration of neuropeptide Y and its precursor in plasma and follicular fluid. *Endocrinology* **127,** 1682–1688.
79. Allen, L. G., Wilson, F. J., and Macdonald, G. J. (1989) Neuropeptide Y-containing nerves in rat gonads: Sex difference and development. *Biol. Reprod.* **40,** 371–378.
80. Harfstrand, A., Eneroth, P., Agnati, L., and Fuxe, K. (1987a) Further studies on the effect of central administration of neuropeptide Y on neuroendocrine function in the male rat relationship to hypothalamic catecholamines. *Regul. Pept.* **17,** 167–179.
81. Wahlestedt, C., Skagerberg, G., Ekman, R., Heilig, M., Sundler, F., and Hakanson, R. (1987) Neuropeptide Y (NPY) in the area of the hypothalamic paraventricular nucleus activates the pituitary-adrenocortical axis in the rat. *Brain Res.* **417,** 33–38.
82. Haas, D. A. and George, S. R. (1987) Neuropeptide Y administration acutely increases hypothalamic corticotropin-releasing factor immunoreactivity: lack of effect in other rat brain regions. *Life Sci.* **41,** 2725–2731.
83. Inoue, T., Inui, A., Okita, M., Sakatani, N., Oya, M., Morioka, H., Mizuno, N., Oimomi, M., and Baba, S. (1989) Effect of neuropeptide Y on the hypothalamic-pituitary adrenal axis in the dog. *Life Sci.* **44,** 1043–1051.
84. Leibowitz, S. F. (1990) Hypothalamic neuropeptide Y in relation to energy balance. *Ann. NY Acad. Sci.* **611,** 284–301.
85. Albers, H. E., Ottenweller, J. E., Liou, S. Y., Lumpkin, M. D., and Anderson, E. R. (1990) Neuropeptide Y in the hypothalamus: effect on corticosterone and single-unit activity. *Am. J. Physiol.* **258,** R376–R382.
86. Liposits et al. (1989)
87. Tsagarakis, S., Rees, L. H., Besser, G. M., and Grossman, A. (1989) Neuropeptide-Y stimulates CRF-41 release from rat hypothalami in vitro. *Brain Res.* **502,** 167–170.

88. Lynch, D. R., Walker, M. W., Miller, R. J., and Snyder, S. H. (1989) Neuropeptide Y receptor binding sites in rat brain: differential autoradiographic localizations with ^{125}I-peptide YY and ^{125}I-neuropeptide Y imply receptor heterogeneity. *J. Neurosci.* **9,** 2607–2619.
89. Zoli, M., Agnati, L. F., Fuxe, K., and Bjelke, B. (1989) Demonstration of NPY transmitter receptor mismatches in the central nervous system of the male rat. *Acta Physiol. Scand.* **135,** 201,202.
90. Quiron, R., Martel, J-C., Dumont, Y., Cadieux, A., Jolicoeur, F., St-Pierre, S., and Fournier, A. (1990) Neuropeptide Y receptors: autoradiographic distribution in the brain and structure–activity relationships. *Ann. NY Acad. Sci.* **611,** 58–72.
91. Plotsky, P. M. (1987) Regulation of hypophysiotropic factors mediating ACTH secretion. *Ann. NY Acad. Sci.* **512,** 205–217.
92. Harfstrand, A., Fuxe, K., Terenius, L., and Kalia, M. (1987b) Neuropeptide Y-immunoreactive perikarya and nerve terminals in the rat medulla oblongata: relationship to cytoarchitecture and catecholaminergic cell groups. *J. Comp. Neurol.* **260,** 20–35.
93. Yokoo, H., Schlesinger, D. H., and Goldstein, M. (1987) The effect of neuropeptide Y (NPY) on stimulation-evoked release of [^{3}H] norepinephrine (NE) from rat hypothalamic and cerebral cortical slices. *Eur. J. Pharmacol.* **143,** 283–286.
94. Shimizu, H. and Bray, G. A. (1989) Effects of neuropeptide Y on norepinephrine and serotonin metabolism in rat hypothalamus in vivo. *Brain Res. Bull.* **22,** 945–950.
95. Corder, R., Pralong, F., Turnhill, D., Saudan, P., Muller A. F., and Gaillard, R. C. (1988) Dexamethasone treatment increases neuropeptide Y levels in rat hypothalamic neurons. *Life Sci.* **43,** 1879–1886.
96. Rivet, J.-M., Castagne, V., Corder, R., and Mormede, P. (1989) Study of the influence of stress and adrenalectomy on central and peripheral neuropeptide Y levels. Comparison with catecholamines. *Neuroendocrinology* **50,** 413–420.
97. Dean, R. G. and White, B. D. (1990) Neuropeptide expression in rat brain: effects of adrenalectomy. *Neurosci. Lett.* **114,** 339–344.
98. Sabol, S. L. and Higuchi, H. (1990) Transcriptional regulation of the neuropeptide Y gene by nerve growth factor: antagonism by glucocorticoids and potentiation by adenosine 3', 5'-monophosphate and phorbol ester. *Mol. Endocrinol.* **4,** 384–392.
99. Assenmacher, I., Szafarczyk, A., Alonso, G., Ixart, G., and Barbanel, G. (1987) Physiology of neural pathways affecting CRH secretion. *Ann. NY Acad. Sci.* **512,** 149–161.
100. Inui, A., Inoue, T., Nakajima, M., Okita, M., Sakatani, N., Olimura, Y., Chihara, K., and Baba, S. (1990) Brain neuropeptide Y in the control of adrenocorticotropic hormone secretion in the dog. *Brain Res.* **510,** 211–215.
101. Koenig, J. I. (1990) Regulation of the hypothalamo-pituitary axis by neuropeptide Y. *Ann. NY Acad. Sci.* **611,** 317–328.
102. Plotsky, P. M., Bruhn, T. O., and Vale, W. W. (1985) Hypophysiotropic regulation of adrenocorticotropin secretion in response to insulin-induced hypoglycemia. *Endocrinology* **117,** 323–329.

103. Caraty, A., Grino, M., Locatelli, A., Guillaume, V., Boudouresque, F., Conte-Devolx, B., and Oliver, C. (1990) Insulin-induced hypoglycemia stimulates corticotropin-releasing factor and arginine vasopressin secretion into hypophysial portal blood of conscious, unrestrained rams. *J. Clin. Invest.* **85,** 1716–1721.
104. Kalra, S. P., Sahu, A., Kalra, P. S., and Crowley, W. R. (1990) Hypothalamic neuropeptide Y: a circuit in the regulation of gonadotropin and feeding behavior. *Ann. NY Acad. Sci.* **611,** 273–283.
105. Williams, G., Gill, J. S., Lee, Y. C., Cardoso, H., Okpere, B. E., and Bloom, S. R. (1989) Increased neuropeptide Y concentrations in specific hypothalamic segions of streptozotocin-induced diabetic rats. *Diabetes* **38,** 321–327.
106. Brady, L. S., Smith, M. A., Gold, P. W., and Herkenham, M. (1990) Altered expression of hypothalamic neuropeptide mRNAs in food-restricted and food-deprived rats. *Neuroendocrinology* **52,** 441–447.
107. Abe, M., Saito, M., Ikeda, H., and Shimazu, T. (1991) Increased neuropeptide Y content in the arcuato-paraventricular hypothalamic neuronal system in both insulin-dependent and non-insulin-dependent diabetic rats. *Brain Res.* **539,** 223–227.
108. Chua, S. C., Jr., Leibel, R. L., and Hirsch, J. (1991) Food deprivation and age modulate neuropeptide gene expression in the murine hypothalamus and adrenal gland. *Mol. Brain Res.* **9,** 95–101.
109. White, J. D., Olchovsky, D., Kershaw, M., and Berelowitz, M. (1990) Increased hypothalamic content of preproneuropeptide Y messenger ribonucleic acid in streptozotocin-diabetic rats. *Endocrinology* **126,** 765–772.
110. Sanacora, G., Kershaw, M., Finkelstein, J. A., and White, J. D. (1990) Increased hypothalamic content of preproneuropeptide Y messenger ribonucleic acid in genetically obese Zucker rats and its regulation by food deprivation. *Endocrinology* **127,** 730–737.
111. Chabot, J-G., Enjalbert, A., Pelletier, G., Dubois, P. M., and Morel, G. (1988) Evidence for a direct action of neuropeptide Y in the rat pituitary gland. *Neuroendocrinology* **47,** 511–517.
112. Takahashi, K., Mouri, T., Murakami, O., Itoi, K., Sone, M., Ohneda, M., Nozuki, M., and Yoshinaga, K. (1988) Increases of neuropeptide Y-like immunoreactivity in plasma during insulin-induced hypoglycemia in man. *Peptides* **9,** 433–435.
113. Schalling, M., Seroogy, K., Hokfelt, T., Chai, S. Y., Hallman, H., Persson, H., Larhammar, D., Ericsson, D., Terenius, L., and Graffi, J. (1988) Neuropeptide tyrosine in the rat adrenal gland—Immunohistochemical and in situ hybridization studies. *Neuroscience* **24,** 337–349.
114. DeQuidt, M. E. and Emson, P. C. (1986c) Neuropeptide Y in the adrenal gland: characterization, distribution and drug effect. *Neuroscience* **19,** 1011–1022.
115. Higuchi, H., Costa, E., and Yang, H. Y. (1988) Neuropeptide Y inhibits nicotine-mediated release of catecholamines from bovine adrenal chromaffin cells. *J. Pharmacol. Exp. Ther.* **244,** 468–474.
116. Hexum, T. D., Majane, E. A., Russett, L. R., and Yang, H. Y. (1987) Neuropeptide Y release from the adrenal medulla after cholinergic receptor stimulation. *J. Pharmacol. Exp. Ther.* **243,** 927–930.

117. Allen, J. M. (1990) Molecular structure of neuropeptide Y. *Ann. NY Acad. Sci.* **611,** 86–98.
118. Minth, C. A. and Dixon, J. E. (1990) Regulation of the human neuropeptide Y gene. *Ann. NY Acad. Sci.* **611,** 99–110.
119. Beroukas, D., Willoughby, J. O., and Blessing, W. W. (1989) Neuropeptide Y-like immunoreactivity is present in boutons synapsing on vasopressin-containing neurons in rabbit supraoptic nucleus. *Neuroendocrinology* **50,** 222–228.
120. Day, T. A., Jhamandas, J. H., and Renaud, L. R. (1985) Comparison between the actions of avian pancreatic polypeptide, neuropeptide Y and norepinephrine on the excitability of rat supraoptic vasopressin neurons. *Neurosci. Lett.* **62,** 181–185.
121. Willoughby, J. O. and Blessing, W. W. (1987) Neuropeptide Y injected into the supraoptic nucleus causes secretion of vasopressin in the unanesthetized rat. *Neurosci. Lett.* **75,** 17–22.
122. Vallejo, J., Carter, D. A., Javier Diez-Guerra, F., Emson, P. C., and Lightman, S. L. (1987) Neonatal administration of a specific neuropeptide Y antiserum alters the vasopressin response to haemorrhage and the hypothalamic content of noradrenaline in rats. *Neuroendocrinology* **45,** 507–509.
123. Edwards, G. L., Cunningham, J. T., Beltz, T. G., and Johnson, A. K. (1989) Neuropeptide Y-immunoreactive cells in the caudal medulla project to the medial preoptic nucleus. *Neurosci. Lett.* **105,** 19–26.
124. Peng, C., Huang, Y-P., and Peter, R. E. (1990) Neuropeptide Y stimulates growth hormone and gonadotropin release from the goldfish pituitary in vitro. *Neuroendocrinology* **52,** 28–34.
125. Krulich, L., Mayfield, M. A., Steele, M. K., McMillen, B. A., McCann, S. M., and Koenig, J. I. (1982) Differential effects of pharmacological manipulations of central alpha-1 and alpha-2 adrenergic receptors on the secretion of thyrotropin and growth hormone in male rats. *Endocrinology* **110,** 796–804.
126. Williams, G., Steel, J. H., Cardoso, H., Ghatei, M. A., Lee, Y. C., Gill, J. S., Burrin, J. M., Polak, J. M., and Bloom, S. R. (1988) Increased hypothalamic neuropeptide Y concentrations in diabetic rat. *Diabetes* **37,** 763–772.
127. Grunditz, T., Hakanson, R., Rerup, C., Sundler, F., and Uddman, R. (1984) Neuropeptide Y in the thyroid gland: neuronal localization and enhancement of stimulated thyroid hormone secretion. *Endocrinology* **115,** 1537–1542.
128. Ahren, B. (1986) Neuropeptide Y and pancreatic polypeptide: effects on thyroid hormone secretion. *Eur. J. Pharmacol.* **126,** 97–102.
129. Huffman, L. and Hedge, G. A. (1986) Neuropeptide control of thyroid blood flow and hormone secretion in the rat. *Life Sci.* **39,** 2143–2150.
130. Crowley, W. R. and Kalra, S. P. (1988) Regulation of luteinizing hormone secretion by neuropeptide Y in rats: hypothalamic and pituitary actions. *Synapse* **2,** 276–281.
131. McDonald, J. K. (1990) Role of neuropeptide Y in reproductive function. *Ann. NY Acad. Sci.* **611,** 258–272.
132. McDonald, J. K. (1991) Reproductive neuroendocrine effects of neuropeptide Y and related peptides, in *Brain-Gut Peptides and Reproductive Function,* Johnston, C. and Barnes, C. D., eds., CRC Press, Boca Raton, FL, pp. 105–154.

Neuropeptide Y in Multiple Hypothalamic Sites Controls Eating Behavior, Endocrine, and Autonomic Systems for Body Energy Balance

B. Glenn Stanley

1. Introduction

To survive, all animals must have sufficient nutrients available to maintain cellular metabolism and function. To ensure that these nutrients are continuously available, complex animals have neuronal systems that regulate food intake, energy storage, and energy expenditure. These regulatory systems control and coordinate the behavioral response (eating behavior), and numerous autonomic and endocrine responses to achieve energy balance. These systems are sensitive to and integrate inputs from a variety of food-related, environmental, humoral, and endocrine factors, which they in turn regulate, thus forming continuous feedback loops. Intensive research is focused on defining these systems in order to reveal general principles of neural integration and control, and to determine how dysfunctions of these mechanisms may contribute to, or even cause, some of the eating and body-weight disorders so prevalent in Western society.

A major goal of research in this area is to identify the neurotransmitters and neuromodulators that act as critical links in these control systems. The clearest evidence that a particular neurotransmitter may participate in the control of eating is that it elicits an eating response when it is injected directly into the brain. Of the scores of the neurotransmitter candidates, only a few have proven

From: *The Biology of Neuropeptide Y and Related Peptides;*
W. F. Colmers and C. Wahlestedt, Eds.

to stimulate eating. These eating-stimulatory neurotransmitters are norepinephrine (NE) or epinephrine (EPI), γ-amino butyric acid (GABA), specific opiate peptides, growth-hormone-releasing hormone, galanin, and the focus of this review, neuropeptide Y (NPY) and other members of the pancreatic polypeptide family *(1)*.

Neuropeptide Y is a member of the 36 amino acid family of pancreatic polypeptides. In mammals, this family consists of pancreatic polypeptide (PP), peptide YY (PYY), and NPY. Although PP and PYY are primarily located in the periphery, NPY is predominantly located within neurons of the central and sympathetic nervous systems *(2–4; see also* chapters by Hendry and by Sundler, this vol.). In fact, NPY is the most abundant candidate peptide-neurotransmitter measured in the brain to date *(5,6)*.

Among the neurotransmitters that stimulate eating, NPY and PYY are in a class by themselves. Like other eating-stimulatory neurochemicals, NPY, PYY, and, to a lesser extent, PP, at low doses can induce satiated animals to eat a normal size meal *(7,8,10)*. However, as shown in Fig. 1, at higher doses, NPY and PYY can induce eating behavior of unparalleled proportions: eating of an intensity and duration that is near the animals' maximum physiological capacity. The initial meal eaten after hypothalamic injection of a high dose of NPY (1 μg) is larger, on average, than that eaten after 24 h of food deprivation, and individual animals occasionally eat more in this meal than they would normally eat in an entire day *(9,11,12)*. Remarkably, there is no evidence for tolerance to this effect. Animals repeatedly injected with NPY or PYY exhibit sustained and dramatic overeating, as well as rapid weight gain with consequent development of obesity *(13,14)*.

Within the context of previous work on mechanisms of eating behavior, the magnitude of NPY's effects was surprising. As noted earlier, acute injections of other neurotransmitters produced considerably smaller effects on eating, and likewise, chronic injections had comparatively minor effects on daily food intake and body-weight gain (e.g., *15–18*). These small effects, in conjunction with the extensive evidence that the controls of eating are extremely complex and multifaceted, involving many different neurotransmitters at different brain sites (e.g., *19*), led to the view that a single neurotransmitter would produce only small and short-lived effects on eating behavior. Therefore, the dramatic effects of NPY focused attention on this peptide, because they suggested that NPY and its

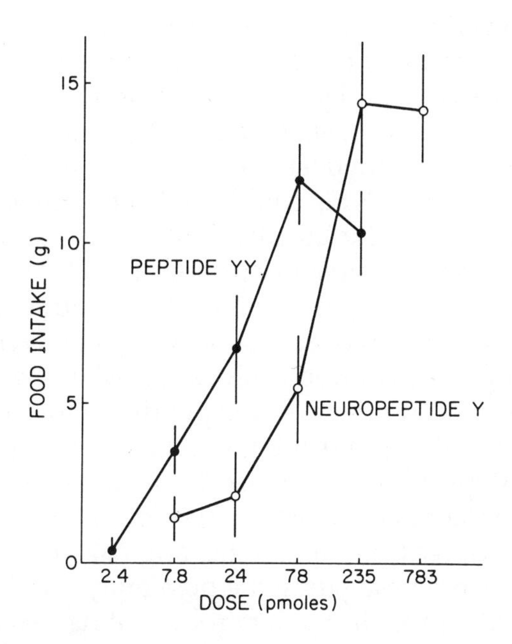

Fig. 1. Increased food intake (mean ± SEM) 60 min postinjection, by rats given PVN injections of PYY (closed circles, n = 7) or NPY (open circles, n = 10). (From *[10]*. Reprinted with permission from *Peptides*.)

structurally related peptides might play particularly pivotal roles in regulating naturally occurring eating behavior and that, like dopamine in Parkinson's disease, they also might play key roles in some disorders of eating behavior and body-weight control. Both possibilities have received support from subsequent research.

2. The First Studies

The discovery that NPY and PP stimulate food intake resulted from a serendipitous observation by John Clark, a postdoctoral fellow in Satya Kalra's laboratory at the University of Florida. Clark was interested in the neuroendocrine functions of PP and, while assessing the effects of third cerebral ventricular injection of human PP on luteinizing-hormone release, he noticed that some of his peptide-injected rats began to eat. He pursued this observation and demonstrated that third ventricular injections of NPY, and to a lesser extent PP, at doses of 2 or 10 µg, stimulated a feeding response of almost 4 g in satiated rats *(7)*.

Based on early reports of this work, Levine and Morley *(20)* tested whether lateral cerebroventricular injections of NPY also would stimulate eating. Neuropeptide Y at doses of 2 and 10 μg was highly effective, increasing food intake by up to 10 g. The rats began to eat about 20 min postinjection, and some eating was still evident up to 2 h later. NPY could elicit eating during either the light or dark portion of the circadian eating cycle, and it also could increase food intake in animals deprived of food for 24 h. Analysis of other behaviors revealed that NPY also produced significant increases in drinking and overall behavioral activity, but it did not affect grooming. Peripheral administration of NPY was ineffective, arguing for a central effect of this peptide. Finally, peripheral administration of the opioid antagonist naloxone or the dopamine antagonist haloperidol suppressed NPY-elicited eating, but the α-adrenergic antagonist phentolamine did not.

While Clark was conducting his original study with PP and NPY, I also began investigating the possibility that hypothalamic NPY might play a role in the control of eating behavior. Earlier work by Sarah Leibowitz and her colleagues had demonstrated that NE and EPI injected into the paraventricular nucleus of the hypothalamus (PVN) elicit eating behavior *(21)*. Other evidence indicated that the PVN contains extremely high concentrations of endogenous NPY *(6)*, whereas yet other studies suggested that NPY might be a cotransmitter in some NE- or EPI-containing neurons *(22,23)*. Since most of my work up to that time had focused on PVN peptide–catecholamine interactions in the control of eating *(24)*, these just-described findings suggested to me that NPY might coexist with NE or EPI in presynaptic neurons innervating the PVN, and that NPY might act in some manner to regulate eating behavior.

To test this, I injected NPY directly into the PVN of satiated rats and measured their consequent food intake. Prior to this, I had similarly tested other peptides, and obtained either no response or a suppressive effect, but never a stimulatory effect on eating. Consequently, I was astounded by the unprecedented size and duration of the eating response to NPY (Fig. 2). For example, in my first test, one subject ate over 30 g (the equivalent of an entire day's intake) within 1 h and then still continued to overeat for several hours afterward. I immediately turned all my efforts toward investigating this phenomenon and shortly thereafter published some of the initial findings *(8,9)*.

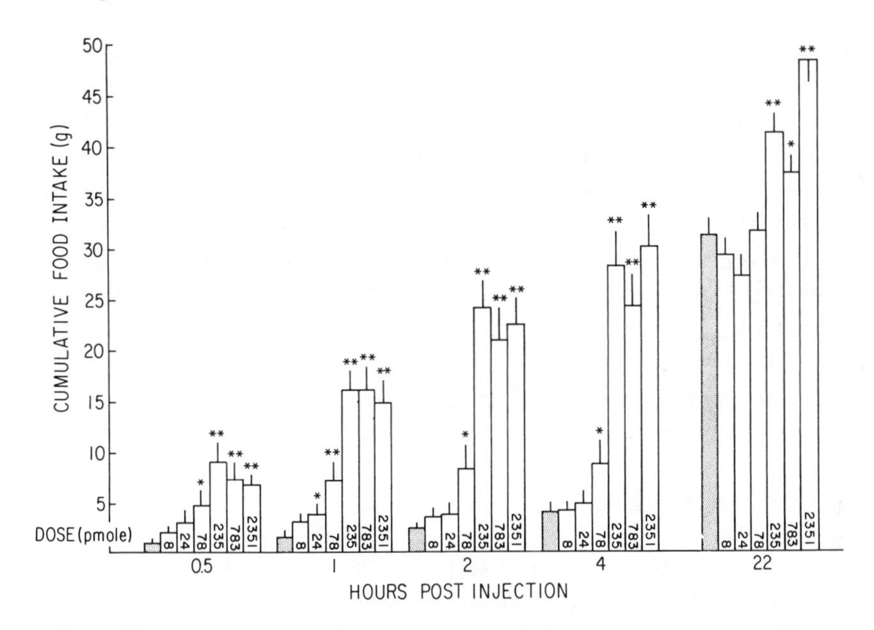

Fig. 2. Increased cumulative food intake (mean ± SEM) in rats 0.5, 1, 2, 4, and 22 h produced by PVN injection of NPY (8, 24, 78, 235, 783, and 2351 pmol/0.3 μL) compared to vehicle (shaded bar). $*p < 0.05$, $**p < 0.01$ relative to vehicle. (From *[9]*. Reprinted with permission from the National Academy of Sciences.)

Since these earliest studies, NPY has been shown to stimulate eating in a wide range of species, including snakes *(25)*, various birds *(26,27)*, and rodents *(28,29)*, as well as in larger mammals, such as pigs *(30)* and sheep *(31)*. Additionally, the effect can be demonstrated over a wide range of developmental periods *(32,170)*. For example, preliminary evidence indicates that rat pups at the earliest stage of development tested (2 d of age) will eat in response to intracerebroventricular (ICV) injections of NPY *(32)*. Thus, NPY appears to be significant to the control of eating in both evolutionary and developmental terms.

The remainder of this chapter will integrate much of the evidence for NPY and related peptides in the control of eating behavior. In relation to this, the focus will be on elucidating: the location of the relevant release sites for endogenous NPY and PYY; the location of the receptors upon which these peptides act to produce eating; the types of NPY receptors and second messengers that mediate its effects; the electrophysiological effects of NPY on hypothalamic neurons; the specific functions of this peptide in control of normal eating and energy expenditure; and its role in disorders of eating behavior and body-weight control. Moreover, it is proposed that synaptic NPY in multiple hypothalamic sites integrates the inputs from a range of metabolic factors associated with decreased available body energy. The released NPY is suggested to serve as a key link between these inputs and systems that control and coordinate several different physiological systems together with eating behavior to yield an overall regulation of body energy balance. A primary locus of the feeding-stimulatory effect of NPY is proposed to be the perifornical hypothalamus (PFH), whereas a primary locus of the related physiological effects is proposed to be the PVN.

3. Sites of Action

Which brain regions contain the receptors activated by exogenous NPY to induce eating behavior, and which brain regions release the endogenous NPY that activates these receptors? There appear to be several brain regions that can mediate the stimulatory effects of NPY and PYY on eating behavior. The most intensely studied are within the hypothalamus, but there is also evidence that NPY can act at both brain stem and cortical sites to stimulate eating behavior and influence metabolism.

3.1. Extrahypothalamic Sites

The involvement of brain stem sites was suggested by studies showing that injections of either NPY or PYY into the fourth ventricle elicit a robust eating response in satiated rats *(33,34)*. Although the precise locus of this effect has not yet been identified, these brain stem sites do appear to be distinct from those in the forebrain, since midbrain microknife cuts attenuate the eating produced by the fourth ventricular injections, but not by the third ventricular injections of these peptides. The involvement of a cortical site was suggested by evidence that NPY injected directly into the sulcal prefrontal cortex, an area known to participate in the control of feeding behavior and related physiological responses, stimulates eating and alters carbohydrate utilization and energy expenditure in rats *(35)*. These findings demonstrate that NPY and structurally related peptides may act in several different brain regions to regulate eating behavior. They also emphasize that the effects of NPY are not restricted to eating behavior, but encompass a spectrum of different, but probably related, physiological functions.

3.2. Hypothalamic Sites

Before proceeding, selected aspects of NPY's location within the hypothalamus will be described (*see* chapters by Hendry, and by McDonald and Koenig, this vol.). The highest concentrations of NPY in the brain are found within hypothalamic nuclei, particularly the PVN, arcuate nucleus, suprachiasmatic nucleus, median eminence, and dorsomedial nucleus, respectively *(6,36)*. Except for the suprachiasmatic nucleus, which receives major NPY innervation from the intergeniculate leaflet of the thalamus *(37)*, the primary source of NPY in these and in most other hypothalamic nuclei appears be from neurons of the arcuate nucleus *(38,39)*, and the brain stem *(40)*. Studies focused on the PVN have shown that it, too, receives major inputs from the arcuate nucleus *(41)*, from brain stem cell groups (C1, C2, and C3), which contain both NPY and EPI, and from group A1, which contains both NPY and NE *(42)*. Concerning the PFH, although there is little specific information, findings indicate that the PFH/lateral hypothalamic area has relatively dense concentrations of NPY-terminal immunoreactivity, particularly overlapping hypothalamic neurons that project to brain stem autonomic nuclei *(43)*. It also may be noteworthy that axons from NPY/catecholamine brain stem neurons traverse the PFH en route to their destination in the PVN *(38,42)*. Considering the suggested roles of the PVN and PFH in mediating NPY's effects, it would be interesting to determine whether the axons traversing the PFH provide synaptic or paracrine NPY input to this region.

3.2.1. The Perifornical Hypothalamus Is a Primary Locus of NPY's Eating-Stimulatory Effect

The most logical locus of NPY's eating-stimulatory effect might appear to be the PVN. This nucleus, as noted earlier, has a well-established role in the regulation of eating behavior *(44)*; it is the focus of the NE and EPI eating-stimulatory effect *(21)*; it receives substantial innervation by neurons containing both NPY and EPI or NE *(42)*; it has one of the highest concentrations of endogenous NPY in the brain *(36)*; and injections of NPY there are much more effective in eliciting eating than injections into either the lateral or the third ventricle *(7–9,20)*.

Although this indirect evidence might suggest that the PVN mediates NPY's effects, more direct evidence questions this hypothesis and suggests instead that an adjacent brain region, the perifornical hypothalamus, is more critical for NPY's eating-stimulatory effect. Evidence against the PVN as the primary locus of NPY's action is that injections there were no more effective than injections into other sites, including the ventromedial and the lateral hypothalamus *(45)*. Indeed, Morley et al. *(46)* found that injections into the anterior portion of the ventromedial nucleus were more effective than injections directly into the PVN. In contrast to the widespread effectiveness of injections within the hypothalamus, injections into sites bracketing this structure were ineffective, demonstrating that NPY does act intrahypothalamically to stimulate eating *(45,46)*. The widespread effectiveness of NPY within the hypothalamus suggests that either the relevant receptor sites are distributed widely throughout the hypothalamus or that the injected NPY acts subsequent to diffusion.

To discriminate between these alternatives, we conducted an extensive cannula-mapping study using an injection system that reliably delivers the NPY in the extremely small volume of 10 nL, rather than in the 300–500 nL volumes typically used. We also demonstrated that virtually all of the NPY administered in these volumes remained within 0.8 mm of the injection, thus severely restricting the impact of diffusion on the anatomical pattern of elicited eating *(47)*. In this study, we mapped NPY's eating-stimulatory effect after injection into each of nearly 50 systematically spaced hypothalamic sites. Unlike our earlier cannula-mapping study using 300-nL injection volumes *(45)*, this work did reveal a site-specific pattern of effects within the hypothalamus. The hypothalamic site most sensitive to NPY was the PFH (the shaded portion of Fig. 3), located at the same anterior/posterior level as the posterior border of the PVN *(47)*. Injections of NPY (0.33 μg) into the PFH yielded mean intakes of over 12 g in 1 h, and over 23 g in 4 h. This was more than twice the eating response obtained <1 mm further medially, in the PVN itself, arguing that the response was the result of a local action of NPY in the PFH rather than diffusion of NPY to the PVN. Likewise, all other sites, including those bracketing the PFH about 1 mm anterior, posterior, medial, lateral, dorsal, and ventral, yielded significantly smaller effects. Injections into even more distant sites had still smaller effects or no effect.

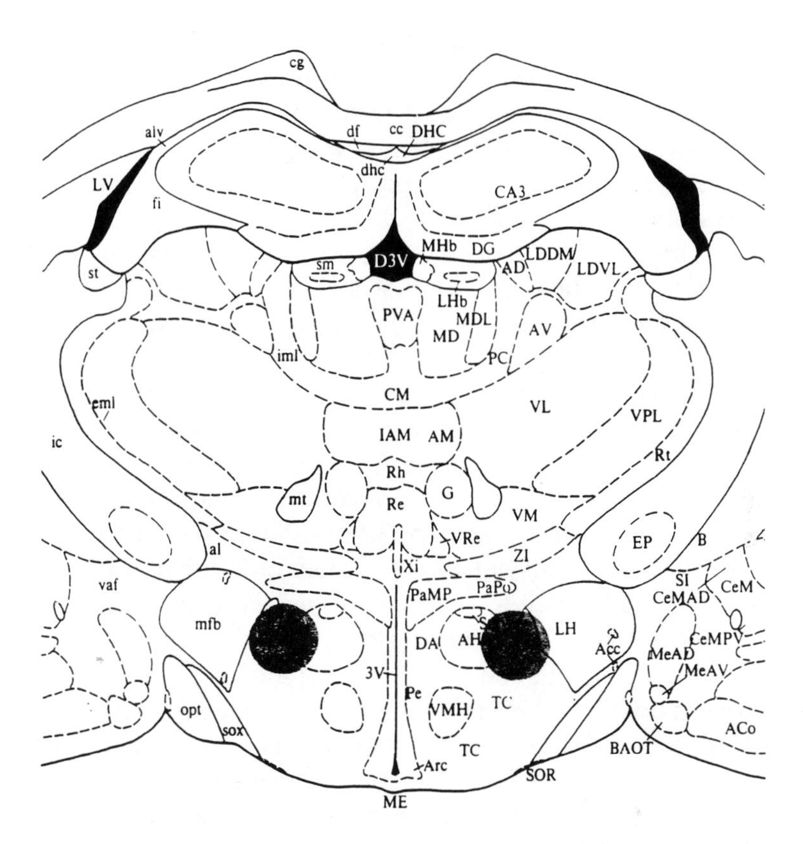

Fig. 3. The brain site that is most sensitive to the eating-stimulatory effect of hypothalamic NPY injection is represented by the shaded portions of a schematic of a coronal section through the rat hypothalamus. (Schematic from *[48]*. Modified with permission from Academic Press.)

The only apparent exception to this pattern of diminished response with increased distance from the PFH was obtained in the dorsomedial nucleus. Injection of NPY into this structure produced mean intakes of as much as 8 g in 1 h, and 15 g in 4 h.

These data suggest that the PFH is a primary locus for the feeding-stimulatory effect of NPY. This area, in contrast to the medial PVN, is proposed to be the primary mediator of the eating-stimulatory effects of hypothalamically injected NPY and to contain the greatest abundance of *feeding-related* NPY-sensitive receptors. Consistent with this suggestion, a recent receptor-binding study reported high concentrations of PYY binding sites in the

PFH/lateral hypothalamic area *(49)*, whereas only low concentrations of PYY and NPY binding were found in most other hypothalamic areas. Our data also point to the dorsomedial nucleus as a possible additional locus for NPY's stimulatory effects. Although the PFH and dorsomedial nucleus appear to be the most sensitive hypothalamic sites for NPY to induce eating, it should be emphasized that the effects are not exclusive to these areas. Smaller, but still significant feeding responses were obtained in other hypothalamic sites. Therefore, although the highest concentrations of eating-related NPY-receptors are proposed to exist within the PFH and dorsomedial nucleus, these receptors are also likely to be present at lower concentrations in other nearby hypothalamic regions.

Consistent with a role for the PFH in mediating NPY's eating-stimulatory effect, previous studies have implicated this area in the control of eating behavior and body-weight gain. Of particular interest is the evidence that specific catecholamine neurotransmitters act in the PFH not to stimulate eating, but to inhibit eating. Specifically, synaptic dopamine and EPI from midbrain and brain stem neurons are believed to act via dopamine and β-adrenergic receptors in the PFH to inhibit eating *(50,51)*. These catecholamines appear to exert a chronic suppressive effect on spontaneous eating behavior, because neurotoxin lesions of the dopamine and EPI neurons projecting to the PFH cause overeating and obesity *(51,52)*. Moreover, catecholamine neurotransmitter activity and catecholamine receptor expression in the PFH are especially sensitive to manipulations, such as food deprivation and gastric loading, that markedly influence natural eating behavior *(53–55)*. Additionally, the PFH is one of the hypothalamic sites at which electrical stimulation is most effective in stimulating eating behavior *(56)*.

3.2.2. Catecholamines and NPY May Interact Antagonistically in the PFH

The findings described above, suggesting that the PFH is an important relay in feeding control systems, support the proposal that this region is a primary locus for NPY's eating-stimulatory effect. Moreover, the virtual identity of the brain sites at which NPY is most effective in feeding *stimulation*, and at which EPI and dopamine are most effective in feeding *suppression*, additionally suggests that these neurotransmitter systems interact within the PFH in an antagonistic manner to control natural eating behavior. More specifically, it is suggested that dopamine and/or epinephrine act within the PFH to reduce NPY's effectiveness in eliciting eating behavior. Consistent with this suggestion, we have recently shown that hypothalamic injection of catecholamine synthesis inhibitors actually potentiates the NPY-elicited eating response *(57)*. This augmented responsiveness to NPY by depletion of endogenous catecholamines suggests that there is a basal release of catecholamines that chronically inhibits neural responsiveness to NPY in satiated animals. Additionally, the injected NPY may release PFH catecholamines, which then act to reduce NPY's effectiveness in stimulating eating. This possibility is supported by a recent study showing that lateral hypothalamic microinjection of NPY increased extracellular concentrations of catecholamines at this site *(58)*. We also have preliminary evidence that drug-induced release of presynaptic catecholamines within the PFH can suppress the NPY-induced eating response *(168)*. Specifically, the catecholamine-releasing drug amphetamine injected into the PFH just prior to injection of NPY markedly reduced the peptide-elicited eating response. This suppression is most likely mediated by synaptic dopamine acting via its receptors in the PFH, because amphetamine's suppressive effect on NPY-induced eating was blocked by PFH injection of the dopamine antagonist haloperidol, but was unaffected by the α and β adrenergic antagonists phentolamine and propranolol. Further support was provided by data showing that NPY-induced eating could also be suppressed by injection of dopamine, but not by injection of epinephrine.

3.2.3. The PVN Most Frequently Exhibits Changes in Endogenous NPY

Besides cannula mapping, other approaches with implications for localizing NPY's effects involve determining which brain sites exhibit alterations of endogenous NPY levels in response to physiological manipulations that influence eating behavior. Interestingly, and in contrast to the cannula-mapping experiments, these studies do implicate the PVN and the arcuate nucleus, a major source of its NPY afferents *(41)*, as likely mediators of NPY's effects. The first of these studies compared tissue concentrations of NPY within various hypothalamic nuclei in rats subjected to different periods of food deprivation and refeeding. Severe food deprivations, lasting 3–4 d, increased tissue levels of NPY in the PVN and in the arcuate nucleus, while allowing the deprived animals to eat normalized NPY levels in the PVN, but not in the arcuate nucleus *(59)*. More recently, shorter periods of deprivation have also been shown to enhance PVN levels of NPY *(60,61)*.

A push–pull cannula study has shown that extracellular levels of NPY, like tissue levels, increase in the PVN during food deprivation and then normalize during refeeding *(63)*. Extracellular levels also increased prior to daily scheduled feeding periods and then rapidly declined while the animals ate. These findings suggest that NPY may be released by terminals in the PVN in response to factors associated with conditions, such as food deprivation, that produce eating behavior. Since increased release was accompanied by increased tissue levels of NPY, it is apparent that the synthesis and/or degradation of this peptide also is highly sensitive to food deprivation. Results of this nature are important because, independent of their implications for the locus of NPY's effects, they demonstrate that endogenous NPY is influenced by, and sensitive to, conditions that normally activate the neural substrates of eating behavior. In combination with the results of injection studies, these findings provide more convincing evidence that NPY is part of a system that acts to initiate corrective responses, such as eating behavior, in response to energy deficits.

This interpretation is reinforced and extended by other studies showing that tissue levels of NPY in the PVN and arcuate nucleus are also increased in response to a variety of other conditions accompanied by increased eating behavior. For example, PVN and arcuate nucleus levels of tissue NPY increase in spontaneously diabetic rats *(64)*, in rats made diabetic with streptozotocin *(65,66)*, in genetically obese Zucker rats *(67)*, and in normal rats just before the onset of the nocturnal-feeding phase of the light–dark cycle *(68)*. These increases appear to result largely from enhanced NPY gene expression within the arcuate nucleus, since preproNPY mRNA levels there are elevated by several conditions associated with overeating. These conditions include: food deprivation *(69,71,73,172)*, diabetes *(70)*, genetic obesity *(69)*, and increased circulating levels of glucocorticoids *(74,135)*. In addition, manipulations that suppress eating also may influence PVN levels of NPY. As noted earlier, rats allowed to eat to satiety exhibited lower levels of tissue and extracellular NPY in the PVN than comparison subjects that were not allowed to eat after deprivation *(59,63)*. These changes may occur rapidly in response to intestinal absorption of nutrients. This was suggested by the findings of Della-Fera et al. *(75)*, who showed that intestinal infusion of nutrients produced an 86% increase in PVN levels of tissue NPY within 7 min. In this case, the rapid increase in tissue NPY may reflect a decrease in the release of this peptide in response to the satiating effects of the infused food. This observation also implies a very high rate of NPY release in hungry animals.

Although the largest and most consistent effects of these manipulations are in the arcuate nucleus and PVN, effects are also sometimes observed in other hypothalamic nuclei and, even very rarely, in extrahypothalamic areas *(67,76)*. In terms of the PFH, it is difficult to ascertain the effects of physiological manipulations on NPY levels in this area with confidence, because few studies have assayed NPY at this specific site. However, some clues may be derived from measures of NPY in the lateral hypothalamus, because "lateral hypothalamus" tissue usually includes substantial portions of the PFH. In this regard, it is interesting that the PFH/lateral hypothalamus frequently exhibits changes in levels of endogenous NPY. Specifically, it has been shown that NPY levels increase in the PFH/lateral hypothalamus in rats that: are eating subsequent to food deprivation *(61)*, have been made diabetic with streptozotocin *(65,77)*, have been maintained on a high fat diet *(76)*, or are in the early portion of the nocturnal feeding cycle *(78)*. These findings support the cannula-mapping data, suggesting that the PFH/lateral hypothalamus is a key site for mediating the eating-stimulatory effects of NPY. A comparison of these studies reveals that the levels of NPY in the PVN and the PFH/lateral hypothalamus do not always change under the same conditions or even in the same direction, suggesting that NPY may have different functions in these two areas. Finally, irrespective of the locus of these effects, changes in endogenous NPY are clearly not restricted to just a few physiological conditions that produce eating behavior, but rather occur in response to a broad spectrum of physiological factors relevant to the control of energy intake, storage, and expenditure.

3.3. Models for Integrating Injection and Release Studies

On the one hand, injection studies indicate that the PFH, not the medial PVN, is most sensitive to NPY's eating-stimulatory effects and is therefore likely to contain the highest concentration of the critical receptors. On the other hand, studies measuring endogenous NPY indicate that levels of this peptide in the arcuate nucleus-PVN system are most sensitive to physiological and behavioral manipulations that increase eating, with the PFH exhibiting changes under fewer and perhaps even under different conditions. There are several plausible explanations for this apparent discrepancy between release sites and sites of action for this peptide.

3.3.1. Model 1: The NPY Released in the PVN Acts in the PFH

A possible explanation for this mismatch is that the NPY released from terminals in the PVN does stimulate eating, but that it produces this effect not in that nucleus, but in the PFH, where it migrates after its release. Agnati et al. *(79)* have suggested that two general types of information transfer exist in the central nervous system—volume transmission and wiring transmission. Wiring transmission refers to neurotransmitters acting in a classical synaptic manner; volume transmission refers to neurotransmitters acting at a distance from their point of release. Zoli et al. *(80)* have recently proposed that NPY can act in a volume transmission mode, at a distance from its site of release. This suggestion was based on their observation that, like some other neurotransmitters, a marked discrepancy exists in the location of endogenous NPY and in the location of NPY receptors. In terms of the relationship between the PVN and PFH in the control of eating behavior, volume transmission seems plausible in view of the close anatomical proximity of these two areas. As shown in Fig. 3, these structures are essentially adjacent. In addition, the possibility that volume transmission of NPY may contribute to control of eating is reinforced by evidence suggesting that many NPY-containing terminals in the hypothalamus exist as free endings, without obvious postsynaptic specializations *(81)*.

However, in terms of the control of eating, a concern exists regarding this hypothesis, since NPY appears to remain relatively localized, at least when it is exogenously administered. As described earlier, 95% of the radiolabeled NPY injected into the PFH remained localized to within 0.5 mm of the injection site, and none was found more than 0.8 mm from the injection site *(47)*. Therefore, if volume transmission is to account for the apparent discrepancy between the PVN as an apparent site of NPY release and the PFH as the apparent site of action, some mechanism for transport of NPY between these sites other than passive diffusion must exist. Furthermore, examination of NPY-containing terminals in the PVN by electron microscopy indicates that most of these endings form classical synaptic connections *(82)*.

3.3.2. Model 2: The PVN Release Is Related to Stress or Endocrine Function

Another possibility is that the increased levels of PVN NPY following manipulations, like food deprivation, may be produced by the physiological stress and increased circulating glucocorticoid levels associated with these manipulations, rather than by their effects on eating mechanisms *per se*. This hypothesis might suggest that the eating induced by NPY injection into the PVN is attributed to diffusion to other brain sites, such as the PFH, and to activation of receptors at those distant sites. The possibility that prolonged stress might elevate NPY levels has recently been noted by Corder et al. *(83)*. Their findings show that the levels of NPY in the infundibular nucleus (i.e., arcuate nucleus), PVN, and ventromedial nucleus were elevated in human patients that had died after prolonged respiratory distress, relative to levels in patients that had died after a briefer illness. They suggested that the prolonged stress might produce the elevated levels of NPY, and noted that many manipulations that produce eating and increase NPY levels in the PVN are also stressful.

In relation to this hypothesis, there is considerable evidence that NPY-containing neurons terminating in the PVN are involved in controlling the release of adrenal glucocorticoids and, conversely, that circulating glucocorticoids can themselves increase levels of endogenous NPY (*see* Sections 8.4., 9., and 10.4.). This is relevant because most conditions that induce alterations in endogenous NPY in the PVN (including food deprivation, diabetes, and genetic obesity) are also associated with stress and/or high levels of circulating corticosterone (CORT). Even the circadian rhythm of PVN NPY parallels the circadian rhythm of circulating CORT levels *(68,84)*, with both peaking near the onset of the dark phase of the light–dark cycle. Thus, it is conceivable that the changes in PVN levels of NPY might have more to do with CORT control than with a direct control of eating.

However, although arcuate-PVN levels of NPY are influenced by circulating glucocorticoids, and NPY in the PVN also regulates glucocorticoid levels, it is unlikely that hypothalamic NPY is solely influenced by stress and glucocorticoids. This conclusion is based on work showing that manipulations that affect eating behavior, but are not stressful, can impact levels of endogenous NPY and, conversely, that some stressful manipulations do not affect levels of NPY. For example, merely manipulating the ratio of fats to carbohydrates in subjects' diets can significantly alter PVN levels of endogenous NPY *(76)*. Likewise, Zucker obese rats have elevated levels of NPY and preproNPY mRNA that do not appear to be produced by the elevated CORT levels normally present in these animals. This conclusion is based on data showing that repeated administration of a glucocorticoid receptor antagonist did not alter the abnormal NPY gene expression in obese Zucker rats *(85)*. Conversely, highly stressful injections of insulin did not alter regional hypothalamic levels of NPY *(86)*. Thus, arcuate-PVN NPY is unlikely to be solely influenced by stress and glucocorticoids.

3.3.3. Model 3: Multiple Sites with Related Functions

A more likely explanation for the contrasting sensitivities of the PFH to eating stimulation by NPY and of the PVN to manipulations that alter endogenous levels of NPY is that both of these brain areas are critically involved in mediating NPY's effects, but that their functions are different. More specifically, it is proposed that the main function of NPY in the PVN is to mediate the autonomic and endocrine responses associated with eating, whereas the main function of NPY in the PFH is to mediate the eating response itself. Furthermore, the activity of NPY in these two areas is proposed to be sensitive to a broad range of factors that are normally involved in initiating and terminating eating behavior, and its associated autonomic and endocrine responses. These influences may include circulating levels of glucose, the hormones insulin, glucagon, and glucocorticoids, and perhaps factors associated with long-term changes in body fat. Although the activity of NPY in the PVN and PFH may be sensitive to a broad range of physiological conditions associated with energy deficits, the pattern and ratio of release between these areas are suggested to vary according to the particular combination of physiological factors (e.g., glucose, CORT, insulin) prevailing at the time. These different anatomical patterns of NPY release are proposed to produce different combinations of behavioral and physiological effects that are appropriate for specific situations. The essence of this suggestion is that under physiological conditions, NPY acting in different brain areas may, in concert with other neurotransmitters, mediate the coordinated output of behavioral, autonomic, and endocrine systems to produce a unified overall effect appropriate to the conditions.

What findings have generated this hypothesis? The possibility that NPY in the PVN and PFH may be influenced in a distinctive manner by different conditions was suggested by studies showing that some conditions that affect NPY levels in the PVN do not affect NPY in the PFH/lateral hypothalamus, and vice versa. For example, food deprivation markedly increased PVN levels of NPY, without affecting the levels in the PFH *(61)*. In contrast, refeeding the deprived animals increased NPY levels in the PFH, while decreasing those in the PVN. Differential responses of NPY in the PVN as compared to the PFH/lateral hypothalamus have also been observed in response to the ratio of fats and carbohydrates in the maintenance diet *(76)* and to the phase of the circadian cycle *(68,78)*.

The suggestion that the PVN also plays a role in NPY feeding, as opposed to exclusive mediation by other sites, particularly the PFH and dorsomedial nucleus, is based largely on evidence that electrolytic lesions of the PVN are effective in reducing the eating response to intracerebroventricular NPY *(87)*. This is important, because the cannula-mapping data by itself does not provide convincing evidence for an involvement of the PVN in NPY-induced eating. Specifically, although injection of NPY into the PVN elicited significant eating, the responses were no larger than those in areas surrounding the PVN. Therefore, the mapping data and the lesion data in combination suggest that both the PFH and the PVN contribute to the eating response produced by NPY.

The idea that NPY may act in several hypothalamic sites to stimulate eating is particularly plausible, considering the evidence that it can act in separate cortical, brain stem, and hypothalamic sites to produce eating *(34,35)*, and that the PVN and PFH form parts of two different hypothalamic systems controlling eating behavior, and its associated autonomic and endocrine responses *(21,50,51,88,89)*. If NPY acts to produce eating by affecting receptors in several hypothalamic sites, then its precise functions in these sites are likely to be different. Consistent with this possibility, there are already some preliminary indications that the eating responses produced by NPY in the PVN and in the PFH are different. For example, it has been shown that in the PVN there is a day–night rhythm of responsiveness to NPY injection, with a peak response in the early portion of the nocturnal eating period *(90)*. In contrast, in the PFH, the eating response to NPY injection does not appear to exhibit an appreciable circadian rhythm *(91)*.

The site-specific differences in the effects of NPY in the PVN and PFH do not appear to be confined to feeding behavior, but instead seem to extend to endocrine and autonomic responses. It has been shown that PVN injections of NPY inhibit sympathetic activity to brown adipose tissue and inhibit gastric acid secretion *(92,93)*. In contrast, injections further laterally in the PFH lateral hypothalamic area have no effect on sympathetic activity of brown adipose tissue and may actually increase gastric acid secretions. Likewise, as will be discussed in more detail in Section 9., it has been shown that NPY injected into the PVN elicits a variety of endocrine effects, including increased levels of circulating insulin, glucagon, and CORT, that appear to be related to control of nutrient disposition *(94–96)*. Interestingly, the NPY-elicited endocrine effects, although correlated with each other, were not correlated with the peptide-elicited eating response measured in the same tests *(95)*, supporting separate anatomical loci for these effects. In contrast to these effects in the PVN, preliminary data indicate that NPY injection into the PFH elicits eating without significantly increasing the levels of any of these hormones *(97)*. Thus, it appears that the PVN may be most sensitive to the autonomic and endocrine effects of NPY, whereas the PFH is most sensitive to its eating-stimulatory effects. In relation to these anatomically different effects, it may be noteworthy that NPY injected into the medial and lateral hypothalamus appears to produce different or even opposite effects on extracellular levels of catecholamine and indolamine neurotransmitters *(58)*.

4. Receptor Subtypes

What are the characteristics of the receptors activated by NPY to produce eating? Evidence suggests that there are at least two subtypes of receptors for NPY: Y1 and Y2 *(98)*. These subtypes are usually distinguished by their differing response profiles to fragments of the NPY molecule. The Y1 subtype requires essentially the entire NPY molecule for complete activation; deleting one or more N-terminal amino acids from the NPY molecule markedly reduces or eliminates its ability to activate this subtype. In contrast, the Y2 subtype is relatively insensitive to N-terminal amino acid deletions; the NPY fragment 13–36 is nearly as effective as the entire molecule in Y2-receptor-dependent effects. Both subtypes are highly sensitive to amino acid deletions at the C-terminal; loss of a single amino acid there usually abolishes NPY's effects. Of late, NPY analogs that are relatively selective for these subtypes have also been developed *(99,100, see also* Grundemar et al., this volume).

To identify the contribution of each of these different receptor subtypes to feeding elicited by NPY, we tested the ability of different NPY fragments to stimulate feeding after injection in the PVN *(101,173)*. Other studies have also employed PVN and icv injections to examine these issues *(62,102,103)*. These studies suggest that the eating response is predominantly mediated through an NPY receptor that is similar to, but not identical with, the Y1 subtype. In agreement with an earlier study testing icv injections *(28)*, we found that PVN injection of the free acid form of NPY, which is ineffective in all other preparations, did not stimulate eating. Although this might suggest that the C terminal is essential for eating stimulation, the recently demonstrated effectiveness of the N-terminal fragment NPY_{1-27} questions this conclusion *(104)*. We tested fragments with N-terminal deletions, such as NPY_{5-36} and shorter fragments, and found that they had either markedly reduced or no effect, consistent with mediation by Y1 receptors. Subsequently, we employed analogs of NPY that are selective for the Y1 and Y2 subtypes and, like Kalra et al. *(62)*, found that the Y1 analog elicited an eating response almost as large as NPY itself, whereas the Y2 analog elicited only a small effect and only at higher doses *(173)*. These two different lines of evidence are consistent with mediation by Y1 receptors; however, it was also found that the NPY fragment 2–36 was substantially more effective than NPY in eliciting eating *(62,101,102,105,173)*. This was unexpected, since in other preparations, this fragment was at best as effective as the parent molecule (*106; see also* chapter by Bleakman et al., this vol.), and, more frequently, was markedly less effective than NPY itself *(49,107,108)*. This suggests that the receptor subtype mediating the eating response is not identical with the Y1 receptor mediating the other effects of NPY. This possibility is supported by the effectiveness of pancreatic polypeptide in eliciting eating *(7)* in comparison to its ineffectiveness in other preparations. Thus, although the primary receptor category mediating the effects of NPY on eating appears to be Y1, there may be further divisions of this subtype that more specifically account for NPY's effects.

5. Electrophysiological Responses

Collectively, the evidence described to this point suggests that the stimulatory effects of NPY on feeding are mediated by Y1-like receptors that are primarily concentrated in the PFH. In contrast, NPY may primarily act in the PVN to produce a variety of endocrine and autonomic effects that are likely to be related to its role in control of eating *(92,93,95,96,109,110)*. The question to be addressed now is: What are the electrophysiological effects of activating these receptors with NPY? In particular, to what extent does the time-course of NPY's electrophysiological effects parallel its behavioral and metabolic effects? In general, NPY's effects on eating and metabolism appear to occur with a moderately long latency and to be extremely prolonged. For example, the minimum latency to eat after hypothalamic injection of NPY is 1–2 min, with an average of about 9–10 min *(9)*. In terms of duration, at high doses, animals will overeat for at least 5 h *(11)*. Likewise, the autonomic and endocrine effects usually appear 5–20 min postinjection and last for several hours thereafter. The major exception to this pattern is the effect of NPY on respiratory quotient. This effect was not apparent at some doses until 20 h postinjection *(110)*.

Although no published studies have assessed the electrophysiological effects of NPY in the PFH, the effects of bath-applied NPY on unit activity in the PVN and in the suprachiasmatic nucleus (SCN) of rats have been examined *(109,111)*. In the PVN, the most frequent effect was inhibition, with a latency of approx 1–2 min and a duration of about 5 min. A smaller number of excitatory responses of approximately the same latency and duration were also obtained in this nucleus. Preliminary data *(165)* suggest that in the PFH the electrophysiological responses to NPY may be similar. In the rat SCN, the responses were more complex, frequently consisting of an initial short-lived excitation followed by a longer-lasting inhibition of up to 60 min. In contrast, in the hamster SCN, the effect of pressure-applied NPY was predominantly excitatory, occurring within seconds of application and lasting up to 60 min *(112)*. Perhaps the most intriguing finding was that these responses exhibit a circadian rhythm. Application of NPY to the SCN during the subjective day altered the activity of these neurons, whereas application during the subjective night did not *(111,112)*.

These electrophysiological findings reveal both parallels and contrasts between the behavioral/metabolic and electrophysiological responses to NPY. For example, the 1–2 min latency of the PVN neural response to NPY may account for the equivalent latency to eat in the most rapidly responding animals. In contrast, the brief (5 min) electrophysiological effects of NPY in the PVN do not appear to account for the long-lasting behavioral, autonomic, and endocrine effects of this peptide. To the extent that these in vitro findings reflect the neural responses of intact animals, they suggest that the apparent electrophysiological responses to NPY may only account for triggering its behavioral and metabolic effects. Other processes may be responsible for NPY's prolonged action. Although the physiological basis for this prolonged action has not yet been established, its existence suggests that NPY may be producing behaviorally relevant, long-lasting intracellular changes that outlast its most obvious electrophysiological effects.

6. Second Messengers

The prolonged eating induced by NPY suggests that this effect may be mediated by intracellular second messengers. This seems particularly likely considering the evidence that second messengers mediate many of NPY's other effects (e.g., *113*). However, there is only one published study to date examining this issue. Chance et al. *(114)* examined the effects of inactivating pertussis-toxin-sensitive G proteins on NPY-induced eating and found that prior administration of pertussis toxin ventrolateral to the PVN abolished the eating behavior induced by injection of NPY. This effect was observed 4 d after the pertussis toxin was administered. By that time, daily food intake had returned to normal levels, arguing that the suppression was unlikely to be the result of general malaise. In addition, NPY inhibited isoproterenol-induced cAMP production in the hypothalamus, an effect that was also blocked by pertussis-toxin pretreatment. These investigators suggested that NPY induces eating by inhibiting cAMP production and that a pertussis-toxin-sensitive G protein is involved in this signal transduction. Future studies will need to evaluate the specificity and locus of these proposed second-messenger effects, and their precise contributions to the duration of the peptide-elicited eating, endocrine, and autonomic effects. Nevertheless, the combination of these finding and those described earlier may suggest that NPY-induced activation of Y1-like receptors in the PFH elicits eating by inhibiting cAMP production via a pertussis-toxin-sensitive G protein and that this contributes to the prolonged duration of NPY's effects.

7. Behavioral Specificity and Eating Patterns

7.1. Behavioral Specificity

Given the evidence for a role of hypothalamic Y1-like receptors in the elicited eating response, the next question is: What are the behavioral effects of activating this system? Does NPY increase eating directly through a brain system normally involved in the control of this behavior, or does it act in an indirect, nonspecific manner? This is an important question, because some evidence suggests that eating behavior and obesity can be induced by nonspecific manipulations that produce stress *(115)*. The possibility of nonspecific effects is of particular concern when NPY is administered icv, because the widespread diffusion of NPY administered by this route produces a multitude of behavioral and physiological effects that are unlikely to be directly related to eating behavior (*116–119*; Heilig, this vol.). Many of these individual responses appear to result from NPY's actions within distinct brain regions (*116,118; see also* McAuley et al., this vol.), suggesting that NPY acts on different neurocircuits in anatomically distinct brain regions to produce its different effects. Thus, when attempting to determine the specificity of NPY's behavioral effects, it is important to focus on studies that employ injections directly into the most sensitive brain regions, rather than on icv injections, which are likely to affect simultaneously many unrelated systems.

The issue of behavioral specificity was examined in the original paper by Clark et al. *(7)*, who found that icv injection of NPY increased food and water intake, and decreased grooming. More recently, Jolicoeur et al. *(119)* have shown that icv injections of NPY increased eating and muscle tone, decreased body temperature and spontaneous motor activity, and induced catalepsy. Likewise, Levine et al. *(120)* showed that although NPY injected icv elicited eating, the animals' behavioral patterns were different from those in animals eating subsequent to food deprivation. Thus, when injected icv, NPY appears to alter several behaviors, which might suggest that its effects on eating are nonspecific. Yet other work employing hypothalamic injections suggests that NPY administered by this route is highly selective in its effects *(8)*. In rats given PVN injection of an intermediate dose of NPY and then allowed access to food, there was a large increase in the time spent eating and a small increase in the time spent drinking, but there was no significant change in any other behavior measured (Fig. 4, bottom panel). Specifically, there were no changes in grooming, sleeping, locomotion, or in levels of general behavioral activity. Likewise, examination of the sequence of behaviors performed after NPY injection revealed that this sequence was similar to that observed in spontaneously eating rats. The rats began to eat approximately 10 min postinjection; they ate a single large meal and, after they stopped eating, they groomed and slept. This sequence of behaviors is characteristic of naturally occurring eating in rats and has been suggested by Antin et al. *(121)* to be a marker of physiological satiety (however, *see* 122). The similarities in the behavioral patterns produced by NPY and those occurring in spontaneously eating rats support the idea that some neurocircuits involved in the generation of natural eating behavior have receptors that are sensitive to NPY.

This is not to suggest that direct hypothalamic injections of NPY exclusively affect eating behavior. First, as will be described in detail in Section 9., PVN injection of NPY produces an array of autonomic and physiological effects on systems that regulate metabolic fuels. Second, in other hypothalamic areas, NPY may influence other behavioral and physiological systems that are

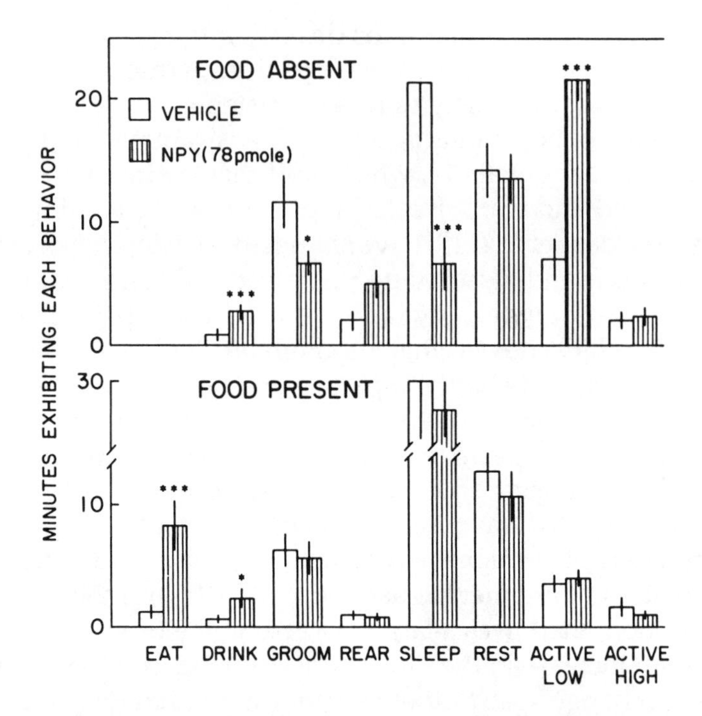

Fig. 4. Minutes (mean ± SEM) exhibiting eating, drinking, grooming, rearing, sleeping, resting, or low and high levels of activity by rats given PVN injection of NPY (78 pmol/0.3 μL) or vehicle. The top panel represents tests conducted without food present; the bottom panel represents tests conducted with food present. $*p < 0.05$, $***p < 0.005$ relative to vehicle. (From *[8]*. Reprinted with permission of *Life Sci.*)

less directly related, or even independent of, the regulation of body energy balance. For example, NPY appears to act in the SCN to modulate circadian rhythms *(116)*, in the preoptic area to suppress sexual behavior *(123)*, and in multiple hypothalamic and extrahypothalamic areas to regulate blood pressure *(118,124; see also* chapters by Heilig, by McDonald and Koenig, and by McAuley et al., this vol.).

Finally, the intense activation of systems that generate eating behavior by NPY may produce "side effects" if the injected animals are prevented from eating. Thus, for example, PVN injection of NPY specifically elicited eating behavior when the rats were given food (Fig. 4, bottom panel), but produced marked changes in other behaviors (increased activity and decreased sleep) when the subjects were denied access to food (Fig. 4, top panel). Similarly, icv NPY decreased blood pressure, heart rate, and respiratory rates in rats without food, but not in rats allowed to eat *(125)*. Since these "side effects" did not occur in animals that were allowed to eat, we suggest that the NPY made the animals "hungry" (i.e., activated eating control mechanisms) and that, in the absence of food, this generated the effects described above. This interpretation is consistent with the evidence for nonspecific components in eating-control mechanisms *(126)* and with a similar study examining the behavioral effect of icv injection of NPY in rats without food *(46)*. In that study, NPY increased behavioral activity, and this increased activity was focused on the subjects' empty food cups. Additional evidence that NPY activates eating-control mechanisms is that NPY, like food deprivation and other natural inducers of eating, increases the reward value of food. It has been shown that icv injection of NPY will induce rats or pigs to perform a previously learned operant response to obtain a food reward *(30,127,169)*.

Besides behavioral specificity, we examined the factors responsible for the termination of eating after PVN injection of an intermediate dose of NPY (78 pmol or 0.33 μg; *8*). At this dose, rats normally stop eating within 1 h of injection, and we investigated whether this was because the peptide itself lost effectiveness or because the subjects became satiated by the ingested food. This was accomplished by determining whether NPY elicited eating in animals that were not fed until 4 h after the injection. The results demonstrated that this dose of NPY still exerts a strong eating-stimulatory effect even 4 h after injection. Thus, NPY has long-lasting effects that are antagonized by ingested food. These are precisely the types of results that should be produced by NPY activation of an endogenous eating-control system.

7.2. Meal Patterns

These initial studies only examined the animals' behavior for 1 h after injection of an intermediate dose of NPY. However, higher doses of NPY produce effects that are much larger and last much longer *(9)*, apparently near the limits of the animals' capacity to eat. These large and prolonged effects led us to question whether this increased eating occurred in discrete meals, and if so, what parameters of eating behavior were most affected? Did the animals eat a few large meals or many small meals? Perhaps more importantly, did the pattern of eating produced by NPY injections resemble a naturally occurring pattern?

These issues have been addressed previously for continuous icv administration of NPY *(128)*, and it was found that the low dose of NPY (117 pmol/h) increased food intake by increasing the average duration of feeding episodes, without increasing their number. The higher doses (470–1175 pmol/h) increased food intake even further, by increasing the number of episodes of eating from a baseline of 1/h up to 10/h, without increasing their duration. To resolve these issues for hypothalamically injected NPY, we have recently examined the pattern of meal-taking induced by PFH injection of NPY, using minute-to-minute measures of eating patterns *(11,12)*. It was found that at low doses (8–24 pmol) of NPY, the increments in food intake were exclusively accounted for by the ingestion of a single meal that averaged approximately two to four times larger in size than spontaneously occurring meals. At higher doses (78 and 235 pmol), NPY also increased the number of meals eaten. The initial meal was enormous, averaging 9.6 g at the highest dose. Despite the enormous quantity of food eaten in this initial meal, for up to 5 h postinjection, the animals continued to eat a series of three additional meals that were still significantly larger than normal. These results demonstrate that across doses, the eating induced by NPY is organized into discrete, well-defined meals, and that this peptide can increase both meal size and meal frequency. These findings are consistent with a role for NPY in organizing spontaneously occurring patterns of eating behavior.

It was suggested earlier that a function of NPY might be to generate eating in response to a deficiency in available energy. A circumstance during which this occurs is food deprivation. If the increased eating produced by food deprivation is mediated by NPY, then the meal patterns following food deprivation should be similar to those following NPY injection. To examine this, we compared the meal patterns produced by PFH injections of NPY to those in rats deprived of food for intervals of up to 24 h. The results indicate that the patterns of eating produced by these two different conditions are similar, particularly at high doses of NPY and longer periods of deprivation. Like the rats injected with a high dose of NPY, the 24-h food-deprived animals ate an enormous first meal and then subsequently ate a series of smaller meals that were still significantly larger than normal *(11)*. Likewise, similarities in eating patterns of food-deprived and NPY-injected rats have been noted in animals performing an operant response for food *(127)*. In contrast, Levine et al. *(120)* found significant differences in the behavioral patterns of food-deprived and icv NPY-injected rats. Thus, although the effects of icv injections are mixed, the similarities between the meal patterns of food-deprived and hypothalamically injected rats suggest that food-deprivation-induced eating may be mediated in part by hypothalamic NPY *(59)*.

Even more compelling are studies examining changes in levels of endogenous NPY in response to food deprivation. As noted earlier, tissue and extracellular concentrations of NPY increase in the PVN of rats that have been food deprived *(59–61,63)*. Additional support was provided by recent work showing that icv injection of NPY antisera caused a concentration-dependent suppression of eating produced by mild food deprivation *(173)*. Further work will be needed to determine whether this suppression was actually due to the inactivation of endogenous NPY, but the findings are consistent with a role for NPY in the generation of food deprivation-induced eating behavior. Thus, biochemical and behavioral results collectively suggest that increased hypothalamic release of NPY may mediate the eating produced by food deprivation.

8. NPY in Relation to Carbohydrate Ingestion

8.1. NPY Preferentially Elicits Consumption of Carbohydrates

In nature and the laboratory, animals are frequently capable of appropriately regulating the consumption of different macronutrients, such as carbohydrates, fats, and proteins *(129,130)*. Evidence also suggests that some eating-stimulatory neurotransmitters preferentially elicit ingestion of specific macronutrients, leading to the suggestion that different neurotransmitters may regulate intake of different macronutrients *(131)*. Therefore, we next determined whether NPY might differentially affect ingestion of these macronutrients. To examine this, NPY was injected into the PVN to determine whether it would induce a macronutrient-selective effect in rats offered a choice of three different diets, consisting of pure carbohydrate, pure protein, or pure fat. Peptide YY was similarly tested. As shown in the top panel of Fig. 5, when offered all three of

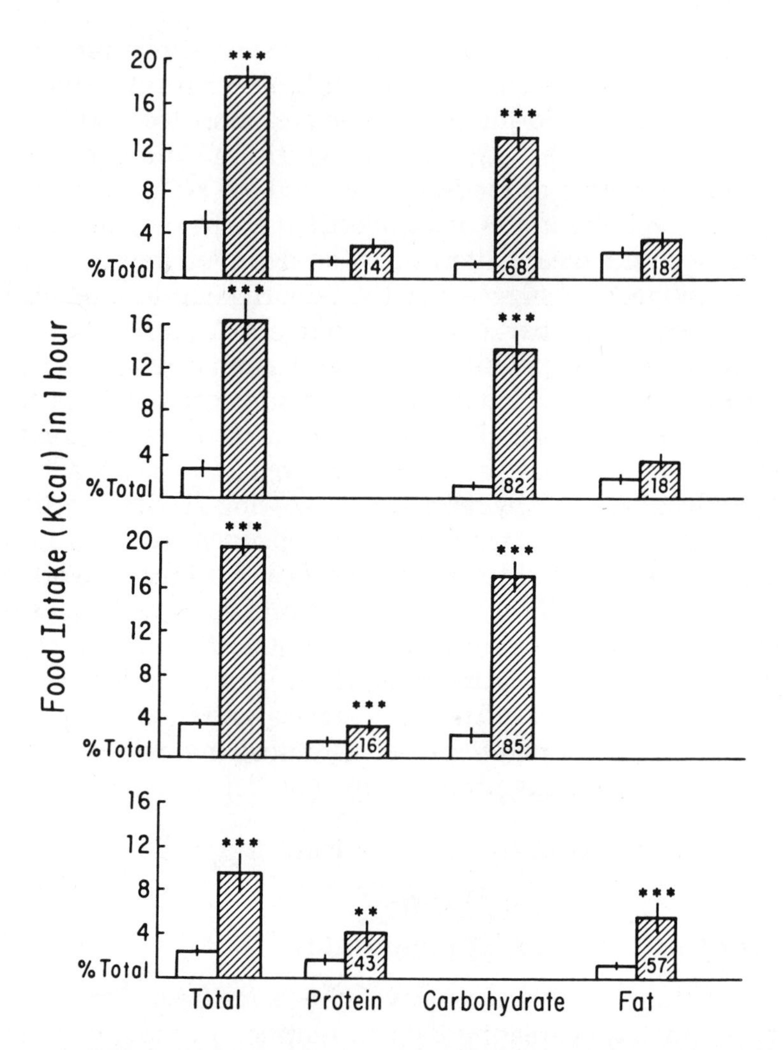

Fig. 5. Effect of PVN injection of NPY (78 pmol, filled bars) or vehicle (open bars) on nutrient intake (mean kcal ± SEM) 1 h postinjection. Ingestion of protein, carbohydrate, and fat as well as total diet intake is shown. Percent total (% total) refers to the numbers within the bars, which indicate the amount of that diet consumed as a percent of the combined totals of all the diets consumed. The top panel shows results when all three diets were available. The second, third, and fourth panels, respectively, shown results when protein, fat, and carbohydrate diets were not available. ***p* < 0.01, ****p* < 0.001 relative to vehicle. (From *[10]*. Reprinted by permission from *Peptides*.) □ Saline; ▨ neuropeptide Y.

the macronutrients, the NPY- and PYY- (not shown) injected rats primarily ate the pure carbohydrate diet *(10)*. Likewise, others have found a carbohydrate-selective effect employing mixed macronutrient diets *(46)*. These results suggested that NPY might play a role in regulating carbohydrate consumption.

The NPY-injected rats were additionally tested with combinations of only two diets present *(10)*, and it was found that the peptide-elicited eating was focused almost exclusively on the carbohydrate diet, as long as that diet was available (Fig. 5, middle panels). Little, if any, of the fat or protein diets were consumed when carbohydrates were available. This led us to question whether NPY would elicit eating if only the fat and protein diets were present. What occurred was that, in the absence of available carbohydrates, the NPY-injected rats did eat substantial quantities of both the fat and the protein diets (*see* Fig. 5, bottom panel), demonstrating that the effect is not exclusive to carbohydrate consumption. However, in this condition, the total food intake (expressed in kilocalories) was only about half that when carbohydrates were present, suggesting that carbohydrates do have a special role in NPY's effects. Peptide YY produced an almost identical pattern of effects in all conditions, implying that it might be acting through similar mechanisms and, in particular, through similar receptors to produce its effect on eating behavior. The Y1 receptor appears to mediate these effects, since PVN injection of Y1-, but not Y2-receptor agonists elicits preferential consumption of carbohydrates *(105)*.

8.2. Nutritive Factors May Be More Critical than Sensory Factors

The reduced response to NPY in the absence of carbohydrate diets suggests that NPY preferentially, but not exclusively, elicits carbohydrate intake. What aspects of the diets are critical to their selection by NPY-injected rats? Our interpretation was that NPY elicits consumption of particular foods based, in part, on how rapidly they can be absorbed and used for energy. Carbohydrates are most rapidly absorbed into the blood *(132)*, and this, we suggest, might account for the preference for this diet. In the absence of carbohydrate, the NPY-injected rats may eat the other diets because they also provide energy, although not as rapidly as carbohydrates.

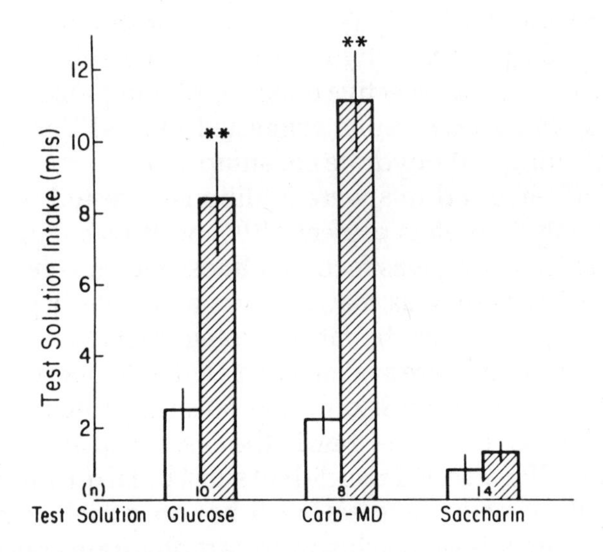

Fig. 6. Effect of PVN injection of NPY (78 pmol, filled bars) or vehicle (open bars) on nutrient intake (mean mL ± SEM) of approximately equally palatable solutions of glucose, carbohydrate-MD, and saccharin. $^{**}p < 0.01$ relative to vehicle. ▨ neuropeptide Y; □ saline.

This hypothesis focuses on the postingestive consequence of a diet, rather than on its pre-absorptive effects (e.g., taste), as the primary factor in determining whether it will be consumed by NPY-injected rats. Therefore, a study was conducted to determine whether taste or caloric consequences might be more influential in diet selection by rats injected with NPY. Neuropeptide Y was injected into the PVN of rats given one of three different liquid diets:

1. Glucose, which is both sweet and nutritive;
2. Saccharin, which is sweet but nonnutritive; and
3. Carbohydrate-MD, a nonsweet nutritive solution.

As shown in Fig. 6, the NPY-injected rats avidly consumed either the sweet or the nonsweet nutritive solutions of glucose or carbohydrate-MD, but did not consume the sweet nonnutritive saccharin solution *(133)*. These data are consistent with our suggestion that the postingestive nutritive consequences of diets are more critical than their orosensory properties for their selection by NPY-injected animals.

In relation to this, others have shown that NPY-induced feeding is highly sensitive to levels of the circulating nutrient glucose. Specifically, iv injections of a physiological dose of glucose produced a reduction in NPY-elicited food intake that was equivalent to the caloric value of the infusion *(134)*. Interestingly, iv injections of fructose were ineffective. Since, unlike glucose, fructose does not cross the blood–brain barrier, these results suggest that the satiating effect of food on the eating behavior induced by NPY may be mediated in part by the actions of nutrients directly on central nervous system neurons. These findings also may suggest a more specific mechanism for the diet preferences exhibited by NPY-injected rats. Specifically, diets may be preferred in proportion to their ability to counteract NPY's activational effects on brain mechanisms of eating behavior. Thus, for example, carbohydrate diets may be preferred, because they rapidly increase blood glucose levels, presumably inhibiting the activity of the brain system(s) activated by NPY.

8.3. Endogenous NPY Is Sensitive to Manipulations that Alter Metabolic Fuels and Their Regulatory Hormones

Measurements of the response of endogenous NPY to manipulations that alter nutrient availability also appear to support the idea that the metabolic effects of food are important in NPY's function. Many studies have shown that hypothalamic NPY is influenced by various treatments associated with altered availability of metabolic fuels. These include food deprivation *(59–61,63)*, onset of the circadian feeding period *(68,78)*, intestinal infusions of nutrients *(75)*, removal of endogenous glucocorticoids *(74,135)*, loss of endogenous insulin *(65,66,70,77,136,137)*, and several forms of genetic obesity *(67,69,70,175)*. Since a common feature of these conditions is altered nutrient utilization, an interesting implication may be that endogenous NPY is affected because it is a neurotransmitter in a neural system that monitors nutrient availability. Additionally, hypothalamic NPY may be directly sensitive to levels of some hormones that regulate the utilization of metabolic fuels. The evidence for a direct hormonal effect is clearest for glucocorticoids. It has been shown that arcuate nucleus NPY-containing neurons have dense concentrations of glucocorticoid receptors

(138,139), and that glucocorticoid administration markedly increases levels of hypothalamic NPY in both intact rats and in neuronal culture *(140)*. It has also been suggested that insulin directly affects endogenous NPY. This was based on results demonstrating that NPY gene expression is enhanced in the arcuate nucleus of diabetic rats *(70)*, and that direct icv injection of insulin can reduce NPY gene expression in this nucleus *(141,172)*. In contrast, Corrin et al. *(86)* found that peripheral administrations of insulin overdoses did not alter hypothalamic concentrations of NPY.

In relation to these findings, we have recently shown that both insulin-deficient diabetic and CORT-deficient rats are markedly less sensitive to the eating-stimulatory effects of hypothalamic NPY injection. Specifically, adrenalectomy or hypophysectomy produced marked reductions in the eating response to PVN injection of NPY, and the response was normalized by CORT replacement therapy *(142)*. We also have preliminary evidence that streptozotocin-diabetic rats are refractory to the feeding-stimulatory effects of NPY injected into the PFH, even when given insulin-replacement therapy *(97)*. These findings demonstrate that some conditions, like insulin and CORT deficiency, that produce changes in the levels of endogenous NPY also alter the sensitivity to the eating-stimulatory effects of this peptide.

8.4. NPY May Regulate Circadian Rhythms of Carbohydrate Consumption

As described earlier, NPY injection produces a marked preference for carbohydrate diets *(10,46)*. Based on this finding, we suggested that hypothalamic NPY might mediate the carbohydrate consumption that normally occurs in the spontaneously feeding rat *(10)*. In the spontaneously eating rat, there is a circadian rhythm of carbohydrate ingestion, with peak intakes occurring just after the onset of the nocturnal period *(143)*. Therefore, we additionally suggested that endogenous NPY activity might be enhanced in this portion of the light–dark cycle and might act, with other neurotransmitters, to generate the carbohydrate intake that occurs at this time *(10)*. Kalra et al. *(128)* also suggested that NPY might underlie nocturnal eating, based on the episodic feeding patterns of rats given icv injections of NPY. There is considerable evidence that supports this proposal. Specifically, tissue levels of endogenous

NPY in the PVN and lateral hypothalamus have been shown to peak at the onset of the nocturnal feeding period *(68,78)*. In relation to this, it may be interesting that there is a parallel circadian rhythm of circulating CORT *(84,144)*, that glucocorticoids enhance hypothalamic levels of NPY and its gene expression within the arcuate nucleus *(74,135,140)*, and that PVN injections of CORT elicit consumption of carbohydrates in adrenalectomized rats *(145)*. It has also been shown that NPY injected into the PVN during this early dark phase of the eating cycle is more effective in eliciting carbohydrate consumption than when it is injected later in the dark phase *(90)*. Based on these findings, Leibowitz and White and their respective associates have recently elaborated the original hypothesis *(68,90,135,142,146)*. They suggest that the normally occurring peak of circulating CORT at the onset of the dark period may act via glucocorticoid receptors on NPY-containing arcuate nucleus neurons to enhance their expression of the NPY gene leading to increased NPY levels within the PVN. This NPY, in concert with PVN NE, was proposed to generate the carbohydrate intake and shift to carbohydrate metabolism that normally occur at that time.

Evidence suggests that this daily pattern of arcuate PVN NPY may be more closely linked to factors that initiate eating, rather than to the light–dark cycle *per se*. Specifically, rats fed only during a scheduled daily 4-h period exhibited a peak of extracellular NPY in the PVN that anticipated the scheduled feeding period (suggesting the existence of a daily peak of NPY), even though this feeding period occurred during the light phase rather than at the onset of the dark phase *(63)*.

In addition to a possible role in regulating circadian rhythms of carbohydrate consumption, endogenous NPY may mediate changes in carbohydrate consumption occurring in more chronic disorders of eating behavior. As noted earlier, concentrations of endogenous NPY or preproNPY mRNA are enhanced in diabetic and in Zucker obese rats, and are decreased in rats deprived of endogenous CORT. In each of these conditions, there is a parallel change in total daily food intake, particularly of carbohydrates *(145,147,148)*. Several investigators have proposed that the altered NPY activity present in these chronic disorders might underlie the changes in food intake and carbohydrate consumption that are present *(77,135,142)*.

9. Endocrine and Autonomic Effects of NPY in Relation to Eating Behavior

Earlier in this chapter, it was proposed that NPY acting primarily in the PVN mediates the coordinated activity of a variety of physiological systems that, in concert with its stimulatory effect on eating behavior, enhance the supplies of energy available for cellular metabolism and storage. Neuropeptide Y has been shown to produce a variety of physiological effects that support and expand this idea.

Specifically, icv or PVN administration of NPY produces a marked increase in circulating levels of insulin, with a small stimulatory effect or, more frequently, no effect on serum glucose levels *(95,96,149,150)*. Since insulin increases glucose utilization, which should produce hypoglycemia, the minimal impact of NPY on glucose levels suggested that NPY also produced counteracting effects. This appears to be the case. Injections of NPY into the PVN have been shown to increase serum levels of glucagon *(96)*, and CORT *(94,95,109)*, while actually decreasing circulating levels of growth hormone *(151)*. Glucagon and CORT both enhance levels of circulating glucose, primarily via gluconeogenesis and glycogenolysis. In contrast, growth hormone increases circulating levels of glucose by decreasing cellular utilization of this nutrient. Thus, the decrease in levels of circulating growth hormone produced by NPY may also increase cellular carbohydrate utilization. Collectively, these endocrine effects of NPY should combine to increase cellular glucose utilization while also maintaining circulating levels of this nutrient. Supporting this hypothesis is the evidence that injections of NPY into the PVN or into the sulcal prefrontal cortex *(35,110)* increase respiratory quotient, which indicates increased carbohydrate metabolism. The NPY-induced increase in eating, particularly of carbohydrates, also should contribute to this effect.

Earlier we noted that NPY might play a role in generating the eating behavior that occurs at the beginning of the nocturnal feeding period. Concerning this, most of the hormones released by NPY have a 24-h rhythm that is consistent with increased NPY activity at the onset of the dark. Specifically, circulating levels of CORT peak near the onset of the feeding period *(84)*; circulating levels of insulin also show a marked increase at this time *(152)*. Conversely, growth-hormone levels are greatest during the inactive sleep periods *(153)*. Thus, NPY may mediate many effects that normally occur during the active feeding period, including increased carbohydrate intake, CORT, and insulin levels, and decreased growth-hormone levels. Taken with the evidence that PVN-tissue NPY is increased during this time *(68)*, these findings support the suggestion that fluctuations in NPY activity are important in generating these normally occurring effects.

Neuropeptide Y appears to have a variety of functions that are probably not directly related to its control of eating. For example, it can influence reproductive behavior and related hormones *(117,154,155)*, and circadian rhythms of behavioral activity *(116)*, as well as blood pressure and heart rate *(117,118)*. Therefore, it is conceivable that the hormonal effects described above are merely coincidental, reflecting the actions of NPY on various independent hormonal control systems rather than on a set of functionally related systems. Although this possibility cannot be dismissed, it seems unlikely because, despite the powerful ability of each of these hormones to affect blood glucose levels, the levels of this metabolic fuel are either unaltered by NPY *(46,95,149,150)* or undergo a small increase *(96)*. It would be a remarkable coincidence that all four of these hormones were released by NPY in precisely the quantities needed to yield constant levels of glucose. In addition, it should be noted that not all feeding-stimulatory neurochemicals produce constant levels of blood glucose. For example, like NPY, PVN injections of NE or EPI induce eating, and increase circulating levels of CORT and insulin. Yet, unlike NPY, these catecholamines also increase circulating levels of glucose *(95)*. Thus, different eating-stimulatory neurotransmitters may produce unique patterns of endocrine response.

In addition to these endocrinological effects, NPY can alter other physiological parameters that may be related to eating. Of particular interest are two related studies showing that icv or PVN administration of NPY markedly reduces the activity of sympathetic innervation of brown fat and decreases the thermogenic activity of this tissue *(93,156)*. Neuropeptide Y injection also increases white adipose tissue lipoprotein lipase activity, which should enhance fat incorporation in this tissue *(156)*. It was suggested that NPY coordinates behavioral and physiological responses to produce a positive energy balance. It has also been shown that PVN injection of NPY produces a significant reduction in gastric acid secretion *(92)*. Since gastric acid secretion normally increases during eating, this would seem to argue against an NPY-mediated control of gastric acid secretion during eating. However, it should be noted that, in this work, there were frequent occurrences of increases in gastric acid secretion produced by injections outside the PVN. Although there were not enough instances of these increases to localize the site of action, the data seem consistent with a site in the area of the PFH, where NPY also appears to act to stimulate eating.

10. Clinical Implications

Given a probable role for NPY in the regulation of "normal" eating, it would be interesting to determine whether it also might be involved in disorders of eating behavior and body-weight regulation. For example, could some cases of obesity or bulimia be caused by excessive NPY or PYY secretion? Likewise, might endogenous NPY contribute to some cases of anorexia nervosa? Also, might the changes in food intake and body weight in disorders, such as diabetes mellitus, Cushing's, and Addison's disease, be produced by altered NPY function? Although the data are fragmentary, there are some findings that support each of these possibilities.

10.1. Obesity

There is a convergence of evidence consistent with the possibility that NPY or PYY might mediate some forms of obesity. One line of evidence is that repeated administration of NPY or PYY produces overeating and obesity. Moreover, the magnitude of the effect is dramatic. As shown in Fig. 7, animals given repeated PVN injections of NPY doubled their daily food intake, exhibited a sixfold increase in their rate of body-weight gain, and tripled body fat over the 10-d injection period *(14)*. Peptide YY repeatedly injected icv for 2 d also increased total food intake over this period *(13)*. The magnitude of these effects demonstrates that NPY and PYY are more powerful than any other known neurochemical inducer of obesity; in fact, they appear to be as powerful in inducing obesity as any known factor. Thus, NPY and PYY may be capable of mediating even the most extreme and rapid weight gains produced experimentally or encountered in nature.

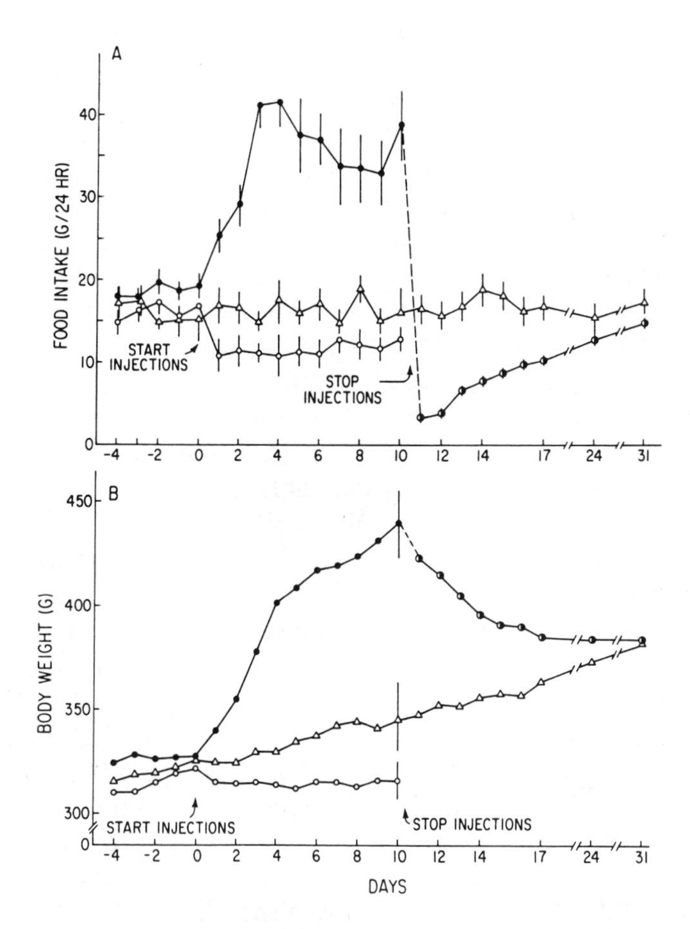

Fig. 7. Daily food intake and body wt (mean ± SEM) of three groups of female rats prior to, during, and after bilateral PVN injections (every 8 h) of either NPY, vehicle, or no injection. (From *[14]*. Reprinted with permission from *Peptides*.) ● Neuropeptide Y; Δ no injection; O saline.

An interesting effect produced by NPY is that the most rapid weight gains occurred initially, when the increments in food intake were smallest. As shown in Fig. 7, the weight gains were most rapid during the initial 4 d of injection, whereas the increases in daily food intake during the first 2 d were actually relatively small. In fact, the rate of body-weight gain during the first 2 d was more rapid than could have been produced just by the increased food intake. This suggested that NPY, besides producing overeating, might in some way be acting to increase metabolic efficiency and/or fat disposition. This is an interesting possibility in view of the evidence that metabolic efficiency can increase during dieting-induced weight loss and that enhanced metabolic efficiency might contribute to obesity *(157)*. Direct examination of this issue has not yet yielded a definitive answer. On the one hand, measures of both total energy expenditure and behavioral activity after PVN injection of NPY did not reveal any effect *(110)*. (This contrasts with NPY injection into sulcal prefrontal cortex, which decreased energy expenditure while increasing food intake *[35]*). On the other hand, NPY has been shown to reduce the thermogenic activity of brown adipose tissue *(156,171)* and produce hypothermia, suggesting that it might reduce energy expenditure.

Additional support for NPY as a mediator of obesity is that hypothalamic expression of this peptide is enhanced in rats made obese by exposure to a highly palatable diet *(174)*, as well as in at least two different forms of genetic obesity *(67,69,85,175)*. For example, obese Zucker rats, compared to their lean litter mates, exhibited higher levels of both tissue NPY and NPY-gene expression in the arcuate nucleus, and higher levels of NPY itself in the PVN *(67,69,85)*. Also, evidence suggests that Zucker fatty rats are insensitive to the eating-stimulatory effects of exogenous NPY *(167)*. These results, in combination with those showing that NPY injection can produce overeating and obesity, suggest that increased hypothalamic NPY in the Zucker obese rats might contribute to their overeating and obesity. This possibility is supported by the similar patterns of eating behavior exhibited by Zucker obese and NPY-injected rats. It has been shown that, like NPY-injected rats, the increased food intake in these obese rats is due to their consumption of abnormally large meals. Also, like NPY-injected rats, the Zucker-obese rats overeat carbohydrate diets *(147)*. Besides these behavioral parameters, these genetically obese animals exhibit several other physiological derangements that also can be produced by injection of NPY. Specifically, like rats given PVN injections of NPY *(94–96,109)*, they exhibit increased levels of circulating insulin and CORT *(159,160)*. Thus, several lines of evidence are consistent with a role for NPY in mediating some forms of obesity.

10.2. Bulimia and Anorexia Nervosa

Two less prevalent eating disorders in humans are bulimia and anorexia nervosa. Bulimia is characterized by binges of eating during which massive quantities of food are consumed and then purged. Anorexia nervosa is characterized by severely restricted eating with extreme and occasionally fatal body-weight loss. The possibility that NPY and/or PYY might play a role in the pathogenesis of bulimia was first suggested by Morley et al. *(13)*, based on their finding that repeated icv injections of PYY caused marked increases in daily food intake. We have recently characterized the pattern of eating produced by NPY in the PFH, and found that, like bulimics, the animals injected with a high dose of this peptide eat a massive meal *(11,12)*. This demonstrates that NPY can produce a pattern of behavior that, in terms of the quantity of food eaten, is like that of bulimics. Thus, NPY is the first neurochemical that can produce bulimic-like behavior.

There is also neurochemical evidence consistent with a role for pancreatic polypeptides in bulimia. Specifically, it has been shown that there are marked increases in the levels of endogenous PYY, but not NPY, in cerebrospinal fluid of humans that are bulimic *(161)*. These increases are not present in all conditions; rather, the marked elevation of PYY was present in bulimic patients that were abstaining from binging, and these levels normalized when these patients were allowed to binge eat. To account for these findings, it was suggested that excessive hypothalamic PYY contributes to the binge eating and that feedback from the ingested food corrects the abnormally high levels of PYY. These finding are perhaps the clearest evidence for an involvement of endogenous brain PP in eating disorders in humans.

In terms of an involvement of NPY and/or other PP in anorexia nervosa, that group of investigators has also shown that the CSF levels of NPY are elevated in underweight anorectic patients and that these elevated levels diminish as the body weight of these patients returns to near normal levels during treatment *(162)*. This evidence may indicate that brain NPY is sensitive to factors related to body weight in humans, and that both NPY and PYY may be involved in control of eating behavior and body-weight control.

10.3. Glucocorticoids, Cushing's Disease, and Addison's Disease

Neuropeptide Y may contribute to the disorders of eating behavior associated with abnormal levels of glucocorticoids. One of these disorders is Cushing's disease, which is caused by an excess in circulating levels of glucocorticoids, resulting in overeating and increased fat deposition. Another is Addison's disease, which is caused by a deficiency in circulating levels of glucocorticoids with consequent anorexia and body-weight loss. We have proposed that excess levels of glucocorticoids in Cushing's disease may enhance the activity of hypothalamic NPY, thereby generating an increase in eating behavior. We have also suggested that the low levels of circulating glucocorticoids in Addison's disease may reduce the activity of endogenous NPY, or the sensitivity to NPY, thereby reducing food intake *(142)*. These possibilities were suggested by the evidence for extensive interactions among circulating glucocorticoids, hypothalamic NPY, and eating behavior. Briefly, PVN injection of NPY induces a marked release of ACTH and glucocorticoids *(94,95,109)*, presumably by acting on CRF-containing neurons of the PVN that express glucocorticoid-binding sites and appear to receive substantial innervation by NPY-containing neurons *(94)*. It has also been shown that adrenalectomy produces a reduction in NPY gene expression within the arcuate nucleus that is reversed by CORT *(74,135)*, and that increased levels of glucocorticoids can enhance NPY levels in vivo or in vitro *(140)*. These effects may be mediated by the abundant glucocorticoid-binding sites on the cell bodies of neurons containing NPY *(138,139)*. Also, it has been shown that the NPY-induced eating response is dependent on glucocorticoids *(128,142)*. Specifically, hypophysectomy or adrenalectomy reduces the eating elicited by NPY, and this response is normalized in animals with CORT replacement *(142)*.

10.4. Diabetes

There is also evidence for a relationship between hypothalamic NPY and circulating insulin that may be involved in the overeating seen in diabetic rats. As has been described earlier, icv and hypothalamic injections of NPY produce an increase in circulating levels of insulin *(96,149,150)*. Conversely, alterations in circulating insulin levels can produce marked changes in hypothalamic NPY levels. Specifically, it has been shown that levels of NPY and/or NPY mRNA are increased in whole hypothalamus, and in specific hypothalamic nuclei of rats made diabetic with streptozotocin *(65,66,70,77)*. Some animal models of genetic diabetes also exhibit increased hypothalamic NPY *(137)*, whereas others exhibit decreased hypothalamic levels of endogenous NPY *(136)*. These changes in

NPY were reversed, at least in part, by insulin replacement *(66,70,137)*. The effect of insulin on NPY may be the result of a direct effect on the brain, since direct third ventricular injection of insulin produces significant changes in NPY mRNA in the arcuate nucleus of the hypothalamus *(141,172)*, whereas peripheral injections do not *(86)*. These findings showing altered levels of NPY in diabetic rats led to the suggestion that the increased hypothalamic NPY in diabetic animals caused the overeating in these animals *(77)*. Although this seems highly plausible given that the hyperphagia in diabetic animals *(148)*, like that produced by NPY injection *(163)*, is characterized by a marked increase in consumption of carbohydrates, this idea is questioned by findings that diabetic rats do not eat in response to PFH injection of NPY *(97)*.

11. Summary and Conclusions

The first decade since the discovery of NPY by Tatemoto et al. *(164)* has yielded dramatic advances in understanding its functions and mechanisms of action. One of the possibilities that has garnered much attention is that NPY may be involved in the control of eating behavior. Much of the evidence for this suggestion has been detailed in this chapter. To review, the key ideas are that NPY, within specific hypothalamic nuclei, is involved in controlling neural systems that stimulate eating behavior, that regulate levels of metabolic fuels, and that control energy expenditure. The primary locus of NPY's control of the autonomic and endocrine systems is proposed to be the PVN, whereas the primary site mediating the feeding-stimulatory effects is proposed to be the PFH. Within these hypothalamic areas, the activity of NPY appears to be sensitive to factors associated with low levels of available metabolic fuels and to alterations in the levels of their regulatory hormones. Neuropeptide Y appears to induce eating via receptors that may be a variant of the Y1 subtype. Activation of these receptors produces a prolonged eating response that may be mediated by G proteins and intracellular second messengers. The eating response itself is focused on ingestion of carbohydrate diets, perhaps because they most rapidly increase levels of usable circulating nutrients. Finally, alterations in the activity of endogenous NPY are likely to contribute to some disorders of eating behavior and body-weight control. It is expected that developments in this area of research over the next decade should not only clarify the role of NPY in control of eating behavior, but may also culminate in the eventual discovery of treatments for disorders, such as bulimia, anorexia nervosa, and obesity.

Acknowledgments

I would like to thank Timothy J. Bartness for reviewing an earlier version of this manuscript. Research was supported by NIH NS 24268.

References

1. Leibowitz, S. F. and Stanley, B. G. (1986) Neurochemical controls of appetite, in *Feeding Behavior: Neural and Humoral Controls* Ritter, R. C., Ritter, S., and Barnes, C. D., eds., Academic, Orlando, FL, pp. 191–234.
2. Kimmel, J. R., Hayden, L. J., and Pollock, H. G. (1975) Isolation and characterization of a new pancreatic polypeptide hormone. *J. Biolog. Chem.* **250,** 9369–9676.
3. Lundberg, J. M., Terenius, L., Hokfelt, T., and Tatemoto, K. (1984) Comparative immunohistochemical and biochemical analysis of pancreatic polypeptide-like peptides with special reference to presence of neuropeptide Y in central and peripheral neurons. *J. Neurosci.* **4,** 2376–2386.
4. Miyachi, Y., Jitsuishi, W., Miyoshi, A., Fujita, S., Mizuchi, A., and Tatemoto, K. (1986) The distribution of polypeptide YY-like immunoreactivity in rat tissues. *Endocrinology* **118,** 2163–2167.
5. Adrian, T. E., Allen, J. M., Bloom, S. R., Ghatei, M. A., Rossor, M. N., Crow, T. J., Tatemoto, K., and Polak, J. M. (1983) Neuropeptide Y distribution in human brain. *Nature* **306,** 584–586.
6. Allen, Y. S., Adrian, T. E., Allen, J. M., Tatemoto, K., Crow, T. J., Bloom, S. R., and Polak, J. M (1983) Neuropeptide Y distribution in the rat brain. *Science* **221,** 877–879.
7. Clark, J. T., Kalra, P. S., Crowley, W. R., and Kalra, S. P. (1984) Neuropeptide Y and human pancreatic polypeptide stimulate feeding behavior in rats. *Endocrinology* **115,** 427–429.
8. Stanley, B. G. and Leibowitz, S. F. (1984) Neuropeptide Y: stimulation of feeding and drinking by injection into the paraventricular nucleus. *Life Sci.* **35,** 2635–2642.
9. Stanley, B. G. and Leibowitz, S. F. (1985) Neuropeptide Y injected in the paraventricular hypothalamus: a powerful stimulant of feeding behavior. *Proc. Natl. Acad. Sci. USA* **82,** 3940–3943.
10. Stanley, B. G., Daniel, D. R., Chin, A. S., and Leibowitz, S. F. (1985) Paraventricular nucleus injections of peptide YY and neuropeptide Y preferentially enhance carbohydrate ingestion. *Peptides* **6,** 1205–1211.
11. Stanley, B. G. and Thomas, W. J. (1990) Patterns of eating behavior elicited by neuropeptide Y injected into the medial perifornical hypothalamus. *Soc. Neurosci. Abstr.* **16,** 773.

12. Stanley, B. G., Thomas, W. J., and Bivens, C. L. (1990) Gorging of food induced by intrahypothalamic injection of neuropeptide Y. Tenth Annual Bristol-Myers Squibb/Mead Johnson Symposium on Nutrition Research, Toronto, Canada.
13. Morley, J. E., Levine, A. S., Grace, M., and Kneip, J. (1985) Peptide YY (PYY), a potent orexigenic agent. *Brain Res.* **341,** 200–203.
14. Stanley, B. G., Kyrkouli, S. E., Lampert, S., and Leibowitz, S. F. (1986) Neuropeptide Y chronically injected into the hypothalamus: a powerful neurochemical inducer of hyperphagia and obesity. *Peptides* **7,** 1189–1192.
15. Grandison, L., and Guidotti, A. (1977) Stimulation of food intake by muscimol and beta endorphin. *Neuropharmacology* **16,** 533–536.
16. McLean, S. and Hoebel, B. G. (1982) Opiate and norepinephrine-induced feeding from the paraventricular nucleus of the hypothalamus are dissociable. *Life Sci.* **31,** 2379–2382.
17. Leibowitz, S. F., Roossin, P., and Rosenn, M. (1984) Chronic norepinephrine injection into the hypothalamic paraventricular nucleus produces hyperphagia and increased body weight in the rat. *Pharmacol. Biochem. Behav.* **21,** 801–808.
18. Vaccarino, F. J., Bloom, F. E., Rivier, J., Vale, W., and Koob, G. F. (1985) Stimulation of food intake in rats by centrally administered hypothalamic growth hormone-releasing factor. *Nature* **314,** 167–168.
19. Leibowitz, S. F. (1986) Brain monoamines and peptides: role in the control of eating behavior. *Fed. Proc.* **45,** 1396–1403.
20. Levine, A. S. and Morley, J. E. (1984) Neuropeptide Y: a potent inducer of consummatory behavior in rats. *Peptides* **5,** 1025–1029.
21. Leibowitz, S. F. (1978) Paraventricular nucleus: a primary site mediating adrenergic stimulation of feeding and drinking. *Pharmacol. Biochem. Behav.* **8,** 163–175.
22. Lundberg, J. M. and Tatemoto, K. (1982) Pancreatic polypeptide family. (APP, BPP, NPY, and PYY) in relation to sympathetic vasoconstriction resistant to α-adrenoceptor blockade. *Acta Physiol. Scand.* **116,** 393–402.
23. Hokfelt, T., Lundberg, J. M., Tatemoto, K., Mutt, V., Terenius, L., Polak, J. M., Bloom, S. R., Sasek, C., Elde, R., and Goldstein, M. (1983) Neuropeptide Y (NPY)- and FMRF-amide neuropeptide-like immunoreactivities in catecholamine neurons of the rat medulla oblongata. *Acta Physiol. Scand.* **117,** 315–318.
24. Stanley, B. G., Leibowitz, S. F., Eppel, N., St-Pierre, S., and Hoebel, B. G. (1985) Suppression of norepinephrine-elicited feeding by neurotensin: evidence for behavioral, anatomical and pharmacological specificity. *Brain Res.* **343,** 297–304.
25. Morris, Y. A. and Crews, D. (1990) The effects of exogenous neuropeptide Y on feeding and sexual behavior in the red-sided garter snake *(Thamnophis sirtalis parietalis). Brain Res.* **530,** 339–341.
26. Kuenzel, W. J., Douglass, L. W., and Davison, B. A. (1987) Robust feeding following central administration of neuropeptide Y or peptide YY in chicks, *Gallus domesticus. Peptides* **8,** 823–828.
27. Denbow, D. M., Duke, G. E., and Chaplin, S. B. (1988) Food intake, gastric secretion, and motility as affected by avian pancreatic polypeptide administered centrally in chickens. *Peptides* **9,** 449–454.

28. Morley, J. E., Hernandez, E. N., and Flood, J. F. (1987b) Neuropeptide Y increases food intake in mice. *Am. J. Physiol.* **253,** R516–R522.
29. Kulkosky, P. J., Glazner, G. W., Moore, H. D., Low, C. A., and Woods, S. C. (1989) Neuropeptide Y: Behavioral effects in the golden hamster. *Peptides* **9,** 1389–1393.
30. Parrott, R. F., Heavens, R. P., and Baldwin, B. A. (1986) Stimulation of feeding in the satiated pig by intracerebroventricular injection of neuropeptide Y. *Physiol. Behav.* **36,** 523–525.
31. Miner, J. L., Della-Fera, M. A., Paterson, J. A., and Baile, C. A. (1989) Lateral cerebroventricular injection of neuropeptide Y stimulates feeding in sheep. *Am. J. Physiol.* **257,** R383–R387.
32. Capuano, C. A., Barr, G. A., and Leibowitz, S. F. (1986) Neuropeptide Y and cholecystokinin: effects on independent feeding in preweanling rats via hypothalamic stimulation. *East. Psychol. Assoc. Abstr.* **57,** 57.
33. Steinman, J. L., Gunion, M. W., and Morley, J. E. (1987) Effects of bilateral partial midbrain transections on feeding and drinking induced by neuropeptide Y (NPY) in rats. *Soc. Neurosci. Abstr.* **13,** 1172.
34. Corp, E. S., Melville, L. D., Greenberg, D., Gibbs, J., and Smith, G. P. (1990) Effect of fourth ventricular neuropeptide Y and peptide YY on ingestive and other behaviors. *Am. J. Physiol.* **259,** R317–R323.
35. McGregor, I. S., Menendez, J. A., and Atrens, D. M. (1990) Metabolic effects of neuropeptide Y injected into the sulcal prefrontal cortex. *Brain Res. Bull.* **24,** 363–367.
36. Chronwall, B. M., DiMaggio, D. A., Massari, V. J., Pickel, V. M., Ruggiero, D. A., and O'Donohue, T. L. (1985) The anatomy of neuropeptide Y-containing neurons in the rat brain. *Neuroscience* **15,** 1159–1181.
37. Harrington, M. E., Nance, D. M., and Rusak, B. (1985) Neuropeptide Y immunoreactivity in the hamster geniculo-suprachiasmatic tract. *Brain Res. Bull.* **15,** 465–472.
38. de Quidt, M. E. and Emson, P. C. (1986) Distribution of neuropeptide Y-like immunoreactivity in the rat central nervous system-II. Immunohistochemical analysis. *Neuroscience* **18,** 545–618.
39. Abe, M., Saito, M., and Shimazu, T. (1990) Neuropeptide Y in the specific hypothalamic nuclei of rats treated neonatally with monosodium glutamate. *Brain Res. Bull.* **24,** 289–291.
40. Sahu, A., Kalra, S. P., Crowley, W. R., and Kalra, P. S. (1988) Evidence that NPY-containing neurons in the brainstem project into selected hypothalamic nucleus: implication in feeding behavior. *Brain Res.* **457,** 376–378.
41. Bai, F. L., Yamano, M., Shiotani, Y., Emson, P. C., Smith, A. D., Powell, J. F., and Tohyama, M. (1985) An arcuato-paraventricular and -dorsomedial hypothalamic neuropeptide Y-containing system which lacks noradrenaline in the rat. *Brain Res.* **331,** 172–175.
42. Sawchenko, P. E., Swanson, L. W., Grzanna, R., Howe, P. R. C., Bloom, S. R., and Polak, J. M. (1985) Co-localization of neuropeptide Y-immunoreactivity in brainstem catecholaminergic neurons that project to the paraventricular nucleus of the hypothalamus. *J. Comp. Neurol.* **241,** 138–153.
43. Gray, T. S., O'Donohue, T. L., and Magnuson, D. J. (1986) Neuropeptide Y innervation of amygdaloid and hypothalamic neurons that project to the dorsal vagal complex in rat. *Peptides* **7,** 341–349.

44. Leibowitz, S. F. (1988) Hypothalamic paraventricular nucleus: interaction between α_2-noradrenergic system and circulating hormones and nutrients in relation to energy balance. *Neurosci. Biobehav. Rev.* **12,** 101–109.
45. Stanley, B. G., Chin, A. S., and Leibowitz, S. F. (1985) Feeding and drinking elicited by central injection of neuropeptide Y: evidence for a hypothalamic site(s) of action. *Brain Res. Bull.* **14,** 521–524.
46. Morley, J. E., Levine, A. S., Gosnell, B. A., Kneip, J., and Grace, M. (1987) Effect of neuropeptide Y on ingestive behaviors in the rat. *Am. J. Physiol.* **252,** R599–R609.
47. Stanley, B. G., Magdalin, W., Serafi, A., Thomas, W. J., and Leibowitz, S. F. (1993) The perifornical area: The major focus of a patchily distributed hypothalamic neuropeptide Y sensitive feeding system(s). *Brain. Res.* (in press).
48. Paxinos, G. and Watson, C. (1986) The Rat Brain in Stereotaxic Coordinates, 2nd ed. Academic, Orlando, FL.
49. Quirion, R., Martel, J.-C., Dumont, Y., Cadieux, A., Jolicoeur, F., St-Pierre, S., and Fournier, A. (1990) Neuropeptide Y receptors: Autoradiographic distribution in the brain and structure-activity relationship. *Ann. NY Acad. Sci.* **611,** 58–72.
50. Leibowitz, S. F. (1975) Amphetamine: Possible site and mode of action for producing anorexia in the rat. *Brain Res.* **84,** 160–167.
51. Leibowitz, S. F., and Brown, L. L. (1980) Histochemical and pharmacological analysis of catecholaminergic projections to the perifornical hypothalamus in relation to feeding behavior. *Brain Res.* **201,** 315–345.
52. Ahlskog, J. E., and Hoebel, B. G. (1973) Overeating and obesity from damage to a noradrenergic system in the brain. *Science* **182,** 166–169.
53. Myers, R. D. and McCaleb, M. L. (1980) Satiety signal from intestine triggers brain's noradrenergic mechanisms. *Science* **209,** 1035–1037.
54. Jhanwar-Uniyal, M. and Leibowitz, S. F. (1986) Impact of food deprivation on α_1- and α_2-noradrenergic receptors in the paraventricular nucleus and other hypothalamic areas. *Brain Res. Bull.* **17,** 889–896.
55. Jhanwar-Uniyal, M., Darwish, M., Levin, B. E., and Leibowitz, S. F. (1987) Alterations in catecholamine levels and turnover in discrete brain areas after food deprivation. *Pharmacol. Biochem. Behav.* **26,** 271–275.
56. Roberts, W. W. (1980) [^{14}C] Deoxyglucose mapping of first-order projections activated by stimulation of lateral hypothalamic sites eliciting gnawing, eating, and drinking in rats. *J. Comp. Neurol.* **194,** 617–638.
57. Kyrkouli, S. E., Stanley, B. G., Hutchinson, R., Seirafi, R. D., and Leibowitz, S. F. (1990) Peptide-amine interactions in the paraventricular hypothalamus: analysis of galanin and neuropeptide Y in relation to feeding. *Brain Res.* **521,** 185–191.
58. Shimizu, H. and Bray, G. A. (1989) Effects of neuropeptide Y on norepinephrine and serotonin metabolism in rat hypothalamus in vivo. *Brain Res. Bull.* **22,** 945–950.
59. Sahu, A., Kalra, P. S., and Kalra, S. P. (1988) Food deprivation and ingestion induce reciprocal changes in neuropeptide Y concentrations in the paraventricular nucleus. *Peptides* **9,** 83–86.
60. Calza, L., Giardino, L., Battistini, N., Zanni, M., Galetti, S., Protopapa, F., and Velardo, A. (1989) Increase of neuropeptide Y-like immunoreactivity in the paraventricular nucleus of fasting rats. *Neurosci. Lett.* **104,** 99–104.

61. Beck, B., Jhanwar-Uniyal, M., Burlet, A., Chapleur-Chateau, M., Leibowitz, S. F., and Burlet, C. (1990) Rapid and localized alterations of neuropeptide Y (NPY) in discrete hypothalamic nuclei with feeding status. *Brain Res.* **528,** 245–249.
62. Kalra, S. P., Dube, M. G., Fournier, A., and Kalra, P. S. (1991) Structure-function analysis of stimulation of food intake by neuropeptide Y: effects of receptor agonists. *Physiol. Behav.* **50,** 5–9.
63. Kalra, S. P., Dube, M. G., Sahu, A., Phelps, C. P., and Kalra, P. S. (1991) Neuropeptide Y secretion increases in the paraventricular nucleus in association with increased appetite for food. *Proc. Natl. Acad. Sci. USA* **88,** 10,931–10,935.
64. Abe, M., Saito, M., Ikeda, H., and Shimazu, T. (1991) Increased neuropeptide Y content in the arcuato-paraventricular hypothalamic neuronal system in both insulin-dependent and non-insulin-dependent rats. *Brain Res.* **539,** 223–227.
65. Williams, G., Gill, J. S., Lee, Y. C., Cardoso, H. M., Okpere, B. E., and Bloom, S. R. (1989) Increased neuropeptide Y concentrations in specific hypothalamic regions of streptozocin-induced diabetic rats. *Diabetes* **38,** 321–327.
66. Sahu, A., Sninsky, C. A., Kalra, P. S., and Kalra, S. P. (1990) Neuropeptide-Y concentration in microdissected hypothalamic regions and *in vitro* release from the medial basal hypothalamus-preoptic area of streptozotocin-diabetic rats with and without insulin substitution therapy. *Endocrinology* **126,** 192–198.
67. Beck, B., Burlet, A., Nicolas, J.-P., and Burlet, C. (1990) Hypothalamic neuropeptide Y (NPY) in obese Zucker rats: implications in feeding and sexual behavior. *Physiol. Behav.* **47,** 449–453.
68. Jhanwar-Uniyal, M., Beck, B., Burlet, C., and Leibowitz, S. F. (1990) Diurnal rhythm of neuropeptide Y-like immunoreactivity in the suprachiasmatic, arcuate and paraventricular nuclei and other hypothalamic sites. *Brain Res.* **536,** 331–334.
69. Sanacora, G., Kershaw, M., Finkelstein, J. A., and White, J. D. (1990) Increased hypothalamic content of preproneuropeptide Y messenger ribonucleic acid in genetically obese Zucker rats and its regulation by food deprivation. *Endocrinology* **127,** 730–737.
70. White, J. D., Olchovsky, D., Kershaw, M., and Berelowitz, M. (1990) Increased hypothalamic content of preproneuropeptide-Y messenger ribonucleic acid in streptozotocin-diabetic rats. *Endocrinology* **126,** 765–772.
71. Brady, L. S., Smith, M. A., Gold, P. W., and Herkenham, M. (1990) Altered expression of hypothalamic neuropeptide Y mRNAs in food-restricted and food-deprived rats. *Neuroendocrinology* **52,** 441–447.
72. Chua, S. C., Leibel, R. L., and Hirsch, J. (1991) Food deprivation and age modulate neuropeptide gene expression in the murine hypothalamus and adrenal gland. *Mol. Brain Res.* **9,** 95–101.
73. O'Shea, R. D. and Gundlach, A. L. (1991) Preproneuropeptide Y messenger ribonucleic acid in the hypothalamic arcuate nucleus of the rat is increased in food deprivation or dehydration. *J. Neuroendocrinol.* **3,** 11–14.
74. Dean, R. G. and White, B. D. (1990) Neuropeptide Y expression in rat brain: Effects of adrenalectomy. *Neurosci. Lett.* **114,** 339–344.

75. Della-Fera, M.A., Koch, J., Gingerich, R. L., and Baile, C. A. (1990) Intestinal infusion of a liquid diet alters CCK and NPY concentrations in specific brain areas of rats. *Physiol. Behav.* **48,** 423–428.
76. Beck, B., Stricker-Krongrad, A., Burlet, A., Nicolas, J.-P., and Burlet, C. (1990) Influence of diet composition on food intake and hypothalamic neuropeptide Y (NPY) in the rat. *Neuropeptides* **17,** 197–203.
77. Williams, G., Steel, J. H., Cardoso, H., Ghatei, M. A., Lee, Y. C., Gill, J. S., Burrin, J. M., Polak, J. M., and Bloom, S. R. (1988) Increased hypothalamic neuropeptide Y concentrations in diabetic rat. *Diabetes* **37,** 763–772.
78. McKibbin, P. E., Rogers, P., and Williams, G. (1991) Increased neuropeptide-Y concentrations in the lateral hypothalamic area of the rat after the onset of darkness: possible relevance to the circadian periodicity of feeding behavior. *Life Sci.* **48,** 2527–2533.
79. Agnati, L. F., Fuxe, K., Zoli, M., Zini, I., Taffano, G., and Ferraguti, F. (1986) A correlation analysis of the regional distribution of central enkephalin and β-endorphin immunoreactive terminals and of opiate receptors in adult and old male rats. Evidence for the existence of two main types of communication in the central nervous system: the volume transmission and the wiring transmission. *Acta Physiol. Scand.* **128,** 201–207.
80. Zoli, M., Agnati, L. F., Fuxe, K., and Bjelke, B. (1989) Demonstration of NPY transmitter receptor mismatches in the central nervous system of the male rat. *Acta Physiol. Scand.* **135,** 201–202.
81. Pelletier, G., Guy, J., Allen, Y. S., and Polak, J. M. (1984) Electron microscope immunocytochemical localization of neuropeptide Y (NPY) in the rat brain. *Neuropeptides* **4,** 319–324.
82. Sawchenko, P. E. and Pfeiffer, S. W. (1988) Ultrastructural localization of neuropeptide Y and galanin immunoreactivity in the paraventricular nucleus of the hypothalamus in the rat. *Brain Res.* **474,** 231–245.
83. Corder, R., Pralong, F. P., Muller, A. F., and Gaillard, R. C. (1990) Regional distribution of neuropeptide Y-like immunoreactivity in human hypothalamus measured by immunoradiometric assay: possible influence of chronic respiratory failure on tissue levels. *Neuroendocrinology* **51,** 23–30.
84. Krieger, D. T. (1974) Food and water restriction shifts corticosterone, temperature, activity and brain amine periodicity. *Endocrinology* **95,** 1195–1201.
85. Pesonen, U., Rouru, J., Huupponen, R., and Koulu, M. (1991) Effects of repeated administration of mifepristone and 8-OH-DPAT on expression of preproneuropeptide Y mRNA in the arcuate nucleus of obese Zucker rats. *Mol. Brain Res.* **10,** 267–272.
86. Corrin, S. E., McCarthy, H. D., McKibbin, P. E., and Williams, G. (1991) Unchanged hypothalamic neuropeptide Y concentrations in hyperphagic, hypoglycemic rats: evidence for specific metabolic regulation of hypothalamic NPY. *Peptides* **12,** 425–430.
87. Kyrkouli, S. E., Sierafi, R. D., Stanley, B. G., and Leibowitz, S. F. (1989) Paraventricular hypothalamic lesions attenuate galanin-induced feeding. *East. Psychol. Assoc. Abstr.* **60,** 24.
88. Swanson, L. W. and Sawchenko, P. E. (1983) Hypothalamic integration: organization of the paraventricular and supraoptic nuclei. *Ann. Rev. Neurosci.* **6,** 269–324.

89. Steffens, A. B., Scheurink, A. J. W., Luiten, P. G. M., and Bohus, B. (1988) Hypothalamic food intake regulating areas are involved in the homeostasis of blood glucose and plasma FFA levels. *Physiol. Behav.* **44,** 581–589.
90. Tempel, D. L. and Leibowitz, S. F. (1990) Diurnal variations in the feeding responses to norepinephrine, neuropeptide Y and galanin in the PVN. *Brain Res. Bull.* **25,** 821–825.
91. Stanley, B. G. and Thomas, W. J. (1993) Feeding responses to perifornical hypothalamic injection of neuropeptide Y in relation to circadian rhythms of eating behavior. *Peptides* (in press).
92. Humphreys, G. A., Davison, J. S., and Veale, W. L. (1988) Injection of neuropeptide Y into the paraventricular nucleus of the hypothalamus inhibits gastric acid secretion in the rat. *Brain Res.* **456,** 241–248.
93. Egawa, M., Yoshimatsu, H., and Bray, G. A. (1991) Neuropeptide Y suppresses sympathetic activity to interscapular brown adipose tissue in rats. *Am. J. Physiol.* **260,** R328–R334.
94. Wahlestedt, C., Skagerberg, G., Ekman, R., Heilig, M., Sundler, F., and Håkanson, R. (1987) Neuropeptide Y (NPY) in the area of the hypothalamic paraventricular nucleus activates the pituitary-adrenocortical axis in the rat. *Brain Res.* **417,** 33–38.
95. Leibowitz, S. F., Sladek, C., Spencer, L., and Tempel, D. (1988) Neuropeptide Y, epinephrine and norepinephrine in the paraventricular nucleus: stimulation of feeding and the release of corticosterone, vasopressin and glucose. *Brain Res. Bull.* **21,** 905–912.
96. Abe, M., Saito, M., and Shimazu, T. (1989) Neuropeptide Y and norepinephrine injected into the paraventricular nucleus of the hypothalamus activate endocrine pancreas. *Biomed. Res.* **10,** 431–436.
97. Stanley, B. G., Yee, S. M., Rosenthal, M. J., and Gunion, M. W. (1991) Eating elicited by perifornical hypothalamic neuropeptide Y injection is abolished in streptozotocin-diabetic rats. *Soc. Neurosci. Abstr.* **17,** 544.
98. Wahlestedt, C., Yanaihara, N., and Håkanson, R. (1986) Evidence for different pre- and post-junctional receptors for neuropeptide Y and related peptides. *Regul. Peptides* **13,** 307–318.
99. Fuhlendorff, J., Gether, U., Askerlund, L., Langeland-Johansen, N., Thogersen, H., Melberg, S. G., Bang Olsen, U., Thastrup, O., and Schwartz, T. W. (1990) [Leu31,Pro34] Neuropeptide Y: A specific Y1 receptor agonist. *Proc. Natl. Acad. Sci. USA* **87,** 182–186.
100. Cox, H. M. and Krstenansky, J. L. (1991) The effects of selective amino acid substitution upon neuropeptide Y antisecretary potency in rat jejunum mucosa. *Peptides* **12,** 323–327.
101. Magdalin, W., Stanley, B. G., Fournier, A., and Leibowitz, S. F. (1989) A structure-activity analysis of neuropeptide Y-induced eating behavior. *Soc. Neurosci. Abstr.* **15,** 895.
102. Jolicoeur, F. B., Michaud, J.-N., Menard, D., and Fournier, A. (1991b) In vivo structure activity study supports the existence of heterogeneous neuropeptide Y receptors. *Brain Res. Bull.* **26,** 309–311.
103. McLaughlin, C. L., Tou, J. S., Rogan, R. J., and Baile, C. A. (1991) Full amino acid sequence of centrally administered NPY required for maximal food intake response. *Physiol. Behav.* **49,** 521–526.

104. Paez, X., Nyce, J. W., and Myers, R. D. (1991) Differential feeding responses evoked in the rate by NPY and NPY_{1-27} injected intracerebroventricularly. *Pharmacol. Biochem. Behav.* **38,** 379–384.
105. Leibowitz, S. F. and Alexander, J. T. (1991) Analysis of neuropeptide Y-induced feeding: Dissociation of Y_1 and Y_2 receptor effects on natural meal patterns. *Peptides* **12,** 1251–1260.
106. Colmers, W. F., Klapstein, G. F., Fournier, A., St-Pierre, S., and Treherne, K. A. (1991) Presynaptic inhibition by neuropeptide Y in rat hippocampal slice in vitro is mediated by a Y2 receptor. *Br. J. Pharmacol.* **102,** 41–44.
107. Rioux, F., Bachelard, H., Martel, J.-C., and St-Pierre, S. (1986) The vasoconstrictor effect of neuropeptide Y and related peptides in the guinea pig isolated heart. *Peptides* **7,** 27–31.
108. Donoso, V., Silva, M., St-Pierre, S., and Huidobro Toro, J. P. (1988) Neuropeptide Y (NPY), an endogenous presynaptic modulator of adrenergic neurotransmission in the rat vas deferens: structural and functional studies. *Peptides* **9,** 545–553.
109. Albers, H. E., Ottenweller, J. E., Liou, S. Y., Lumpkin, M. D., and Anderson, E. R. (1990) Neuropeptide Y in the hypothalamus: effect on corticosterone and single-unit activity. *Am. J. Physiol.* **258,** R376–R382.
110. Menendez, J. A., McGregor, I. S., Healey, P. A., Atrens, D. M., and Leibowitz, S. F. (1990) Metabolic effects of neuropeptide Y injections into the paraventricular nucleus of the hypothalamus. *Brain Res.* **516,** 8–14.
111. Shibata, S. and Moore, R. Y. (1988) Neuropeptide Y and vasopressin effects on rat suprachiasmatic nucleus neurons *in vitro*. *J. Biol. Rhythms* **3,** 265–276.
112. Mason, R., Harrington, M. E., and Rusak, B. (1987) Electrophysiological responses of hamster suprachiasmatic neurones to neuropeptide Y in the hypothalamic slice preparation. *Neurosci. Lett.* **80,** 173–179.
113. Perney, T. M. and Miller, R. J. (1989) Two different G-proteins mediate neuropeptide Y and bradykinin-stimulated phospholipid breakdown in cultured rat sensory neurons. *J. Biol. Chem.* **264,** 7317–7327.
114. Chance, W. T., Sheriff, S., Foley-Nelson, T., Fischer, J. E., and Balasubramaniam, A. (1989) Pertussis toxin inhibits neuropeptide Y-induced feeding in rats. *Peptides* **10,** 1283–1286.
115. Rowland, N. and Antelman, S. M. (1976) Stress-induced hyperphagia and obesity in rats: A possible model for understanding human obesity. *Science* **191,** 310–312.
116. Albers, H. E., and Ferris, C. F. (1984) Neuropeptide Y: Role in light-dark cycle entrainment of hamster circadian rhythms. *Neurosci. Lett.* **50,** 163–168.
117. Harfstrand, A. (1987) Brain neuropeptide Y mechanisms. Basic aspects and involvement in cardiovascular and neuroendocrine regulation. *Acta Physiol. Scand.* **[Suppl.] 565,** 1–83.
118. Westfall, T. C., Martin, J., Chen, X., Ciarleglio, A., Carpentier, S., Henderson, K., Knuepfer, M., Beinfeld, M., and Naes, L. (1988) Cardiovascular effects and modulation of noradrenergic neurotransmission following central and peripheral administration of neuropeptide Y. *Synapse* **2,** 299–307.
119. Jolicoeur, F. B., Michaud, J.-N., Rivest, R., Menard, D., Gaudin, D., Fournier, A., and St-Pierre, S. (1991) Neurobehavioral profile of neuropeptide Y. *Brain Res. Bull.* **26,** 265–268.

120. Levine, A. S., Kuskowski, M. A., Grace, M., and Billington, C. J. (1991) Food deprivation-induced feeding: a behavioral evaluation. *Am. J. Physiol.* **260,** R546–R552.
121. Antin, J., Gibbs, J., Holt, J., Young, R. C., and Smith, G. P. (1975) Cholecystokinin elicits the complete behavioral sequence of satiety in rats. *J. Comp. Psychol. Psychol.* **87,** 784–790.
122. Deutch, J. A. (1990) Food intake: gastric factors, in *Handbook of Behavioral Neurobiology: Neurobiology of Food and Fluid Intake.* Stricker, E. M., ed., Plenum, New York, pp. 151–182.
123. Clark, J. T., Kalra, S. P., and Kalra, P. S. (1986) Neuropeptide Y (NPY)-induced suppression of male sexual behavior: Neural loci. *1st Internatl. Congr. Neuroendocrinol.* **1,** 107.
124. Tseng, C. J., Mosqueda-Garcia, R., Appalsamy, M., and Robertson, D. (1989) Cardiovascular effects of neuropeptide Y in rat brainstem nuclei. *Circulation Res.* **64,** 55–61.
125. Harfstrand, A. (1986b) Intraventricular administration of neuropeptide Y (NPY) induces hypotension, bradycardia and bradypnea in the awake unrestrained male rat. Counteraction by NPY-induced feeding behavior. *Acta Physiol. Scand.* **128,** 121–123.
126. Stricker, E. M. (1990) Homeostatic origins of ingestive behavior, in *Handbook of Behavioral Neurobiology: Neurobiology of Food and Fluid Intake.* Stricker, E. M., ed., Plenum, New York, pp. 151–182.
127. Jewett, D. C., Levine, A. S., Cleary, J., Schall, D. W., and Thompson, T. (1989) The effect of intracerebroventricular (ICV) NPY on food reinforced behavior. *Soc. Neurosci. Abstr.* **15,** 894.
128. Kalra, S. P., Dube, M. G., and Kalra, P. S. (1988) Continuous intraventricular infusion of neuropeptide Y evokes episodic food intake in satiated female rats: effects of adrenalectomy and cholecystokinin. *Peptides* **9,** 723–728.
129. Richter, C. P. (1943) Total self-regulatory function in animals and human beings. *Harvey Lect.* **38,** 63–103.
130. Blundell, J. E. (1983) Problems and processes underlying the control of food selection and nutrient intake, in *Nutrition and the Brain.* Wurtman, R. J. and Wurtman, J. J., eds., Raven, New York, pp. 164–221.
131. Leibowitz, S. F., Weiss, G. F., Yee, F., and Tretter, J. R. (1985) Noradrenergic innervation of the paraventricular nucleus: specific role in control of carbohydrate ingestion. *Brain Res. Bull.* **14,** 561–567.
132. Steffens, A. B. (1969) Rapid absorption of glucose in the intestinal tract of the rat after ingestion of a meal. *Physiol. Behav.* **4,** 829–832.
133. Liberatore, L., Deindorfer, B., Stanley, B. G., and Leibowitz, S. F. (1986) Eating elicited by hypothalamic injection of neuropeptide Y: The role of sensory and nutritive factors. *East. Psychol. Assoc. Abstr.* **57,** 60.
134. Rowland, N. E. (1988) Peripheral and central satiety factors in neuropeptide Y-induced feeding in rats. *Peptides* **9,** 989–992.
135. White, B. D., Dean, R. G., and Martin, R. J. (1990) Adrenalectomy decreases neuropeptide Y mRNA levels in arcuate nucleus. *Brain Res. Bull.* **25,** 711–715.
136. Williams, G., Ghatei, M. A., Diani, A. R., Gerritsen, G. C., and Bloom, S. R. (1988b) Reduced hypothalamic somatostatin and neuropeptide Y concentrations in the spontaneously-diabetic chinese hamster. *Horm. Metabol. Res.* **20,** 668–670.

137. Williams, G., Lee, Y. C., Ghatei, M. A., Cardoso, H. M., Ball, J. M., Bone, A. J., Baird, J. D., and Bloom, S. R. (1989) Elevated neuropeptide Y concentrations in the central hypothalamus of the spontaneously diabetic BB/E Wistar rat. *Diabetic Med.* **6,** 601–607.
138. Hisano, S., Kagotain, Y., Tsuruo, T., Daikoku, S., Chihara, K., and Whitnall, M. H. (1988) Localization of glucocorticoid receptor in neuropeptide Y-containing neurons in the arcuate nucleus of the rat hypothalamus. *Neurosci. Lett.* **95,** 13–18.
139. Harfstrand, A., Cintra, A., Fuxe, K., Aronsson, M., Wikstrom, A.-C., Okret, S., Gustafsson, J. A., and Agnati, L. F. (1989) Regional differences in glucocorticoid receptor immunoreactivity among neuropeptide Y immunoreactive neurons of the rat brain. *Acta Physiol. Scand.* **135,** 3–9.
140. Corder, R., Pralong, F., Turnill, D., Saudan, P., Muller, A. F., and Gaillard, R. C. (1988) Dexamethasone treatment increases neuropeptide Y levels in rat hypothalamic neurons. *Life Sci.* **43,** 1879–1886.
141. Schwartz, M. W., Marks, J. L., Sipols, A. J., Baskin, D. G., Woods, S. C., Kahn, S. E., and Porte, D. (1991) Central insulin administration reduces neuropeptide Y mRNA expression in the arcuate nucleus of food-deprived lean (Fa/Fa) but not obese (fa/fa) Zucker rats. *Endocrinology* **128,** 2645–2647.
142. Stanley, B. G., Lanthier, D., Chin, A. S., and Leibowitz, S. F. (1989) Suppression of neuropeptide Y-elicited eating by adrenalectomy or hypophysectomy: reversal with corticosterone. *Brain Res.* **501,** 32–36.
143. Tempel, D. L., Shor-Posner, G., Dwyer, D., and Leibowitz, S. F. (1989) Nocturnal patterns of macronutrient intake in freely feeding and food-deprived rats. *Am. J. Physiol.* **256,** R541–R548.
144. Krieger, D. and Hauser, H. (1978) Comparison of synchronization of circadian corticosteroid rhythms by photoperiod and food. *Proc. Natl. Acad. Sci. USA* **75,** 1577–1581.
145. Tempel, D. L. and Leibowitz, S. F. (1989) PVN steroid implants: Effect on feeding patterns and macronutrient selection. *Brain Res. Bull.* **23,** 553–560.
146. Leibowitz, S. F. (1990) Hypothalamic neuropeptide Y in relation to energy balance. *Ann. NY Acad. Sci.* **611,** 284–301.
147. Enns, M. P. and Grinker, J. A. (1983) Dietary self-selection and meal patterns of obese and lean Zucker rats. *Appetite* **4,** 281–293.
148. Koopmans, H. S. and Pi-Sunyer, F. X. (1986) Large changes in food intake in diabetic rats fed high-fat and low-fat diets. *Brain Res. Bull.* **17,** 861–871.
149. Moltz, J. H. and McDonald, J. K. (1985) Neuropeptide Y: direct and indirect action on insulin secretion in the rat. *Peptides* **6,** 1155–1159.
150. Kuenzel, W. J. and McMurtry, J. (1988) Neuropeptide Y: Brain localization and central effects on plasma insulin levels in chicks. *Physiol. Behav.* **44,** 669–678.
151. Rettori, V., Milenkovic, L., Aguila, M. C., and McCann, S. M. (1990) Physiologically significant effect of neuropeptide Y to suppress growth hormone release by stimulating somatostatin discharge. *Endocrinology* **126,** 2296–2301.
152. LeMagnen, J. (1981) The metabolic basis of dual periodicity of feeding in rat. *Behav. Brain Sci.* **4,** 561–607.
153. Takahashi, Y. (1979) Growth hormone secretion related to the sleep and waking rhythm, in *The Functions of Sleep* Drucker-Colin, R., Shkurovich, M., and Sterman, M. B., eds., Academic, New York, pp. 113–145.

154. Clark, J. T., Kalra, P. S., and Kalra, S. P. (1985) Neuropeptide Y stimulates feeding but inhibits sexual behavior in rats. *Endocrinology* **117,** 2435–2442.
155. Harfstrand, A., Fuxe, K., Agnati, L. F., Eneroth, P., Zini, I., Zoli, M., Andersson, K., von Euler, G., Terenius, L., Mutt, V., and Goldstein, M. (1986) Studies on neuropeptide Y-catecholamine interactions in the hypothalamus and in the forebrain of the male rat. Relationship to neuroendocrine function. *Neurochem. Int.* **8,** 355–376.
156. Billington, C. J., Briggs, J. E., Grace, M., and Levine, A. S. (1991) Effects of intracerebroventricular injection of neuropeptide Y on energy metabolism. *Am. J. Physiol.* **260,** R321–R327.
157. Brownell, K. D., Greenwood, M. R. C., Stellar, E., and Shrager, E. E. (1986) The effects of repeated cycles of weight loss and regain in rats. *Physiol. Behav.* **38,** 459–464.
158. Brief, D. J., Sipols, A., Ginter, K., Amend, D., and Woods, S. C. (1987) Intraventricular neuropeptide Y injections stimulate food intake in lean, but not obese Zucker rats. *Soc. Neurosci. Abstr.* **13,** 586.
159. Zucker, L. M. and Antoniades, H. N. (1972) Insulin and obesity in the Zucker genetically obese rat "fatty." *Endocrinology* **90,** 1320–1330.
160. Fletcher, J. M., Haggarty, P., Wahle, K. W. J., and Reeds, P. J. (1986) Hormonal studies of young lean and obese Zucker rats. *Horm. Metabol. Res.* **18,** 530–536.
161. Berrettini, W. H., Kaye, W. H., Gwirtsman, H., and Allbright, A. (1988) Cerebrospinal fluid peptide YY immunoreactivity in eating disorders. *Neuropsychobiol.* **19,** 121–124.
162. Kaye, W. H., Berrettini, W., Gwirtsman, H., and George, D. T. (1990) Altered cerebrospinal fluid neuropeptide Y and peptide YY immunoreactivity in anorexia and bulimia nervosa. *Arch. Gen. Psychiat.* **47,** 548–556.
163. Stanley, B. G., Anderson, K. C., Grayson, M. H., and Leibowitz, S. F. (1989b) Repeated hypothalamic stimulation with neuropeptide Y increases daily carbohydrate and fat intake and body weight gain in female rats. *Physiol. Behav.* **46,** 173–177.
164. Tatemoto, K., Carlquist, M., and Mutt, V. (1982) Neuropeptide Y—a novel brain peptide with structural similarities to peptide YY and pancreatic polypeptide. *Nature* **296,** 659–662.
165. Aramakis, V. B., Ashe, J. H., Juranek, J., Lomeli, L. M., Taneja, A., and Stanley, B. G. (1992) Differential action of neuropeptide Y and subtype agonists on single unit activity in the paraventricular hypothalamus. *Soc. Neurosci. Abstr.* **18,** 988.
166. Beck, B., Burlet, A., Nicolas, J.-P., and Burlet, C. (1992) Unexpected regulation of hypothalamic neuropeptide Y by food deprivation and refeeding in the Zucker rat. *Life Sci.* **50,** 923–930.
167. Brief, D. J., Sipols, A. J., and Woods, S. C. (1992) Intraventricular neuropeptide Y injections stimulate food intake in lean, but not obese Zucker rats. *Physiol. Behav.* **51,** 1105–1110.
168. Gillard, E. R., Dang, D. Q., and Stanley, B. G. (1992) Dopamine in the perifornical hypothalamus attenuates neuropeptide Y-induced feeding. *Soc. Neurosci. Abstr.* **18,** 743.

169. Jewett, D. C., Cleary, J., Levine, A. S., Schaal, D. W., and Thompson, T. (1992) Effects of neuropeptide Y on food-reinforced behavior in satiated rats. *Pharmacol. Biochem. Behav.* **42,** 207–212.
170. Pich, M. E., Messori, B., Zoli, M., Ferraguti, F., Marrama, P., Biagini, G., Fuxe, K., and Agnati, L. F. (1992) Feeding and drinking responses to neuropeptide Y injections in the paraventricular hypothalmic nucleus of aged rats. *Brain. Res.* **575,** 265–271.
171. Roscoe, A. K. and Myers, R. D. (1991) Hypothermia and feeding induced simultaneously in rats by perfusion of neuropeptide Y in preoptic area. *Pharmacol. Biochem. Behav.* **39,** 1003–1009.
172. Schwartz, M. W., Sipols, A. J., Marks, J. L., Sanacora, G., White, J. D., Scheurink, A., Kahn, S. E., Baskin, D. G., Woods, S. C., Figlewicz, D. P., and Porte, D. (1992) Inhibition of hypothalamic neuropeptide Y gene expression by insulin. *Endocrinology* **130,** 3608–3616.
173. Stanley, B. G., Magdalin, W., Seirafi, A., Nguyen, M. M., and Leibowitz, S. F. (1992) Evidence for neuropeptide Y mediation of eating produced by food deprivation and for a variant of the Y1 receptor mediating this peptide's effect. *Peptides* **13,** 581–587.
174. Wilding, J. P. H., Gilbey, S. G., Mannan, M., Aslam, N., Ghatei, M. A., and Bloom S. R. (1992) Increased neuropeptide Y content in individual hypothalamic nuclei, but not neuropeptide Y mRNA, in diet-induced obesity in rats. *J. Endocrinol.* **132,** 299–304.
175. Williams, G., Shellard, L., Lewis, D. E., McKibbin, P. E., McCarthy, H. D., Koeslag, D. G., and Russell, J. C. (1992) Hypothalamic neuropeptide Y disturbances in the obese (cp/cp) JCR: LA corpulent rat. *Peptides* **13,** 537–540.

Neuropeptide Y in Relation to Behavior and Psychiatric Disorders

Some Animal and Clinical Observations

Markus Heilig

1. Introduction

1.1. Psychiatry, Biology, and Animal Studies

Before proceeding to review and discuss "hard" observations on NPY in relation to behavior and psychiatric disorders, allow for some introductory remarks of a more theoretical nature. Basic scientists sometimes have difficulties fully comprehending the degree of controversy evoked in clinical settings by the reduction of mental processes to biological phenomena. A modern school of philosophy of science (*see*, e.g., *1*) claims such different frameworks as, e.g., those of neurobiology and psychoanalytic psychology to be theoretically impossible to reconcile or "incommensurable." I beg to differ. The postulated incommensurability is destructive, artificial, and unnecessary. Release and action of neurotransmitters, or the flow of primitive id impulses and the activation of specific ego defenses, can all be seen as different ways of modeling inherently monistic processes of the human mind. These different models may be useful at different levels of description or within different domains, and they may or may not be possible to map onto each other. Much can be won when they are integrated. Finding and applying isomorphisms between psychological and biological models is likely to enrich both. Animal models of anxiety,

From: *The Biology of Neuropeptide Y and Related Peptides;*
W. F. Colmers and C. Wahlestedt, Eds.

used in the work described below and widely used by other investigators, are based on a conflict model of anxiety that, nomenclature aside, is virtually identical to that used by psychoanalytic psychology (*see*, e.g., 2). Integration, it turns out, can be productive.

1.2. Neuropeptides

Monoaminergic transmission has been of obvious interest to psychiatry for almost four decades. This field now seems to witness important breakthroughs. Monoaminergic receptors are cloned and their subtypes characterized in manners never possible before *(3–5)*, making it possible to search for drugs with more precise targets than ever. Neuropeptides, meanwhile, are newcomers in a psychiatrist's vocabulary. In addition, because of their molecular nature, they are relatively more difficult to study. Yet there are several reasons to believe that studying peptidergic systems will turn out to be important for our understanding of the function and malfunction of the brain. One of these reasons is the phylogenetic conservation of genes for a number of neuropeptides. Genetic material being retained under selectional pressure suggests that the corresponding gene product is important for the function of the organism. As an example, the gene for NPY, the most abundant neuropeptide of the brain, seems indeed to be highly conserved through the phylogenetic scale (*see* Section 1.3.1., *see also* Chapter 1).

Further, peptide messengers often act as long-term modulators of neuronal function. NPY seems to adhere to this rule. Psychiatric diseases, as for instance illustrated by the psychomotor inhibition of major depression, are states, not singular events. Underlying pathophysiological mechanisms are therefore more likely to be found among systems employing long-acting, state-determining compounds. Drugs designed for treatment of psychiatric disorders may therefore also be better targeted at these long-acting, "gain-setting" systems.

Finally, difficulties posed by the chemical properties of neuropeptides are appreciable. Neuropeptide analogs active on central receptors on peripheral administration can, however, be found (*see*, e.g., *6* on anxiolytic properties of a class of synthetic CCK-B receptor antagonists). The findings presented below suggest that the accomplishment of a similar task with respect to NPY could offer a valuable new therapeutic principle.

1.3. NPY

Shortly following its isolation *(6a)*, numerous observations pointed to NPY as a candidate for regulating aspects of behavior and being involved in certain psychiatric disorders. The following observations are covered in detail elsewhere in this book, but will here be collected and restated from that perspective.

1.3.1. Genetics

The sheer amount of NPY present in the brain makes the peptide interesting. The synthesis of the 97-a.a. preproNPY requires a considerable energy expenditure on the part of the neuron and would thus constitute a clear selectional disadvantage were the end product not important for the organism. Yet, the peptide is present in amounts that on a molar basis exceed those of other known neuropeptides (*see*, e.g., *7*). Also, as mentioned above, genes for NPY-like peptides seem to have been highly conserved throughout evolution and have been replicated, giving rise to a family of related peptides. Gene products that on the basis of sequence homology and/or crossreactivity, belong to the NPY superfamily are present in a number of lower species. In protostomian invertebrates, PP immunoreactivity is found in neurons and endocrine cells, whereas NPY immunoreactivity is lacking *(8)*. In three fish species, the angler fish *(9,10)*, the daddy sculpin *(11)*, and the Pacific salmon *(12)*, members of the NPY superfamily are found. These have been sequenced, and all three display a much higher degree of sequence homology with PYY and NPY than with any PP. These findings indicate that the "ancestor gene" of the NPY superfamily has its origin before the evolution of protostomia and deuterostomia. The gene duplications giving rise to different members of the PP-family are more recent events, taking place after the divergence of evolutionary lines leading to fish and mammals (that is, some 350 million years ago). Thereafter, the gene for NPY seems to have been highly conserved, indicating a functional importance of the peptide.

1.3.2. NPY Affects the Regulation of Circadian Rhythms

One of the core symptoms of major depression is a dysregulation of circadian rhythms. A number of physiological as well as behavioral mechanisms in the mammalian brain follow circadian (approx 24 h) rhythms. These are kept up by a dual mechanism: In the absence of light/darkness stimuli, internal pacemaker activity generates cycles slightly longer or shorter than 24 h (so-called "free-running rhythms"). These are normally reset or "entrained" by the external light/darkness cycle. The principal endogenous pacemaker identified in the mammalian brain is located in the hypothalamic suprachiasmatic nucleus (SCN). The SCN is innervated by two visual pathways that may mediate the entrainment of the endogenously generated cycle: the retinohypothalamic tract, a direct projection from retinal ganglion cells, and a secondary projection from the thalamic VLGN *(13)*. NPY is present in the latter pathway. When NPY (or aPP) is injected into the SCN of hamsters kept under constant light, the free-running rhythm of the animals is shifted. The direction of the shift is dependent on the time of injection. Administration of NPY during the 12-h interval preceding the daily onset of locomotor activity produces an earlier onset of the activity phase ("phase advance"), whereas injection during the 12 h following the activity phase produces opposite effects ("phase delay"). The length of the activity phase is not affected. These effects are identical to those produced by pulses of darkness during daylight time. The effects seem to be highly localized, since they are not reproduced by intracerebroventricular (icv) injection of NPY *(14,15)*. Interestingly, the ability of fetal SCN transplants to restore the generation of endogenous, free-running rhythms in SCN-lesioned animals seems to correlate with the presence of vasopressin, VIP, and/or NPY in the integrated graft *(16)*. Finally, in a hamster hypothalamic slice preparation, the firing rate of SCN neurons has been reported to increase on application of NPY *(17)*.

1.3.3. NPY Affects the Hypothalamic–Pituitary–Adrenal (HPA) Axis

Another component of the depressive syndrome is a dysregulation of the HPA axis, with hypercortisolism and abolished diurnal variation in the secretion of cortisol. NPY has complex effects on glucocorticoid regulation. Injection of NPY into the hypothalamic PVN was early reported to produce an increase in plasma adrenocorticotropic hormone (ACTH) and corticosterone levels both in anesthetized and in conscious, freely moving rats *(18)*. When the dose range was extended downward *(19)*, icv injection of low NPY doses suppressed corticosterone secretion, whereas higher doses increased both ACTH and corticosterone levels. NPY-induced increase in ACTH and corticosterone release may be mediated through a modulation of corticotropin-releasing factor (CRF) secretion and action, since NPY fibers innervate CRF cells of the PVN *(20)* and NPY administration increases hypothalamic CRF levels *(21)*. In fact, NPY seems to interact with CRF transmission at two levels. The release of CRF seems to be stimulated by NPY *(22)*. Although these authors suggested this effect to be direct, mediation through an α_2-adrenergic pathway has been suggested by others *(23)*. In addition, the postsynaptic effect of CRF seems to be potentiated by NPY. In the dog, the ACTH response to NPY is partially blocked by a CRF antagonist, whereas a subthreshold dose of NPY potentiates the effects of exogenous CRF *(24)*.

1.3.4. NPY Affects Food Intake

Another core symptom of major depression is a dysregulation of food intake. Ordinarily, this dysregulation is expressed as an anorexia and followed by weight loss, but in subgroups of "atypically" depressed, mostly young subjects, it is instead expressed as overeating. Rats (and a number of other species) injected icv with NPY or hPP eat excessively. The effect of NPY is probably mediated through the hypothalamic PVN. The magnitude of NPY-induced feeding is higher than of that induced by any pharmacological agent previously tested and is extremely long lasting (*see*, e.g., *25–27*). NPY-induced stimulation of feeding has been reproduced in a number of species (e.g., *28,29*). Among the three basic macronutrients (fat, protein, and carbohydrate), the intake of carbohydrates is preferentially stimulated *(30)*. Drinking is affected in a much less consistent manner, which in addition varies with species (e.g., *28,30,31*). No tolerance is developed toward the orexigenic effect of NPY, and when administration of the peptide was repeated over 10 days, a marked increase in the rate of weight gain was observed *(32)*. Following starvation, the concentration of NPY in the hypothalamic PVN increases with time and returns rapidly to control levels following food ingestion *(33)*. This indicates that effects of NPY on food intake may be representative of inherent physiological mechanisms. The effects of NPY on feeding are discussed in a separate chapter. In addition, some further psychiatric implications of these findings are in addition discussed below (Section 3.5.).

1.3.5. NPY Coexists and Cooperates with Norepinephrine (NE)

In a number of variations on the theme, central NE systems have been implicated in the pathogenesis of major depressive illness *(34,35)*. A fact that may have received smaller attention is that NE systems of the brain also have been implicated in the pathophysiology of anxiety. α_2-adrenoceptor antagonists induce anxiety, or signs thereof, whereas opposite, seemingly anxiolytic effects are produced by agonists. These effects have been observed in humans *(36,37)*, as well as in animals *(38–40)*. Also, antidepressant drugs affecting noradrenergic transmission are of benefit in the treatment of anxiety states *(41,42)*.

In the peripheral nervous system (PNS), NPY is present in virtually all noradrenergic nerve endings and in a majority of medullary adrenal cells. In most commonly employed peripheral model systems, NPY has a threefold action: It potentiates postsynaptic, vasoconstrictive actions of NE, produces some direct postsynaptically mediated vasoconstriction of its own, and inhibits, presynaptically, the release of NE (e.g., 43). In the brain, a partial coexistence of NPY and NE, as well as NPY and epinephrine (E), is present *(44)*. It should, however, be noted that numerous central NPY neurons do not contain any of the biogenic amines, whereas coexistence of NPY and GABA *(45)* and/or NPY and somatostatin *(46)* is abundant in the cerebral cortex, basal ganglia, and the amygdaloid complex. Functionally, NPY produces NA-like effects on hormone release *(47)*, and upregulates α_2-adrenergic binding-sites in the rat brain stem *(48)*.

1.3.6. NPY Is Present in Brain Areas Important for "Emotionality"

The regional distribution of NPY has been determined quantitatively in a number of species, including humans *(7,49)*, nonhuman primates *(50)*, and the rat *(51–53)*. Although numeric data vary up to one order of magnitude, because of differences in extraction procedures and antibodies used, the relative contents of different regions are in good agreement between the studies quoted above. Thus, in the rat, hypothalamic areas, nucleus accumbens, septum, and the periaqueductal gray matter contain the highest levels of NPY. Moderately high amounts are found in cortical regions, including the hippocampus, in the thalamus, and in the basal ganglia. NPY immunoreactivity is low or absent in the pons and the cerebellum. In humans, high levels of NPY are found in the basal ganglia, nucleus accumbens, amygdala, and the hypothalamus. Intermediate amounts are detected in several cortical areas, including the hippocampus, in the septum, and in the periaqueductal gray matter. Lowest levels of NPY are found in the thalamus, globus pallidus, substantia nigra, pons, and cerebellum.

1.3.7. Summary

The observations reviewed above suggested a role for NPY in the pathophysiology of a specific psychiatric disorder: major depressive illness. This suggestion initiated our own line of study. Anxiety being a symptom almost invariably accompanying depression, it came as no surprise that results obtained during the course of the study also indicated a link between NPY and anxiety. The present chapter will concentrate on presenting both animal and human data, obtained by ourselves as well as others, to elucidate the role of NPY in depression and anxiety. In addition, the literature reporting on NPY in connection with other psychiatric and neurological diseases will be reviewed.

2. NPY in Relation to Animal Behavior, Anxiety, and Depression

2.1. Initial Animal Experiments: Spontaneous Activity in the Home Cage and the Open Field, and Effects on Stress-Induced Gastric Ulcers

These early experiments were performed to study behavioral correlates of EEG effects reported by others to be produced by NPY (*see* Section 2.7.). The specific hypothesis to be tested was that NPY may produce sedative effects. This was examined by studying effects of exogenous NPY on basic behavioral parameters of activity in rats. Since the initial observations seemed to support our hypothesis, this hypothesis was also approached in a different manner: Sedative drugs are known to protect against stress-induced gastric ulcers (*see* Section 2.7.). Therefore, the effects of pretreatment with NPY on gastric ulcer formation in a stress model were also studied (cf *54,55*).

NPY was injected into the lateral cerebral ventricle of awake, freely moving rats. Neurological status of the animals was assessed in order to rule out possible toxic effects of the peptide. No gross neurological deficits were detected in rats treated with 2.0 nmol NPY, a dose in the high range of the dose–response curve for other effects of NPY. Locomotor activity in the home cage was recorded using an automated capacitive instrument. This behavioral parameter is believed to mirror the spontaneous locomotor activity of the animals in a familiar environment, as opposed to that observed when faced with novelty (such as that in the open field). NPY greatly suppressed home cage activity independently of the time-point during the light/dark cycle at which it was administered. The effect lasted at least throughout one light/dark cycle, abolishing the normal circadian variation in activity, and was fully reversible after 5 d of recovery. In addition, it was noted that NPY-treated rats started to eat within approx 20 min after administration of NPY.

Open field testing was performed in order to study behavior of the animals when faced with a novel environment. The number of crossings (i.e., the animal crossed a dividing line of the open field), rearings (the animal lifted both its forepaws), and groomings (episodes of repetitive cleaning of the animals fur with any of its paws) was scored. NPY decreased all activity parameters in the open field in a dose-dependent manner. The apparent ED_{50} for this effect was approx 1.0 nmol. On three successive days of treatment, no tolerance to the activity-suppressing effect of NPY was developed. Treated animals returned to control activity levels after 5 d of recovery, showing the effect to be reversible. In addition, a wagging walk and piloerection were observed in NPY-treated animals both in and outside the open field. No other qualitative effects of NPY on open-field behavior could be detected.

When animals are repeatedly exposed to a testing environment, habituation occurs over time. The exploratory drive decreases, and so does spontaneous locomotor activity in the open field. As expected, this was observed in our control animals. By the second day of testing, these had decreased their activity in the open field and remained on that level on subsequent test sessions. NPY-treated animals habituated as rapidly, and to the same extent as controls.

Following pretreatment with NPY, there was a tendency for increased plasma levels of corticosterone on exposure to the novelty stimulus of the open field. This tendency, however, did not reach statistical significance.

As a stress model, water restraint was used. Following food deprivation, animals were restrained in plastic tubes and placed in 19°C for 75 min. After sacrifice, the stomachs were removed and everted, and scores were made of the number and length of any gastric erosion. The total amount of gastric ulceration produced by water-restraint stress was reduced by approx 50% by 2.0 nmol of NPY. Maximal plasma levels (approx 0.86 mg/L) of corticosterone were present after the stress exposure, and these were not additionally affected by NPY treatment.

2.2. NPY, Depressed Patients, and Antidepressant Treatments

In order to address indirectly the hypothesis that NPY may be involved in the pathophysiology of depressive illness, we examined the effects of different antidepressant treatments on cerebral tissue levels of NPY in experimental animals. Since the initial animal findings seemed to support our hypothesis, this notion was further examined by comparing the concentrations of NPY in the CSF of depressed patients with those of healthy controls (cf *56–58*).

2.2.1. Antidepressant Drug Treatment

Imipramine, zimeldine, or saline were given orally (by gastric lavage) twice daily for 3 weeks. The animals were sacrificed 1–2 h after the last treatment. The brains were dissected out, divided into five major regions, extracted, and assayed for NPY-like immunoreactivity. In frontal cortical tissue, imipramine and zimeldine elevated the concentration of NPY by 40 and 60%, respectively. The concentration of NPY in parietal cortical tissue was not affected. In the hypothalamus, imipramine increased the concentration of NPY by approx 65%, whereas zimeldine was without effect.

2.2.2. Electroconvulsive Shock

Electroconvulsive shocks or control treatments were administered once daily for 13 or 14 d. For the induction of seizures, current was delivered through corneal electrodes. Each shock elicited tonic-clonic seizures, which lasted approx 30 s. One control group was handled in the same manner, but no current was delivered. A second control group was handled identically, but received low-current shocks, insufficient for the induction of seizures. A third control group, finally, was not handled at all and kept in a separate room. In both frontal and parietal neocortical tissue, as well as in the hippocampus, ECS increased the concentration of NPY by approx 100% compared to any of the three control groups. There were no differences in tissue NPY concentrations between the different control groups. No effects of ECS on tissue concentrations of NPY were seen in the olfactory bulbs, striatum, hypothalamus, pons, or cerebellum.

2.2.3. CSF Concentrations of NPY in Depressed Patients

Cerebrospinal fluid (CSF) samples were drawn from healthy volunteers and from hospitalized depressed patients. The healthy subjects had no present somatic or psychiatric disease, nor had they any personal or family history of psychiatric illness or drug abuse. Apart from oral contraceptives used by some of the females, no drugs were used by the members of this group. The depressed patients were diagnosed as suffering from major depression following the diagnostic criteria of DSM III. At the time of the lumbar puncture, they had received no antidepressant medication for at least 2 weeks. CSF samples were drawn in the morning. Preceding the lumbar puncture, the occurrence and severity of depressive symptoms were assessed using the Hamilton depression rating scale (HAM-D; *59*). The CSF content of NPY was measured using radioimmunoassay.

Depressed patients had reduced CSF concentrations of NPY in the CSF by 30% when compared with the control subjects (395.3 ± 27.1 vs 582.1 ± 36.2 ng/L, mean ± SEM; $p = 0.006$). No sex differences were observed in any of the groups. No correlation was observed between the total severity of depression as assessed by the total scores on HAM-D and the CSF concentrations of NPY in the individual patients.

2.3. NPY and Anxiety: Animal Anxiety Models and Human Data

Drugs capable of producing sedation often produce anxiolysis when administered in lower doses. Also, anxiety is one of the core symptoms of major depressive illness. In order to examine possible anxiolytic effects of NPY, different animal models were used (cf *60*). On the basis of animal data, data from the population of depressed patients described above were reanalyzed with respect to a possible correlation between anxiety scores and NPY levels in CSF (cf *61*).

2.3.1. The Elevated X-Maze (Montgomery's Conflict Test, MT)

This test is based on principles described by Montgomery *(61a)* and developed further by others *(38–40)*. It is based on the conflict between the exploratory drive of the rat and its fear of elevated open areas. An elevated (1 m above the ground) plus-formed maze was used. Two opposing arms were surrounded by black 10 cm high Plexiglas™ walls (closed arms), whereas the other arms were devoid of walls (open arms). The number of entries made into, as well as the cumulative time spent in the open and the closed arms of the maze, respectively, were recorded during 5-min sessions.

This test has been validated for detection of anxiolytic and anxiogenic drug effects. Normal rats display a preference for the "safe" enclosed arms of the maze, and this preference is increased by anxiogenic, and decreased by anxiolytic drugs *(38–40)*.

In our experiments, saline-treated controls showed the expected preference for the closed arms of the maze, both when the number of entries into, and time spent in arms were assessed. This preference was counteracted by NPY in a dose-dependent manner in the dose range 0.2–5.0 nmol, and was abolished at the highest dose level. Thus, an anxiolytic effect was present (Fig. 1). Also, although the overall activity, as measured by the total number of entries into any arm of the maze, was not affected by the lower doses of NPY, the highest dose level (5.0 nmol) decreased it by approx 50%, indicating significant sedation.

2.3.2. Punished Drinking Test (Vogel's Conflict Test, VT)

A modified Vogel's drinking conflict model was also used to assess possible anxiolytic effects of NPY *(63,64)*. This test is based on the creation of a conflict between the thirst drive of a water-deprived animal and its fear of a mild electric shock. After partial water deprivation, the animals were tested in an operant chamber. When approaching the drinking spout, the rat was allowed to drink for 30 s. On each further attempt to drink, an electric shock was administered. The number of shocks accepted during a ten-min session was recorded.

This test has also been validated for the detection of anxiogenic and anxiolytic drug effects. The number of shocks accepted in order to obtain the drinking solution is decreased by anxiogenic and increased by anxiolytic drugs *(63,64)*.

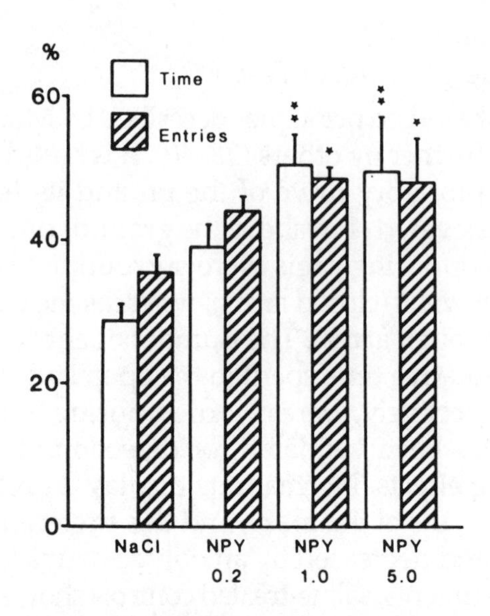

Fig. 1. Behavior during a 5-min session in the Montgomery's conflict test 1 h after icv injection of saline, NPY (0.2–5.0 nmol), or NPY_{13-36} (0.4–2.0 nmol). Bars represent percentage time spent in (open bars) and entries into (hatched bars) the open arms of the maze, ± SEM. $F(41,5) = 5.27$ and 3.50, $p < 0.001$ and 0.010 with respect to treatment effect for time and entries, respectively. * and ** indicate difference vs the saline-treated controls at $\alpha = 0.05$ and 0.01, respectively, on Tukey's HSD test.

The number of shocks accepted in order to obtain the drinking solution was markedly increased by NPY. At the lower two doses, this increase was approximately fivefold. At the highest dose, it was threefold. The effect of NPY was fully antagonized by a dose of the α_2-adrenergic antagonist idazoxan, which did not affect behavior in this model when given alone.

The number of accepted shocks in the Vogel test can also be influenced by drug effects on pain sensitivity or thirst. Therefore, we tested these separately. Neither shock threshold nor drinking motivation was affected by a dose of NPY that under identical conditions produced highly significant "anxiolytic-like" effects in the VT.

2.3.3. Anxiety Levels in Depressed Human Subjects

These were were scored according to the Hamilton Depression Scale (item 10, psychological anxiety, and 11, somatic anxiety). The correlations between scores, as well as their sum, and the CSF concentration of NPY were calculated. Significant negative corre-

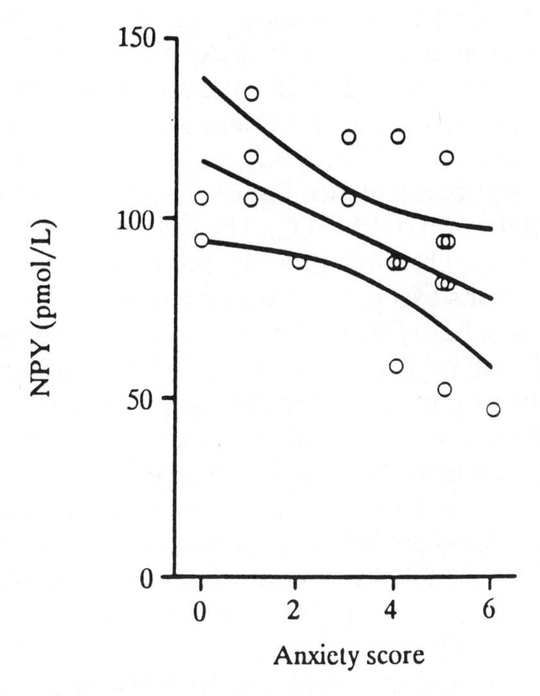

Fig. 2. Correlation between NPY in the CSF of depressed patients and the sum of HAM-D anxiety scores. A 90% confidence interval is indicated. $R = -0.56$; $p = 0.02$.

lations were found between the anxiety scores in the HAM-D and the CSF NPY concentrations of depressed patients (Fig 2.). In the analysis, healthy control subjects were not included. By definition, these did not have a DSMIII diagnosis of major depression and were therefore not scored according to HAM-D. By definition of the inclusion criteria, however, the control subjects were free from psychiatric symptoms, including anxiety. If the controls are included under the assumption that they were free of anxiety symptoms, the correlation becomes 0.64, and the p value < 0.001.

2.4. Effects of NPY on Central Monoaminergic Transmission

In order to study the mode of action by which NPY exerts its behavioral effects, we examined the effects of NPY (0.2–5.0 nmol icv) and its C-terminal 13–36 fragment (NPY_{13-36}, 0.4–10 nmol icv) on brain monoamines and their metabolites in rats (cf *65*).

NPY was injected into awake, freely moving rats. One hour after injection, the animals were decapitated, and the following brain regions were dissected out: frontal cortex, parietal cortex, basal ganglia, hypothalamus, and brain stem. After homogenization and acidic extraction, DA, DOPAC, NA, 5-HT, and 5-HIAA were analyzed using high-performance liquid chromatography with electrochemical detection (HPLC-EC). The NA metabolite MHPG was analyzed using gas chromatographic-mass spectrometry (GC-MS). In all brain regions except the parietal cortex, NPY increased the concentrations of both DA and DOPAC in a dose-dependent manner. In the frontal cortex, the effects of NPY were reproduced by NPY_{13-36}. In the subcortical regions, NPY_{13-36} reproduced the effect of NPY on DOPAC levels, but differed from the parent molecule in leaving DA levels unaltered. NPY decreased tissue levels of NA at higher doses, this in some cases being preceded by increased NA levels at lower doses of NPY. These effects were most pronounced in the parietal cortex and the striatum. In addition, MHPG levels were elevated in a dose-dependent manner by NPY in the frontal cortex. The effects of NPY_{13-36} on NA and MHPG resembled those of NPY. The 5-HT systems were largely unaffected by NPY.

2.5. Interactions Between NPY and α_2-Adrenergic Receptors

Interactions between NPY and α-2 adrenergic receptors were suggested by binding data reported by others (*see* Section 1.3.5.). Also, the behavioral effects of NPY were reminiscent of those produced by the α_2-adrenergic agonist clonidine *(66)*. Therefore, we examined whether the behavioral effects of NPY could be antagonized by different adrenergic blockers (cf *60,67a*).

For pharmacological testing, the open-field activity model was used. Animals were injected with various doses of NPY after pretreatment with phentolamine (nonselective α-adrenergic blocker), idazoxan (α_2-selective), or prazosine (α_1-selective). Since the initial experiments seemed to support the hypothesis that NPY-induced sedation was mediated via an interaction with α_2-adrenoceptors, idazoxan was also tested as pretreatment in one of the animal anxiety models, the VT described above. NPY-induced sedation was partially antagonized in a dose-dependent manner by phentolamine, in doses that did not affect behavior when administered alone. Idazoxan blocked the effect of NPY completely, also in a dose that was inert when given alone. Prazosine was ineffective in a number of doses. NPY-induced anxiolysis in the Vogel test (*see* Section 2.3.2.) was blocked by pretreatment with idazoxan.

2.6. Functional and Neurochemical Evidence for a Heterogeneity of Central NPY Receptors

Observations in peripheral model systems suggested that a heterogeneity of NPY receptors may exist (*see* Section 2.7.). According to these observations, the intact NPY_{1-36} molecule would activate both a Y1- and a Y2-receptor type, whereas NPY_{13-36} would preferentially activate the Y2 subtype of NPY receptors. The following studies were performed in order to examine whether a similar subdivision exists in the brain and, if so, to determine which receptor subtype is involved in the various behavioral effects of NPY (cf *60,65,67a,67b*).

The effects of NPY_{13-36} on open-field activity were examined and compared to those of NPY. NPY_{13-36} did not reproduce the sedative effect of NPY. Instead, the fragment produced an increase in both crossings and rearings by at most approx 35%. This increase followed an inverted-U-shaped dose–response curve. The increase in activity produced by NPY_{13-36} was not affected by pretreatment with phentolamine. The exploratory pattern in animals receiving NPY_{13-36} was one of a marked initial hyperactivity, followed by a rapid decrease in activity. In addition, rats treated with NPY_{13-36} showed no signs of eating when observed in the time between injection and open field testing.

Effects of NPY_{13-36} on conflict behavior were studied in the Vogel test, and the fragment did not affect conflict behavior. The effects of NPY_{13-36} on brain monoamines were also studied and compared to those of NPY. NPY_{13-36} largely reproduced the effects of NPY DA and DOPAC in cortical tissue, but not in subcortical areas.

Finally, the effects of both NPY and NPY_{13-36} on open-field activity were examined in the genetically aberrant spontaneously hypertensive (SH) rat. The rationale for this was that this strain may differ from normal animals in the balance between Y1- and Y2-mediated effects of NPY (*see* Section 2.7.). Contrary to its effects in normal (Sprague Dawley) rats, NPY increased open field activity in SH rats. This effect was dose dependent, and was present both in morning and evening sessions. In the commonly employed control strain, WKy, NPY suppressed locomotor activity in a manner similar to that observed in "normal" rats. This suppression was only detectable in evening sessions, since the basal activity of WKy rats was already low or absent in morning sessions. In preliminary experiments, NPY_{13-36} reproduced the activity-stimulating effects of NPY in the SH rats, although with an approximately fivefold lower potency.

2.7. Discussion

2.7.1. Sedation

The effects of centrally administered NPY on neurological status, home cage activity, and open-field behavior of laboratory rats can be summarized as a dose-dependent induction of a reversible overall activity suppression. This activity suppression is not accompanied by gross neurological deficits. Apart from sedation, a number of alternative interpretations of these findings should be considered, including ataxia or hypotension. Both from the studies of others (e.g., 27) and from observations obtained during the course of our own studies, it is, however, apparent that neurological function and overall motor ability of rats treated centrally with NPY are not markedly impaired. In fact, under certain circumstances (such as when food is present or in a conflict situation), NPY-treated animals express higher locomotor activity than controls. On the other hand, NPY produces effects on the EEG that are typical of sedation *(68)*. Therefore, the author and collaborators favor the interpretation that NPY-induced activity suppression is an expression of centrally mediated sedation.

The observation that NPY protected against stress-induced injury further supported the notion that NPY may produce sedation, since such an action is typical of sedative drugs (*see*, e.g., *69*). Some caution is, however, appropriate with respect to effects of central NPY on gastric function. Like the effects of NPY on ACTH/glucocorticoid secretion, these seem to be highly site dependent. Microinjections of NPY into the hypothalamic PVN markedly reduce the rate of gastric acid secretion, whereas injections into a number of other hypothalamic loci increase it. Both effects are abolished by vagotomy *(70)*. Accordingly, in preliminary experiments, we have recently found that stress-induced gastric erosion formation can be inhibited or enhanced by NPY, depending on injection site within the hypothalamus (Heilig, unpublished data).

In one respect, NPY seems to differ from established sedatives. These drugs usually decrease plasma glucocorticoid levels (e.g., *71*). In our line of experiments, NPY did not affect plasma corticosterone after icv administration, and, if anything, there was a tendency toward increased corticosterone *(55)*. Although this tendency did not reach statistical significance, in a related paper *(18)*, we have shown that localized administration of NPY into the hypothalamic PVN increases both ACTH and corticosterone secretion.

2.7.2. Anxiolysis

Drugs producing sedation are commonly capable of producing anxiolysis as well. We have therefore tested NPY for possible anxiolytic effects in two animal conflict models. The two models employed, the elevated x-maze and the punished drinking test, share the conflict element, but otherwise differ widely. In the punished drinking test, we controlled for interference from possible effects of NPY on pain sensitivity and thirst. In both models, NPY produced effects indistinguishable from well-known anxiolytics. Neither pain sensitivity nor thirst, two possibly confounding variables, were affected by NPY. We therefore conclude that the behavioral effects of NPY observed in these two anxiety models are representative of an anxiolytic action of the peptide.

As commonly seen with sedative–anxiolytic drugs, the anxiolytic action of NPY seems to be exerted at doses lower than those required to produce sedation. In the dose range 0.2–1.0 nmol icv NPY did not affect locomotor activity in the open field, or overall activity in the MT. The same dose range produced maximal anxiolysis in both anxiety models. At 4.0–5.0 nmol, both open-field locomotor activity and overall activity in the MT were reduced by 50% or more, indicating the presence of sedation; in addition, sedation was observed also in the VT, although this test does not provide an independent quantitative measure of this parameter.

In recent experiments *(72)*, we have replicated the finding of an anxiolytic action by NPY in another animal model of anxiety, the Geller-Seifter test. In these experiments, Wistar rats were used (as opposed to Sprague-Dawley animals). Interestingly, no sedative effects were seen at 5 nmol NPY icv in these experiments, and the "anxiolytic" action of NPY increased with dose over the dose range 0.2–5.0 nmol. It is thus possible that the ratio of sedative:anxiolytic dose of NPY may vary between strains of experimental animals.

2.7.3. NPY in Relation to Depression and Anxiety

Decreased CSF levels of NPY have been found in patients with major depression *(58)*. Chronic oral treatment with the antidepressant drugs imipramine and zimeldine markedly elevated the frontal cortical concentrations of NPY. In addition, treatment with imipramine increased hypothalamic levels of NPY. A number of other brain regions were not affected *(56)*. ECS similarly elevated tissue concentrations of NPY in both the frontal and the parietal neocortex, as well as in the hippocampus, a region not separately assessed in the study employing antidepressant drugs. Rats given shocks of insufficient strength for the development of generalized convulsions, as well as sham-treated animals, did not differ from untreated controls *(57)*. It is to be noted that for these three studies (humans, as well as drug treated and ECS treated rats), the same RIA was used. These findings have been replicated in part by others *(73)*. Finally, lithium has been reported to increase cortical and striatal NPY concentrations *(74)*.

Zimeldine is thought to inhibit preferentially the reuptake of 5-HT. Imipramine is rapidly and almost completely metabolized to des-methyl-imipramine (desipramine) after oral administration in the rat, and desipramine is thought to preferentially inhibit the reuptake of NA. The actions of ECS as well as lithium treatment are obviously numerous. In neither case is the mechanism of antidepressant action known. Thus, different treatments known to have an antidepressant effect in common seem to share an ability to increase levels of NPY in the rat brain. This observation, together with the finding of decreased levels of NPY in the CSF of depressed patients *(58)*, makes it tempting to hypothesize that NPY may be involved in the pathophysiology of depressive illness and in the mechanism of action of various antidepressant treatments. More specifically, it could be claimed that major depression involves a deficit of central NPY, whereas antidepressant treatments bring NPY levels back to normal.

Clearly, however, the situation is not nearly so simple. In an early report *(75)*, elevated CSF concentrations of NPY were found in depressed patients, in contrast to the later findings of decreased levels cited above. Further, normal concentrations of NPY in CSF of depressed patients have been reported by others *(76)*. The only obvious difference between the studies has been the use of different antibodies in the RIAs used for analysis of the material. HPLC characterization of the immunoreactive "NPY" detected by the antibody used by Widerlöv et al. *(58)* is shown in Fig. 3, whereas the corresponding data for the antibody used by Berrettini et al. *(76)* are present in the original paper. From a comparison of these profiles, it is evident that the molecular nature of the immunoreactive material detected differs. Notably, the antibody used by Berrettini et al. *(76)* is reported to detect material present as a single peak that coelutes with authentic NPY on HPLC. The antibody used by Widerlöv et al. *(58)* detects several peaks. Thus, the role of NPY in the pathophysiology of depressive illness, if any, does not seem to be a simple deficit. Rather, an altered processing or metabolism of NPY or its precursor may take place in depression, giving rise to different patterns of immunoreactive fragments in CSF of depressed patients and healthy subjects. These fragments will be recognized to different extents depending on which part of the sequence the specific antibody happens to have affinity to. In fact, HPLC analysis shows that the NPY-immunoreactive material from the CSF of depressed patients differs qualitatively from that of controls (*see* Fig. 3).

In order to clarify further the possible role of NPY in depression, valid animal models of depressive illness are desirable. Behavioral deficits regarded as a model of human depression are induced by surgically removing the olfactory bulbs in the laboratory rat *(77)*. In a preliminary experiment, the concentration of NPY in the cerebral cortex of bulbectomized rats was decreased *(78)*. Again, however, seemingly conflicting data are present. In a study by Kataoka and coworkers *(79)*, bulbectomy was reported to increase levels of NPY in the amygdala and the hypothalamus of rats, and this increase was brought back to normal by antidepressant drugs. An HPLC characterization of the immunoreactive material is not available. Therefore, the two seemingly contrary findings may again reflect an alteration in the processing of NPY, as detected by the two different antibodies used. Finally, the ability of drugs to inhibit mouse-killing behavior in bulbectomized rats has been used when screening for antidepressant properties of drugs. In this model, NPY potentiated subthreshold doses of NA, again functionally indicating an "antidepressant" action of the peptide *(80)*.

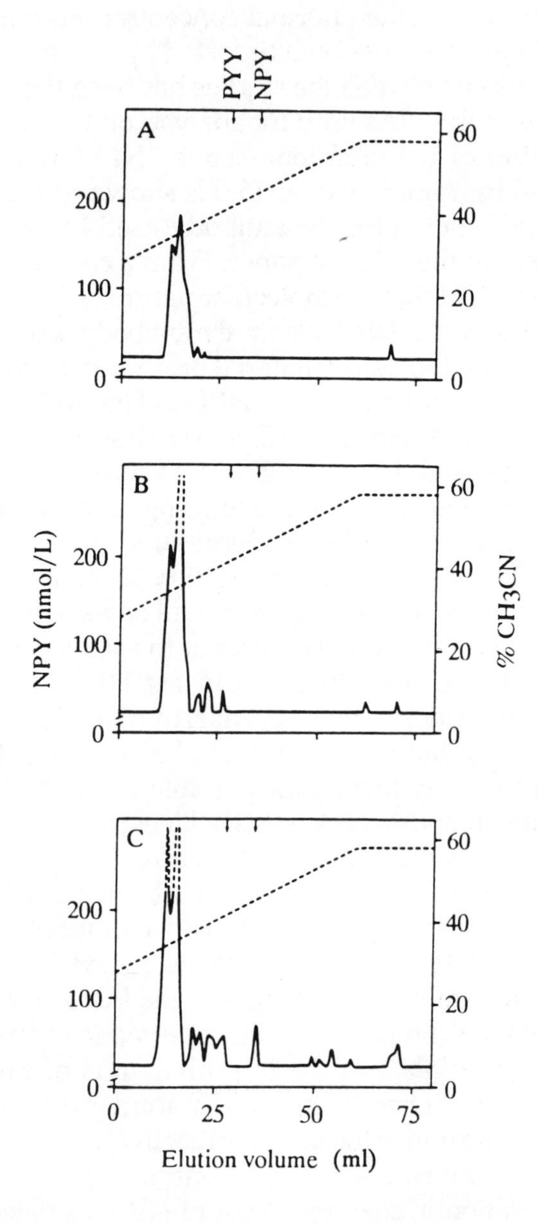

Fig. 3. HPLC profiles of material detected by the NPY antibody in pooled CSF samples from A. controls (125 ± 11 pmol/L); B. depressed patients with "normal" total CSF levels of NPY (125 ± 10 pmol/L); C. depressed patients with reduced NPY levels in the CSF (97 ± 2 pmol/L). The elution positions of synthetic NPY and PYY are indicated by arrows.

Another intriguing line of evidence is emerging to support an important role for NPY in depression. Cocaine withdrawal produces a depressive syndrome in humans. This syndrome is in fact indistinguishable from major depression and responds to conventional antidepressant treatments. Behavioral signs of a depression-like syndrome are also seen in experimental animals during cocaine withdrawal. In fact, cocaine withdrawal has been proposed as an animal model of a central feature of the depressive syndrome, anhedonia (i.e., an inability to perceive pleasure or internal reward) *(81)*.

As described below (*see* section on NPY–DA interactions), NPY given into the nucleus accumbens was found to be rewarding, an effect that was mediated through dopaminergic mechanisms. Conversely, in an elegant study, sustained administration of cocaine was shown to produce a decrease of both NPY and its mRNA. These persisted long after administration had been terminated *(82)*. Thus, a deficit of active NPY in the period of withdrawal from cocaine could account for the inability to experience pleasure, which is at the very core of the depressive syndrome.

2.7.4. Anatomical Sites of Action

The icv administration route of NPY employed in our animal studies may possess a certain advantage: If NPY analogs capable of passing the blood–brain barrier are synthesized and used therapeutically, they will most likely produce a "cerebrosystemic" effect much like that produced by icv administration. Unfortunately, using this route of administration does not clarify the anatomical site(s) at which the observed behavioral effects are produced. With respect to endocrine, feeding-inducing, and memory-modulating effects of NPY, sites within the hypothalamus (hormone secretion and feeding) and the hippocampus (memory) have been suggested on the basis of localized administration, as well as knowledge of anatomical localization of these functions (*see*, e.g., *18,26,83*). With respect to the nonspecific sedation observed in our experiments, it can be speculated on the basis of EEG findings *(68)* that the cerebral cortex is affected. Such a cortical involvement could, however, represent a direct effect on cortical neurons or, e.g., effects on ascending pathways from brain stem noradrenergic projections.

With respect to the anxiolytic and the hypothesized antidepressant effect of NPY, it is similarly unclear whether the effect is, in a broad sense, pre- or postsynaptic. In addition, the possible target areas of interest are not known. Two distinct and testable hypotheses can be put forward against the background of current literature. It has been suggested that an increased activity in the principal brain NA nucleus, the LC, promotes anxiety, whereas decreased LC activity results in anxiolysis *(84)*. Activation of α_2-adrenoceptors on LC cell bodies lowers the firing rate of these neurons *(85)*. Since both the sedative and anxiolytic effects of central NPY seem to be mediated through interactions with α_2-adrenoceptors (*see* Section 2.7.5.), it is tempting to speculate that an α_2-adrenoceptor mediated decrease of LC activity is responsible for at least the anxiolytic action of NPY.

On the other hand, α_2-adrenoceptors are also present in other areas of possible interest for the pathophysiology of anxiety, and some of these receptors are also believed to be postjunctional to NA neurons. A structure of particular interest in this context is the amygdaloid complex. Lesions of the amygdala produce anxiolysis in animal conflict models *(86)*, an effect that is reproduced by direct injection of NA into the nucleus *(80)*. Also, the amygdala seems to be a key structure in the "antidepressant" action of both antidepressant drugs *(87)* and electroconvulsive treatment *(88)*, as observed in the forced swim test; also, as mentioned above, antidepressant drugs as well as NA injected into the medial amygdala inhibit mouse-killing behavior in olfactory bulbectomized rats, another depression model. As mentioned above, NPY potentiates this "antidepressant" effect of subthreshold doses of NA when both drugs are injected locally into the medial amygdala *(80)*, thus providing additional support for the hypothesis that this structure may be the anatomical locus of some behavioral effects induced by central NPY. In recent experiments, the anxiolytic effect of NPY was reproduced by bilateral localized injection of the peptide (50–100 pmol/side) into the central nucleus of the amygdala *(88a)*.

2.7.5. Receptor Mechanisms: Interactions with α-Adrenoceptors

Effects on both the EEG pattern and locomotor activity similar to those produced by NPY are seen after activation of central α_2-adrenergic receptors with clonidine *(66)*. Furthermore, low doses of clonidine have been reported to produce anxiolytic-like effects in animal anxiety models *(40)*. In a number of in vitro models, NPY seems to mimic the effects of α_2-adrenoceptor activation, or potentiate the actions of α_2-agonists. At the same time, these in vitro effects of NPY are blocked by highly selective α_2-adrenergic antagonists. Such effects have been shown in preparations of the medulla oblongata, the hypothalamus, and frontoparietal cortex, both in synaptosomal and slice preparations *(89–92)*.

NPY was early reported to upregulate the number of α_2-adrenergic binding sites in membranes from the medulla oblongata of normal rats *(48)*. Additional effects of NPY on α_2-adrenoceptor protein synthesis, receptor conformation, and receptor–effector coupling have been suggested by this group *(93)*. Recently, NPY has been directly shown to increase hyperpolarizing effects of α_2-agonists selectively on locus coeruleus neurons, thus decreasing the firing rate of these cells *(94)*.

On the basis of the findings summarized above, we hypothesized that NPY-induced sedation and anxiolysis may be the result of an enhancement of α_2-adrenergic neurotransmission. This hypothesis seems substantiated by our experimental findings. Thus, the nonselective α_2-adrenergic receptor antagonist phentolamine, as well as the selective α_2-adrenergic antagonist idazoxan, antagonized the sedative action of NPY, whereas the α_1-antagonist prazosine was ineffective in this respect. Although not equally extensively studied, the anxiolytic-like effect of NPY was also blocked by idazoxan.

Finally, NPY-induced upregulation of α_2-binding sites is absent in the SH rat *(95)*. Accordingly, we have been able to show that NPY does not produce sedation in this strain. Similar observations have also been made by others, who also observed that the anxiolytic action of NPY was retained in the absence of sedation (Merlo Pich, submitted to *Regulatory Peptides*). Interestingly, it has recently been shown that SH rats exhibit lower levels of anxiety in behavioral tests *(96)*.

2.7.6. Receptor Mechanisms: Heterogeneity of Central NPY Receptors

Using different fragments of NPY and PYY, evidence for at least two pharmacologically different populations of NPY receptors has been obtained in experiments on the peripheral sympathetic neuroeffector junction *(97)*. Thus, both the modulatory and the direct vasoconstrictive effects of NPY on blood vessels seem to require the full sequence of NPY or the related PYY. Both these effects have been suggested to represent a postsynaptic action of the peptides. On the other hand, the inhibitory effect of NPY (or PYY) on the contractile response of the rat vas deferens to low-frequency electrical stimulation is reproduced by the 13–36 a.a. fragment of NPY (or PYY). This inhibitory action is thought to reflect a presynaptic inhibition of NA release. The putative post- and presynaptic NPY receptors thus suggested were termed NPY-Y1 and Y2, respectively *(98)*.

At present, NPY_{13-36} remains the most selective pharmacological tool available for a preferential activation of the putative NPY-Y2 receptor. We have used NPY_{13-36} in parallel with the intact NPY in order to probe the receptor mechanism underlying the behavioral effects of NPY. NPY_{13-36}, in the dose range 0.4–10.0 nmol icv, did not reproduce the sedative or the anxiolytic effects of the intact NPY molecule. Instead, the fragment stimulated locomotor activity of rats in the open field. Contrary to the effects of intact NPY, this activity stimulant effect of NPY_{13-36} was not blocked by α-adrenergic brocade. In addition, NPY_{13-36} fully reproduced the effects of NPY on both DA and DOPAC levels within the frontal cortex. Within other areas studied (striatum, hypothalamus, and brain stem), however, NPY_{13-36} consistently reproduced only the effects of NPY on tissue concentrations of DOPAC, but not of DA.

It could be argued that an antagonistic or partially agonistic action of NPY_{13-36} on a homogeneous population of NPY receptors would be compatible with our data. Also, the behavioral effects of central NPY_{13-36} could be nonspecific and not related to NPY receptors at all. Recently, however, the postulated Y1 and Y2 receptors have been increasingly well characterized both in terms of binding and second-messenger mechanisms (*see*, e.g., *99–102*), and the fragment NPY_{13-36} or similar C-terminal fragments seem indeed to be preferential agonists on the NPY-Y2 receptor. It therefore seems justified to conclude that the sedative and anxiolytic effects of NPY are mediated through Y1 receptors. In fact, the occurrence of NPY_{12-36} in brain tissue extracts *(102)* further supports the notion that NPY in vivo indeed is processed to molecules with different profiles of action and that a disturbance of this processing may be the mechanism by which NPY is involved in the pathophysiology of depressive illness (*see* Section 2.7.3.).

In the peripheral model system, the rat vas deferens, an intrinsic activity of PYY_{13-36} roughly equal to that of NPY and PYY, has been reported. The apparent affinity of PYY_{13-36}, as reflected by the pD2 value, was approx 50% of that of PYY (8.16 vs 7.53) *(97)*. The relative intrinsic activities and affinities of NPY_{13-36} and NPY are similar to those of PYY_{13-36} in this system (Wahlestedt, personal communication). In the behavioral studies, we have therefore examined effects of NPY_{13-36} at doses twice as high as those of NPY. Presently, we are lacking a specific bioassay system in which to study central effects of NPY-Y2 receptor activation. It is therefore difficult to estimate the relative affinity and intrinsic activity of putative Y2-agonists, including NPY and NPY_{13-36} itself, on this receptor in vivo. Various in vitro estimates of this parameter are given in, e.g., *102* and *99*, but it is difficult to know how these translate to the in vivo situation. Recent experiments suggest that the selective Y_1-receptor agonist, [Leu^{31},Pro^{34}]NPY, is anxiolytic after intra-amygdala injection, while 4-fold higher doses of NPY_{13-36} do not fully reproduce this effect *(103)*.

2.7.7. Effects of NPY on Brain Monoaminergic Systems

Although extensive interactions between NPY and central monoaminergic system seem to be present, they are not easily interpreted in relation to the behavioral effects of NPY.

2.7.8. NA Systems

We have found a general tendency for NPY to decrease NE concentrations after high icv doses, whereas lower doses tend to increase NE. In addition, fronto-cortical levels of MHPG were elevated by NPY in a dose-dependent manner. Icv injection of 0.47 nmol NPY after α-methyl-*p*-tyrosine (α-MPT, 300 mg/kg) has been reported to increase NA levels in the hippocampus, hypothalamus, and brain stem of the rat *(103)*. After similar pretreatment, biphasic effects on hypothalamic NA fluorescence have been reported in the dose range 0.075–0.750 nmol icv. Fluorescence was increased by low, and decreased by high doses of NPY within this dose range *(19)*. These data are thus largely in agreement with our results. Assuming that synthesis of NE is not substantially affected by NPY, it would therefore seem that

NPY at lower doses decreases the release of NA, whereas higher doses act oppositely. Indeed, an increased release of NA produced by NPY is suggested by increased tissue levels of MHPG in the frontal cortex after treatment with NPY.

Hypotheses postulating an involvement of NA in both depression (*see*, e.g., *34,35*) and anxiety *(84)* have been presented, and in this context the ability of NPY to affect central NA systems is interesting. As discussed above, our findings indicate that NPY administered icv produces anxiolytic effects at lower doses, whereas sedation is produced at higher dose levels. It could therefore be speculated that biphasic effects of NPY on central NE release may be of importance for explaining the different effects produced by low and high doses of NPY. It must, however, be kept in mind that postmortem assessment of tissue concentrations of NE and MHPG only provides an indirect reflection of the processes occurring in vivo. Recently, data obtained using in vivo brain microdialysis have been published. These data seem to indicate that effects of NPY on central NE release may vary not only with dose, but also with anatomical location. After a single dose of NPY, signs of suppressed NE release were obtained from the medial hypothalamus, whereas the opposite was the case when lateral hypothalamus was sampled *(105)*. Clearly, therefore, different effect profiles of low and high doses of icv NPY could also represent different anatomical sites being differentially affected.

2.7.9. DA Systems

Under steady-state conditions, NPY in our studies produced a dose-dependent increase of both DA and its major metabolite, DOPAC, in the frontal cortex, the basal ganglia, the hypothalamus, and the brain stem. This parallel increase in DA and DOPAC is compatible with an increase in both synthesis and release of DA. Similar data, although only for the basal ganglia, have been reported by others *(106)*, who also reported additive effects of NPY and somatostatin. In addition, after pretreatment with α-MPT (300 mg/kg), NPY (0.47 nmol icv) has been reported to cause higher striatal levels of DA compared to rats in which the α-MPT pretreatment was followed by an icv injection of saline. Other brain regions, including the cerebral cortex and the hypothalamus, were not affected by NPY in this study *(103)*. This finding was interpreted by the authors as a result of a reduced DA turnover induced by NPY. Our findings of both DA and DOPAC increasing after NPY administrations are in agreement with those obtained by others *(106)* and make this interpretation unlikely. Ultimately, studies of in vivo release will again be required to determine the matter. As for NE, the in vivo microdialysis study quoted above *(105)* in addition indicates that effects of NPY on DA systems may be highly region specific.

If it is accepted that NPY in general seems to activate central DA neurons, this may represent a primary effect or a secondary, compensatory mechanism. A primary activation of the dopaminergic system would be expected to produce a behavioral activation, manifested, e.g., by an increase of locomotor activity *(107)*, and a desynchronization of the EEG pattern *(108)*. Instead, effects to the opposite are produced by NPY, both on locomotor activity (e.g., *55*) and on the EEG *(68)*. Interestingly, the effects of NPY on brain tissue concentrations of DA and DOPAC, as well as on locomotor activity and EEG pattern, are similar to those seen after blockade of central DA receptors with haloperidol *(107–109)*. Therefore, the observed effects of NPY on the DA systems of the brain could represent a compensatory mechanism, secondary to a functional blockade of dopaminergic transmission.

On the other hand, preliminary evidence suggests that NPY could indeed potentiate DA transmission, at least in certain brain areas. Central reward mechanisms have been shown to depend largely on DA transmission in the nucleus accumbens (*see*, e.g., *110*). It has recently been shown that microinjection of NPY into the nucleus accumbens produces place preference (i.e., after having been conditioned to associate two compartments of an apparatus with injections of NPY or saline, animals given a free choice spend significantly more time in the compartment associated with NPY). Such a place preference is usually taken as evidence for a substance being rewarding, and is, e.g., produced by amphetamine and cocaine. The place preference produced by NPY was blocked by *cis*-flupenthixol, a dopaminergic antagonist *(111)*.

The interactions between central NPY and DA seem to be reciprocal. Chemical deafferentiation of the DA input to the striatum produced by 6-OH-DA has been reported to result in increased staining of striatal tissue by NPY antibodies, representing increased tissue levels of NPY *(112)*. When push–pull perfusion of the conscious rat caudate nucleus was used, amphetamine-induced DA release was followed by a marked suppression of NPY release. This suppression was paralleled by a similar suppression of somatostatin release *(113)*. Similar results have been obtained in the arcuate nucleus. In this structure, inhibition of DA synthesis by α-MPT, as well as long-term blockade of DA receptors by haloperidol, produced an increased NPY staining *(114)*. Also, the dopaminergic D2 receptor agonist bromocriptin decreased preproNPY mRNA levels in the arcuate nucleus, whereas the dopaminergic antagonist haloperidol had the opposite effect *(115)*. The effects of dopaminergic transmission on NPY levels may, however, be different in different brain regions. Although repeated (twice daily for 6 d) administration of *d*-amphetamine, thought to augment dopaminergic transmission, produced decreased levels of NPY in the striatum, the PVN, and the dorsomedial nucleus of the hypothalamus, concentrations were actually increased in the medial preoptic nucleus *(116)*.

Also, findings seemingly opposite to those obtained from the caudate nucleus have been reported in the nucleus accumbens, where intact DA innervation was required for maintaining expression of NPY *(117)*. The latter findings are intriguing in view of the rewarding properties of intraaccumbens NPY, long-lasting NPY deficits reported in cocaine-withdrawing animals *(82)*, and depressive symptoms displayed by both humans and experimental animals during cocaine withdrawal (*see*, e.g., *81*). If, in contrast to the caudate, the NPY–DA interactions in nucleus accumbens would constitute a positive feedback loop, a disruption of this loop would not show an intrinsic tendency to recover.

2.8. Future Directions

It seems justified to conclude that a role of NPY in mechanisms of anxiety and depression is strongly suggested by the observations reviewed above. Clearly, however, several central questions remain unanswered at this moment. Conflicting data are present with regard to CSF levels of NPY in depression. Similarly, conflicting data are present from experimental animals treated with antidepressant drugs. It may therefore be adequate to state that, although a role indeed is suggested for NPY in anxiety and depression, we do not know what that role is. Reconciling the conflicting data will probably require a characterization of the processing that the NPY precursor undergoes in different conditions, rather than measuring crude concentrations of immunoreactive material.

Another area where it is yet difficult to extract a pattern is interactions of NPY–monoaminergic systems, and the relation of these interactions to behavioral effects of NPY. Interactions with monoaminergic transmitter systems will need to be characterized in vivo, and with respect to specific anatomical locations before they can be better understood.

The most pressing need, however, is the development of selective ligands for NPY receptors. Substituted peptide analogs, the commonly employed approach, may prove immensely important for the basic sciences. In the particular context of NPY and psychiatric disease, however, the important task seems to be a much more difficult one: to develop selective, nonpeptidergic, stable, and lipid-soluble analogs for central NPY receptors. Only then will it ultimately become possible to put the hypothesis that activation of these receptors may represent a new therapeutic principle to a final test.

3. NPY and Other CNS Pathology

3.1. Memory, Alzheimer's Dementia (AD), and Senile Dementia of the Alzheimer Type (SDAT)

A modulation of memory processing by central NPY has been reported in the mouse *(118)*. NPY given into the third ventricle within 3 min after training sessions improved the retention of both passive and active avoidance tasks. The improvement of passive avoidance retention required lower doses of NPY, and differed also in having a bell-shaped dose–response curve. Also, recall of a task poorly stored owing to weak training was improved by NPY given 1 h before testing. Once again, the dose–response curve was bell shaped. Finally, NPY counteracted the amnestic effects of both anisomycin (an inhibitor of protein synthesis) and scopolamine (a cholinergic muscarinic receptor antagonist). The effects of NPY on memory retention were not the result of an improved acquisition and were independent of the orexigenic action of the peptide. Peripherally administered NPY was ineffective.

The effects of NPY on memory processing seem to vary with anatomical site. Memory retention is improved by local injections of NPY into the rostral portion of the hippocampus and the septum, whereas injections into the caudal portion of the hippocampus and the amygdala act to the opposite. Injections of an NPY antibody produced effects opposite to those produced by the peptide in all these four areas, indicating a physiological role of endogenous NPY in memory processing *(83)*.

The C-terminal 17-a.a. fragment of NPY (NPY_{20-36}) has been used as a tentative NPY-Y2 agonist in order to determine receptor mechanisms underlying the effects of NPY on memory. NPY_{20-36} reproduced the effects of the intact NPY molecule on memory, thus indicating these to be Y2 mediated. For comparison, the stimulatory effects of NPY on food intake were not reproduced by the fragment, suggesting that these are mediated by Y1 receptors *(119)*. Flood and Morley conclude that the effects of NPY on memory are produced presynaptically.

It could be argued that presynaptic location of Y2 receptors only has been demonstrated in the periphery, and there only functionally. Electrophysiological observations, however, seem to support the conclusion made by Flood and Morley. Presynaptic inhibition of excitatory input to granule cells of the CA1 area has been demonstrated in the hippocampal slice preparation *(121–123)*. This finding has been replicated by others *(124)*, who in addition found an NPY-mediated reduction in the amplitude of postsynaptic potentials normally produced by stimulation of the CA1 input. It should be noted that "presynaptic" in this context should be interpreted in a broad sense, including presynaptic NPY heteroreceptors on, e.g., glutamatergic terminals.

A direct excitatory postsynaptic action of NPY on the granule cells of the dentate gyrus has been reported *(125)*. In addition, it was recently reported that NPY potentiates the increase in firing of rat CA3 dorsal hippocampus pyramidal neurons, which is produced in vivo by microiontophoretic application of *N*-methyl-D-aspartate *(126)*. These data may seem to contradict those reviewed above. Both sets of data are, however, compatible with NPY presynaptically inhibiting the release of glutamate (or a related excitatory transmitter), while postsynaptically potentiating its effect. Such an interaction would be a parallel to what has been described with respect to NPY and NE in the peripheral sympathetic junction.

Finally, like a number of other memory-enhancing drugs, but as opposed to two α_2-adrenoceptor agonists, NPY enhanced memory retention in aged mice with the optimal dose being the same as in young animals *(153)*. Although a role for NPY in memory processing seems to be firmly established by the observations reviewed above, a possible involvement of the peptide in disease states affecting memory has not been convincingly demonstrated, and conflicting data are abundant in the literature.

With the extensive coexistence of NPY and somatostatin in mind, considerable interest was produced by the finding of a dissociation between these two peptides in SDAT. Although decreased levels of somatostatin have been found in the temporal cortex of SDAT patients, NPY levels remained normal in that area. In contrast to somatostatin, tissue levels of NPY were not correlated with the count of neuritic plaques in this study *(127)*, despite the fact that NPY immunoreactivity has been found in neuritic plaques of the demented brain by the same group *(128)*. A similar dissociation between somatostatin and NPY in cortical tissue has also been reported by others, who in addition found a parallel increase in the levels of both peptides in the substantia innominata *(129)*. Normal numbers of NPY neurons have been found in the ventral striatum of AD patients *(130)*. In agreement with these data, unaltered CSF levels of NPY have been found in patients with SDAT, whereas somatostatin levels were decreased *(131)*.

In contrast, reductions in neocortical and hippocampal NPY immunoreactivity have been found by Beal and associates, who also reported NPY levels in the substantia innominata and in the striatum to be unaltered *(132,133)*. Morphological examination of the AD hippocampus shows a severe loss of NPY neurons and axons in areas most affected by the disease process *(134)*. Identical morphological abnormalities were seen in the SDAT neocortex when neurons were labeled with antibodies against NPY or somatostatin *(135)*.

It has been suggested that the discrepancies present in the literature could be caused by a lack of correction for the atrophy present in AD brains. Employing such a correction, decreased numbers of both somatostatin and NPY-immunoreactive neurons, as well as decreased measures of fiber lengths, have been obtained from both frontal and temporal cortex of AD patients *(136)*.

Finally, excitotoxic lesions of the Nc. Basalis Meynert have been suggested as an animal model of human dementia. Increased cortical levels of both NPY and somatostatin have been found in this model *(137)*.

3.2. Huntington's Chorea (HC)

The choreic striatum is severely atrophic, resulting in a loss of most neuronal markers. As an exception to this rule, medium-sized aspiny neurons containing NPY, somatostatin, and NADPH-diaphorase are selectively spared, resulting in increased relative tissue concentrations of NPY and somatostatin *(138–140)*. The mechanism behind this selective sparing is not known. In the equally atrophic amygdala of HC patients, NPY neurons are not spared *(141)*.

An excitotoxin hypothesis for the pathogenesis of HC has been postulated. The report that quinolinic acid, but not other excitotoxins, produces NPY- and somatostatin-sparing striatal lesions closely resembling those found in HC *(142)* was met with considerable interest. These authors later examined the receptor specificity of this effect. They reported that among several excitotoxic compounds, only those selective for glutamate AA1 receptors (the "NMDA" type) produce the "NPY-sparing" lesion *(143)*. Other groups, however, have failed to replicate the original finding of quinolinic acid lesions sparing NPY-somatostatin neurons *(144,145)*. Recently, a similar capability of the endogenously present excitatory transmitter candidate, L-homocysteic acid, to produce NPY and somatostatin sparing striatal lesion has been suggested by Beal and associates *(146)*. Clearly, this matter remains unresolved for the moment. Attempts at linking specific symptoms of HC with specific neuropeptides have been made, but have this far not provided evidence for specific links *(147)*.

3.3. Parkinson's Disease

Despite a marked reduction in the concentration of somatostatin in both the neocortex and the hippocampus of patients with severe Parkinson's disease, no significant change in the concentration of NPY has been found in either region *(148)*.

3.4. Schizophrenia

Normal NPY levels have been reported in the CSF of schizophrenic patients *(58,76)*. In postmortem amygdala *(149)* and hypothalamus *(150)* from schizophrenic patients, normal tissue concentrations of NPY, as well as somatostatin, have also been reported. In contrast, reduced concentrations of NPY have been reported in the temporal cortex *(150)* of schizophrenic patients.

3.5. Eating Disorders

Elevated CSF content of NPY has been reported in both underweight amenorrheic anorectics and the same amenorrheic patients studied again within 6 weeks after weight restoration. An inverse relationship between CSF NPY and caloric intake in healthy volunteer women was also found. Thus, the increase in CSF levels of NPY may be a secondary, compensatory response to decreased food intake. Also, since NPY regulates a number of endocrine parameters, including the secretion of LH, FSH, and ACTH, several symptoms of anorexia could be secondary to such an increase in NPY secretion *(151,152)*. These findings are in agreement with animal observations mentioned above and reported in detail elsewhere in this book (*see* Chapter 11).

4. Concluding Remarks

Among CNS diseases, current evidence seems to suggest a role for NPY in depression and anxiety. As a preliminary hypothesis, it could be postulated that altered processing and/or turnover of NPY might both produce some central symptoms of these disease states and account for some apparent discrepancies present in the literature. In addition, it seems likely that NPY contributes to producing some characteristic symptoms of eating disorders, although its role in this case seems to be secondary to changes in food intake. Finally, a role for NPY in memory processing seems convincing, but present data do not convincingly support the notion that NPY participates in the pathophysiology of diseases affecting memory.

Over the decade since its isolation, a vast literature on NPY has rapidly accumulated. An important part of that literature consist of studies examining the relation of NPY to various CNS diseases. This literature is reviewed and discussed above. Ultimately, however, the traditional approach that tries to link simple changes in specific transmitters with specific psychiatric diseases may not be the most productive one. Considering the massive parallelism of the brain, exemplified by coexistence of multiple transmitters within single neurons and by anatomical organization principles of the CNS, new multidimensional models may be required before the involvement of any single transmitter in any psychiatric disease is understood.

Acknowledgments

This chapter is an attempt to integrate data originating from numerous original publications. Every attempt has been made to credit the authors of these papers by properly referencing the relevant citations. With regard to the work on NPY, depression, and anxiety, I would especially like to underscore the importance of George F. Koob, Karen T. Britton, Erik Widerlöv, Robert Murison, Rolf Ekman, Claes Wahlestedt, Jörgen Engel, Bo Söderpalm, Christer Alling, Laszlo Vecsei, and Emilio Merlo Pich. I initially planned to credit each of them with specific aspects of the work. I rapidly found, however, that a net of ideas and experiments woven in several intellectually fruitful environments (Institute of Physiological Psychology, Bergen, Norway; Department of Psychiatry and Neurochemistry, Lund, Sweden; Department of Pharmacology, Göteborg, Sweden, Department of Neuropharmacology, Scripps Clinic and Research Foundation, La Jolla, California) was not easily disentangled. Then of course each of these persons had their network of interactions, and so on.

References

1. Kuhn, T. S. (1970) *Structure of Scientific Revolutions.* University of Chicago Press, Phoenix.
2. Freud, S (1957; original 1905) *Jokes and Their Relation to the Unconscious.* Vol. 8 of *Standard Edition of the Complete Psychological Works of Sigmund Freud.* Hogarth, London.
3. Christie, M., Machida, C. A., Neve, K. A., and Civelli, O. (1988) Cloning and expression of a rat D2 dopamine receptor cDNA. *Nature* **336(6201),** 783–787.
4. Zhou, Q. Y., Grandy, D. K., Thambi, L., Kushner, J. A., Van Tol, H. H., Cone, R., Pribnow, D., Salon, J., Bunzow, J. R., and Civelli, O. (1990) Cloning and expression of human and rat D1 dopamine receptors. *Nature* **347(6288),** 76–80.
5. Sokoloff, P., Giros, B., Martres, M. P., Bouthenet, M. L., and Schwartz, J. C. (1990) Molecular cloning and characterization of a novel dopamine receptor (D3) as a target for neuroleptics. *Nature* **347,** 146–151.
6. Hughes, J., Boden, P., Costall, B., Domeney, A., Kelly, E., Horwell, D. C., Hunter, J. C., Pinnock, R. D., and Woodruff, G. N. (1990) Development of a class of selective cholecystokinin type B receptor antagonists having potent anxiolytic activity. *Proc. Natl. Acad. Sci. USA* **87(17),** 6728–6732.
7. Adrian, T. E., Allen, J. M., Bloom, S. R., Ghatei, M. A., Rossor, M. N., Roberts, G. W., Crow, T. J., Tatemoto, K., and Polak, J. M. (1983) Neuropeptide Y distribution in human brain. *Nature* **306,** 584–586.

8. Yui, R., Iwanaga, T., Kuramoto, H., and Fujita, T. (1985) Neuropeptide immunocytochemistry in protostomian invertebrates, with special reference to insects and molluscs. *Peptides* 6 **Suppl. 3,** 411–415.
9. Andrews, P. C., Hawke, D., Shively, J. E., and Dixon, J. E. (1985) A nonamidated peptide homologous to porcine peptide YY and neuropeptide YY. *Endocrinology* **116,** 2677–2681.
10. Noe, B. D., McDonald, J. K., Greiner, F., Wood, J. G., and Andrews, P. C. (1986) Anglerfish islets contain NPY immunoreactive nerves and produce the NPY analog aPY. *Peptides* **7,** 147–154.
11. Conlon, J. M., Schmidt, W. E., Gallwitz, B., Falkmer, S., and Thim, L. (1986) Characterization of an amidated form of pancreatic polypeptide from the daddy sculpin *(Cottus scorpius)*. *Regul. Peptides* **16,** 261–268.
12. Kimmel, J. R., Plisetskaya, E. M., Pollock, H. G., Hamilton, J. W., Rouse, J. B., Ebner, K. E., and Rawitch, A. B. (1986) Structure of a peptide from coho salmon endocrine pancreas with homology to neuropeptide Y. *Biochem. Biophys. Res. Commun.* **141,** 1084–1091.
13. Moore, R. Y. and Card, J. P. (1985) Visual pathways and the entrainment of circadian rhythms. *Ann. NY Acad. Sci.* **453,** 123–133.
14. Albers, H. E., Ferris, C. F., Leeman, S. E., and Goldman, B. D. (1984) Avian pancreatic polypeptide phase shifts hamster circadian rhythms when microinjected into the suprachiasmatic region. *Science* **223,** 833–835.
15. Albers, H. E. and Ferris, C. F. (1984) Neuropeptide Y: role in light-dark cycle entrainment of hamster circadian rhythms. *Neurosci. Lett.* **50,** 163–168.
16. Lehman, M. N., Silver, R., Gladstone, W. R., Kahn, R. M., Gibson, M., and Bittman, E. L. (1987) Circadian rhythmicity restored by neural transplant. Immunocytochemical characterization of the graft and its integration with the host brain. *J. Neurosci.* **7,** 1626–1638.
17. Mason, R., Harrington, M. E., and Rusak, B. (1987) Electrophysiological responses of hamster suprachiasmatic neurones to neuropeptide Y in the hypothalamic slice preparation. *Neurosci. Lett.* **80,** 173–179.
18. Wahlestedt, C., Skagerberg, G., Ekman, R., Heilig, M., Sundler, F., and Håkanson, R. (1987) Neuropeptide Y (NPY) in the area of the hypothalamic paraventricular nucleus activates the pituitary-adrenocortical axis in the rat. *Brain Res.* **417,** 33–38.
19. Härfstrand, A., Eneroth, P., Agnati, L., and Fuxe, K. (1987) Further studies on the effects of central administration of neuropeptide Y on neuroendocrine function in the male rat. *Regul. Peptides* **17,** 167–179.
20. Liposits, Z., Sievers, L., and Paull, W. K. (1988) Neuropeptide-Y and ACTH-immunoreactive innervation of corticotropin releasing factor (CRF)-synthesizing neurons in the hypothalamus of the rat. An immunocytochemical analysis at the light and electron microscopic levels. *Histochemistry* **88,** 227–234.
21. Haas, D. A. and George, S. R. (1987) Neuropeptide Y administration acutely increases hypothalamic corticotropin-releasing factor immunoreactivity: lack of effect in other rat brain regions. *Life Sci.* **41,** 2725–2731.
22. Tsagarakis, S., Rees, L. H., Besser, G. M., and Grossman, A. (1989) Neuropeptide-Y stimulates CRF-41 release from rat hypothalami in vitro. *Brain Res.* **502,** 167–170.

23. Haas, D. A. and George, S. R. (1989) Neuropeptide Y-induced effects on hypothalamic corticotropin-releasing factor content and release are dependent on noradrenergic/adrenergic neurotransmission. *Brain Res.* **498,** 333–338.
24. Inoue, T., Inui, A., Okita, M., Sakatani, N., Oya, M., Morioka, H., Mizuno, N., Oimomi, M., and Baba, S. (1989) Effect of neuropeptide Y on the hypothalamic-pituitary-adrenal axis in the pig. *Life Sci.* **44,** 1043–1051.
25. Clark, J. T., Kalra, P. S., Crowley, W. R., and Kalra, S. P. (1984) Neuropeptide Y and human pancreatic polypeptide stimulate feeding behavior in rats. *Endocrinology* **115,** 427–429.
26. Stanley, B. G. and Leibowitz, S. F. (1984) Neuropeptide Y: stimulation of feeding and drinking by injection into the paraventricular nucleus. *Life Sci.* **35,** 2635–2642.
27. Levine, A. S. and Morley, J. E. (1984) Neuropeptide Y: a potent inducer of consummatory behavior in rats. *Peptides* **5,** 1025–1029.
28. Parrott, R. F., Heavens, R. P., and Baldwin, B. A. (1986) Stimulation of feeding in the satiated pig by intracerebroventricular injection of neuropeptide Y. *Physiol. Behav.* **36,** 523–525.
29. Morley, J. E., Hernandez, E. N., and Flood, J. F. (1987) Neuropeptide Y increases food intake in mice. *Am. J. Physiol.* **253,** R516–R522.
30. Stanley, B. G., Daniel, D. R., Chin, A. S., and Leibowitz, S. F. (1985) Paraventricular nucleus injections of peptide YY and neuropeptide Y preferentially enhance carbohydrate ingestion. *Peptides* **6,** 1205–1211.
31. Morley, J. E. and Flood, J. F. (1989) The effect of neuropeptide Y on drinking in mice. *Brain Res.* **494,** 129–137.
32. Stanley, B. G., Kyrkouli, S. E., Lampert, S., and Leibowitz, S. F. (1986) Neuropeptide Y chronically injected into the hypothalamus: a powerful neurochemical inducer of hyperphagia and obesity. *Peptides* **7,** 1189–1192.
33. Sahu, A., Kalra, P. S., and Kalra, S. P. (1988) Food deprivation and ingestion induce reciprocal changes in neuropeptide Y concentrations in the paraventricular nucleus. *Peptides* **9,** 83–86.
34. Schildkraut, J. J. (1965) The catecholamine hypothesis of affective disorder: A review of supporting evidence. *Am. J. Psychiat.* **122,** 509–522.
35. Siever, L. J. and Davies, K. L. (1985) Overview: Towards a dysregulation hypothesis of depression. *Am. J. Psychiat.* **142,** 1017–1031.
36. Holmberg, G. and Gershon, S. (1961) Autonomic and psychic effects of yohimbine hydrochloride. *Psychopharmacology* **2,** 93–106.
37. Hoen-Saric, R., Merchant, A. F., Keyser, M. L., and Smith, V. K. (1981) Effects of clonidine on anxiety disorders. *Arch. Gen. Psychiat.* **39,** 735–742.
38. Handley, S. L. and Mithani, S. (1984) Effects of alpha-adrenoceptor agonists and antagonists in a maze-exploration model of 'fear'-motivated behaviour. *Naunyn-Schmiedeberg's Arch. Pharmacol.* **327,** 1–5.
39. Pellow, S., Chopin, P., File, S. E., and Briley, M. (1985) Validation of open:closed arm entries in an elevated plus-maze as a measure of anxiety in the rat. *J. Neurosci. Methods* **14,** 149–167.
40. Söderpalm, B. and Engel, J. A. (1988) Biphasic effects of clonidine on conflict behaviour: involvement of different alpha-receptors. *Pharmacol. Biochem. Behav.* **30,** 471–477.

41. Kahn, R. J., McNair, D. M., Lipman, R. S., Covi, L., Rickels, K., Downing, R., Fisher, S., and Frankenthaler, L. M. (1986) Imipramine and chlordiazepoxide in depressive and anxiety disorders. II. Efficacy in anxious outpatients. *Arch. Gen. Psychiat.* **43,** 79–85.
42. Modigh, K. (1987) Antidepressant drugs in anxiety disorders. *Acta Psychiat. Scand.* **76 (Suppl. 335),** 51–71.
43. Lundberg, J. M., Terenius, L., Hökfelt, T., Martling, C. R., Tatemoto, K., Mutt, V., Polak, J., Bloom, S., and Goldstein, M. (1982) Neuropeptide Y (NPY)-like immunoreactivity in peripheral noradrenergic neurons and effects of NPY on sympathetic function. *Acta Physiol. Scand.* **116,** 477–480.
44. Everitt, B. J., Hökfelt, T., Terenius, L., Tatemoto, K., Mutt, V., and Goldstein, M. (1984) Differential co-existence of neuropeptide Y (NPY)-like immunoreactivity with catecholamines in the central nervous system of the rat. *Neuroscience* **11,** 443–462.
45. Hendry, S. H., Jones, E. G., DeFelipe, J., Schmechel, D., Brandon, C., and Emson, P. C. (1984) Neuropeptide-containing neurons of the cerebral cortex are also GABAergic. *Proc. Natl. Acad. Sci. USA* **81,** 6526–6530.
46. Chronwall, B. M., Chase, T. N., and O'Donohue, T. L. (1984) Coexistence of neuropeptide Y and somatostatin in rat and human cortical and rat hypothalamic neurons. *Neurosci. Lett.* **52,** 213–217.
47. Kalra, S. P. and Crowley, W. R. (1984) Norepinephrine-like effects of neuropeptide Y on LH release in the rat. *Life Sci.* **35,** 1173–1176.
48. Agnati, L. F., Fuxe, K., Benfenati, F., Battistini, N., Härfstrand, A., Tatemoto, K., Hökfelt, T., and Mutt, V. (1983) Neuropeptide Y in vitro selectivity increases the number of alpha 2-adrenergic binding sites in membranes of the medulla oblongata of the rat. *Acta Physiol. Scand.* **118,** 293–295.
49. Dawbarn, D., Hunt, S. P., and Emson, P. C. (1984) Neuropeptide Y: regional distribution chromatographic characterization and immunohistochemical demonstration in post-mortem human brain. *Brain Res.* **296,** 168–173.
50. Khorram, O., Roselli, C. E., Ellinwood, W. E., and Spies, H. G. (1987) The measurement of neuropeptide Y in discrete hypothalamic and limbic regions of male rhesus macaques with a human NPY-directed antiserum. *Peptides* **8,** 159–163.
51. Allen, Y. S., Adrian, T. E., Allen, J. M., Tatemoto, K., Crow, T. J., Bloom, S. R., and Polak, J. M. (1983) Neuropeptide Y distribution in the rat brain. *Science* **221,** 877–879.
52. Chronwall, B. M., DiMaggio, D. A., Massari, V. J., Pickel, V. M., Ruggiero, D. A., and O'Donohue, T. L. (1985) The anatomy of neuropeptide-Y-containing neurons in rat brain. *Neuroscience* **15,** 1159–1181.
53. de Quidt, M. E. and Emson, P. C. (1986) Distribution of neuropeptide Y-like immunoreactivity in the rat central nervous system—II. Immunohistochemical analysis. *Neuroscience* **18,** 545–618.
54. Heilig, M. and Murison, R. (1987) Intracerebroventricular neuropeptide Y protects against stress-induced gastric erosion in the rat. *Eur. J. Pharmacol.* **137,** 127–129.
55. Heilig, M. and Murison, R. (1987) Intracerebroventricular neuropeptide Y suppresses open field and home cage activity in the rat. *Regul. Peptides* **19,** 221–231.

56. Heilig, M., Wahlestedt, C., Ekman, R., and Widerlöv, E. (1988) Antidepressant drugs increase the concentration of neuropeptide Y (NPY)-like immunoreactivity in the rat brain. *Eur. J. Pharmacol.* **147,** 465–467.
57. Wahlestedt, C., Blendy, J. A., Kellar, K. J., Heilig, M., Widerlöv, E., and Ekman, R. (1990) Electroconvulsive shocks increase the concentration of neocortical and hippocampal neuropeptide Y (NPY)-like immunoreactivity in the rat. *Brain Res.* **507,** 65–68.
58. Widerlöv, E., Lindström, L. H., Wahlestedt, C., and Ekman, R. (1988) Neuropeptide Y and peptide YY as possible cerebrospinal markers for major depression and schizophrenia, respectively. *J. Psychiat. Res.* **22 (1),** 69–79.
59. Hamilton, M. (1967) Development of a rating scale for primary depressive illness. *Brit. J. Soc. Clin. Psychol.* **6,** 278–296.
60. Heilig, M., Söderpalm, B., Engel, J. A., and Widerlöv, E. (1989) Centrally administered neuropeptide Y (NPY) produces anxiolytic-like effects in animal anxiety models. *Psychopharmacology* **98,** 524–529.
61. Widerlöv, E., Heilig, M., Ekman, R., and Wahlestedt, C. (1988) Possible relationship between neuropeptide Y (NPY) and major depression evidence from human and animal studies. *Nord Psykiat Tidsskr* **42,** 131–137.
61a. Montgomery, K. C. (1958) The relation between fear induced by novel stimulation and exploratory behaviour. *J. Comp. Physiol. Psychol.* **48,** 254–260.
62. Pellow et al. (1984)
63. Vogel, J. R., Beer, B., and Clody, D. E. (1971) A simple and reliable conflict procedure for testing anti-anxiety agents. *Psychopharmacology* **21,** 1–7.
64. Engel, J. A., Hjort, S., Svensson, K., Carlsson, A., and Liljequist, S. (1984) Anticonflict effect of the putative serotonine receptor agonist 8-hydroxy-2-(Dl-n-propylamino)tetralin (8-OH-DPAT). *Eur. J. Pharmacol.* **105,** 365–368.
65. Heilig, M., Vecsei, L., Wahlestedt, C., Alling, C., and Widerlöv, E. (1990) Effects of centrally administered neuropeptide Y (NPY) and $NPY_{13\text{-}36}$ on the brain monoaminergic systems of the rat. *J. Neural Trans. (Gen. Sect.)* **79,** 193–208.
66. Drew, G. M., Gower, A. J., and Marriott, A. S. (1979) Alpha-2-adrenoceptors mediate clonidine-induced sedation in the rat. *Brit. J. Pharmacol.* **67,** 133–142.
67a. Heilig, M., Wahlestedt, C., and Widerlöv, E. (1988) Neuropeptide Y (NPY) induced activity suppression in the rat: evidence for NPY receptor heterogeneity and for interaction with alpha-adrenergic receptors. *Eur. J. Pharmacol.* **157,** 205–213.
67b. Heilig, M., Vecsei, L., and Widerlöv, E. (1989) Opposite effects of centrally administered neuropeptide Y (NPY) on locomotor activity of spontaneously hypertensive (SH) and normal rats. *Acta Physiol. Scand.* **137,** 243–248.
68. Fuxe, K., Agnati, L. F., Härfstrand, A., Zini, I., Tatemoto, K., Pich, E. M., Hökfelt, T., Mutt, V., and Terenius, L. (1983) Central administration of neuropeptide Y induces hypotension, bradypnea and EEG synchronization in the rat. *Acta Physiol. Scand.* **118,** 189–192.
69. Keim, K. L. and Sigg, E. B. (1977) Plasma corticosterone and brain catecholamins in stress: Effects of psychotropic drugs. *Pharmacol. Biochem. Behav.* **6,** 79–85.
70. Humphreys, G. A., Veale, W. L., and Davison, J. S. (1987) Effect of intrahypothalamic microinjection of neuropeptide Y on gastric acid secretion in the rat. *Proc. West Pharmacol. Soc.* **30,** 317–319.

71. File, S. E. and Peet, L. A. (1980) The Sensitivity of the Rat Corticosterone Response to Environmental Manipulations and to Chronic Chlordiazepoxide Treatment. *Physiol. Behav.* **25,** 753–758.
72. Heilig M., McLeod S., Koob G. F., and Britton (1992) Anxiolytic-like action of neuropeptide Y, but not other peptides, in an operant conflict test. *Regulatory Peptides* **41,** 61–69.
73. Theodorsson, E., Mathe, A. A., and Stenfors, C. (1990) Brain neuropeptides: tachykinins, neuropeptide Y, neurotensin and vasoactive intestinal polypeptide in the rat brain: modifications by ECT and indomethacin. *Prog. Neuro-Psychopharmacol. Biol. Psychiat.* **14,** 387–407.
74. Mathé, A. A., Jousisto-Hanson, J., Stenfors, C., and Theodorsson, E. (1990) Effect of lithium on tachykinins, calcitonin gene-related peptide, and neuropeptide Y in rat brain. *J. Neurosci. Res.* **26,** 233–237.
75. Ekman, R., Widerlöv, E., Walléus, H., and Lindström, L. H. (1984) Elevated levels of NPY-like material in CSF in depression. ISPNE XVth International Congress in Vienna July 15–19, 73.
76. Berrettini, W. H., Doran, A. R., Kelsoe, J., Roy, A., and Pickar, D. (1987) Cerebrospinal Fluid Neuropeptide Y in Depression and Schizophrenia. *Neuropsychopharmacology* **1(1),** 81–83.
77. Jesberger, J. A. and Richardson, J. S. (1985) Animal models of depression: parallels and correlates to severe depression in humans. *Biol. Psychiat.* **20,** 764–784.
78. Widerlöv, E., Heilig, M., Ekman, R., and Wahlestedt, C. (1989) Neuropeptide Y—possible involvement in depression and anxiety, in *Nobel Conference on NPY* (Mutt, V., Fuxe, K., and Hökfelt, T., eds.), Raven, New York, pp. 331–342.
79. Kataoka, Y., Ushio, M., Koizumi, S., Niwa, M., and Veki, S. (1988) Changes in the concentration of brain neuropeptide Y (NPY)-like immunoreactivity following olfactory bulbectomy and effects of antidepressants in rats. *Soc. Neurosci. Abstr.* **114,** 11.
80. Kataoka, Y., Sakurai, Y., Mine, K., Yamashita, K., Fujiwara, M., Niwa, M., and Ueki, S. (1987) The involvement of neuropeptide Y in the antimuricide action of noradrenaline injected into the medial amygdala of olfactory bulbectomized rats. *Pharmacol. Biochem. Behav.* **28,** 101–103.
81. Markou, A. and Koob, G. F. (1991) Postcocaine Anhedonia. An Animal Model of Cocaine Withdrawal. *Neuropsychopharmacology* **4(1),** 17–26.
82. Wahlestedt, C., Karoum, F., Jaskino, G., Wyatt, R. J., Larhammar, D., Ekman, R., and Reis, D. J. (1991) Cocaine-induced reduction of brain neuropeptide Y synthesis dependent on medial prefrontal cortex. *PNAS* **88(6),** 2078–2082.
83. Flood, J. F., Baker, M. L., Hernandez, E. N., and Morley, J. E. (1989) Modulation of memory processing by neuropeptide Y varies with brain injection site. *Brain Res.* **503,** 73–82.
84. Redmond, D. E. and Huang, Y. H. (1979) New evidence for a locus coeruleus-norepinephrine connection with anxiety. *Life Sci.* **25,** 2149–2162.
85. Svensson, T. H., Bunney, B. S., and Aghajanin, G. K. (1975) Inhibition of both noradrenergic and serotonergic neurons in the brain by the alpha adrenergic agonist clonidine. *Brain Res.* **92,** 291–306.
86. Shibata, K., Kataoka, Y., Yamashita, K., and Ueki, S. (1986) An important

role of the central amygdaloid nucleus and mammillary body in the mediation of conflict behavior in rats. *Brain Res.* **372(1),** 159–162.
87. Duncan, G. E., Breese, G. R., Criswell, H., Stumpf, W. E., Mueller, R. A., and Covey, J. B. (1986) Effects of antidepressant drugs injected into the amygdala on behavioural responses of rats in the forced swim test. *J. Pharmacol. Exp. Ther.* **238(2),** 758–762.
88. Kawashima, K., Araki, H., Uchiyama, Y., and Aihara, H. (1987) Amygdaloid catecholaminergic mechanisms involved in suppressive effects of electroconvulsive shock on duration of immmobility in rats forced to swim. *Eur. J. Pharmacol.* **141(1),** 1–6.
88a. Heilig, M., McLeod, S., Brot, M., Koob, G. F., Britton, K. T. (1992) Anxiolytic-like action of neuropeptide Y: Mediation by Y1-receptors in amygdala, and dissociation from food intake effects. In press, *Neuropsychopharmacology.*
89. Martire, M., Fuxe, K., Pistritto, G., Preziosi, P., and Agnati, L. F. (1986) Neuropeptide Y enhances the inhibitory effects of clonidine on 3H-noradrenaline release in synaptosomes isolated from the medulla oblongata of the male rat. *J. Neural. Transm.* **67,** 113–124.
90. Martire, M., Fuxe, K., Pistritto, G., Preziosi, P., and Agnati, L. F. (1989) Neuropeptide Y increases the inhibitory effects of clonidine on potassium evoked 3H-noradrenaline but not 3H-5-hydroxytryptamine release from synaptosomes of the hypothalamus and the frontoparietal cortex of the male Sprague-Dawley rat. *J. Neural. Transm.* (Gen Sect) **78,** 61–72.
91. Yokoo, H., Schlesinger, D. H., and Goldstein, M. (1987) The effect of neuropeptide Y (NPY) on stimulation-evoked release of ^{3}H-norepinephrine (NE) from rat hypothalamic and cerebral cortical slices. *Eur. J. Pharmacol.* **143,** 283–286.
92. Tsuda, K., Yokoo, H., and Goldstein, M. (1989) Neuropeptide Y and galanin in norepinephrine release in hypothalamic slices. *Hypertension* **14,** 81–86.
93. Fuxe, K., Agnati, L. F., Härfstrand, A., Janson, A. M., Neumeyer, A., Andersson, K., Ruggeri, M., Zoli, M., and Goldstein, M. (1986) Morphofunctional studies on the neuropeptide Y/adrenaline costoring terminal systems in the dorsal cardiovascular region of the medulla oblongata. Focus on receptor-receptor interactions in cotransmission. *Prog. Brain Res.* **68,** 303–320.
94. Illes, P. and Regenold J. T. (1990) Interaction between neuropeptide Y and noradrenaline on central catecholamine neurons. *Nature* **344,** 62,63.
95. Agnati, L. F., Fuxe, K., Benfenati, F., Battistini, N., Härfstrand, A., Hökfelt, T., Cavicchioli, L., Tatemoto, K., and Mutt, V. (1983) Failure of neuropeptide Y in vitro to increase the number of alpha 2-adrenergic binding sites in membranes of medulla oblongata of the spontaneous hypertensive rat. *Acta Physiol. Scand.* **119,** 309–312.
96. Söderpalm, B. (1989) The SHR exhibits less 'anxiety' but increased sensitivity to the anticonflict effect of clonidine compared to normotensive controls. *Pharmacol. Toxicol.* **65,** 381–386.
97. Wahlestedt, C., Yanaihara, N., and Håkanson, R. (1986) Evidence for different pre- and post-junctional receptors for neuropeptide Y and related peptides. *Regul. Peptides* **13,** 307–318.
98. Wahlestedt, C., Edvinson, L., Ekblad, E., and Håkanson, R. (1988) Effects of neuropeptide Y at sympathetic neuroeffector junctions: Existence of Y1

and Y2 receptors., in *Neuronal Messengers in Vascular Function* (Nobin, A. and Owman, C., eds.), Elsevier, Amsterdam, pp. 231–244.

99. Sheikh, S. P., Håkanson, R., and Schwartz, T. W. (1989) Y1 and Y2 receptors for neuropeptide Y. *FEBS Lett.* **245,** 209–214.
100. Michel, M. C., Schlicker, E., Fink, K., Boublik, J. H., Gothert, M., Willette, R. N., Daly, R. N., Hieble, J. P., Rivier, J. E., and Motulsky, H. J. (1990) Distinction of NPY receptors in vitro and in vivo. I. NPY-(18-36) Discriminates NPY receptor subtypes in vitro. *Am. J. Physiol.* **259,** E131–E139.
101. Aakerlund, L, Gether, U., Fuhlendorff, J., Schwartz ,T. W., and Thastrup, O. (1990) Y1 receptors for neuropeptide Y are coupled to mobilization of intracellular calcium and inhibition of adenylate cyclase. *FEBS Lett.* **260,** 73–78.
102. Kitamura, K., Kangawa, K., Tanaka, K., and Matsuo, H. (1990) Isolation of NPY-25 (neuropeptide Y[12-36]), a potent inhibitor of calmodulin, from porcine brain. *Biochem. Biophys. Res. Commun.* **169,** 1164–1171.
103. Heilig, M., McLeod, S., Brot, M., Koob, G. F., and Britton, K. T. (1992) Anxiolytic-like action of neuropeptide Y: Mediation by Y1-receptors in amygdala, and dissociation from food intake effects. *Neuropsychopharmacology* (in press).
104. Vallejo, M., Carter, D. A., Biswas, S., and Lightman, S. L. (1987) Neuropeptide Y alters monoamine turnover in the rat brain. *Neurosci. Lett.* **73,** 155–160.
105. Shimizu, H. and Bray, G. A. (1989) Effects of neuropeptide Y on norepinephrine and serotonin metabolism in rat hypothalamus in vivo. *Brain Res. Bull.* **22,** 945–950.
106. Beal, M. F., Frank, R. C., Ellison, D. W., and Martin, J. B. (1986) The effect of neuropeptide Y on striatal catecholamines. *Neurosci. Lett.* **71,** 118–123.
107. Iversen, S. D. (1977) Brain dopamine systems and behaviour, in *Handbook of Psychopharmacology*, vol. 8 (Iversen, L. L., Iversen, S. D., and Snyder, S. H., eds.), Plenum, New York, pp. 333–384.
108. Monti, J. M. (1982) Catecholamines and the sleep-wake cycle. I. EEG and behavioural arousal. *Life Sci.* **30,** 1145–1157.
109. Carlsson, A. (1978) Mechanism of action of neuroleptic drugs, in *Psychopharmacology: A Generation of Progress* (Lipton, M. A., DiMascio, A., and Killam, K. F., eds.), Raven, New York, pp. 1057–1070.
110. Koob, G. F. and Bloom, F. E. (1988) Cellular and Molecular Mechanisms of Drug Dependence. *Science* **242,** 715-723.
111. Josselyn, S. A. and Beninger, R. J. (1991) Intra-accumbens neuropeptide Y produces reward: A conditioned place-preference that is blocked by pretreatment with a dopamine antagonist. *Soc. Neurosci. Abs.* 420.15.
112. Kerkerian, L., Bosler, O., Pelletier, G., and Nieoullon, A. (1986) Striatal neuropeptide Y neurones are under the influence of the nigrostriatal dopaminergic pathway: immunohistochemical evidence. *Neurosci. Lett.* **66,** 106–112.
113. Tatsuoka, Y., Riskind, P. N., Beal, M. F., and Martin, J. B. (1987) The effect of amphetamine on the in vivo release of dopamine, somatostatin and neuropeptide Y from rat caudate nucleus. *Brain Res.* **411,** 200–203.
114. Li, S. and Pelletier, G. (1986) The role of dopamine in the control of neuropeptide Y neurons in the rat arcuate nucleus. *Neurosci. Lett.* **69,** 74–77.
115. Pelletier, G. and Simard, J. (1991) Dopamine regulation of pre-proNPY mRNA levels in the rat arcuate nucleus. *Neurosci. Lett.* **127(1),** 96–98.

116. Tessel, R. E., DiMaggio, D. A., and O'Donohue, T. L. (1985) Amphetamine-induced changes in immunoreactive NPY in rat brain, pineal gland and plasma. *Peptides* **6,** 1219–1224.
117. Salin, P., Kerkerian, L., and Nieoullon, A. (1990) Expression of neuropeptide Y immunoreactivity in the rat nucleus accumbens is under the influence of the dopaminergic mesencephalic pathway. *Exp. Brain Res.* **81(2),** 363–371.
118. Flood, J. F., Hernandez, E. N., and Morley, J. E. (1987) Modulation of memory processing by neuropeptide Y. *Brain Res.* **421,** 280–290.
119. Flood, J. F. and Morley, J. E. (1989) Dissociation of the effects of neuropeptide Y on feeding and memory: evidence for pre- and postsynaptic mediation. *Peptides* **10,** 963–966.
120. Citation withdrawn.
121. Colmers, W. F., Lukowiak, K., and Pittman, Q. J. (1985) Neuropeptide Y reduces orthodromically evoked population spike in rat hippocampal CA1 by a possibly presynaptic mechanism. *Brain Res.* **346,** 404–408.
122. Colmers, W. F., Lukowiak, K., and Pittman, Q. J. (1987) Presynaptic action of neuropeptide Y in area CA1 of the rat hippocampal slice. *J. Physiol. (Lond.)* **383,** 285–299.
123. Colmers, W. F. and Pittman, Q. J. (1989) Presynaptic inhibition by neuropeptide Y and baclofen in hippocampus: sensitivity to pertussis toxin treatment. *Brain Res.* **498,** 99–104.
124. Haas, H. L., Hermann, A., Greene, R. W., and Chan-Palay, V. (1987) Action and location of neuropeptide tyrosine (Y) on hippocampal neurons of the rat in slice preparations. *J. Comp. Neurol.* **257,** 208–215.
125. Brooks, P. A., Kelly, J. S., Allen, J. M., Smith, D. A., and Stone, T. W. (1987) Direct excitatory effects of neuropeptide Y on rat hippocampal neurons in vitro. *Brain Res.* **408,** 295–298.
126. Monnet, F. P., Debonnel, G., and de Montigny, C. (1990) Neuropeptide Y selectively potentiates *N*-methyl-D-aspartate-induced neuronal activation. *Eur. J. Pharmacol.* **182,** 207,208.
127. Dawbarn, D., Rossor, M. N., Mountjoy, C. Q., Roth, M., and Emson, P. C. (1986) Decreased somatostatin immunoreactivity but not neuropeptide Y immunoreactivity in cerebral cortex in senile dementia of Alzheimer type. *Neurosci. Lett.* **70,** 154–159.
128. Dawbarn, D. and Emson, P. C. (1985) Neuropeptide Y-like immunoreactivity in neuritic plaques of Alzheimer's disease. *Biochem. Biophys. Res. Commun.* **126,** 289–294.
129. Allen, J. M., Ferrier, I. N., Roberts, G. W., Cross, A. J., Adrian, T. E., Crow, T. J., and Bloom, S. R. (1984) Elevation of neuropeptide Y (NPY) in substantia innominata in Alzheimer's type dementia. *J. Neurol. Sci.* **64,** 325–331.
130. Lehericy, S., Hirsch, E. C., Cervera, P., Hersh, L. B., Hauw, J. J., Ruberg, M., and Agid, Y. (1989) Selective loss of cholinergic neurons in the ventral striatum of patients with Alzheimer disease. *Proc. Natl. Acad. Sci. USA* **86,** 8580–8584.
131. Atack, J. R., Beal, M. F., May, C., Kaye, J. A., Mazurek, M. F., Kay, A. D., and Rapoport, S. I. (1988) Cerebrospinal fluid somatostatin and neuropeptide Y. Concentrations in aging and in dementia of the Alzheimer type with and without extrapyramidal signs. *Arch. Neurol.* **45,** 269–274.
132. Beal, M. F., Mazurek, M. F., Chattha, G. K., Svendsen, C. N., Bird, E. D., and

Martin, J. B. (1986) Neuropeptide Y immunoreactivity is reduced in cerebral cortex in Alzheimer's disease. *Ann. Neurol.* **20,** 282–288.
133. Beal, M. F., Mazurek, M. F., and McKee, M. A. (1987) The regional distribution of somatostatin and neuropeptide Y in control and Alzheimer's disease striatum. *Neurosci. Lett.* **79,** 201–206.
134. Chan-Palay, V., Lang, W., Haesler, U., Köhler, C., and Yasargil, G. (1986) Distribution of altered hippocampal neurons and axons immunoreactive with antisera against neuropeptide Y in Alzheimer's-type dementia. *J. Comp. Neurol.* **248,** 376–394.
135. Nakamura, S. and Vincent, S. R. (1986) Somatostatin- and neuropeptide Y-immunoreactive neurons in the neocortex in senile dementia of Alzheimer's type. *Brain Res.* **370,** 11–20.
136. Davies, C. A., Morroll, D. R., Prinja, D., Mann, D. M., and Gibbs, A. (1990) A quantitative assessment of somatostatin-like and neuropeptide Y-like immunostained cells in the frontal and temporal cortex of patients with Alzheimer's disease. *J. Neurol. Sci.* **96,** 59–73.
137. Arendash, G. W., Millard, W. J., Dunn, A. J., and Meyer, E. M. (1987) Long-term neuropathological and neurochemical effects of nucleus basalis lesions in the rat. *Science* **238,** 952–956.
138. Ferrante, R. J., Kowall, N. W., Beal, M. F., Richardson, E. P., Jr., Bird, E. D., and Martin, J. B. (1985) Selective sparing of a class of striatal neurons in Huntington's disease. *Science* **230,** 561–563.
139. Ferrante, R. J., Kowall, N. W., Beal, M. F., Martin, J. B., Bird, E. D., and Richardson, E. P., Jr. (1987) Morphologic and histochemical characteristics of a spared subset of striatal neurons in Huntington's disease. *J. Neuropathol. Exp. Neurol.* **46,** 12–27.
140. Dawbarn, D., De Quidt, M. E., and Emson, P. C. (1985) Survival of basal ganglia neuropeptide Y-somatostatin neurones in Huntington's disease. *Brain Res.* **340,** 251–260.
141. Zech, M., Roberts, G. W., Bogerts, B., Crow, T. J., and Polak, J. M. (1986) Neuropeptides in the amygdala of controls, schizophrenics and patients suffering from Huntington's chorea: an immunohistochemical study. *Acta Neuropathol. (Berl)* **71,** 259–266.
142. Beal, M. F., Kowall, N. W., Ellison, D. W., Mazurek, M. F., Swartz, K. J., and Martin, J. B. (1986) Replication of the neurochemical characteristics of Huntington's disease by quinolinic acid. *Nature* **321,** 168–171.
143. Beal, M. F., Kowall, N. W., Swartz, K. J., Ferrante, R. J., and Martin, J. B. (1989) Differential sparing of somatostatin-neuropeptide Y and cholinergic neurons following striatal excitotoxin lesions. *Synapse* **3,** 38–47.
144. Davies, S. W. and Roberts, P. J. (1987) No evidence for preservation of somatostatin-containing neurons after intrastriatal injections of quinolinic acid. *Nature* **327,** 326–329.
145. Boegman, R. J., Smith, Y., and Parent, A. (1987) Quinolinic acid does not spare striatal neuropeptide Y-immunoreactive neurons. *Brain Res.* **415,** 178–182.
146. Beal, M. F., Kowall, N. W., Swartz, K. J., and Ferrante, R. J. (1990) Homocysteic acid lesions in rat striatum spare somatostatin-neuropeptide Y (NADPH-diaphorase) neurons. *Neurosci. Lett.* **108,** 36–42.
147. Albin, R. L., Reiner, A., Anderson, K. D., Penney, J. B., and Young, A. B.

(1990) Striatal and nigral neuron subpopulations in rigid Huntington's disease: implications for the functional anatomy of chorea and rigidity-akinesia. *Ann. Neurol.* **27,** 357–365.
148. Allen, J. M., Cross, A. J., Crow, T. J., Javoy-Agid, F., Agid, Y., and Bloom, S. R. (1985) Dissociation of neuropeptide Y and somatostatin in Parkinson's disease. *Brain Res.* **337,** 197–200.
149. Beal, M. F., Svendsen, C. N., Bird, E. D., and Martin, J. B. (1987) Somatostatin and neuropeptide Y are unaltered in the amygdala in schizophrenia. *Neurochem. Pathol.* **6,** 169–176.
150. Widerlöv, E., Gottfries, C. G., Lindström, L., and Ekman, R. (1988) Brain and CSF neuropeptide alterations in schizophrenia. *Neurochem. Internat.* 13 suppl. **1,** 66.
151. Kaye, W. H., Berrettini, W. H., Gwirtsman, H. E., Gold, P. W., George, D. T., Jimerson, D. C., and Ebert, M. H. (1989) Contribution of CNS neuropeptide (NPY, CRH, and beta-endorphin) alterations to psychophysiological abnormalities in anorexia nervosa. *Psychopharmacol. Bull.* **25,** 433–438.
152. Kaye, W. H., Berrettini, W., Gwirtsman, H., and George, D. T. (1990) Altered cerebrospinal fluid neuropeptide Y and peptide YY immunoreactivity in anorexia and bulimia nervosa. *Arch. Gen. Psychiat.* **47,** 548–556.
153. Flood, J. F. and Morley, J. E. (1990) Pharmacological enhancement of long-term memory retention in old mice. *J. Gerontol.* **45,** B101–B104.

Index